Pesticide Residues in Coastal Tropical Ecosystems

Pesticide Residues in Coastal Tropical Ecosystems

Distribution, fate and effects

Edited by
Milton D. Taylor,
Stephen J. Klaine,
Fernando P. Carvalho,
Damia Barcelo, and
Jan Everaarts*

CRC Press
Taylor & Francis Group
Boca Raton London New York

CRC Press is an imprint of the
Taylor & Francis Group, an **informa** business
A TAYLOR & FRANCIS BOOK

CRC Press
Taylor & Francis Group
6000 Broken Sound Parkway NW, Suite 300
Boca Raton, FL 33487-2742

First issued in paperback 2020

© 2003 Milton D. Taylor, Stephen J. Klaine, Fernando P. Carvalho,
Damia Barcelo and Jan Everaarts
CRC Press is an imprint of Taylor & Francis Group, an Informa business

No claim to original U.S. Government works

ISBN-13: 978-0-367-45468-5 (pbk)
ISBN-13: 978-0-415-23917-2 (hbk)

Visit the Taylor & Francis Web site at
http://www.taylorandfrancis.com

and the CRC Press Web site at
http://www.crcpress.com

Typeset in Baskerville by
HWA Text and Data Management, Tunbridge Wells

British Library Cataloguing in Publication Data
A catalogue record for this book is available from the British Library

Library of Congress Cataloging in Publication Data
Pesticide residues in coastal tropical ecosystems : distribution, fate and effects /
edited by Milton D. Taylor ... [et al.].
     p. cm.
     Includes bibliographical references and index.
     1. Pesticides – Environmental aspects – Tropics.   2. Coastal ecology –
Tropics.   I. Taylor, Milton D.

QH545.P4 P44796 2002
628.5′29′0913–dc21   202072711

Contents

Figures

Tables

Contributors

Many of the book's authors or their representatives were among the coordinated research program attendees at the International Symposium on Marine Pollution held in Monaco, 5–9 October 1998. Standing from left to right are Cai Fulong (PRC), Lily Varca (Philippines), Jahan (Bangladesh), Ajai Mansingh (Commonwealth Caribbean), Dang Duc Nhan (Vietnam), Tania Tavares (Brazil) Luz Angela Castro (Colombia), Elba M. de la Cruz (Costa Rica), F. González-Farias (Mexico), S.P. Kale (India), Damia Barcelo (Spain), Stephen J. Klaine (USA), Gonzalo Dierksmeier (Cuba) and Abdul Rani Abdullah (Malaysia). Seated or kneeling in the front row are Fernando Carvalho (IAEA), Shem O. Wandiga (Kenya) and Zhong Chuangguang (PRC).

Abdul Rani Abdullah Department of Chemistry, University of Malaya, 50603, Kuala Lumpur, Malaysia.

Cristina M. Bajet Pesticide Toxicology and Chemistry Laboratory, National Crop Protection Center, University of the Philippines at Los Baños, College, Laguna 4031, Philippines.

Damia Barcelo CSIC, CID, Department of Environmental Chemistry C Jordi Girona Barcelona, Spain

Cai Fulong Third Institute of Oceanography, State Oceanic Administration, Xiamen, People's Republic of China.

Luisa E. Castillo Central American Institute for Studies in Toxic Substances, Universidad Nacional, Costa Rica.

Fernando P. Carvalho Instituto Tecnológico e Nuclear, Dpto Protecção Radio-lógica, E.N. 10 P-2685-953 Sacavém, Portugal.

Luz Angela Castro Centro Control Contaminación del Pacífico (CCCP), Sección Contaminación Marina. Isla del Morro, Tumaco – Nariño, Colombia S.A.

Chen Shunhua School of Life Sciences, Zhongshan University, Guangzhou 510274, People's Republic of China.

Elba M. de la Cruz Central American Institute for Studies in Toxic Substances, Universidad Nacional, Costa Rica.

Kathy M. Dalip Pesticide and Pest Research Group, Department of Life Sciences, University of the West Indies, Mona, Kingston 7, Jamaica. Current Affiliation: Senior Entomologist, Caribbean Agriculture Research and Development Institute, University of the West Indies, Mona, Kingston 7, Jamaica.

Dang Duc Nhan Vietnam Atomic Energy Commission, 59 Ly Thuong Kiet, Hanoi, Vietnam.

Gonzalo Dierksmeier Plant Protection Research Institute, 110 No. 514 esq. 5taB Playa, Habana City, Cuba.

Jan Everaarts* Netherlands Institute for Sea Research AB Den Burg-Textel The Netherlands

E. Frempong Department of Biological Sciences, University of Science and Technology, Kumasi, Ghana.

F. González-Farias Center for Nutrition Research and Development (CIAD), Mazatlan Unit in Aquaculture and Environmental Management. P.O. Box 711, Mazatlan, Sin. México 82000. Permanent address: National Autonomous University of México, Institute of Marine Science and Limnology, Mazatlan Marine Station. P.O. Box 811, Mazatlan, Sin. México 82000.

* deceased

R. Hernández Plant Protection Research Institute, 110 No. 514 esq. 5taB Playa, Habana City, Cuba.

P.N. Kaigwara University of Nairobi, Department of Chemistry, P.O. Box 30197, Nairobi, Kenya.

S.P. Kale Nuclear Agriculture and Biotechnology Division, Bhabha Atomic Research Centre, Mumbai 400 085, India.

Stephen J. Klaine Department of Environmental Toxicology, Clemson Institute of Environmental Toxicology, Clemson University, Pendleton, SC USA 29670.

J.O. Lalah University of Nairobi, Department of Chemistry, P.O. Box 30197, Nairobi, Kenya.

Ajai Mansingh Pesticide and Pest Research Group, Department of Life Sciences, University of the West Indies, Mona, Kingston 7, Jamaica. Current Affiliation: Natural Products Institute, University of the West Indies, Mona, P.O. Box 12, Kingston 7, Jamaica.

K. Martinez Plant Protection Research Institute, 110 No. 514 esq. 5taB Playa, Havana City, Cuba.

M.A. Matin Institute of Food and Radiation Biology, Atomic Energy Research Establishment, GPO Box 3787, Dhaka-1000, Bangladesh

Foster L. Mayer United States Environmental Protection Agency (USEPA), Environmental Research Lab, Gulf Breeze, FL USA 32561.

J. Mitra Nuclear Agriculture and Biotechnology Division, Bhabha Atomic Research Centre, Mumbai 400 085, India.

A.M. Mohan Rao Environmental Assessment Division, Bhabha Atomic Research Centre, Mumbai 400 085, India.

Pura Moreno Plant Protection Research Institute, 110 No. 514 esq. 5taB Playa, Havana City, Cuba.

P.K. Mukherjee Nuclear Agriculture and Biotechnology Division, Bhabha Atomic Research Centre, Mumbai 400 085, India.

N.B.K. Murthy Nuclear Agriculture and Biotechnology Division, Bhabha Atomic Research Centre, Mumbai 400 085, India.

S. Osafo Acquaah Department of Chemistry, University of Science and Technology, Kumasi, Ghana.

G.G. Pandit Environmental Assessment Division, Bhabha Atomic Research Centre, Mumbai 400 085, India.

K. Raghu Nuclear Agriculture and Biotechnology Division, Bhabha Atomic Research Centre, Mumbai 400 085, India.

Dwight E. Robinson Pesticide and Pest Research Group, Department of Life Sciences, University of the West Indies, Mona, Kingston 7, Jamaica. Current Affiliation: Natural Products Institute, University of the West Indies, Mona, P.O. Box 12, Kingston 7, Jamaica.

Milton D. Taylor Department of Environmental Toxicology, Clemson Institute of Environmental Toxicology, Clemson University, Pendleton, SC USA 29670.

S.O. Wandiga University of Nairobi, Department of Chemistry, P.O. Box 30197, Nairobi, Kenya.

Pen Yefang School of Life Sciences, Zhongshan University, Guangzhou 510274, People's Republic of China.

Liao Yuanqi The Plant Protection Research Institute, Guangdong Academy of Agricultural Sciences, Guangzhou, People's Republic of China.

Mark F. Zaranyika Chemistry Department, University of Zimbabwe, P.O. Box MP 167 Mount Pleasant, Harare, Zimbabwe.

Zhao Xiaokui School of Life Sciences, Zhongshan University, Guangzhou 510274, People's Republic of China.

Zhong Chuangguang School of Life Sciences, Zhongshan University, Guangzhou 510274, People's Republic of China.

Foreword

The second half of the twentieth century saw tremendous increases in world food production brought about by revolutionary changes in agriculture. With the introduction and widespread acceptance of DDT as a nearly universal insecticide for agriculture in 1944, crop losses to insects declined rapidly and food production accelerated through a paradigm shift from small family farms to industrialized agriculture. By the late 1960s when insect resistance to DDT and the rising concern for environmental contamination caused by DDT and other organochlorine pesticides became apparent, other pesticide chemistries such as the organophosphates, carbamates, organotins, and second generation pyrethroids had become available. These compounds filled the pest control gap as organochlorine use gradually declined from the combination of resistance and environmental regulatory pressures. The Green Revolution of the 1980s introduced high-yield hybrid seed varieties, which in combination with chemical fertilizers, resulted in dramatic increases in crop yields; world rice and wheat yields tripled between 1950 and the mid-1980s. With the recent development of genetically modified transgenic crops, there has been a further shift to large-scale plantation farming. The shift to monoculture-based industrial farming has brought about the introduction of many new and novel pesticide chemistries, primarily because monocultures are more susceptible to pest population explosions than the old-style mixed crop planting that would sustain and promote insect predator populations.

Because of the environmental problems associated with the organochlorines, pesticide regulatory agencies were established, initially in developed countries located in temperate climes, and began issuing regulations to control pesticide production and use. These regulations required data about the efficacy and non-target effects of individual pesticides. Consequently, most of the data generated for regulatory purposes was based on data obtained from field research in temperate climates. Thus, guidelines for using pesticides in the tropics have been based on research conducted in Europe and the United States. While some tropical countries recognized the potential problems associated with this fact, they had little recourse but to adopt regulations copied from their more affluent northern neighbors. Over the past 20 years, many developing countries have adopted some form of environmental protection, either through establishment of a regulatory agency or agencies or through adoption of a body of law to regulate the pesticide industry. As a result, some tropical countries are taking a more proactive approach to regulating pesticides by requiring licensing data for individual pesticides to be generated in soils, climates, and under conditions reflective of local conditions. This is an encouraging sign of environmental responsibility. There are groups in some countries, pushing for national long-term pesticide residue monitoring programs for foodstuffs, surface waters, or biota. However, these efforts are likely to be stillborn until and unless the international community, viz. the developed countries, assists by providing stable long-term funding to support these efforts.

This book grew out of a research program to determine the distribution, fate, and effects of pesticides in the tropical marine environment as envisioned and led by Fernando Carvalho at the Marine Environmental Laboratory of the International Atomic Energy Agency in Monaco. He recruited respected scientists from 17 countries and charged them to accomplish a specific set of objectives. These included measuring current levels of pesticide residues in coastal environments, characterizing the cycling and fate of pesticides, evaluating the effects of their residues on marine biota, assessing the risk associated with pesticides in coastal tropical ecosystems, and advising on measures to protect the tropical marine environment. With funding provided by the Swedish International Development Authority, a coordinated research program (CRP) was established and through joint activities including courses on aquatic toxicology, ecological risk assessment, and quality assurance and quality control; training workshops; and participation in international scientific meetings, the participating scientists were forged into a dynamic working group. As their research yielded new information about the extent of pesticide residues in coastal environments, they became familiar with the results of other groups working in their respective countries. The genesis of this book came from discussions between themselves and other colleagues. As the idea evolved, it became apparent that more than a summary of research results would be needed to convey the potential for pesticides to affect the tropical marine environment. Contributing colleagues were asked to include some basic information about their country, e.g., location, extent, geographic features, etc.; brief histories of pesticide use with import and export data where available; summaries of their pesticide regulations and descriptions of the regulatory bodies in their country; descriptions of pesticide research in their country including residue levels in surface waters, coastal waters, sediments and biota; and synopses of efforts in their respective countries to limit the growth of pesticide use, either through Integrated Pest Management efforts, public education programs, or through other means. Readers will discover that individual chapters are not rigidly structured as some authors tended to emphasize one area more than another. As editors, we felt that this less-rigid format made for a more interesting book and provided some information about a country's focus or concerns either through inclusion or omission. Our hope is that the resulting book will bring together as much of the current peer-reviewed, government, and "grey" literature on pesticide use, fate, and effects in the tropics as possible and serve as a state-of-the-science summary for generating the next set of research projects by scientists and serve as a basis for management decisions regarding coastal tropical ecosystems.

Obviously, this project would never have come to fruition without the continued efforts of its contributors to respond to editorial suggestions and supply updated information for their respective chapters. The editors would also like to thank the students of Professor Klaine's laboratory in the Department of Environmental Toxicology, Clemson University, for their assistance in reviewing page proofs of the book.

M. D. Taylor
June 2002
Pendleton, South Carolina, USA

Abbreviations

Abbreviations for the following terms and organizations are used throughout this book. However, abbreviations for terms and organizations listed within individual chapters are applicable only to that chapter.

AChE	acetylcholinesterase
a.i.(s)	active ingredient(s)
B	billion
BCF	bioconcentration factor
cm	centimeters
d	day(s)
DPM	disintegrations per minute
dw	dry weight
EIL	economic injury level
EC	emulsifiable concentrate
EPA	Environmental Protection Agency
FW	flowable
G	granular
GDP	gross domestic product
ha(s)	hectare(s)
HYV	high yield variety
h	hour(s)
IPM	Integrated Pest Management
IAEA	International Atomic Energy Agency
L	liter(s)
kg	kilogram
LOEC	lowest observed effect concentration
MRLs	maximum residue levels
m	meter(s)
T	metric tonnes
μg	micrograms
mm	millimeters
M	million

ng	nanogram
NOAA	National Oceanographic and Atmospheric Administration
NOEC	no observed effect concentration
OC	organochlorine pesticides
OP	organophosphate/organophosphorus pesticides
ppb	part per billion
ppm	part per million
PCB	polychlorinated biphenyl
POP(s)	persistent organic pollutants
SC	soluble concentrate
Σ	total (used to denote: sum of all isomers or sum of a class)
FAO	United Nations Food and Agriculture Organization
UNEP	United Nations Environment Programme
UNESCO	United Nations Educational, Scientific, and Cultural Organization
WP	wettable powder
ww	wet weight
WHO	World Health Organization
y	year(s)

Chapter I

Introduction[1]

*Fernando Carvalho, Stephen J. Klaine and
Milton D. Taylor*

Since the 1940s the use of pesticides has grown steadily at about 11 percent a year, reaching five million T in 1995. Pesticides and fertilizers play a central role in agriculture and contribute to an enhanced food production worldwide. Agrochemical use, mainly in developed countries, is being reduced while organic (*no* synthetic chemicals) farming methods are being reinvented. However, in most countries, agrochemicals remain an essential component of agricultural practice and will remain so for the foreseeable future. Pesticides, especially insecticides and fungicides, are more heavily applied to tropical cash crops, e.g. banana, coffee, cotton, and vegetables, than to crops in temperate regions. For example, the application of pesticides to banana plantations in Costa Rica reached 45 kg a.i. per ha, whereas the comparable average application of pesticides in Japan to crops is 10.8 kg. Only a minor fraction of applied pesticides actually reaches the target pest species – less than 0.1 percent; the excess pesticide moves through the environment potentially contaminating soil, water, and biotic matrices. It is critical, especially in the tropics where use patterns lead to much higher loads of pesticides than in temperate zones, to characterize the fate and non-target toxicity of these pesticides to confidently assess the risk associated with their use.

Agricultural fields are generally located in coastal plains and river valleys so it is not surprising that streams and rivers receive the majority of agriculture runoff and carry the residues into estuaries and coastal seas. For example, the USA's Mississippi River may have transported an estimated 430 T of atrazine in 1989 from mid-western corn and soybean plantations to the Gulf of Mexico. The effect of these pesticide residues on human and environmental health is of great concern. Recent studies on the estrogen-like behavior of some DDTs and PCBs in humans implicate these compounds in the etiology of breast cancer. Furthermore, the growing awareness of the environmental persistence of organic pollutants (pesticides and industrial organic chemicals) has led governments to agree on an international Convention on Persistent Organic Pollutants (POPs), phasing out a number of noxious substances, including DDT, PCBs, and other chlorinated hydrocarbons.

Nevertheless, the use of hundreds of old and new compounds as crop-protection chemicals continues worldwide. It is urgent that we develop strategies to facilitate the coexistence of productive, healthy, and economically viable agriculture with the preservation of natural resources. To accomplish this, more research is needed to assess the impact of pesticide burden on ecosystems. Then, chemical management strategies could be developed to minimize the risk associated with those burdens. Exacerbating the general plethora of data is the fact that most studies on environmental cycling, fate, and effects of pesticides have been carried out in temperate climates (North America and western Europe). Much less information has been available on the behavior of these chemicals in tropical ecosystems where the use of POPs continues.

To help fill this information gap, the IAEA set up a coordinated research program (CRP) on the 'Distribution, Fate, and Effects of Pesticides on Biota in the Tropical Marine Environment', funded by the Swedish International Development Authority. Eighteen laboratories in seventeen countries including Bangladesh, Brazil, China, Colombia, Costa Rica, Cuba, Ecuador, India, Jamaica, Kenya, Malaysia, Mexico, the Philippines, Spain, the Netherlands, the United States, and Vietnam took part in the program. The CRP's technical objectives were to measure the current levels of pesticide residues in coastal environments; to characterize the cycling and fate of pesticides using radiolabeled compounds and nuclear techniques; to evaluate the effects of residues on marine biota; to assess the risk associated with pesticide residues in coastal tropical ecosystems; and to advise on measures to protect the tropical marine environment. A large number of requests for participation and information were received at the IAEA, which attested to the project's relevance among countries and their interest to act on associated environmental problems.

Eighteen laboratories in 17 IAEA Member States agreed to participate in the project, which was led by the IAEA's Marine Environment Laboratory (MEL) in Monaco. The first Research Coordination Meeting (RCM) was held at MEL in June 1994. It reviewed pesticide usage and data on contamination from pesticide residues in the participating countries, identified the compounds to be targeted by the research, and determined the applicable methodologies for meeting project objectives. Furthermore, the equipment and training needs of the participating laboratories were defined to organize IAEA technical support to the laboratories. Since then, annual research coordination meetings have been organized to present and discuss the project results achieved in every country. Between the meetings, ongoing liaison and data exchange among participants was maintained through periodic circular notes.

Major joint activities were implemented to meet common needs or objectives. These included courses (lectures) on pesticide chemistry, aquatic toxicology, and ecological risk assessment; training workshops on pesticide analysis (one organized in Costa Rica and another in Malaysia) for analytical training for CRP participants; and inter-comparison exercises to test the accuracy of laboratory results. Research using compounds labeled with ^{14}C was introduced in 12 laboratories and gas chromatography analytical techniques are now used in 14 participating laboratories. Most of the laboratories adopted quality assurance procedures, including regular participation in the inter-comparison exercises and the use of certified reference materials to ensure the quality of data in the analyses of organochlorine (OC) pesticides and radiolabeled compounds. Research results have been presented at various scientific forums, including the International Symposium on the Environmental Behavior of Crop Protection Chemicals and the International Symposium on Marine Pollution.

Among the least expensive and best research tools that the laboratories applied were the radiotracer techniques. Pesticides labeled with radioisotopes were used in model ecosystems to help researchers investigate the persistence of pesticides, their degradation pathways, and pesticide transfer in marine food chains. Large numbers of samples could be processed rapidly at low cost and accurately measured with standard liquid scintillation equipment. Several participating laboratories applied this technique to obtain previously unavailable data on pesticide behavior in tropical marine environments.

Each CRP participating institute planned to carry out field investigations to monitor pesticide residues in the coastal ecosystems in their respective countries. Study areas covered diverse ecosystems, e.g. Manila Bay in Philippines, the Zhujiang (Pearl) River Delta in The People's Republic of China, Cartagena Bay in Colombia, several watersheds around Kingston Harbor in Jamaica, Baia de Todos os Santos in Brazil, the Red River Valley and Delta in northern Vietnam, and the Indian Coast of Kenya. As expected, pesticide residues identified in coastal environments in these regions followed the trends of local pesticide usage. Nevertheless, volatile pesticides, transported by atmospheric processes and redeposited worldwide, were measured in areas far from their original application sites. Through the CRP, the capacity of laboratories to measure pesticide residues in marine samples, e.g. DDT residues in mussels, was significantly improved. Results of the CRP suggest that the persistent OC pesticides are present throughout the tropics, although fortunately only in trace amounts in some areas. When detected, organophosphate (OP) pesticides generally seem to be present in concentrations below the level of OCs.

Using ^{14}C labeled compounds and liquid scintillation counting, many CRP participants investigated the behavior of various pesticides in the aquatic environment. Their research focused on a selection of compounds, i.e. DDT, endosulfan, lindane, chlorpyrifos, and parathion, which have been found in tropical coastal environments. The studies were designed to obtain information on several aspects related to the fate, persistence, and bioaccumulation of these pesticides in marine waters. Typically, studies were carried out using laboratory microcosms and model ecosystems according to the common methodologies adopted by the CRP participating institutes.

The transport, dispersion, and, ultimately, the biological effects of pesticides in lagoon systems depend upon the persistence of these chemicals under tropical conditions and their bioaccumulation and biodegradation. It is generally believed that sunlight and the elevated temperatures in the tropics would contribute to a rapid breakdown of these compounds. However, preliminary experimental results indicate that photolysis plays a minor role in the breakdown of these compounds in comparison with chemical hydrolysis. Results from the CRP indicated that aqueous half-lives of dissolved pesticides at 32°C are in the range of 1.4 to 10 days for chlorpyrifos, nine to 46 days for parathion, and 130 to 155 days for DDT, depending upon salinity. Further, experiments involving water:sediment systems demonstrated that half-lives of pesticides sorbed onto lagoon sediments are 10 to 100 times longer than the half-lives for the same compounds in the overlying waters, despite the larger microbial biomass in the sediment compared to the water. Therefore, the rapid sorption of pesticides onto sediment particles may increase their persistence. It is likely that the largest reservoirs of these compounds will be found in the sediments of lagoons, wetlands, and river deltas that receive runoff from adjacent agricultural fields.

Several CRP participating institutes in Bangladesh, China, India, the Philippines, Vietnam, Malaysia, Jamaica, and Mexico carried out similar experiments to examine the fate of DDT and chlorpyrifos in aquaria, simulating the conditions of the tropical marine environment and using local species of marine biota. Their results indicated that the accumulation of those pesticides from water by mussels, clams, shrimp, and fish was very rapid, occurring over a time scale of minutes to hours. Furthermore, the biological concentration factor of lipophilic compounds, e.g. DDT and its metabolites, was generally very high.

Considerable progress was achieved by the laboratories using radiotracer techniques in pesticide research. It included the development of miniaturized experimental systems to investigate compound degradation and volatilization. Data showed that the persistence and bioaccummulation of pesticides in the marine environment was related to molecular configuration and chlorine content. It was evident that in tropical environments OCs such as DDT degraded very slowly even though half-lives were shorter than reported for temperate environments. Further, some OPs may survive long enough in the tropical environment to disperse into estuarine and coastal ecosystems and, thus, impact aquatic biota. Nevertheless, OPs generally degraded much faster than OCs in the marine environment.

Toxicity assays on marine species under tropical conditions have been carried out by several institutes in the Philippines, Jamaica, Costa Rica, and Mexico. The aim was to assess the sensitivity of common tropical species that were likely to be exposed to pesticide residues, e.g. tilapia and farmed shrimp, under conditions of tropical ecosystems. One common feature of tropical coastal lagoons, which are surrounded by mangrove forests, is the elevated concentration of humic substances. Humic substances are formed by the gradual decomposition of leaf litter from mangrove trees. Due to the hydrophobicity of OC and OP compounds, a possible association of these pesticides with particulate and dissolved humics may modify their overall fate and their bioavailability. This hypothesis was tested using marine mussels exposed to pesticides dissolved in seawater either without humic substances or with particulate humics already containing bound pesticides. Results showed that the accumulation by the mussels of pesticides dissolved in seawater increased rapidly in the first 12 hours, followed by a slower increase thereafter. Pesticides bound by the humic substances were also accumulated in mussel tissues, although to a lesser extent than directly from seawater. Therefore, it appeared that previous binding of pesticides by humics acted to reduce the bioaccumulation of those compounds by lagoon fauna. However, further research is still needed to elucidate the exact mechanisms that govern the distribution, fate, and bioavailability of pesticide residues in tropical lagoons.

One overall goal of the CRP was to characterize the existing risk to both humans and coastal marine ecosystems and suggest strategies to reduce that risk – if it is presently unacceptable – in the future. Several case studies within the CRP generated sufficient pesticide residue data and toxicity data to begin to assess ecological risk in particular ecosystems. In some ecosystems, particularly estuarine and coastal lagoon systems receiving agricultural drainage, pesticide residues in sediments and biota approached the acutely toxic level. More often, however, residues were below lethal values and represent potential food chain problems. Coincidentally, these ecosystems were ideal habitat for fish, shrimp, and oyster farming. Hence, these residues have found their way into the general population through the aquaculture industry. In addition, these residues presented a measurable risk to the stability of the ecosystems. This instability could arise through the loss of a critical species or trophic level or through the deterioration of general water quality, e.g. dissolved oxygen, caused by the microbial degradation of these contaminants. The preservation of these systems and associated resources would require the implementation of integrated management plans for coastal zones in order to harmonize the interests of agriculturalists, aquaculturalists, and the fishing community.

The CRP's development facilitated the assessment of pesticide contamination in key coastal areas. Furthermore, experimental data on the cycling of pesticides were obtained to provide a more complete understanding of the impact and fate of their residues in tropical coastal environments and to suggest the best methods for implementing environmental management strategies. Other immediate benefits of the CRP to participating countries included improved capability to measure pesticide residues in environmental samples and to conduct research on pesticide cycling in tropical regions. Overall, the project has enhanced awareness of the need to reduce environmental contamination traced to persistent organic compounds.

Ultimately, comprehensive watershed-based ecological risk assessments are needed to provide the foundation for developing land and water resource management strategies to ensure the stability of tropical ecosystems and provide resources for the necessary future development of agriculture, aquaculture, business and industry, recreation facilities, and domestic dwellings in tropical countries. Existing pesticide residue and toxicity data from CRP participants and others would be incorporated along with land-use data for the watershed to develop cause-and-effect relationships for various activities. However, practical solutions to treat contaminated waste-water and surface runoff from agricultural fields, aquaculture facilities, industrial and commercial business sites, and residential areas must still be developed, investigated, and tested. Tentative solutions have been suggested in a few cases, e.g. the use of natural or constructed wetlands around agriculture fields to remove pesticide residues from runoff. This and other suggestions must be tested on pilot-scale demonstration projects in the tropics, e.g. banana plantations, rice paddies, cotton fields, etc. Such projects would be the next logical step forwards based on the work already performed by the world's laboratories through this IAEA-sponsored research program.

The purpose of this book is to document the productivity of the CRP, place it into perspective with other available information on pesticide disposition in tropical marine environments, and facilitate the evolution of ecosystem risk assessments in these environments. As such, this book is organized geographically to assist readers with particular spatial interests. Most chapters attempt to describe local environmental regulations, the history of pesticide use and production in the country, pesticide licensing, registration and use regulations, results of pesticide residue studies in both riverine and coastal ecosystems, pesticide toxicity results with indigenous species, and finally descriptions of local and national efforts to educate pesticide users about both worker safety and integrated pest management. The primary exception to the rule, e.g. describing studies of coastal and marine ecosystems, is the chapter from Zimbabwe, a landlocked nation. Zimbabwe was included because of its extensive shoreline along Lake Kariba and the continued use of DDT to combat the tsetse fly. However, several chapters contain information regarding studies of pesticide levels in large freshwater bodies and some rivers. These data are included because they allow some interesting comparisons between freshwater pesticide residue levels and levels of the same pesticides found in the upper reaches of estuaries. Some chapters also compile research and policy needs for individual countries. For a number of country chapters, data on PCB residue levels is included, especially when PCBs represent a serious potential contaminant hazard for the country. Finally, a chapter on risk assessment lays the framework for the use of these data to describe pesticide risk in coastal watersheds.

There is not enough space in this book to adequately describe the existing pesticide fate and toxicity data. The reader is directed to the extensive list of references provided in each chapter. The appendix contains a list of all pesticides mentioned in this book along with their individual CAS or Merck numbers and their chemical activity. For those interested in the Quality Assurance and Quality Control procedures and checks applicable to the IAEA and the CRP participants, the reader is directed to consult articles listed under 'Further reading'.

NOTE

1   Primarily taken from the article titled 'Results of an International Research Project: tracking pesticides in the tropics', by F.P. Carvalho, D.D. Nhan, C. Zhong, T. Tavares, and S.J. Klaine (1998) that appeared in the *IAEA Bulletin* 40(3):24–30.

FURTHER READING

Carvalho, F.P., Villeneuve, J.P. and Cattini, C. 1999a. Determination of organochlorine compounds, petroleum hydrocarbons, and sterols in a sediment sample, IAEA-383. Results of an intercomparison exercise. *International Journal of Environmental Analytical Chemistry* 75(4):315–29.

Carvalho, F.P., Villeneuve, J.P. and Cattini, C. 1999b. The determination of organochlorine compounds and petroleum hydrocarbons in a seaweed sample: results of a worldwide intercomparison exercise. *Trends in Analytical Chemistry* 18(11):656–64.

Carvalho, F.P., Villeneuve, J.P. and Coquery, M. 1999. Analytical intercomparison exercises and harmonization within environmental laboratories from developing countries. *International Journal of Environmental Analytical Chemistry* 74(1–4):263–74.

Coquery, M., Carvalho, F.P., Azemard, S. and Horvat, M. 1999. The IAEA worldwide intercomparison exercises (1990–1997): determination of trace elements in marine sediments and biological samples. *Sci. Tot. Environ.* 238:501–8 Special Issue SI SEP 30 1999.

Parr, R.M., Fajgelj, A., Dekner, R., Ruiz, H.V., Carvalho, F.P. and Povinec, P.P. 1998. IAEA analytical quality assurance programmes to meet the present and future needs of developing countries. *Fresenius Journal of Analytical Chemistry* 360(3–4):287–90.

Villeneuve, J.P., de Mora, S.J., Cattini, C. and Carvalho, F.P. 2000. Determination of organochlorinated compounds and petroleum hydrocarbons in sediment sample IAEA-408. Results from a world-wide intercalibration exercise. *Journal of Environmental Monitoring* 2(5):524–8.

Chapter 2

Pesticides in the marine environment of Ghana

S. Osafo Acquaah and E. Frempong

INTRODUCTION

The Republic of Ghana – formed from the merger of the British colony of the Gold Coast and the Togoland trust territory in 1957 – was the first country in colonial Africa to gain its independence. It is located in western Africa roughly between Lat. 5°N and 11°N, Long. 1°E and 3°30'W. It borders the Gulf of Guinea in the south, Burkina Faso (Upper Volta) along its north and northwest border, the Republic of Côte d'Ivoire in the west and the Togolese Republic (Togo) in the east. Ghana has a total land area of 238,540 km² divided among 10 administrative regions: Ashanti, Brong-Ahafo, Central, Eastern, Greater Accra, Northern, Upper East, Upper West, Volta, and Western (Figure 2.1).

Ghana's climate is warm and comparatively dry along the southeast coast but hot and humid in the southwest and hot and dry in the north. There are two distinct rainy seasons in the south: April to July and September to November; however, in the north, the rainy seasons tend to merge lasting from April through September. Annual rainfall ranges from about 1,100 mm in the north to about 2,100 mm in the southeast. The harmattan, a dry, northeasterly desert wind, blows from December to March, lowering the humidity and creating hot days and cool nights in the north. In the south the effects of the harmattan are felt primarily in January.

A tropical rainforest belt, broken by heavily forested hills along with many streams and rivers, extends northward from the Gulf of Guinea, near the Côte d'Ivoire frontier. This area, known as the Ashanti, produces most of the country's cocoa, minerals, and timber. North of the rainforest, Ghana is covered by low bush, park-like savanna, and grassy plains. Ghana's dominant feature is 8,482 km² Lake Volta, which backs up behind Akosombo Dam on the Volta River, and is one of the world's largest artificially created lakes. This lake generates electricity, provides inland transportation, and is a potentially valuable resource for irrigation and fish farming in Ghana. It also provides an additional 1,125 km of arterial and feeder waterways beyond the 168 km of perennial navigation – for launches and

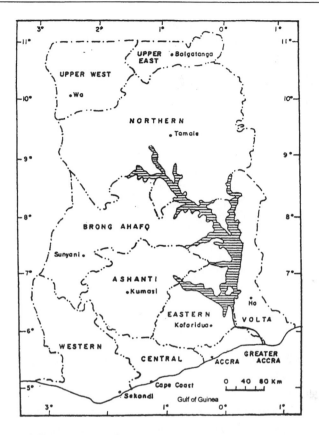

Figure 2.1 Map of Ghana showing the ten administrative regions and the marine coastline on the Gulf of Guinea

lighters – on the Volta, Ankobra, and Tano rivers. Ghana's coastline stretches for 539 km along the Gulf of Guinea and is mostly a low, sandy shore backed by plains and scrub. It is intersected by several rivers and streams, most of which are navigable only by canoe.

The country is well endowed with natural resources, e.g. gold, timber, industrial diamonds, bauxite, manganese, fish, rubber, and hydropower, and has twice the per capita output of the poorer countries in West Africa. Gold, timber, and cocoa production are its major sources of foreign exchange while the domestic economy continues to revolve around subsistence agriculture. Agriculture, including natural resource extraction, accounts for 47 percent of Ghana's GDP and employs 54.7 percent of the work force, mainly small landholders. Ghana's land use includes 12 percent in arable land, 7 percent in permanent crops, 22 percent in permanent pastures and 35 percent in forests and woodlands with the remaining 24 percent in other uses (1993 est.). There are also 60 km² of irrigated farmland (1993 est.). The recent drought in the north has severely affected agricultural activities in that region. However, throughout the country deforestation, overgrazing, soil erosion, poaching, and habitat destruction threaten wildlife populations. Water pollution and inadequate supplies of potable water also pose serious problems for the approximately 19.5 million inhabitants of Ghana.

Ghana is party to a number of international agreements affecting the environment including the Biodiversity, Climate Change, Desertification, Endangered Species, Environmental Modification, Law of the Sea, Nuclear Test Ban, Ozone Layer Protection, Ship Pollution, Tropical Timber 83, Tropical Timber 94, and Wetlands agreements. Ghana has also signed but not yet ratified the Marine Life Conservation agreement.

The effect of pesticides on public health and on the environment including the tropical marine environment of Ghana are of special interest to the nation. The marine environment of Ghana is polluted from a variety of sources: industrial waste effluent of several industries, e.g. breweries, textile manufacturers, fish processors, paint manufacturers, poultry and livestock feedmills, cement manufacturing companies, and oil refineries, located in the coastal industrial towns of Accra, Tema, Sekondi, Takoradi, and Cape Coast; discharges from the inland Lake Volta and the rivers Densu, Tano, Ankobra, and Pra; domestic sources such as sewage; agricultural activities; mining; quarrying; and others.

In the past 15 years, there has been a growing interest and increased investment in agriculture to boost food production in Ghana. This has led to increased use of pesticides for the control or eradication of agricultural pests. Most of the pesticides used for agricultural and other purposes are non-specific and can affect both target and non-target organisms if they reach the marine ecosystems of Ghana. This chapter presents the status of pesticide pollution in the tropical marine environment of Ghana. Although some of the problems associated with the pollution of Ghana's marine environment have been recognized, very little research work has been done to assess the magnitude of the pollution problem and the nature of the pollutants.

PESTICIDE LEGISLATION IN GHANA

For many years, there was no organization or agency responsible for regulating the importation, manufacture, distribution, handling, and use of pesticides in Ghana. Early attempts by some government organizations such as the Ministry of Health, Ministry of Agriculture, and the Ghana Standards Board to control the importation and distribution and use of pesticides in the country were not successful due to lack of coordination and logistics. The Environmental Protection Council, now referred to as the Environmental Protection Agency (EPA), was established in 1974 as an advisory and research organization to coordinate the activities of other government bodies concerned with issues related to the environment. It adopted procedures that required all importers and formulators of pesticides to seek clearance for the importation of pesticides (Environmental Protection Council, 1991). In this regard, chemicals are screened by the agency before they are imported into the country. The EPA is assisted and supported by the Ghana Standards Board

(GSB), Ghana Customs, Excise and Preventive Services, Factories Inspectorate Department, and the Plant Quarantine and Regulatory Services. This screening provides a form of control over the industrial chemical trade within the country and the importation of chemicals into the country.

In 1994, the EPA Act (Act 490) was promulgated and, eventually the Pesticide Control and Management Bill (PCMB) was passed in 1996. The PCMB ensures an effective monitoring of the importation, distribution, storage, sale, and use of pesticides and other hazardous substances in the country. The Ghana Government enacted the PCMB to protect farmers, consumers, and the environment from harmful chemicals by ensuring quality in the manufacture of these chemicals and judicious use of these substances (particularly pesticides). Through this PCMB, the government hopes to ensure that end users are able to buy efficacious pesticides and apply them to best advantage; protect pesticide applicators, handlers, and those in contact with pesticides against the toxic effects of pesticides, especially irreversible effects; protect the general population against hazardous residues in foodstuffs; and protect the environment, especially non-target wildlife, against undesirable effects or hazards from applied pesticides.

The legislation is divided into four parts and the essential provisions of each part are as set out below. Part I addresses the registration of pesticides and provides for the compulsory registration, by Ghana's EPA, of all pesticides unless a pesticide is exempted from registration under Clause 2. Under such circumstances an unregistered pesticide may be imported for experimental or research purposes but not for general distribution, imported in the event of a national emergency, or imported only for purposes of a direct transhipment through Ghana (and it is permitted for entry into the country of destination). Part I also outlines the method for companies to apply for registration of pesticides and the mode and basis of classification of pesticides. Pesticides are classified as restricted, suspended, or banned and are subject to the 'Prior Informed Consent Procedure', which enables the national pesticide law to be linked with internal regulations at the local district and regional levels in order to meet obligations under international trade protocols for control of pesticides and other chemical substances.

Ghana's EPA may approve and register a pesticide under such conditions as it may determine and may only register a pesticide if it is satisfied that the pesticide is safe and effective for the use for which it is intended and that the pesticide has been tested for efficacy and safety under local conditions. In addition, it has the power to ban or suspend the registration of any pesticide or cancel the registration where it is satisfied that the pesticide is not effective or may cause hazards to people, animals, crops, or the environment (Environmental Protection Council, 1991). There is also a provision for the provisional clearance of some pesticides, valid for no more than two years, for use before registration under certain circumstances. The EPA also has the authority to cancel such provisional clearance status. Applicants are informed in writing of refusals to register a pesticide and the grounds for the refusal. A pesticide registration remains valid for a period not exceeding two years and is renewable subject to review. Ghana's EPA maintains a registry of pesticides with the names and particulars of registered and provisionally cleared pesticides and publishes, annually, in the *Gazette* a list of registered pesticides, their classification, and amendments made to their classifications.

Part II of the legislation sets forth requirements for the licensing of dealers in pesticides, which are defined as importers, exporters-manufacturers, distributors, advertisers, and sellers of pesticides. Dealers are required to obtained licenses as a prerequisite for dealing in pesticides and to keep records of their activities, retain them for 10 years, and make them available to the EPA when so required. The EPA also sets out safeguards for the use of pesticides and has the power to restrict or prohibit the use of a registered pesticide in designated areas within specified periods.

The legislation prescribes appropriate containers, packaging, and labels for pesticides; stipulates that certain acts in relation to the adulteration of registered pesticides are prohibited; sets forth certain standards for the chemical component of registered pesticides; and prohibits the false advertisement of registered or provisionally cleared pesticides. Furthermore, it establishes a Pesticides Technical Committee of the GSB to perform pesticide control and management functions under the Bill as directed by the Board.

Part III details the enforcement provisions of the PCMB and provides for inspectors and persons authorized by District Assemblies to perform the functions of properly implementing the provisions of this law. It also provides penalties for failing to comply with provisions of the law.

The final part of this landmark legislation contains several general provisions that enjoin customs officers to assist in enforcement, provided that the Minister for the Environment in consultation with the Minister responsible for Food and Agriculture makes regulations for the full implementation of the law, and allowed a 6-month grace period for registration of pesticides and issuance of licenses before sanctions could be imposed (Environmental Protection Council, 1991). Ghana's EPA is working with other appropriate bodies to establish the necessary machinery for executing provisions of the law.

PESTICIDE USE IN GHANA

For the past several decades, pesticides have been used in Ghana in the public health sector for disease vector control and in agriculture to control and eradicate crop pests. However, there has been a rapid rise in the quantity of pesticides used in agriculture over the past ten years and this growth trend is expected to continue for the next ten or more years. Agricultural pesticides are used in the cocoa growing industry (cacao tree cultivars from the genus *Theobroma*), for cotton farming, in vegetable production, and for other mixed-crop farming systems involving maize, cassava (perennial shrubs of the genus *Manihot* grown for the edible tubers), cowpeas (the vine-like herb *Vigna unguiculata* L. grown for its edible seeds), sugarcane, rice, etc. The majority of pesticides used in agriculture are employed in the forest region located in the Ashanti, Brong Ahafo, western, and eastern regions of Ghana.

Organochlorine pesticides are widely used by farmers because of their cost effectiveness and their broad spectrum activity. Lindane is widely used in Ghana on cocoa plantations, on vegetable farms, and for the control of stemborers in maize (Ministry of Agriculture, 1990). Endosulfan is widely used in cotton growing areas, on vegetable farms, and on coffee plantations. Most of the farms in Ghana's vegetable growing areas are situated along rivers that also serve as the water supply source for farming and for drinking water.

Pesticides are also employed to control ectoparasites of farm animals in Ghana. In a study encompassing 30 organized farms and 110 kraals distributed throughout the 10 regions of Ghana, Awumbila and Bokuma (1994) found that 20 different pesticides were in use with the OC lindane being the most widely distributed and used pesticide, accounting for 35 percent of those used on farms and 85 percent of those used by herdsmen. Of the 20 pesticides, 45 percent were OPs, 30 percent were pyrethroids, 15 percent were carbamates, and 10 percent were OCs (Awumbila and Bokuma, 1994). The favorite method of application was by hand dressing and no post application interval was used before slaughter or sale of milk for human consumption (Awumbila, 1996). In this regard, there is the potential for significant risk to human health, both for the applicators and the end-users of agricultural products. Clarke *et al.* (1997) studied the knowledge, attitudes, and practices of 123 farm workers on three irrigation project areas in the Accra Plains regarding the safe handling and use of pesticides. They found moderate levels of knowledge of pesticide absorption routes and of potential symptoms following exposure. However, farm workers' knowledge of personal protective measures was poor to moderate and their use of personal protective equipment was minimal – due primarily to financial constraints. They concluded that training of agriculture and health workers in safety precautions, recognition, and management of pesticide poisoning was a matter of urgency.

In the public health arena, pesticides, primarily temephos (Osei-Atweneboana *et al.*, 2001), have been used by the Onchocerciasis Programme in the Volta Basin for control of black flies (*Simulium* spp. Diptera: Simuliidae), which transmit Onchocerciasis (African river blindness, a disease caused by the pathogenic nematode, *Onchocerca volvulus*) to humans and for control of domestic pests, e.g. cockroaches, various flies, mosquitoes, ectoparasites including ticks, and other insects. Blindness caused by Onchocerciasis is a serious and ongoing problem in this region of Africa such that the systematic treatment of most rivers across West Africa is likely to continue for the foreseeable future. However, Osei-Atweneboana *et al.* (2001) have found evidence for mild resistance development (slight to five-fold) to temephos in black fly populations from two sites in south-west Ghana – Sutri Rapids on the Tano River and Sekyere-Heman on the Pra River – where the larvicide had not been applied. They speculate that the likely cause of resistance could have been local selection of black flies exposed to agrochemical runoff from cocoa, coffee, and palm oil plantations flanking the rivers.

No detailed records have been kept on the volume and types of pesticides used in the country, except for a few chemicals. However, analysis of pesticide trade flow patterns, recorded by Ghana's Statistical Service, between Ghana and other countries for the period January to December 1992 and January to June 1993 indicate a total of 2,589,254 kg and 1,264,872 kg of pesticides respectively, were imported during this period (Boateng, 1993a; 1993b). Analysis of external trade statistics compiled from custom bills of entry completed by importers and exporters or their agents show that pesticides are mostly imported rather than exported in Ghana.

Analysis of the data available for the period January to December 1992 showed the following distribution of the various types of pesticides: insecticides (60.8 percent), herbicides (24.2 percent), fungicides (8.9 percent), rodenticides (1.5 percent), and other pesticide types (4.6 percent) (Boateng, 1993a). A survey conducted between 1992 and 1994 found that the following pesticides are most commonly used by farmers (in percentage of farmers using a pesticide): copper (II) hydroxide (kocide) 29.0 percent, mancozeb (dithane) 11.0 percent, fenitrothion (sumithion) 6.0 percent, dimethoate (perfekthion) 11.0 percent, pirimiphos methyl (actellic) 11.0 percent, λ-cyhalothrin (karate) 22.0 percent, and endosulfan (thiodan) 10.0 percent. Lindane is widely distributed by the Ministry of Agriculture and was not included in the survey. The study also indicated that insecticides constitute about 67 percent of pesticides used by farmers while fungicides made up about 30 percent and herbicides and other pesticide types were about 3 percent of total use (Osafa Acquaah and Frempong, 1995). Table 2.1 lists some of the pesticides used in agriculture in Ghana based on a survey conducted from 1992 to 1994 in the Ashanti, Brong Ahafo, eastern, and western regions of Ghana.

PESTICIDE RESIDUES IN BIOTA AND NON BIOTIC MATRICES IN THE INLAND, COASTAL, AND MARINE ECOSYSTEMS OF GHANA

OC residues in the interior of Ghana

Osafo Acquaah (1997) measured lindane and endosulfan in river water and fish tissues collected from rivers passing through regions of intense cocoa production and other farming activities in the Ashanti Region between 1993 and 1995. Water and fish samples from 1995 contained both lindane and endosulfan – these were the only two pesticides being investigated but chromatogram peaks corresponding to the DDTs and other OCs were observed – with much lower levels of both pesticides present in water samples. For the Oda, Kowire, and Atwetwe rivers, mean concentrations found in water samples for lindane and endosulfan were 19.4, 12.4, 16.4 and 17.9, 20.5, and 21.4 ng L^{-1}, respectively (Osafo Acquaah, 1997). Concentrations in all the fish varied by species and month of sampling but were generally higher for lindane and much higher for endosulfan.

Ntow (2001) measured OC pesticide residues in water; sediment; the tomato crop, which constitutes more than 90 percent of the major vegetables grown in the

Table 2.1 Trade names of agrochemicals, classified by use, used by farmers from the Ashanti, Brong Ahafo, eastern, and western regions of Ghana

Insecticide	Herbicide	Fungicide	Other
Actellic (primiphos methyl)[a]	Gesaprim (atrazine)	Baygon (propoxur)	Grofol (foliar 20-30-10 NPK + micronutrients)
Cymbush (cypermethrin)	Bellater (atrazine + cyanazine)	Caacobre (copper (II) oxide)	Pracol (ampicillin trihydrate –antibacterial)
Desis (deltamethrin)	Garlon (triclopyr, a pyridine herbicide)	Champion (copper (II) hydroxide)	
Diazinon	Gramoxone (paraquat)	Cobox (copper oxychloride)	
Dimethoate 40 EC	Ronstar (oxadiazon)	Dithane (mancozeb)	
Dursban (chlorpyrifos)	Roundup (glyphosate)	Kocide (copper (II) hydroxide)	
Endosulfan	Stam F34T (propanil)	Gastoxin (aluminum phosphide)	
Fenitrothion	Basta 20 SL (glufosinate)		
Fenom C (profenophos + cypermethrin)	Sturnmate (thiobencarb)		
Primigram (metolachlor + atrazine)	Bladex (cyanazine)		
Nogos (dichlorvos)	Bladex (cyanazine)		
Ofunack (pyridiphenthion)			
Perfekthion (dimethoate)			
Phostoxin (aluminium phosphide)			
Ripcord (cypermethrin)			
Secto (lindane + synergized pyrethroids)			
Sumithion (fenitrothion)			
Thiodan (endosulfan)			
Trebon (etofenprox)			
Gammalin 20 (lindane)			
Callifan 50CE (endosulfan)			
Furadan (carbofuran)			
Unden (propoxur)			
Karate (λ-cyhalothrin)			
Totals 24	10	7	2

Note:

a Common names for a.i.(s) of commercial products are given in parenthesis where known.

district; and human fluids – blood samples and mothers' breast milk – for 208 samples collected from the environs of Akumadan, a vegetable farming community located 95 km northwest of Kumasi in the Ashanti Region. Aldrin, dieldrin, endrin, 2,3,5-TCB, p,p'-DDD, and p,p'-DDT were not detected in any of the samples. However, endosulfan sulfate, α-endosulfan, β-endosulfan, and lindane were detected in water samples taken from four area streams and a community standpipe at mean levels (frequency of detection in parentheses) of 30.8 ng L^{-1} (78 percent), 62.3 ng L^{-1} (60 percent), 31.4 ng L^{-1} (60 percent), and 9.5 ng L^{-1} (76 percent), respectively (Table 2.2). Sediment samples from the four streams contained all seven OCs, each appearing in 88 percent or more of the samples analyzed. Lindane was detected at the highest level (3.2 μg kg^{-1}) followed by HCB (0.9 μg kg^{-1}), heptachlor epoxide (0.63 μg kg^{-1}), and p,p'-DDE (0.46 μg kg^{-1}). HCB and p,p'-DDE were the only OCs detected in blood and milk and both were at lower levels in milk than those reported in industrialized countries. Ntow (2001) estimated the daily intake of DDE and HCB by nursing infants and found that consumption of both pesticides were below WHO/UNEP acceptable daily intakes.

The Ghanaian coastline

The Ghanaian coastal zone is a part of the coastal area of the Gulf of Guinea and is subject to some level of erosion. The degree of erosion depends on the location and geology of the area of concern. The coastal zone, especially around the ports of Accra, Tema, and Takoradi, has been the major area of industrial development in Ghana. Almost 60 percent of all industries in the country are located in the Accra-Tema metropolis, which covers less than 1 percent of the total area of Ghana. Along the entire coastline of Ghana, discharges into the environment are, to a large extent, untreated and unregulated, thus increasing the risk of pollution and modification to the marine environment especially in areas of high population density (Calamari, 1985; Calamari and Naeve, 1994).

The marine area of Ghana

The marine area of Ghana covers about 203,720 km^2 based on the 550 km stretch of Ghana's coastline and encompasses the area between the landward low-water mark of the coastal zone and the seaward boundary of the 200 nautical mile Exclusive Economic Zone, which is 370.4 km from shore. This area is equivalent to 85 percent of the total land area of the country.

Impact of inland drainage and land-based pollution sources on coastal lands and waters

The Volta River system which, apart from its large tributaries, includes Volta Lake and the Kpong Headpond is the dominant inland drainage system in Ghana. Some other major rivers, including the Tano, Ankobra, Pra, and Densu, also drain into the sea through coastal lagoons and estuaries. These rivers may carry pollutants

Table 2.2 Organochlorine pesticide residues in water, sediment, tomato, and human fluids in a vegetable farming community located in the Ashanti Region of Ghana

Pesticide	Water ng L⁻¹	Water % of samples	Sediment µg kg⁻¹ dwᵃ	Sediment % of samples	Tomato µg kg⁻¹ fwᵇ	Tomato % of samples	Blood µg kg⁻¹	Blood % of samples	Milk µg kg⁻¹	Milk % of samples
HCB	BDᶜ	10	0.90	90.5	<0.10ᵈ	51.3	30	55	1.75	95
Lindane	9.5	76	3.20	95.2	<2.50	23.7	NDᵉ	0	ND	0
p,p´-DDE	BD	8	0.46	88.1	<0.10	7.9	380	85	17.15	80
Heptachloor epoxide	BD	12	0.63	97.6	1.65	55.3	ND	0	ND	0
α-endosulfan	62.3	64	0.19	95.2	<0.05	23.7	ND	0	ND	0
β-endosulfan	31.4	60	0.13	88.1	<0.01	11.8	ND	0	ND	0
Endosulfan sulfate	30.8	78	0.23	97.6	<0.01	35.5	ND	0	ND	0

Source: Adapted from Ntow, 2001.

Notes:
a dw indicates dry weight.
b fw indicates fresh weight.
c BD indicates compound detected but below detection limit of 100 ng L⁻¹.
d (<) indicates compound detected but below detection limits.
e ND indicates compound not detected.

originating from point sources such as municipal sewers and industrial discharges into the coastal zone. The principal sources of land-based pollution are municipal, industrial, and agricultural discharges, which are usually untreated and unregulated. Agricultural practices contribute to the release of pollutants, mainly pesticides and fertilizers, into the coastal zone through land runoff, river loading, and atmospheric transport. In drainage basins, where the levels of pollutants such as pesticides, trace metals, and nutrients are high as a result of industrial, agricultural, or domestic activities, the continuous discharge of such surface waters may cause localized pollution of coastal environments, especially in lagoons and estuaries.

Marine pollution and pesticide residue research

Within the framework of the United Nations Environmental Program (UNEP) Regional Seas Program, four action plans covering African countries have been signed. The action plans covering West and Central Africa, of which Ghana is a part, comprise projects on marine pollution monitoring and research with components for analysis of metals and OCs in biota; oil pollution monitoring on beaches and in coastal waters; and bacteriological quality control of bathing waters (UNEP, 1985).

Many chlorinated insecticides, e.g. DDT, aldrin, dieldrin, and endosulfan, have been used in Ghana for more than three decades in agriculture, vector control, and other facets of public health. So far, however, no studies have looked for the presence of OCs in biota. Biney (1982) identified the main sources of pollution of the coastal environment of Ghana as sewage of both industrial and domestic origin and oil, in the form of tar balls. Of the 16 lagoons he investigated, 12 were found to be polluted and two were grossly polluted. The grossly polluted lagoons, the Korle and Chemu estuaries, receive both industrial and domestic wastes. The estuaries were generally slightly polluted. The major problem along the beaches was pollution by domestic sewage including refuse. Sewage pollution of beaches was found to be commonly associated with 'high' population areas including the urban centers of Accra, Tema, Cape Coast, and Takoradi.

RECOMMENDATIONS

In view of the paucity of studies on OCs in biota, the following recommendations are made:

Policy

1   Existing legislation, regulations, and standards concerning the marine environment and the coastal zone must be enforced. There is a need to strengthen and broaden the current role of the Environmental Protection Agency to ensure enforcement of existing laws.

2   The legal regime for the protection of the marine environment must be further developed. Comprehensive laws dealing with pollution of surface and underground waters by pesticides must be enacted. Environmental impact assessments of proposed projects in Ghana must be given due consideration.

Research

1   Studies should be carried out on the extent and effect of pollution with particular attention to pesticides in marine and coastal environments. The objective would be to collect sufficient data to properly develop and manage the marine environment. This would include measuring the level of contaminants, studying contaminant accumulation that might lead to biological impacts, and recording baseline data on the distribution of flora and fauna on beaches, in coastal lagoons and estuaries, and in other near-shore ecosystems.

2   Conduct studies on the impact of inland drainage and land-based pollution sources on coastal lands and waters. The objective would be to develop the scientific basis for legislative provisions for appropriate abatement and control measures. These studies should include the characterization of industrial effluent to identify the most hazardous pollutants that would require immediate control measures to be implemented, an estimation of the rate of input of pollutants into the coastal zone from land-based sources, the distribution pathways of pollutants in coastal waters, and long-term studies on the biological impact of pollution discharges into the coastal zone.

Education

1   Training programs should be established to increase public education on sound environmental practices. Furthermore, agricultural management practices that minimize excessive use of pesticides and chemical fertilizers should be developed and instituted. Training programs for pesticide handlers and applicators should be implemented.

2   The training of more environmental scientists and managers, both locally and overseas, should be accelerated.

Monitoring

An effective monitoring system must be established to monitor estuarine and marine pesticide levels, oil pollution, industrial and sewage pollution, and to conduct research including analysis for metals and OC pesticides in biota. Additionally, pollution monitoring of beaches and coastal waters should be instituted and bacteriological quality control of bathing waters should begin.

Management

Relevant institutions must be strengthened and restructured through provisions for skilled staff, state-of-the-art equipment, and logistical support, as necessary. This would ensure reliable data gathering, standard setting, and monitoring as well as effective execution and delivery of environmental projects and programs.

REFERENCES

Awumbila, B. 1996. Acaricides in tick control in Ghana and methods of application. *Tropical Animal Health and Production* 28(2):S50–S52.

Awumbila, B. and Bokuma, E. 1994. Survey of pesticides used in the control of ectoparasites of farm animals in Ghana. *Tropical Animal Health and Production* 26(1):7–12.

Biney, C. 1982. Preliminary survey on the state of pollution of the coastal environment of Ghana. *Oceanal Acta.* 4(suppl.).

Boateng, O. 1993a. External Trade Statistics January–December 1992. Accra, Ghana: Statistical Services, pp. 116–77.

Boateng, O. 1993b. External Trade Statistics January–June 1993. Accra, Ghana: Statistical Services, pp. 79–80.

Calamari, D. and Naeve, H. (eds) 1994. Review of pollution in the African aquatic environment. *Committee on Inland Fisheries Association (CIFA) Technical Paper* 25. Rome: FAO.

Calamari, D. 1985. Review of the state of aquatic pollution of west and central African inland waters. *Committee on Inland Fisheries Association (CIFA) Occasional Papers* 12:25.

Clarke, E.E.K., Levy, L.S., Spurgeon, A. and Calvert, I.A. 1997. The problems associated with pesticide use by irrigation workers in Ghana. *Occup Med Oxf.* 47(5):301–8.

Environmental Protection Council. 1991. *Ghana environmental action plan*, Volume1. Accra, Ghana: Environmental Protection Council.

Ministry of Agriculture. 1990. *Technical Bulletin* GDB/MI. Accra, Ghana: Ministry of Agriculture.

Ntow, W.J. 2001. Organochlorine pesticides in water, sediment, crops, and human fluids in a farming community in Ghana. *Arch Envir Contam Toxicol.* 40(4):557–63.

Osafo Acquaah, S and Frempong E. 1995. *Organochlorine Insecticides in African Agroecosystems.* Vienna: IAEA. IAEA TECDOC-931, pp. 111–18.

Osafo Acquaah, S. 1997. Lindane and endosulfan residues in water and fish in the Ashanti Region of Ghana. In: *Environmental Behaviour of Crop Protection Chemicals.* Proceedings of a symposium held 1–5 July 1996 in Vienna by the IAEA/FAO. Vienna, Austria: IAEA, pp. 471–8.

Osei-Atweneboana, M.Y., Wilson, M.D., Post, R.J. and Boakye, D.A. 2001. Temephos-resistant larvae of *Simulium sanctipauli* associated with a distinctive new chromosome inversion in untreated rivers of south-western Ghana. *Medical and Veterinary Entomology* 15(1):113–16.

UNEP. 1985. Industry and environment: Agrochemicals and their impact on the environment. *UNEP Technical Publications* 8(3) July/August/September.

Chapter 3

Pesticide use in Zimbabwe

Impact on Lake Kariba, a tropical freshwater ecosystem

Mark F. Zaranyika

INTRODUCTION

Zimbabwe is situated within the African tropics (Lat. 15° to 22°S and Long. 26° to 34°E), occupies an area of 390,580 km² and has a population of about 11 M. The Zambezi River forms the boundary with Zambia to the north, and the Limpopo River forms the boundary with South Africa to the south. The eastern highlands that form the rim of the African Plateau (before descent to the Mozambique Coastal Plain), constitute the greater part of the border with Mozambique. In the west, the boundary with Botswana follows the eastern limit of the Kalahari Desert.

Zimbabwe is a landlocked country. Its GDP for 1990 and 1994 was Z$14,643 M and Z$39,775 M, respectively. The agriculture sector contributes about 13 percent of GDP, while the export of agricultural products contributes 50 percent of the country's total annual export earnings (Central Statistical Office, 1990). The use of pesticides plays a major role in maintaining these high levels of agricultural production. As in most tropical countries in Africa, pesticides are also extensively used in the public health arena to control diseases such as malaria (a nonhemorrhagic fever caused by protozoans of the genus *Plasmodium Marchiafava* and *Celli* and vectored by *Anopheles Meigen* spp. (Diptera: Culicidae) mosquitoes), African trypanosomiasis (African sleeping sickness, a nonhemorrhagic fever caused by protozoans of the genus *Trypanosoma Gruby* and vectored by tsetse flies, *Glossina* spp. (Diptera: Glossinidae), and typhoid (a bacterial illness caused by *Salmonella typhi* spread in contaminated food and water). Of late, there has been concern about the possible effects the use of pesticides has on tropical environments, including tropical marine and fresh water ecosystems. The effect of pesticides on the environment depends on several factors such as climate, in particular temperature and rainfall; soil type and the nature of the vegetative cover; biotic activity; light intensity; agricultural practices; and mode of introduction of the pesticide into a particular environmental compartment. These factors determine the persistence of a pesticide in a specific environment, and in this respect, OC pesticides as a group have been found to be the most persistent.

OC pesticides have been used extensively since the early 1960s to control the tsetse fly and malaria vectors in southern east-central Africa, i.e. Zimbabwe,

Mozambique, and South Africa (Ford, 1971). DDT has been used for this purpose since 1962, while dieldrin was used during the period 1962 to 1967 (Ford, 1971; Mpofu, 1987; Whitwell et al., 1974). Endosulfan and BHC are currently used, especially for aerial spraying (Chapman, 1976).

DDT was used in Zimbabwe for more than four and a half decades, from 1946 to 1983 (Chikuni, 1996). In addition to its use to control the tsetse fly and malaria vectors, DDT was used extensively for the control of agricultural pests such as the maize stalkborer Busseola fusca Fuller (Lepidoptera: Noctuidae), cotton cutworm Agrotis spp. (Lepidoptera: Noctuidae), and cotton bollworm Helicoverpa armigera Habner (Lepidoptera: Noctuidae). DDT was used in agriculture until 1983 when this use was banned. However, it is still registered with the Ministry of Health's Hazardous Substances Control Board as a 'hazardous substance class 1', i.e. a chemical that can endanger humans and domestic and wild animals, and its procurement and use are restricted to cover tsetse and mosquito control only (Chikuni, 1996). Other OC pesticides registered for use in agriculture in Zimbabwe include dieldrin, endosulfan, BHC, aldrin, chlordane, dicofol, and chlorthal-dimethyl.

This chapter discusses the use of pesticides in Zimbabwe and how this has impacted on Lake Kariba, a tropical freshwater ecosystem. Lake Kariba is one of the world's largest man-made lakes. It was constructed in the mid-1950s, started to fill in 1958, and reached full capacity in 1963. Situated in south-central Africa (between Lat. 16°30' to 18°S, and Long. 27° to 39°E), the lake is politically shared by Zambia and Zimbabwe, with the international border bisecting the lake longitudinally (Figure 3.1). Its physical dimensions are given in Table 3.1.

Geographically Lake Kariba is part of the middle Zambezi region and lies in a rift valley (the Gwembe Valley), overlooked on both sides by steep escarpments. The mean maximum temperature is 30.4°C, while the minimum annual mean temperature is 18.2°C. Rainfall around the lake region is generally low; the annual mean for the period 1951 to 1986 was 734 mm (Leggett et al., 1991). Generally the wet season occurs in the months of December to March, with occasional storms

Table 3.1 The physical dimensions of Lake Kariba, Zimbabwe at the normal operating water level (see Figure 3.1)[a]

Water level (above sea level)	485 m
Length	277 km
Mean breadth	19.4 km
Mean depth	29.18 m
Maximum depth[b]	93 m
Surface area	5364 km²
Volume	156.5 km³
Theoretical renewal time of the water mass	3 years approx.

Notes:
a Source Balon and Coche (1974).
b Occasional deeper 'holes' not considered.

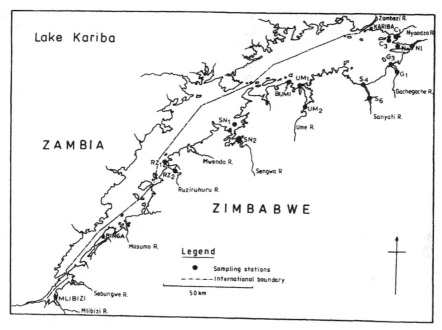

Figure 3.1 Lake Kariba with inflows from Zimbabwean rivers
Source: Wessels *et al.* (1972).

in October, April, and May. There is no rainfall from June to September. Evidence
of a pesticide residue build-up in Lake Kariba has been reported by several workers
(Billings and Phelps, 1972; Whitwell *et al.*, 1974; Greichus *et al.*, 1978; Wessels *et
al.*, 1980). Thus, in 1980, Wessels *et al.* were prompted to warn that 'in view of the
extensive fishery development taking place on the lake, the problem of residues in
human food may become a serious matter'. The background and extent of the
problem are the subject of this chapter.

IMPORTATION, MANUFACTURE, AND
REGULATION OF PESTICIDES IN ZIMBABWE

The use of pesticides in agriculture:
importation and regulation of pesticides in
Zimbabwe

The use of pesticides in Zimbabwe is regulated in terms of the Pesticide Regulations
of 1977, under the Fertilizer, Farm Feeds and Remedies Act (Chapter 111) of
1952 (Government of Zimbabwe, 1952) and the Hazardous Substances and Articles
Act (Chapter 322) of 1972 (Government of Zimbabwe, 1972). The Fertilizer, Farm
Feeds and Remedies Act (Chapter 111) is administered by the Ministry of Lands,
Agriculture and Rural Resettlement. This act prohibits the sale or distribution of

pesticides unless they have been registered with the Plant Protection Research Institute. Registration is carried out under the Hazardous Substances and Articles Act (Chapter 322) which is administered by the Ministry of Health and Child Welfare.

Before registration, pesticides are classified on the basis of their acute oral lethal dose (LD_{50}) and persistence after application. The classification (poison group) is indicated by a green, amber, red, or purple triangle on the label for LD_{50} values of greater than 2,001, 501 to 2,000, 101 to 500, or 0.1 to 100 mg kg^{-1}, respectively. Pesticides can only be imported into the country after they have been registered. It is also a requirement that all imported pesticides be registered in the country of origin.

The complete list of registered pesticide products in Zimbabwe is very long (approximately 600), but compared to the total number of pesticides available worldwide (>40,000), this number is small. Several companies are involved in the formulation and marketing of pesticides in Zimbabwe. The major formulators are (with the number of formulations registered by each company shown in brackets): Agricura (99), Zimbabwe Fertilizer Co. (ZFC) (71), Windmill (66), Bayer (62), Shell (48), Ciba-Geigy (34), Spray-quip (26), Hoechst (16), Omnichem (15), and Wellcome Environmental (17) (Mathuthu, 1993). There are several minor pesticide formulating companies including Rhone Poulenc (11), Technical Services (6), Fercochem. (4), Oxyco (4) and T.S.A. (2). Some formulations of Agricura and ZFC are made totally from local raw materials. The other companies, to a large extent, merely import the active ingredients from which they make their formulations.

A list of pesticides commonly used for crop pests in Zimbabwe is given in Tables 3.2 and 3.3. The formulations commonly marketed in Zimbabwe were selected following a market survey, which involved visits to major outlets that sell pesticides, e.g. the Farmers Cooperation, Agricultural Buying and Veterinary Services, wholesale centers, and supermarkets, in addition to interviews with farmers (Mathuthu, 1993). The pesticides listed in Tables 3.2 and 3.3 are those formulations that are most commonly used around the country. The tables show the brand name and a.i. of the pesticide, its poison group, and the chemical class of the compound. The poison group indicates the degree of toxicity of the pesticide and is indicated by the symbols P (Purple), R (Red), A (Amber), and G (Green), with P indicating the most toxic pesticides and G the least toxic.

The Agricultural Chemicals Industry Association (ACIA) represents all manufacturers and distributors of agrochemical and animal health products in Zimbabwe. The ACIA is a member of the International Group of National Associations of Agrochemical Manufacturers (GIFAP). Through GIFAP, the ACIA has endorsed the FAO's code of conduct on the distribution and use of agrochemicals (Mbanga, 1996).

Table 3.2 Pesticides used in Zimbabwe: grouped according to chemical classes of the compounds[a]

Brand name	Active ingredient	Poison group[b]	Type of compound
Alfacron 50 W.P.	azamethiphos 50%	R	OP
Kaptasan F	captan 31.35% + fenitrothion 1%	G	OP
Steladon	chlorfenvinphos 30%	P	OP
Dursban 4E	chlorpyrifos 40.8%	R	OP
Fly Bait	dichlorvos 0.5%	A	OP
Diazinon DFF	diazinon 86.88%	P	OP
Diaz 30	diazinon 30%	R	OP
Rogor C.E.	dimethoate 36%	R	OP
Dimethoate 40 E.C.	dimethoate 40%	R	OP
Disystem 5% granule	disulfoton 5%	P	OP
Altomix 7.75G	disulfoton 7.5% + cyproconazole 0.25%	P	OP
Lebaycide 50%	fenthion 50%	R	OP
Folithion 60 E.C.	fenitrothion 55.49%	R	OP
Kontakil	fenitrothion 60%	R	OP
Ant-Killer	fenitrothion 17.5%	A	OP
Anthio 33 E.C.	formothion 33%	R	OP
Fyfanon 1000 E.C.	malathion 86%	A	OP
Kilathion 100 E.C.	malathion 83.5%	A	OP
A.B.C. powder (dust)	malathion and other	G	OP
Damfin 2P	methacrifos 2%	G	OP
Kudzivirira Mbesa	malathion 1	G	OP
Malathion 1% Dust	malathion 1%	G	OP
Malathion 5% Dust	malathion 5%	G	OP
Malathion 25% W.P.	malathion 25%	G	OP
Malathion 50% E.C.	malathion 50%	G	OP
Python 21	malathion 2.2%	G	OP
Metasystox R 25% E.C.	oxydemeton-methyl 25%	R	OP
Wellcome grainguard	pirimiphos-methyl 48.8%	G	OP
Shumba 2% dust	pirimiphos-methyl 2%	G	OP
Bolstar 720 E.C.	sulprofos 72%	R	OP
Sprayquip stalkborer 2% granules	trichlorfon 2.5%	G	OP
Aldrin 40% W.P.	aldrin 40%	P	OC
Anti-Kil	chlordane 30%	A	OC
Razor	chlorthal-dimethyl 36%	G	OC
Dicofol 40 E.D.	dicofol 40%	A	OC
Kelthane	dicofol 18.5%	G	OC
Dieldrex 50 W.P.	dieldrin 50%	P	OC
Thionex 1% granules	endosulfan 1%	G	OC
Thiodan 1% granules	endosulfan 1%	G	OC
Thiodan 20 E.C.	endosulfan 20%	P	OC
Multi Benhex	γ-BHC 12% + total BHC 75%	R	OC
Gamatox house spray	γ-BHC 5.0%	A	OC
Bexadust (L)	γ-BHC 0.6%	G	OC
Agri seed dress 75%	lindane 1%	R	OC
Temik 15G	aldicarb 15%	P	carbamate
Carbaryl 85 S	carbaryl 85%	A	carbamate
Carbaryl 85 W.P.	carbaryl 85%	A	carbamate
Harakiri	carbaryl 0.3%	G	carbamate

continued...

Table 3.2 continued

Brand name	Active ingredient	Poison group[b]	Type of compound
Cypam	E.P.T.C. 77%	G	carbamate
Baygon residual spray	propoxur 2.0% + dichlorvos 0.5%	G	carbamate
Decis 2.5 E.C.	deltamethrin 2.5%	R	pyrethroid
Agrithrin Super 5 E.C.	esfenvalerate 5%	R	pyrethroid
ICON 10 W.P.	λ-cyhalothrin	G	pyrethroid
Bymo insect killer	pyrethrins 0.125%	G	pyrethrin
Wellcome	permethrin 25%	G	pyrethroid
G-17	pyrethrins 2.25%	G	pyrethroid
Dusting powder	pyrethrins 0.20%	G	pyrethroid
Garden insecticide concentrate	pyrethrins 1.5%	G	pyrethroid
Killem insect aerosol	tetramethrin 0.2% + Δ-phenothrin 0.08%	G	pyrethroid
Gramoxone	paraquat 24.75%	P	heterocyclic
Fungazil 75% S.P.	imazalil 75%	R	heterocyclic
Thiram 80% W.P.	thiram 80% (disulphide)	R	carbamate (fungicide)
Tritifix	MCPA/amine 41.5%	A	phenoxy acid
Copper fungicide	copper oxychloride 88%	A	inorganic compound
Copper oxychloride 50 E.C.	copper oxychloride 50%	A	inorganic
Dormex	cyanamide 49%	A	inorganic amide
Agri Dust	dusting sulphur 65% + copper oxychloride 6.5% +malathion 5%	A	inorganic compound inorganic salt OP
Arsenal	imazapyr	G	heterocyclic
Cosan wettable sulphur	sulphur 80%	G	inorganic compound
Lime sulphur	polysulphide sulphur 24.8%	G	inorganic compound
Racumin rat poison	coumatetralyl Na+	R	heterocyclic
Basagran	bentazone 48%	A	heterocyclic
Bladex 5 S.C.	cyanazine 50%	A	heterocyclic
Citrocyclin 90	tetracycline + hydrochloride 90%	A	heterocyclic
Funginex	triforine 18.7%	G	heterocyclic
Fumigas 10	ethylene oxide 10%	P	organic compound
Agrifume EDB 4.5	ethylene dibromide 42.2%	P	organohalide
Agrithrin 20 E.C.	fenvalerate 20%	A	organic acid derivative
Daconate 6	MSMA 48%	A	organic acid derivative
MSMA	MSMA 48.4%	A	organic acid derivative
Snail and slug killer	metaldehyde 2%	G	organic acid derivative
Norax ready mixed	warfarin 0.0375%	A	organic compound
NABU	sethoxydim 20%	G	organic compound
Alachlor	alachlor 48%	P	acetanilide
Ronstar FLO	oxadiazon 240 g/l	A	organic amine
Weedkiller M	M.C.P.A. 400 g/l potassium salt	A	organic salt
Atrazine 5 G	atrazine 6.25%	G	organic acid derivative
Bayer Diuron 80 W.P.	diuron 80%	G	dimethylurea
Bayleton 5% W.P.	triadimefon 5%	G	organic derivative
Bayton 15%	triadimenol 15%	G	alcohol
Benlate	benomyl 50%	G	carboxylic acid derivative
Mitac	amitraz 20%	A	organic acid derivative

continued...

Table 3.2 continued

Brand name	Active ingredient	Poison group[b]	Type of compound
Cotogard 500 F.W.	fluometuron 25% + prometryn 25%	G	urea derivative triazine
Gesagra 500 F.W.	metolachlor 25% + triazine 23.5%	G	acetamide
Gibberellic acid	gibberellic acid 32%	G	organic acid
Karathane 2% dust	dinocap 2%	G	nitrophenol
Dithane M-45 W.P.	mancozeb 80%	G	organic acid derivative
Dithane M-45	mancozeb 80%	G	organic acid derivative
Orchard oil	mineral oil 99.7%	G	petroleum oil
Orchex oil 695	mineral oil 99.25%	G	petroleum oil
Pilot S.C.	quizalofop-ethyl	G	organic compound
Ronstar Flo	oxadiazon 36%	G	organic amine
Roundup	glyphosate 41%	G	phosphoglycine
Rovral 250 S.C.	iprodione 25%	G	carboxamide
Rovral	iprodione 50%	G	carboxamide
Sprayquip tak	n-decanol 79%	G	petroleum oil
Stomp 500E	pendimethalin 50%	G	nitrobenzamine
TCA 90 grass killer	sodium trichloro-acetate 90%	G	organic salt
Tordon 101 mixture	2,4-D amine salt 39.6% + picloram 10.2%	P	phenoxy acid carboxcylic acid derivative
Gesaprim 500 F.W.	atrazine 47.0%	G	atrazine
Gesagard 500 F.W.	prometryn 50%	G	triazine
Gardomil 500 F.W.	terbuthylazine 36.7% + metolachlor 12.5%	G	triazine acetamide
Tetradifon 8 E.C.	tetradifon 8%	G	sulfone

Notes:
a Reproduced with permission, Table 23 in SADC ELMS Report Series 35 (1993).
b Poison group (see text for LC$_{50}$s corresponding to each group): A (amber) = toxic; G (green) = non-toxic; R (red) = highly toxic; P (purple) = extremely toxic.

Table 3.3 List of commonly used dipping chemicals in Zimbabwe[a]

Brand name	Active ingredient	Poison group	% a.i.	Class of compound
Fendona	alphacypermethrin	G	5	pyrethroid
Paracide	alphacypermethrin	G	7	pyrethroid
Triatix D	amitraz	G	12.5	amidine
Barricade	cypermethrin	A	15	pyrethroid
Ectopor	cypermethrin	G	2	pyrethroid
Grenade	cyhalothrin	G	5	pyrethroid
Ektoban	cypermethrin	G	2.5	pyrethroid
Decatix	deltamethrin	G	5	pyrethroid
Sumitik	fenvalerate	A	20	pyrethroid
Bayticol	flumethrin	G	20g/L	pyrethroid
Drastic Deadline	flumethrin	G	10g/L	pyrethroid

Note:
a Adapted from SADC ELMS Report Series 35 (1993).

Pesticide pollution from agriculture

Zimbabwe is a landlocked country where elevation and rainfall are highly correlated. Rainfall varies from below 300 mm annually in the low-lying areas in the south and southeast, to more than 1,500 mm in the mountains bordering Mozambique. Rain falls from November to March, and only about one-third of the country is suitable for intensive agriculture. Figure 3.2 shows the agro-ecological regions of Zimbabwe (ENDA, 1991). Traditionally tobacco has been the primary agricultural commodity, although cotton, tea, citrus, livestock, wheat, sugar, and maize are also important. Figure 3.3 shows that all rivers within Zimbabwe originate from the high veld – veld or veldt is the extensive grassland region of eastern and southern Africa – where most of the intensive agriculture is practiced. These rivers drain to the Zambezi River and Lake Kariba in the north, and to the Limpopo River in the south. The Zambezi and Limpopo rivers are the two major rivers flowing to the Indian Ocean.

A major climatic factor in the dispersal of pesticides from agricultural use in Zimbabwe is the fact that rain is usually in the form of short, heavy tropical storms which result in high erosive runoff during the periods that most pesticides are applied in agriculture, i.e. between November and January. This high erosive runoff leads to silting behind dams, so that much of the applied pesticides find their way directly into river and lake sediments (Zaranyika and Makhubalo, 1996). Evidence of a build-up of OC pesticide residues in Lake Kariba sediments has been reported (Zaranyika et al., 1994).

Recently, smallholder vegetable production has rapidly expanded in Zimbabwe. Sibanda et al. (2000) found these small farmers use some cultural control methods and occasionally botanical pesticides but for the most part they rely on conventional synthetic pesticides for controlling the range of serious pests and diseases that affects nonindigenous vegetables. Synthetic pesticides are usually applied using lever-operated knapsack sprayers, although occasionally less orthodox application methods are employed. The primary concerns based on these practices are due to shortcomings in protective clothing for applicators, large deviations from recommended doses (based on the adage that if a little is good, then more is better), and excessive runoff to the soil. Both of the latter concerns can lead to a build-up of pesticide residues in streams and lakes.

PESTICIDE USE IN THE CONTROL OF DISEASE VECTORS

Tsetse fly infestation in eastern central Africa

Tsetse fly infestation in Africa was reviewed by Ford (1971). In southern east-central Africa, infestation is mainly by *Glossina morsitans* Westwood (Diptera: Glossinidae) and *G. pallidipes* Austen. Figure 3.4 shows the areas infested. *G. morsitans* infests the Brachystegia-Fulbernardia woodlands of Mozambique and Zimbabwe below

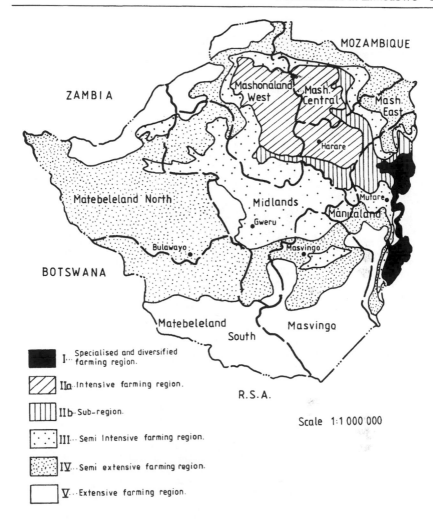

Figure 3.2 Agro-ecological regions of Zimbabwe (from ENDA-Zimbabwe, 1991)

1,200 m above sea level, and the Colophospernum Mopane woodlands in the Zambezi valley. *G. pallidipes* is also found throughout these areas inhabiting thicket or forest-edge areas. Until eliminated by insecticides from Zululand (du Toit, 1959), it extended further south in Africa than any other species. Figure 3.5 shows the distribution of *G. morsitans* and *G. pallidipes* in Zimbabwe, southern Mozambique and South Africa in 1959.

Tsetse fly in Zimbabwe

Zimbabwe forms a single natural geographical system, centered upon the watershed that separates the Zambezi from the Limpopo and Sabi-Lundi river systems (see

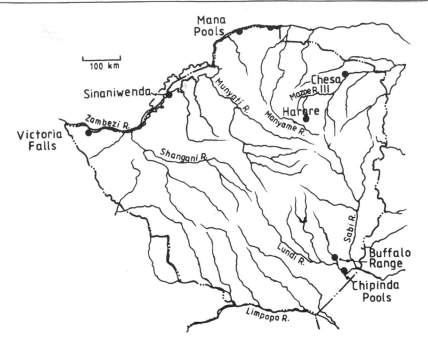

Figure 3.3 Zimbabwe with its major drainage systems
Source: Billings and Phelps (1972).

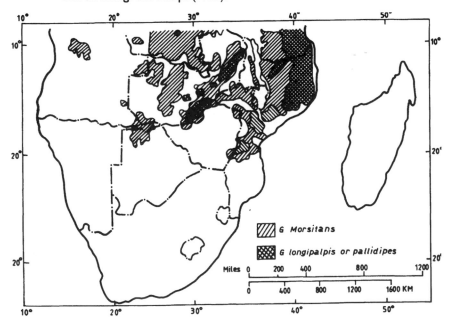

Figure 3.4 Distribution of *G. morsitans* and *G. pallidipes* in S.E. central Africa (adapted
from Ford, 1971). With permission of Oxford University Press.

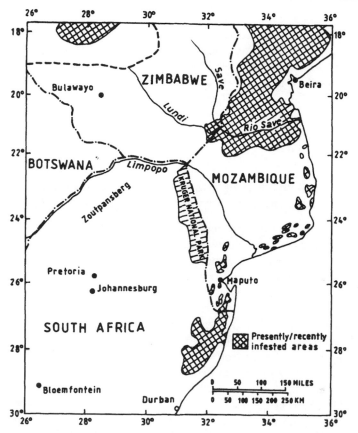

Figure 3.5 Distribution of *G. morsitans* in Zimbabwe, southern Mozambique and South Africa in 1959 (adapted from Ford, 1971). With permission of Oxford University Press.

Figure 3.3) before they descend from the African Plateau into the Mozambique Coastal Plain. The eastern highlands form the rim of the plateau in the east. In the southeast, between the Sabi and the Limpopo, descent to the coastal plain is more gradual. This region lies in an arid zone of low rainfall (200 to 400 mm per year), centered over the Limpopo, but extending chiefly northwards. Here, temperatures are high. In the northeast, the Nyanga Mountains slope down to the Zambezi Valley.

Glossina were found below about 900 m above sea level in the Limpopo Valley, and below about 1,200 m in the Zambezi Valley. The economically important portion of Zimbabwe lies above these altitudes, so that in the north and south are lowlands, both of which were once tsetse infested. The Limpopo Basin lost its infestation during the Great Rinderpest Panzootic (1889 to 1896), except for the Mozambique Plain (Ford, 1971). The Zambezi Valley, at least on the Zimbabwe side, also lost most of its tsetse infestation, except for small pockets.

Pollution of the Zambezi River (including Lake Kariba) and the Mozambique Coastal Plain by OC pesticide residues is related, in the main, to efforts by the three countries – Zimbabwe, South Africa, and Mozambique to prevent the recovery and spread of the tsetse fly in the region. Prior to 1961, control of the spread of the tsetse fly had been carried out by means of brush clearing and game destruction (Robertson and Kluge, 1968). Ever since 1962, the three countries have combined their tsetse control campaigns by carrying out annual applications of persistent pesticides to dry-season resting and refuge sites within the infested areas (Robertson et al., 1972).

Tsetse control using insecticides in the south-eastern Zimbabwe–Mozambique border region

The use of persistent insecticides in the reclamation of the Zimbabwe–Mozambique border region between the Rio Save and the Limpopo River from *G. morsitans* was reviewed by Robertson et al. (1972). Between 1953 and 1962, an extensive and rapid westerly and southwesterly advance of *G. morsitans* occurred in the lower Lundi drainage basin in the Zimbabwe–Mozambique border region, west of the Save River. By mid-1962, the tsetse had advanced to within 80 km of the Krugger National Park in South Africa. The advance seriously threatened to extend cattle trypanosomiasis over vast areas on the Nuanetsi and Limpopo basins, and, thus, became a matter of vital concern to the three countries, Zimbabwe (then southern Rhodesia), Mozambique, and South Africa. Therefore, joint spraying operations to control the advance were started. The operations involved ground application of persistent insecticide to tsetse resting and refuge sites. To ensure that the pesticide deposits remained lethal for as long as possible, the spraying was (and is) done in the dry season, during July to the end of September. In the first two years, the insecticide was applied by means of motorized machines, but, from 1964 onwards, hand-operated spraying machines were used. These sprayers were (and are) fitted with special nozzles capable of throwing a variable jet of spray up to a distance of 7.6 m and with special regulating valves capable of giving a constant output pressure of 2.07 bars. A team of eight spray operators, four of whom are in action at a given time, carry out the spraying operation. In the field, the team of spray operators is guided by maps made from aerial photographs of the application area. Each spray operator normally covers a swathe about 14 m in width.

The type, formulation, and quantity of pesticide used on the Zimbabwe side of the Zimbabwe–Mozambique border and the year and area sprayed are shown in Tables 3.4 and 3.5. The increased application rate of the pesticide, from 50 L km^{-2} in 1962 to 144 L km^{-2} in 1966 as shown in Table 3.4, primarily reflects the increasing density of suitable tsetse habitats as the work progressed from the periphery of the tsetse advance to areas of firmly established tsetse flies. During the same period, 1962 to 1966, spray operations were also conducted on the Mozambique side of the border. The use of DDT was begun in 1968. The 1970 and 1971 campaigns involved some respraying of areas that had been sprayed previously.

Table 3.4 Spraying operations using 3.1 percent dieldrin emulsion in the south-east Zimbabwe–Mozambique border region 1962–67

Year	No. of teams	Quantity used (L)	Approx. area sprayed (km²)
1962	2	46,273	900
1963	2	45,569	1,160
1964	3	75,645	1,300
1965	5	174,907	2,070
1966	6	219,026	1,530
1967	9	1,613,80	2,130[a]

Note:
a 830 km² in Zimbabwe and 1,300 km² in Mozambique.

Table 3.5 Spraying operations in the south-east Zimbabwe–Mozambique border region 1968–71

Year	Territory	Insecticide	Area treated (km²)	Quantity (L)
1968	Zimbabwe	5% DDT suspension	450	81,146
	Mozambique	5% DDT suspension	2,375	233,029
1969	Zimbabwe	5% DDT suspension	266	105,694
	Mozambique	5% DDT suspension	1,805	306,423
	Mozambique	Dieldrin 3.1%	156	43,528
1970	Zimbabwe	5% DDT suspension	207	60,689
	Mozambique	5% DDT suspension	2,396	384,933
1971	Mozambique	5% DDT suspension	715	164,680

Tsetse control using insecticides in the Zambezi valley

Tsetse control spraying in the Zambezi Valley began in 1966, with the ground spraying of the Gokwe-Sanyati and Urungwe areas between 1966 and 1968. After this, spraying campaigns were discontinued as a result of intensification of the 'War of Liberation', but were resumed after independence in 1980. Figure 3.6 shows the areas sprayed in the periods 1982 to 1984 and 1988. In addition to the areas shown in Figure 3.6, a study of the operational maps of the Tsetse and Trypanosomiasis Control Branch of the Department of Veterinary Services shows that DDT sprays were concentrated in the Binga area of the Zambezi Valley prior to 1985. Between 1985 and 1990, spraying was conducted mainly east of Ruzirukuru River in the areas drained by the Sengwa, Ume, Sanyati, and Gachegache rivers.

The quantity of DDT used for tsetse control has been declining since 1981 (see Figure 3.7). In 1981, about 300 T of DDT were used and that figure had dropped to less than 10 T by 1990 (a personal communication from W. Shereni, Head of the Tsetse and Trypanosomiasis Control Unit, Department of Veterinary Services, Ministry of Lands, Agriculture, and Rural Resettlement for Zimbabwe; unreferenced, see Acknowledgments). Two factors have contributed to the drop in the quantity of DDT used. The first is eradication of the tsetse flies from some locations

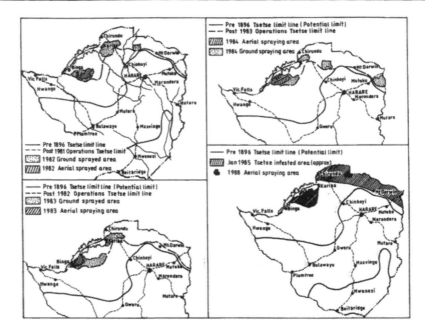

Figure 3.6 Areas sprayed with DDT in the Zambezi valley between 1982 and 1985, and in 1988
Source: Department of Veterinary Sciences, Zimbabwe.

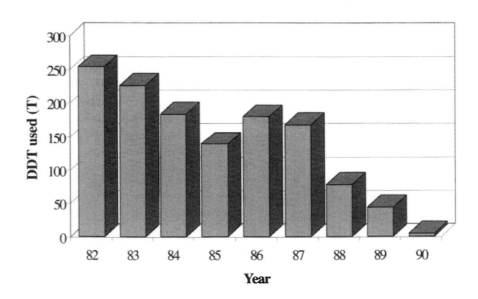

Figure 3.7 Annual use of DDT for tsetse control in Zimbabwe 1981–90

and the resultant reduction in the area infested, and the second factor is the use of alternative insecticides. These alternative insecticides include endosulfan (Chapman, 1976), and synthetic pyrethroids such as deltamethrin, α-cypermethrin and λ-cyhalothrin (Holloway, 1990). Trial aerial sprays with ultra low volume applications of endosulfan were carried out in 1974 and 1975 in the Gokwe District. The quantity of endosulfan and the areas sprayed are shown in Table 3.6. Trials with deltamethrin sprayed on odor-baited targets began in 1988 (Vale *et al.*, 1988).

USE OF PESTICIDES TO CONTROL MALARIA VECTORS

Malaria control in Zimbabwe began in the late 1940s to a limited degree, responding only to epidemics (Mpofu *et al.*, 1988). The program has since been expanded and now covers about two-thirds of the country (Mpofu, 1987). DDT has been used in Zimbabwe since 1972, replacing hexachlorocyclohexane (HCH) to which vector mosquitoes had developed resistance. As in many African countries, indoor spraying is employed, the insecticide being targeted onto the inside surfaces of dwelling huts, the roof thatch, and eaves (Taylor *et al.*, 1981). The sprays are designed to achieve a target of 2 g a.i. per m^2 (Mpofu *et al.*, 1988). Figure 3.8 shows the areas sprayed during 1981 to 1986 and the 1989/90, 1990/91, and 1992/93 seasons, respectively. Substitution of DDT with deltamethrin, a biodegradable pyrethroid, began in the 1989/90 spraying season. However, although deltamethrin was effective and nonpersistent, it was found to be prohibitively expensive, and, in the 1992/93 spraying season, the Ministry of Health, which administers the program, reverted to the use of DDT.

There is evidence that vector control programs are the main source of DDT pollution in Zimbabwe (Chikuni *et al.*, 1997). In a study analyzing PCB and DDT and its metabolites levels in human milk from mothers in seven areas of Zimbabwe, Chikuni *et al.* (1997) found that the Kariba area had the highest mean levels of $\sum$ DDT at 25,259 ng g^{-1} milk fat (range was 2,257 to 101,724 ng g^{-1} milk fat) by a factor of >2.5 over the Nyanga fruit-growing area – possible sources in the Nyanga area include background DDT residues from previous agricultural use and wind drift from nearby areas of rampant DDT aerial spraying. The Kariba area was nearly 16-fold higher compared to a rural area, Esigodini, with the lowest mean $\sum$ DDT concentration, 1,607 ng g^{-1} milk fat. However, 100 percent of milk samples were positive for *p,p'*-DDE and 98 percent positive for *p,p'*-DDT. Most of the

Table 3.6 Use of endosulfan in the Gokwe District of Zimbabwe

Year	Area (km²)	Total quantity of insecticide (L)	Rate (g ha⁻¹)
1974	239	11,500	14
1975	732	24,000	14

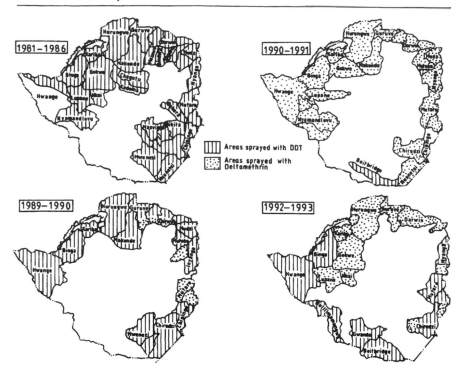

Figure 3.8 Areas sprayed for malaria control between 1981 and 1993 (reproduced by permission of the Blair Research Institute, Ministry of Health and Child Welfare, Zimbabwe)

spraying in the Lake Kariba area for mosquito and tsetse flies is done during the rainy season, so washouts from streams and rivers assist to distribute the DDT, which eventually finds its way into Lake Kariba (Zaranyika *et al.*, 1994). The lake provides food, drinking water, and irrigation water for small vegetable gardens so that the main source of DDT to humans in this area is diet related. The ratio of DDT to DDE was 0.60 in the Lake Kariba area compared to ratios between 0.15 and 0.28 in the other six areas, confirming the continued use of DDT in vector control, especially around the environs of Lake Kariba.

IMPACT OF PESTICIDES ON THE ENVIRONMENT IN ZIMBABWE

Impact on wildlife

The impact on wildlife of the persistent OC pesticides used in the control of tsetse flies has been of major concern in Zimbabwe, and several surveys have been conducted in order to assess its effect on birds and other wildlife. Billings and Phelps (1972), Whitwell *et al.* (1974), Wessels *et al.* (1980), and Phelps *et al.* (1986) carried out such surveys to assess the impact of the use of pesticides on wildlife.

The surveys by Billings and Phelps (1972) and Whitwell *et al.* (1974) were carried out following the tsetse fly control campaigns in 1962 to 1967 with dieldrin, and in 1968 to 1972 with DDT. The surveys were conducted by analyzing birds' eggs and other biological samples collected from various part of the country (see Tables 3.7 and 3.8 and Figure 3.2).

The pesticides tested for in the studies included DDT, DDE, DDD, BHC, dieldrin, aldrin, and endosulfan. Although the surveys did not find evidence of a heavy build-up of pesticides in the terrestrial environment as a result of the anti tsetse spraying, it did find evidence of a build-up of pesticides in the terrestrial and aquatic environments from agriculture. The study by Billings and Phelps (1972), involving analyses of eggs, embryos, and body fat of crocodiles, and livers of water-buck, impala, elephant, darter *Anhinga rufa* Lacépede and Daudin (Aves: Pelecani-formes: Anhingidae), and black flycatcher *Melaenornis pammelaina* Stanley (Aves: Passeriformes: Muscicapidae), found the highest levels of OC pesticide residues on agricultural land, while only traces of DDD, DDE, and DDT were found in the eggs of crocodile and the liver of elephants from game reserves where there is no agriculture (see Table 3.7). The link between agriculture and the incidence of pesticide residues in wildlife was further supported by the fact that practically no

Table 3.7 Pesticide residues (mg kg⁻¹ dw) in animal tissue from different localities in Zimbabwe[a]

Tissue	Locality	DDD	DDE	DDT	DDT	Dieldrin
Croc.[b] egg	Sinamwenda	0.69	0.54	0.40	11.64	0.01
Croc. egg	Sinamwenda	0.49	0.17	0.24	0.9	0.03
Croc. egg	Sinamwenda	0.41	0.24	0.30	0.95	0.09
Croc. egg	Sinamwenda	0.42	0.27	0.30	0.99	0.02
Croc. egg	Buffalo Range	0.20	0.59	0.47	1.26	0.04
Croc. egg	Buffalo Range	0.26	0.69	0.51	1.46	0.05
Croc. egg	Buffalo Range	0.18	0.56	0.45	1.19	0.07
Croc. egg	Buffalo Range	0.30	0.70	0.57	1.57	0.08
Croc. egg	Nyanyadzi	0.16	0.55	0.57	1.28	0.17
Croc. egg	Nyanyadzi	0.14	0.48	0.28	0.90	0.14
Croc. embryo	Nyanyadzi	0.37	0.13	0.23	0.73	0.12
Croc.abd.[c] fat	Chesa TTL[d]	0.79	3.68	2.10	6.57	1.33
Barbel abd. fat	Chipinda	3.41	11.4	1.88	16.64	ND[e]
Darter liver	Chipinda	ND	6.43	ND	6.43	ND
Darter abd. fat	Chipinda	ND	23.2	ND	23.2	ND
Flycatcher liver	Harare	1.53	29.6	12.3	43.4	ND
Waterbuck liver	Victoria Falls	0.09	0.03	0.06	0.18	ND
	Victoria Falls	0.10	0.03	0.06	0.19	ND
	Victoria Falls	0.12	0.03	0.08	0.23	ND

Notes:
a   Adapted from Billings and Phelps, 1972.
b   Croc. abbreviation for crocodile.
c   Abbreviation abd. represents abdominal.
d   TTL represents Tribal Trust Land (the former name for communal land).
e   ND represents residue not detected.

Table 3.8 Pesticide residues in biological samples from various localities in Zimbabwe (μg g^{-1} dw)[a]

Area	Date	Species	Type of sample	DDT	DDD	DDE	∑DDT	BHC	Dieldrin	Aldrin	Endosulfan
Harare	1972	Fiscal Shrike *lanius*	young chick	–[b]	–	5.1	5.5	T[c]	–	–	–
	10/72	Masked weaver *Ploceus velatus*	composite sample of 2 chicks and eggs	0.1	0.3	4.3	5.2	–	–	–	–
	3/73	Red bishop *Euplectes orix*	eggs	0.2	–	2.9	3.3	T	T	–	–
Lake Chivero	7/72	Egyptian goose *Aplopochen aegyptiacus*	egg	–	–	–	–	–	–	–	–
	9/72	Blackheaded heron *Ardea melanocephala*	egg	–	–	4.7	5.1	0.1	–	0.1	–
Rusape	1972	Black goshawk *Accipter melanoleucus*	chicks	–	1.5	29.7	32.4	0.4	3.9	–	–
Headlands	1972	African goshawk *Accipter tachiro*	egg	–	–	0.5	0.5	T	–	–	–

Area	Date	Species	Type of sample	DDT	DDD	DDE	ΣDDT	BHC	Dieldrin	Aldrin	Endosulfan
Kariba (Charara cleared area)	1/73	Reed cormorant Phalacrocorax africanus	composite sample of 2 eggs preserved boiled	B[d]	B	3.4	3.7	T	0.16	T	B
	7/71	Darter Anhinga rufa	egg from oviduct preserved in formalin	0.4	0.3	2.7	0.5	T	B	B	B
	7/71	8 Reed cormorants 4 Darters	composite sample of 8 livers preserved in formalin	B	B	0.9	1.0	B	B	B	B
Matopos	1970	Black eagle Aquila verreauxi	addled egg preserved frozen	T	T	T	T	T	T	B	B
	1971	Black eagle	addled egg preserved frozen	T	B	T	T	T	B	B	B
	1971	Black eagle	addled egg preserved frozen	T	B	T	T	T	B	B	B
Gonarezhou 1971		Wahlberg's eagle	egg	T	–	0.3	0.3	–	0.1	–	B
	1971	Chanting goshawk Melierax musicus	egg	–	–	0.2	0.2	–	–	–	2.2

continued...

Area	Date	Species	Type of sample	DDT	DDD	DDE	ΣDDT	BHC	Dieldrin	Aldrin	Endosulfan
	1971	Wahlberg's eagle	egg	–	–	0.2	0.2	T	–	0.2	–
	1971	Chanting goshawk	egg	–	–	–	–	T	–	–	–
	1972	Black vulture *Torgos tracheliotus*	egg	–	–	1.6	1.7	–	–	–	–
	1972	Black vulture	egg	–	–	0.4	0.5	–	–	–	–
	1972	African hawk eagle *Hieraaetus spilogaster*	egg	–	–	0.4	0.5	–	–	–	–
	1972	Black-breasted snake eagle *Circaetus pectoralis*	egg	–	–	0.6	0.7	T	0.8	–	–
	1972	Hooded vulture *Necrosyrtes monachus*	egg	–	–	0.5	0.6	T	0.3	–	–
	1972	Giant eagle owl *Bubo lacteus*	egg	0.7	T	1.3	2.1	–	–	–	–
Copper Queen	1971	Mopane bark	control	500		T	500				–
			control	539		T	539				–
			control	549		T	549				–

Notes: a Adapted from Whitwell *et al.*, 1974.   b Em dash (–) indicates residue not detected.   c T indicates trace.   d B indicates benzene.

residues were found in eggs of birds from Hatopos, an area of very little agricultural activity near the border with Botswana to the west. The study by Whitwell *et al.* (1974), involving analysis of birds' eggs from Lake Chivero (or McIlwaine, as it used to be called) outside the tsetse control area, showed that eggs of the black-headed heron, *Ardea melanocephala* Virgors and Children (Aves: Ciconiiformes: Ardeidae), whose diet includes aquatic animals, contained pesticide residues. At Lake Kariba, pesticide residues were found in fish-eating birds taken from basins that receive drainage from agricultural areas. Similar findings were reported by Wessels *et al.* (1980) following a study involving analysis of crocodile eggs from Lake Kariba, see Table 3.9. This study found increased residue levels in one basin of the lake, generally in accordance with land use practice in the area drained by the major tributaries running into the basin. The survey conducted by Phelps *et al.* (1986) from 1981 to 1982 is the most recent. These workers reported that chlorinated hydrocarbon residues were widespread in Zimbabwe. As with the earlier survey by Whitwell *et al.* (1974), this survey involved analysis of crocodile eggs collected from several localities. All eggs analyzed showed residues of DDT and its metabolites (see Table 3.10), the levels of which were found to be related to the type of land use in the specific area from which they were collected. Toxaphene was detected in crocodile eggs from cattle ranching areas, while polychlorinated biphenyls were recorded near industrialized areas.

Impact on Lake Kariba and its environs

All the studies discussed in the preceding section found evidence of contamination of Lake Kariba by OC pesticide residues. A further study by Phelps *et al.* (1989), involving analysis of crocodile fat samples collected from seven localities on the shoreline of the lake, found levels of DDT as high as 80 μg g^{-1} (see Table 3.11).

Further evidence of the pollution of Lake Kariba by OC pesticide residues was reported by Kiibus and Berg (1991), Berg *et al.* (1992), Douthwaite (1992), and by Zaranyika *et al.* (1994). The Kiibus and Berg (1991) and Berg et al (1992) study was conducted by sampling and analyzing fish, mussels, snails, prawns, and birds from different localities and trophic levels of the lake in an effort to find some pattern in the distribution of the residues, mainly DDT and its metabolites. They found that DDT seemed to be both bioaccumulating and biomagnifying in the lake. They also showed that the levels of DDT were generally high compared to levels found in lakes outside the tsetse control areas (see Table 3.12). The algae feeder, redbreast tilapia *Tilapia rendalli* Boulenger, had 1,900 ng g^{-1} fat $\sum$ DDT while the predatory tigerish *Hydrocynus forskahlii* Cuvier (Characiformes: Alestiidae, African tetras) had levels of 5,000 ng g^{-1} fat $\sum$ DDT (Berg *et al.*, 1992). Highest levels of $\sum$ DDT were found in bottom-dwelling species, i.e. the mussel *Corbicula africana* (Bivalvia: Corbiculidea) at 10,100 ng g^{-1} fat $\sum$ DDT, and in benthos feeding fish, e.g. *Labeo altivelis* Peters (Cypriniformes: Cyprinidae) at 5,700 ng g^{-1} fat $\sum$ DDT (Berg *et al.*, 1992).

Douthwaite (1992) carried out surveys to assess the effects of DDT treatments applied for tsetse control on white-headed black chat *Thamnolaea arnoti* Tristram

Table 3.9 OC insecticide residues in 15 crocodile eggs (μg g^{-1} dw) from Lake Kariba, Zimbabwe[a]

Collection site	Nest and egg number		α-BHC	β-BHC	Dieldrin	p,p'-DDE	p,p'-DDD	p,p'-DDT	$\sum$ DDT
Mwenda River mouth	M6	i	−[b]	−	−	1.33	0.38	0.29	2.0
	M7	i	−	−	−	1.97	0.69	0.44	3.1
	M7	iii	−	−	−	1.98	0.66	0.56	3.2
Sengwa River mouth	Is.[c]7	i	−	−	−	2.09	0.75	0.46	3.3
	Is.13	i	−	−	−	0.53	0.20	0.23	0.96
	Is.27	i	−	−	−	1.22	0.82	0.29	2.33
	Is.27	iii	−	−	−	1.13	0.45	0.43	2.01
	Is.34	i	−	−	−	0.61	0.28	0.28	1.17
	Is.35	i	−	−	−	9.00	0.80	0.98	10.78
Spencer Creek	K1	i	0.18	9.63	1.19	4.09	0.95	1.10	6.14
	K3(1)	i	0.07	1.01	−	0.67	0.50	0.69	8.0
Crocodile farm	K3(4)	i	0.22	6.38	−	1.42	2.56	−	−
Gwai mouth		i	0.01	−	−	6.85	3.00	1.09	10.94
Deka mouth		i	−	−	−	14.2	3.25	4.50	21.9
Zambezi River mouth		i	5.63	24.5	−	14.0	2.00	2.18	18.18

Notes:
a   Adapted from Wessels et al., 1980.
b   En dash (−) indicates residue not detected.
c   Is. represents island.

Table 3.10 OC insecticide and PCB residues in crocodile eggs from Zimbabwe (μg g⁻dw)[a]

Source	HCB	α-BHC	β-BHC	p,p'-DDE	p,p'-DDD	p,p'-DDT	∑DDT	PCB
Sengwa River								
clutch 10/1981	0.003	0.005	0.002	3.26	0.87	1.08	5.21	0.029
	0.003	0.004	0.003	3.82	1.23	1.40	6.45	0.034
	ND[b]	0.002	0.003	3.94	0.88	1.15	4.97	
Sengwa River								
clutch 10/1981	0.002	0.002	0.003	1.96	0.61	0.67	3.24	
	0.002	0.007	0.002	1.86	0.59	0.69	3.14	
Mpalangena River								
10/1981	ND	ND	0.210	5.02	0.72	0.74	6.48	
	0.004	0.002	0.218	3.11	0.72	0.66	4.49	
Chundu Island								
10/1981	0.004	0.003	0.013	4.60	0.51	0.61	5.72	
	0.003	0.044	0.083	5.43	2.63	1.97	10.03	
Kariba crocodile								
farm 9/1981	0.003	0.046	0.250	16.29	5.68	3.94	25.91	0.063
Lake Chivero								
10/1981	0.003	0.011	1.262	6.22	1.29	1.24	8.75	1.530
	ND	0.018	1.548	10.31	2.05	1.89	14.25	1.356
	ND	0.010	1.398	6.21	1.40	1.14	8.75	1.023
	ND	0.011	1.211	5.99	1.21	1.11	8.31	1.220
	0.003	0.009	1.664	5.89	1.21	1.22	8.32	1.351
	ND	0.011	1.506	6.49	1.49	1.33	9.28	1.351
Ngezi Park								
10/1981	0.003	0.004	0.045	3.08	0.69	0.51	4.28	0.053
	ND	0.003	0.061	3.76	0.90	0.60	6.26	0.038
Kyle Park	0.003	0.004	0.154	3.13	0.66	0.48	4.27	0.248
	0.004	0.005	0.202	3.38	0.64	0.52	4.54	0.270
	ND	0.006	0.538	8.01	1.85	1.53	11.39	0.185
	ND	0.004	0.280	4.46	0.78	0.68	5.92	0.143
	ND	0.005	0.213	8.95	2.69	2.72	14.36	0.143
Runde River I								
10/1981	0.002	0.007	0.148	3.26	0.71	0.52	4.49	0.221
	ND	0.003	0.092	4.25	0.70	0.53	5.48	0.153
Runde River II	ND	0.004	0.043	3.07	0.23	0.34	3.64	0.110

Notes:
a   Adapted from Phelps *et al.*, 1986.
b   ND indicates residue not detected.

(Aves: Passeriformes: Turdidae) populations in northwest Zimbabwe in the Zambezi Valley between 1987 and 1990. The survey was carried out in woodlands that had been sprayed with DDT at the rate of 200 g ha⁻¹. In separate studies, population drops of 88 percent over 33 months and 74 percent over 9 months were reported. The author concluded that tsetse spraying operations have had a severe, and possibly prolonged, impact on the white-headed black chat population of northwest Zimbabwe. Zaranyika *et al.* (1994) analyzed sediment samples from seven of the major river bays on the Zimbabwean side of the lake (see Figure 3.1). The results obtained (see Table 3.13) confirmed that there was contamination of most bays by DDT and its metabolites, endosulfan, aldrin, dieldrin, endrin, and heptachlor.

Table 3.11 Levels of residues of DDT and its metabolites in fat of crocodiles (pg g^{-1} dw fat[a]) from seven localities on the shoreline of Lake Kariba, Zimbabwe[b]

Locality	Sample No./Sex[c]	Bodymass (kg)	o,p-DDE	p,p'-DDE	o,p-TDE	p,p'-TDE	p,p'-DDT	∑ DDT
Kasese River	1/F	5.97	0.28	1.91	0.46	0.50	–[d]	3.15
	2/M	58.04	0.28	17.08	–	2.31	2.26	21.93
	3/F	15.87	0.08	3.65	–	0.42	0.09	4.24
	4/F	48.53	0.27	9.87	–	0.73	1.65	12.52
Cutty Sark	6/M	4.54	0.15	3.08	0.56	0.93	0.30	5.02
Rifa River	7/F	6.80	2.23	46.12	–	6.43	20.31	75.09
	8/M	4.08	0.97	44.58	–	1.84	5.94	53.31
Banana Farm	9/F	7.26	2.66	47.54	–	7.25	10.26	67.71
	10/M	6.80	2.50	45.26	1.32	5.87	9.31	64.26
	11/M	19.50	3.86	49.33	–	11.62	18.90	83.71
Charara River	12/F	16.30	0.32	14.41	–	1.10	1.49	17.32
	13/M	8.62	0.65	34.19	–	1.63	2.04	38.51
Antelope Island	14/F	5.66	0.20	20.31	–	1.49	1.62	23.62
	15/M	55.32	0.69	23.39	–	2.28	4.12	30.48
Nyaodza River	16/F	58.04	8.23	21.97	–	13.50	12.47	88.71
	17/F	35.82	0.66	20.04	1.26	4.95	4.37	31.28
	18/F	17.69	0.90	36.65	–	3.91	3.94	45.40
	19/F	82.99	0.52	14.73	–	3.13	3.03	21.41
	20/F	6.80	0.24	16.69	–	1.24	1.30	19.17
	21/M	14.51	0.15	4.73	–	0.58	0.39	5.04
	22/M	4.99	0.16	6.87	0.46	1.02	0.68	9.19
	23/M	7.26	0.51	33.81	–	1.86	2.20	38.39

Notes:
a Fat samples extracted using hexane. The hexane was evaporated off before weighing and the fat was then redissolved for final analysis.
b Adapted from Phelps et al., 1989.
c M for male; F for female.
d En dash (–) indicates residue not detected.

Table 3.12 Average values (ng g^{-1} dw) of DDT, HCH, aldrin and DDT/$\sum$ DDT for fish from Lake Kariba, Lake Chivero, and Mazvikadei Dam, Zimbabwe[a]

Species	Source	No.	DDT	HCH	Aldrin	DDT/$\sum$ DDT
Red-breasted Tilapia	Kariba	20	360–	25–	3–4	0.17–
(*Tilapia rendalli*)			2,100	64		0.73
	Mazvidadei	8	2100	46	87	0.12
	Chivero	22	790	640	0	0.27
Manyame Labeo						
(*Labeo altivelis*)	Kariba	8	5,700	44	17	0.04
	Chivero	9	1,100	1,100	0	0.23
Tigerfish						
(*Hydrocynus forskahlii*)	Kariba	14	5,000	47	15	0.2
	Chivero	7	1,000	1,300	0	0.17

Note:
a   Adapted from Kiibus and Berg, 1991.

Table 3.13 $\sum$ DDT, $\sum$ Drins, $\sum$ DDE, endosulfan, heptachlor, and $\sum$ DDE/$\sum$ DDT ratios in sediments from the Charara (C), Nyaodza (N), Gachegache (G), Sanyati (S), Ume (UN), Sengwa (SN), and Ruzirukuru (Rz) River bays around Lake Kariba, Zimbabwe (in ng g^{-1} dw)

Sampling point	$\sum$ DDT	$\sum$ DDE	$\sum$ DDE/$\sum$ DDT	$\sum$-Drins	Endosulfan	Heptachlor
C$_3$	ND[a]	ND		1.67	ND	ND
C$_1$	112.60	112.600	1.00	0.017	24.20	ND
S$_4$	12.30	1.260	0.10	ND	ND	0.876
S$_6$	ND	ND		ND	ND	4.900
UM$_1$	ND	ND		ND	1.91	40.02
SN$_1$	65.70	10.500	0.16	34.70	2.23	0.882
SN$_2$	ND	ND	ND	51.50	16.10	0.019
RZ$_1$	13.64	9.340	0.68	2.27	ND	2.660
RZ$_2$	16.60	ND		32.70	25.50	ND
G$_3$	5.62	ND		20.50	12.00	3.590
G$_1$	20.04	0.392	0.12	ND	167.80	ND
N$_3$	13.59	1.730	0.13	63.70	ND	ND

Source: adapted from Zaranyika *et al.*, 1994

Note:
a   ND indicates residue not detected.

IMPACT OF PESTICIDES ON OTHER LAKE ECOSYSTEMS AND COASTAL ZONE ECOSYSTEMS IN EASTERN SOUTH CENTRAL AFRICA

In this chapter I have given a detailed description of the impact of pesticides on the Lake Kariba ecosystem and the background for the problem. This has been possible because of the large amount of research work that has been published on the Kariba ecosystem. One study has been published on pesticide residues in Lake McIlwaine, located about 35 km southwest of Harare, Zimbabwe (Mhlanga and Madziva, 1990). However, very little has been published on the other lake systems and the coastal zones in eastern south-central Africa. This is largely due to the political unrest that has characterized the region since the early 1960s, when the use of pesticides for the control of the tsetse fly began. This position is changing, and peace seems to have finally come to the region, especially Mozambique. Therefore, we can expect tsetse and malaria control programs to intensify in that country. Already moves to increase agricultural activity are under way in Mozambique. Consequently the use of pesticides is bound to increase. There is a need to extend the work which has been carried out on the Lake Kariba ecosystem to other lake ecosystems, including Lake Kahora Bassa in Mozambique and Lake Malawi, as well as the marine coastal regions in Mozambique.

ACKNOWLEDGEMENTS

I wish to thank W. Shereni, Head of Zimbabwe's Tsetse and Trypanosomiasis Control Unit for his permission to include information about current use levels of DDT in the manuscript. This work was supported in part by a grant from the Swedish Agency for Research Cooperation with Developing Countries (SAREC), and in part by a grant from the Research Board of the University of Zimbabwe.

REFERENCES

Berg, H., Kiibus, M. and Kautsky, N. 1992. DDT and other insecticides in the Lake Kariba ecosystem, Zimbabwe. *Ambio* 21(7):444–50.

Balon, E.K. and Coche, A.G. 1974. *Lake Kariba: A Man-made Tropical Ecosystem in Central Africa*. The Hague: Dr. W. Junk Publishers. pp. IX, 85, 87.

Billings, K.J. and Phelps, R.J. 1972. Pesticide levels from animals in Rhodesia. *Trans Rhod Sci Assoc.* 55:6–9.

Central Statistical Office. 1990. *Quarterly Digest of Statistics*. Harare, Zimbabwe: Central Statistical Office.

Chapman, N.G. 1976. Aerial spraying of tsetse flies (*Glossina* spp.) in Rhodesia with ultra low volumes of endosulfan. *Trans Rhod Sci Assoc.* 57(2):12–21.

Chikuni, O. and Nhachi, C.F.B. 1996. Residues of organochlorine pesticides in human milk. In: Nhachi, F.B. and Kasilo, M.J. (eds) *Pesticides in Zimbabwe: Toxicity and health implications*. Harare, Zimbabwe: University of Zimbabwe Publishers, pp. 73–80.

Chikuni, O., Nhachi, C.F.B., Nyazema, N.Z., Polder, A., Nafstad, I. and Skaare, J.U. 1997. Assessment of environmental pollution by PCBs, DDT and its metabolites using human milk of mothers in Zimbabwe. *Sci Total Environ.* 199(1–2):183–90.

Douthwaite, R.J. 1992. Effects of DDT treatments applied for tsetse fly control on white-headed black chat (*Thamnolaea arnoti*) populations in Zimbabwe, part I: population changes. *Ecotoxicology* 1:17–30.

du Toit, R.M. 1959. The eradication of tsetse fly (*Glossina pallidipes*) from Zululand, Union of South Africa. *Adv Vet Sci.* 5:227.

ENDA-Zimbabwe Environmental Unit Cartographic Section. 1991. Harare, Zimbabwe: Government Printers.

Ford, J. 1971. *The Role of Trypanosomiases in African Ecology: A Study of the Tsetse Fly Problem.* Oxford, England: Clarendon Press.

Government of Zimbabwe. 1952. Fertilizer, Farm Feeds and Remedies Act (Chapter 111). Harare, Zimbabwe: Government Printers.

Government of Zimbabwe. 1972. Hazardous Substances and Articles Act (Chapter 322). Harare, Zimbabwe: Government Printers.

Greichus, Y.A., Greichus, A., Draayer, H.A. and Harshal, B. 1978. Insecticides, polychlorinated biphenyls and metals in African lake ecosystems: II Lake McLlwaine, Rhodesia. *Bull Environ Contam Toxicol.* 19(4):444–53.

Holloway, H.T.P. 1990. Alternatives to DDT for use in ground spraying control operations against tsetse flies (Diptera: Glossinidae). *Trans Zimbabwe Sci Assoc.* 64(4):33–40.

Kiibus, H. and Berg, H. 1991. Pesticides in the food web of Lake Kariba, Zimbabwe. Department of Systems Ecology, University of Stockholm Report Nr 5-10691. Stockholm, Sweden: University of Stockholm.

Mathuthu, A.S. 1993. Table 23. In: *Agrochemical Pesticides in the SADC Region: Use and Concerns.* Maseru, Lesotho: Southern African Development Community (SADC) Environment and Land Management Sector (ELMS) Coordination Unit. SADC ELMS Report Series 35, pp. 162–63.

Mbanga, T. 1996. Pesticides and the Agricultural Industry Association. In: Nhachi, F.B. and Kasilo, M.J. (eds) *Pesticides in Zimbabwe: Toxicity and Health Implications.* Harare, Zimbabwe: University of Zimbabwe Publishers, pp. 12–6.

Mhlanga, A.T. and Madziva, T.J. 1990. Pesticide residues in Lake McIlwaine, Zimbabwe. *Ambio.* 19(8):368–72.

Mpofu, S.M. 1987. DDT and its use in Zimbabwe. *Zimbabwe Sci News* 12:31–6.

Mpofu, S.M., Taylor, P. and Govere, J. 1988. An evaluation of the residual lifespan of DDT in malaria control. *J Am Mosq Control Assoc.* 4(4):529–35.

Phelps, R.J., Forcardi, S., Fossi, C., Leonzio, C. and Renzoni, A. 1986. Chlorinated hydrocarbons and heavy metals in crocodile eggs from Zimbabwe. *Trans Zimbabwe Sci Assoc.* 63(2):8–15.

Phelps, R.J., Toet, M. and Hutton, J.M. 1989. DDT residues in the fat of crocodiles from Lake Kariba. *Trans Zimbabwe Sci Assoc.* 64(2):9–14.

Robertson, A.G. and Kluge, E.B. 1968. The use of insecticide in arresting an advance of *Glossina morsitans* West in the South-East Lowveld of Rhodesia. *Proc Trans Rhod Sci Assoc.* 53:17–33.

Robertson, A.G., Kluge, E.B., Kritzinger, D.A. and de Sousa, A.E. 1972. The use of residual insecticides in reclamation of the Rhodesia–Mozambique border region between the Sabi/Save and Limpopo rivers from *Glossina morsitans* Westwood. *Proc Trans Rhod Sci Assoc.* 55(1):34–62.

Sibanda, T., Dobson, H.M., Cooper, J.F., Manyangarirwa, W. and Chiimba, W. 2000. Pest management challenges for smallholder vegetable farmers in Zimbabwe. *Crop Protection* 19(8–10):807–15.

Taylor, P., Crees, M.J. and Hargreaves, K. 1981. Duration of *Anopheles arabiensis* control in experimental huts sprayed with DDT and decamethrin. *Trans Zimbabwe Sci Assoc.* 61(1):1–13.

Vale, G.A., Lovemore, D.F., Flint, S. and Cockbill, G.F. 1988. Odour baited targets to control tsetse flies, *Glossina* spp. (Diptera: Glossinidae), in Zimbabwe. *Bull Entomol Res.* 78:31–49.

Wessels, C.L., Tannock, J., Blake, D. and Phelps, R.J. 1980. Chlorinated hydrocarbon insecticide residues in *Crocodilus nicoticus laurentius* eggs from Lake Kariba. *Trans Zimbabwe Sci Assoc.* 60(3):11–7.

Whitwell, A.C., Phelps, R.J. and Thompson, W.R. 1974. Further records of chlorinated hydrocarbon pesticide residues in Rhodesia. *Arnoldia (Rhodesia)* 6(37):1–7.

Zaranyika, M.F., Mambo, E. and Makhubalo, J.M. 1994. Organochlorine pesticides in the sediments of selected river bays in Lake Kariba, Zimbabwe. *Sci Total Environ.* 142(3): 221–6.

Zaranyika, M.F. and Makhubalo, J.M. 1996. Organochlorine pesticide residues in inland waters in Zimbabwe. In: Nhachi, F.B. and Kasilo, M.J. (eds) *Pesticides in Zimbabwe: Toxicity and Health Implications.* Harare, Zimbabwe: University of Zimbabwe Publishers, pp. 89–106.

Chapter 4

Pesticides in Kenya

S.O. Wandiga, J.O. Lalah, and P.N. Kaigwara

INTRODUCTION

The republic of Kenya lies on the eastern side of the African continent, between Lat. 4°40'N and 4°40'S and between Long. 33°50'W and 41°45'E (NEAP, 1994). The equator bisects the country in almost two equal parts. The climate of Kenya is controlled by movement of the intertropical convergence zone (ITCZ), whose influence is then modified by the altitudinal differences that give rise to Kenya's varied climatic regimes (NEAP, 1994). The country's equatorial location and its position adjacent to the Indian Ocean also influence the local climate. Kenyan soils are grouped into various units (NEAP, 1994) based largely on their physical and chemical properties. These play a major role in explaining vegetation types and their distribution patterns. Kenya may be divided into four major agroecological zones (AEZ) namely the highlands, savannah, coastal, and arid and semi-arid lands (ASAL) (Figure 4.1). The zones have distinct humidity ranges, mean annual temperatures, rainfall patterns, and altitudes that largely dictate their respective ecological potentials. Kenya's population was estimated at 27.5 million in 1995 and was growing at a rate of 2.9 percent per annum (NDP, 1997). Its economy is predominantly agriculture and agroforestry-based, contributing 26 percent to the gross domestic product (GDP) in 1997 (NDP, 1997). Agricultural activities are concentrated in the highlands (high potential), savannah, and coastal (medium potential) AEZs (NEAP, 1994).

PESTICIDE REGULATION IN KENYA

History of pesticide usage

Control of the general use and handling of pesticides in Kenya goes back to the colonial era. The earliest recorded legislation dates from 6 September 1921 when the Public Health Act, Cap 242, was passed by the colonial government. Sixteen years later, a second Act of Parliament dealing with Cattle Cleansing, Cap 358, was passed on 27 April 1937. This Act prescribed various preparations for destroying

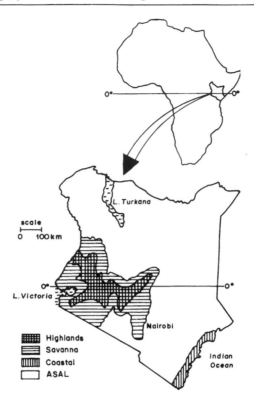

Figure 4.1 The agroecological zones of Kenya

ticks. These preparations are still retained in law though several amendments have modified the original prescriptions.

At the height of Kenya's struggle for independence, when the colonial government declared emergency rules, it also adopted a Voluntary Precaution Scheme for the agricultural industry. Compliance with the scheme was on a voluntary basis. This Scheme led to the proclamation of the Poisonous Substances Ordinance of 1954. The ordinance was based on the United Kingdom Act of 1952, which provided for the protection of employees against risk of poisoning by certain substances used in agriculture and incidental and connected matters.

On the eve of Kenyan independence, the Pharmacy and Poisons Act of Parliament was passed by Westminster on 1 May 1957. The aim of this Act was to incorporate provisions in the law to provide for the control of the profession of pharmacy and the trade in drugs and poisons. Included in this Act was the control of veterinary drugs and poisons with additional rules on the selling and labeling of poisons, including pesticides.

The now independent Kenya Parliament passed an Act on 11 May 1965 for the prevention of adulteration of food, drugs, chemical substances, and incidental and connected matters. In this, the Food, Drug and Chemical Substances Act,

Cap 254, pesticides were given particular attention, and the term 'chemical substances' was defined to refer to any substance or mixture of substances prepared, sold, or represented for use as:

- a germicide,
- a disinfectant,
- an insecticide,
- a rodenticide,
- an antiseptic,
- a pesticide,
- a vermicide,
- a detergent.

For the first time, it also set tolerance levels (in ppm) for pesticides in foodstuffs. This law has neither been amended since then nor has its implementation been effective. The protection of workers in the workplace has not been left outside the ambit of legal protection. The Factories Act, Cap 514, regulates factory working conditions with an aim of maximizing health protection for workers.

Other legislative laws passed by Parliament that have a bearing on pesticide use, distribution, and control include the Agriculture Act, Cap 318; the Fertilizers and Animal Foodstuffs Act, Cap 345; the Forest Act, Cap 385; the Plant Protection Act, Cap 324; and the Water Act, Cap 389. Although in some of these Acts pesticides are not specifically mentioned, it is clear that to fulfill the Act's objectives the control of pesticides may be invoked.

The practice in Kenya has been for Parliament to pass sectorial laws for the regulation and control of environmental matters. There is no umbrella law covering all aspects of environmental matters. Such a bill is on the drawing board and it is hoped that when passed by Parliament it will go a long way toward regulating environmental issues, including pesticides. The major deficiency in the present patchwork of laws is their scattered aims and ineffective implementation. Penalties prescribed for offenders have also been overtaken by economic realities. Kenya, therefore, needs an environmental act that will assist it to better manage its environment. Recent attempts to better regulate the use of pesticides are described in the next section.

Current pesticide regulations

Except for the Poisonous Substances Ordinance, 1954, the rest of the Acts mentioned in the previous section are still in force. A replacement ordinance, called the use of Poisonous Substances Ordinance, which will regulate the protection of people against the risks from exposure to poisonous substances, has been drafted but has not yet been presented to Parliament.

The most comprehensive law regulating pesticides is the Pest Control Products Act, which came into law on 19 May 1983. It was established to regulate the importation, exportation, manufacture, and distribution of products used for the control of pests, and of the organic function of pesticides on plants and animals. Pest Control Product was defined as 'a device, product, organism, substance, or a thing that is manufactured for directly or indirectly controlling, preventing, destroying, attracting, or repelling any pest'. The Act established a Pest Control Products Board (PCPB), which became operational in October 1984. PCPB's mandate as contained in the Act is described below under the respective categories:

Regulatory

1   To register and approve for use all pest control products.
2   To regulate the sale and distribution of pest control products through licensing of imports and exports.
3   To inspect and license all facilities used for the manufacture, storage, and distribution of pesticides.
4   To analyze any pesticides for efficiency before recommending for use.

Technical

1   To receive and evaluate data from manufacturers and importers on the merits of pest control products.
2   To undertake, as appropriate, short and long term research to evaluate the impact of pesticides on the environment.
3   To collect information from international organizations such as FAO, WHO, EPA, UNEP, etc. that are relevant to pesticide use and regulation.

Training and information

1   To educate and inform users and the general public on matters concerning the safety and danger of using pesticides. Other functions that fall under the Training and Information category are advising relevant authorities on aspects of pesticide management, training government extension agents and other interested personnel on pesticide management, and advising the government on the status of approved pesticides. Since its inception in 1986, the Board has also banned or restricted the use of a number of pesticides (Table 4.1).

PESTICIDE USE AND DISTRIBUTION

Past and current usage patterns

Agriculture has been the mainstay of Kenya's economy. This dependence on agricultural production has led to widespread pesticide use during the last four decades. Lindane was introduced in Kenya in 1949, toxaphene in 1950, DDT in 1956, and dieldrin in 1961 (Kaine, 1976). Other compounds in use during the

Table 4.1 Pesticides banned or in restricted use in Kenya (from the Pest Control Products Board)

Banned pesticides in Kenya	
Common name	Former use of the pesticide
Dibromochloropropane (DBCP)	Soil fumigant
Ethylene dibromide (EDB)	Soil fumigant
2,4,5-T phenoxy herbicide	Herbicide
Chlordimeform	Acaricide/insecticide
All isomers of HCH	Insecticide
Chlordane	Insecticide
Captafol	Fungicide
Heptachlor	Insecticide
Toxaphene (camphechlor)	Acaricide
Endrin	Insecticide
Parathion(methyl and ethyl)	Insecticides
Restricted pesticides in Kenya	
Common name	Permitted use
Lindane	Seed dressing only
Aldrin; dieldrin	Termites in building industry – no longer available in Kenya
DDT	Public health only for control of mosquitoes in mosquito breeding grounds – no longer available in Kenya

1950s included dinitrocresol (DNC) and the OP compounds, TEPP (tetraethyl pyrophosphate) and schradan (Keating, 1983).

The livestock industry has been adversely affected by diseases such as East Coast fever or theileriasis (an acute disease of cattle transmitted by ticks and caused by *Theileria parva* Theiler) and anaplasmosis (a peracute to chronic infectious disease of ruminants frequently caused by blood-feeding insects such as ticks). Acaricides such as chlorfenvinphos have been used to combat the disease vectors. DDT has been instrumental in reducing the incidence of malaria, with the consequence of many lives saved, by controlling malaria's vector, the *Anopheles* mosquito.

Foxall (1983) reported that K Sh400 million worth of pesticides was being used annually in Kenya. These consisted of 50 percent fungicides, 20 percent insecticides, 20 percent herbicides, and 10 percent acaricides, rodenticides, molluscides, and nematicides, combined. Currently about K Sh2.5 billion worth of pesticides is used in Kenya annually (Mwaisaka, 1999). In 1987, PCPB reported an increase in imports between 1984 and 1986 from K Sh350 million (1984) to K Sh410 million (1985) and then to K Sh580.2 million (1986). Between 1985 and 1987, pesticides worth K Sh1,732.3 million were imported (Mwanthi and Kimani, 1993), while for the period 1987 through 1990 a total of 31,234 T (PCPB, 1994) was imported into the country. The bulk of imported pesticides was consumed locally with less than 3 percent exported to neighboring countries. About 20 percent were imported in technical grade form and were formulated locally while the rest were imported as ready to use formulated products. For example, malathion (technical) is used locally for the formulation of 2 percent malathion dust and technical carbofuran (furadan) is used for the preparation of 3G, 5G, and 10G granular formulations. Examples of formulated products imported ready for use include furadan 350 ST and marshal 250 FC (carbosulfan). Tables 4.2 to 4.5 show 1986 through 1995 pesticide imports into Kenya in monetary, metric, and percentage terms. By 1997, the Pest Control Products Board (PCPB) had registered 370 formulations, representing 217 active ingredients for use in Kenya. About 22 percent of the volume imported were highly hazardous, 20 percent moderately hazardous, 45 percent slightly hazardous, and the remainder were unclassified (Ohayo-Mitoko, 1997).

A decline in the volume of imports is noticeable between 1988 and 1990. This was probably due to the ban and restriction of some OC pesticides. Munga (1985) reported that 70 T of DDT had been used annually for agricultural pest control on maize and cotton while other OCs, e.g. lindane, aldrin, and dieldrin, were used for seed dressings. DDT was last imported into Kenya in 1985, aldrin and dieldrin in 1992 (PCPB). OCs still in use in Kenya include endosulfan and lindane.

Approximately 33 percent of Kenyan farmers, primarily large farm operators, use pesticides. On most small farms, which are mostly subsistence-level farms, there is minimal use of pesticides. Cash crops, such as coffee, use about 50 percent of imported pesticides while horticultural crops require another ~25 percent (Kanja, 1988). Other important crops that require a significant quantity of pesticides are cotton, sugarcane, maize, and tea. Herbicides, as a substitute for mechanical or hand weeding, are also used by coffee, maize, barley, wheat, sugarcane, and tea farmers.

The pesticide industry in Kenya

The Kenyan pesticide industry comprises companies that manufacture a.i.(s) used in pesticide formulation, formulators contracted to manufacturers of a.i.(s) used

Table 4.2 Importation of different groups of pesticides into Kenya (1986–95) (value of cost and freight in M Kshs, adapted from the Pest Control Products Board)

Year	Insecticides and acaricides	Herbicides	Others	Fungicides	Total
1995	707.0	312.1	74.4	682.6	776.4
1994	479.3	286.5	84.5	432.8	1,283.1
1993	428.7	272.2	64.1	441.8	1,206.8
1992	505.0	228.5	101.7	457.1	1,292.3
1991	202.2	146.8	41.8	223.8	614.6
1990	260.3	159.4	55.6	169.2	644.5
1989	208.1	154.2	30.7	328.8	721.8
1988	158.9	145.2	28.5	329.9	662.5
1987	182.3	173.4	28.1	357.3	741.1
1986	134.9	121.3	42.6	281.3	580.1

Table 4.3 Importation of different groups of pesticides into Kenya (1986–95) (adapted from Pest Control Products Board and quantity expressed in T finished product)

Year	Insecticides and Acaricides	Herbicides	Others	Fungicides	Total
1995	1,413.3	870.6	2,323.0	501.9	5,108.8
1994	1,049.9	747.4	1,671.8	563.3	4,032.4
1993	839	882	1,503	309	3,533
1992	1,670	1,122	2,634	1,164	6,590
1991	1,072	844	1,568	570	4,054
1990	1,572	1,134	1,330	857	4,893
1989	1,571	1,148	4,327	665	7,711
1988	1,089	2,108	4,259	801	8,257
1987	1,206	1,311	715	697	10,371
1986	1,076	112	654	808	9,597

Table 4.4 Importation of different groups of pesticides into Kenya (1986–95) (expressed as a percentage of the total monetary value of imports)

Year	Insecticides and acaricides	Herbicides	Fungicides	Other	Total
1995	39.8	17.6	38.4	4.2	100
1994	37.4	22.3	33.7	6.6	100
1993	35.5	22.6	36.6	5.3	100
1992	39.0	17.7	35.4	7.9	100
1991	32.9	23.9	36.4	6.8	100
1990	40.4	24.7	26.3	6.6	100
1989	28.8	21.4	45.5	8.6	100
1988	24.0	21.9	49.8	4.3	100
1987	24.6	23.4	48.2	4.3	100
1986	23.2	20.9	48.6	7.3	100

Table 4.5 Importation of some pesticides into Kenya (1986–92)

Year	Malathion (technical) (T)	Carbofuran (technical) (T)	Furadan (technical) (L)	Carbosulfan (technical) (L)
1992	10.0	23.0	15,000	20,000
1991	13.0	10.0	21,000	–[a]
1990	18.5	12.0	16,000	–
1989	16.0	7.0	10,000	20,000
1988	9.0	14.0	–	–
1987	15.0	30.0	2,000	–
1986	20.0	8.0	–	–

Note:

a En dash (–) indicates the product was not imported.

in the formulation of agricultural chemicals and related products, and contracted representatives of manufacturers of agricultural chemicals and related products not otherwise represented in Kenya. Most firms are overseas-based companies, except for the Pyrethrum Board of Kenya (PBK), which extracts pyrethrins from the pyrethrum plant *Chrysanthemum cinerariaefolium* Trev. (Compositae). Kenya has been the world's largest producer of pyrethrum products, exporting ground flowers for the mosquito coil market in addition to refined extract for inclusion in aerosols. To increase toxicity and consequently lower production costs of pyrethrum-containing insecticides, it is combined with synergists, such as piperonyl butoxide, which in themselves are not toxic (Casida, 1973; Vickery and Vickery, 1979). Other companies and organizations are also involved in the distribution and use of pesticides and related products. They include locally formed companies and cooperative societies like the Kenya Farmers Association (KFA). Some manu-facturers do not have facilities in Kenya but market their pesticides through an appointed agent(s).

Historically firms in the agrochemical industry have been responsible for pesticide distribution in Kenya. The principal importers before 1963 included Pest Control Ltd (founded in England), Murphy Chemicals (a subsidiary of May and Baker), and Shell Chemical Industries (a subsidiary of Shell Oil). The primary pesticide distributors included the Kenya Farmers Association (KFA) and BEA Corporation (owners of Mitchell Cotts and Simpson and Whitelaw Seed Merchants) (Rocco, 1999). The pesticides were used mostly on plantations, estates, and large farms owned by companies or individuals. After 1963 (post independence), most of the large farms were subdivided, and consequently the distribution of pesticides involved more farmers and became more complex, i.e. through cooperative societies. Representatives of overseas pesticide manufacturers are now involved in the importation of pesticides. Additionally they serve as the principal distributors, supplying pesticides directly to the large-scale and estate farmers and providing continuous supply to stockist shops throughout Kenya. The government regulates this sector through the PCPB.

Other important groups include the Agrochemicals Association of Kenya (AAK), the Kenya Safe Use Project and the Kenya Environment Secretariat. The AAK was established in 1958 as the Pesticide Chemicals Association of East Africa and was formed when the participants saw the need for a joint approach following discussions with the Ministry of Agriculture. At that time, the government was trying to establish certain standards for local formulations, particularly dusting powders. After the demise of the East African Community in 1977, the name was changed to the Pesticide Chemicals Association of Kenya (Rocco, 1999). Then in 1997, to reflect the broadening interests of its members, the name was changed again to the Agrochemicals Association of Kenya. In 1987, the Association started a training program on the safe use of pesticides. This encouraged the International Group of National Associations of Agrochemical Manufacturers (GIFAP) to start the Kenya Safe Use Project in 1991 (Rocco, 1999).

In the public health sector, pesticides have offered control of vector-borne diseases such as malaria, African sleeping sickness, bilharziasis (an infection by parasitic flukes of the genus *Schistosoma* Sambon), and fascioliasis (an infection caused by liver flukes of the genus *Fasciola* L.) through pesticide spray programs aimed at controlling the vectors including mosquitoes, tsetse flies, and water snails. WHO programs to eradicate these pests in areas like Mwea Tabere settlement scheme (an area set aside for rice growing and human settlement), Kano Plain, and Lambwe Valley have rendered them habitable. Historically dieldrin, DDT, and endosulfan were used for the control of mosquitoes and tsetse flies, but due to their detrimental effects on non-target organisms, the less persistent OP, carbamate, and pyrethroid insecticides are now used. Pirimiphos methyl is currently being used to control adult mosquitoes outdoors and permethrin is used for household residual sprays and for treating bed mosquito nets, curtains, and fabrics for protection against mosquitoes and other biting insects. Cyhalothrin-λ is also used in public health for control of houseflies, mosquitoes, and cockroaches. Niclosamide and trifenmorph have been used at Mwea Tabere to control the water snail *Biomphalaria pfeifferi* Krauss (Gastropoda: Planorbidae), which is a vector of bilharziasis. Household pests such as flies, cockroaches, fleas, rats, and mice have been controlled using various products (see Table 4.6).

Storage pests

Problems associated with storage insect pests on maize in Kenya have existed ever since the crop was first introduced. This is because the high temperatures and relative humidity in most regions of the country strongly favor the growth and development of these pests (Asman, 1966). The infestation trend of harvested crops can be broken into three phases depending on the species attacking the crop and the storage environment. The first phase occurs when the grain is maturing in the field and is characterized by infestations by primary pests, e.g. *Sitophilus zeamais* Motschulsky (Coleoptera: Curculionidae) and *Sitotroga cereallela* Olivier (Lepidoptera: Gelechiidae) (Floyd, 1971; Ayertey, 1978), which attack whole grain. *Ephesta cautella* Walker (Lepidoptera: Pyrilidae) is absent from the grain during this phase. Once grain is shelled and placed in the warehouse, *E. cautella* becomes important in close association with other secondary pests, particularly *Tribolium castaneum* Herbst (Coleoptera: Tenebrionidae), *Corcyra cephalonica* Stainton (Lepidoptera: Galeriidae), and *Oryzeaphilus surinamensis* L. (Coleoptera: Silvinidae) (Delima, 1973). Secondary pests are those that feed on grain already damaged by primary pests and also on fragments of grain. The third phase, in which *E. cautella* is less important, occurs when control operations are less than optimal and comprises infestations by *Rhizopertha dominica* Fabricius (Coleoptera: Bostrichidae), *Cryptolestes* spp. (Coleoptera: Cucujidae), and *Tenebroides mauritanicus* L. (Coleoptera: Trogositidae) (Graham, 1970a). Attempts to control the pests have relied heavily on the use of pesticides including DDT, γ-BHC, pyrethrins, and malathion, but they have achieved limited

Table 4.6 Some pesticides used in Kenya (adapted from the Pest Control Products Board)

Pesticide common name	Type: use
λ-Cyhalothrin	Insecticide: for use on cotton, horticulture, ornamentals.
Carbosulfan (carbofuran or furadan)	Insecticide: control of maize stalkborer, coffee berry borer, cotton pests, aphids, thrips, lister scale, soil pests (e.g. termite grubs and nematodes) in coffee nurseries. Seed dressing in beans and maize for the control of soil borne and early foliar pests.
Cypemethrin	Insecticide: for use on cotton, vegetables, citrus, and other fruits and army worm and locust control.
Chlorpyrifos	Insecticide: for use on cotton, locust and army worm control, soil pests and larvicide for public health.
Carbofuran	Systemic insecticide/nematicide: soil pests, nematodes, early foliar feeding pests on coffee, bananas, pineapples, pyrethrum, nurseries, maize. Applied with mechanical granular applicators.
Glyphosate	Herbicide: post-emergence systemic control of weeds in coffee, tea plantations, sugarcane, pasture destruction, reduced tillage.
Copper hydroxide or 50% metallic copper	Fungicide: for the control of Coffee Berry Disease (CBD), leaf rust, bacterial blight on coffee, and horticultural crops.
Chlorfenvinphos	Acaricide: for the control of all species of ticks found in East Africa (vectors of East Coast fever), also fleas, lice on cattle, goats, sheep.
Amitraz (N-methylbis (2,4-xylylimino-methyl)amine)	Acaricide: for veterinary use to control ticks and other ectoparasites on cattle – 0.025% aqueous dispersion applied as dip or spray at 7 d intervals.
Coumatetralyl	Rodenticide: for the control of rats and mice.
Bacillus thuringiensis Berliner var. kurstaki 16	Biological insecticide: for control of lepidopterous larvae and other pests on vegetables; for the control of giant looper, green looper, leaf skeletonizers, and jelly grub in coffee.
Pyrethrin/permethrin/ piperonyl butoxide/ dichlorvos	Insecticide: aerosol for the control of crawling and flying insects, cockroaches, ants, flies, mosquitoes.

success (McFarlane, 1969). Presently bromophos, dichlorvos, pirimiphos-methyl, and permethrin are the primary insecticides used, although control is incomplete and the pests still cause significant losses. Pests attacking grain stored on the farm are controlled by residual chemical sprays on storage structures and insecticidal dusting of cob maize (Anonymous, 1974). In contrast, the primary method employed at centralized storage facilities has been fumigation (McFarlane and

Sylvester, 1969). For immediate control of pest outbreaks or surface infestations, pesticides are sprayed directly on the grain surface and on storage fabrics, thereafter providing residual protection (McFarlane and Sylvester, 1969).

Lalah and Wandiga (1996) found that after 51 weeks of storage, 34 to 60 percent of the initial radiolabelled malathion dust remained on stored beans *Phaseolus vulgaris* and maize *Zea mays* irrespective of the storage method used, i.e. the open basket storage model or the modern wooden box model. Half-life of the pesticide ranged from 194 to 261 d for maize and 259 to 405 d for beans in open baskets or closed boxes, respectively.

Acaricides

Ticks cause the greatest loss of livestock and are the most important vectors of disease agents in domestic animals (Kaine, 1976; Keating, 1983). Several different acaricides with varying application rates, residual action periods, stripping rates, stability, and safety have been used to combat them (Keating, 1983). The use of acaricides in Kenya has been orchestrated to avoid resistance development by restricting the number of available acaricides. Sodium arsenite was the only acaricide in use in Kenya between 1912 and 1949 for vector control of serious livestock diseases such as East Coast fever (Keating, 1983). The first resistance to arsenic was reported in the blue tick *Boophilus decolaratus* (Acari: Ixodidae) in 1953. Lindane (benzene hexachloride, BHC or hexachlorocyclohexane, and HCH) was introduced in 1949 and resistance to BHC was first noted in 1954 in *B. decolaratus* (Keating, 1983). The development of tick strains resistant to arsenic and HCH led to the increased use of toxaphene, a chlorinated camphene, which was introduced in 1950 (Keating, 1983). By 1956, toxaphene was the major acaricide in use, due to its stability in dip washes and its prolonged residue effect. Other OCs, i.e. DDT and dieldrin, were introduced in 1956 and 1961, respectively, but because tick resistance developed, the OC acaricides were banned in 1976 (Keating, 1983). A further disadvantage of OC acaricides was that they accumulated in body fat and were secreted in milk from dairy animals.

OP compounds, such as dioxathion and coumaphos were introduced in 1959 (Keating, 1983). In 1961, resistance to toxaphene was noted in two strains of the red-legged tick (*Rhipicephalus evertsi* (Acari: Ixodidae)) (Anonymous, 1961) and in *B. decolaratus* in 1962 (Anonymous, 1962) leading to increased use of OP compounds. The OPs were often used during this period in combination with arsenic, HCH, and toxaphene (as toxaphene still effectively controlled one of the most important species of ticks *R. appendiculatus* Neumann (Acari: Ixodidae), the vector of East Coast Fever (ECF)). *R. appendiculatus* eventually developed resistance to toxaphene leading to its ban (Kenya, 1976). Then in 1976, two OPs, dioxathion and quintiofos, and a carbamate, carbaryl, were gazetted and recommended for use. Acaricides that are still in use in Kenya include carbaryl, quintiofos, chlorfenvinphos, coumaphos, and several formamidines. However, amitraz is the most widely used. Synthetic pyrethroids are currently undergoing efficacy tests and some, e.g.

cypermethrin, have been recommended for use. Most farmers treat their cattle in cattle dips because this is more economical and they find it easy to maintain the correct chemical concentration in the dip.

Pest–natural predator imbalance

Control of storage pests in Kenya, since the establishment of the Cereals and Produce Board silos for post harvest storage has been accomplished with the use of recommended pesticides (McFarlane, 1969; Hall, 1970; Delima 1973). The pesticides are broad-spectrum and should have controlled all species present. However, some species including the moth, *E. cautella*, have survived following spraying operations (Graham, 1966). Occasionally upsurges of *E. cautella* populations have been observed following pesticide treatment indicating possible development of resistance to commonly used chemicals (Graham, 1970a). Graham (1966) found that after fumigation with methyl bromide, the *E. cautella* population attained its third generation peak in 140 days. During this period, about 3,000 adults were caught in flight traps each day. Thereafter, the moth population declined to a level of one to 10 moths per day because of pressure from the predatory mite *Blattisocius tarsalis* Berlese (Acari: Ascidae), a destructive feeder of *E. cautella* eggs. He noted that in the absence of *B. tarsalis* and chemical sprays, *E. cautella* populations were high and often occurred in combination with another pest *T. castaneum*. Muhihu (1996) observed that although pesticides are considered effective, in that they substantially reduce pest numbers, complete eradication is not possible. Graham (1970b) postulated that where an insecticide combined with mite control was necessary, some pesticides appeared to be more toxic to the mite than to the moth and that the continued use of the insecticide led to increased importance of *E. cautella* as a pest in maize stores. Incomplete eradication by malathion of *E. cautella* has been observed in other countries (Graham, 1970b) and other pests, e.g. *T. casteneum*, have also developed resistance to malathion (Champ and Dyte 1976).

Biological control has been an important component of pest control in Kenya. The common coffee mealybug (CCM) *Planococcus (Pseudococcus) kenyae* Le Pelley (Hemiptera: Pseudocóccidae) was imported into Kenya from Uganda. The first epidemic occurred in 1923 and continued until 1951 when it was reduced to a minor pest (Hill, 1975; Abasa, 1981). Because insecticides had proven ineffective, parasites were used to fight the CCM. These included *Anagyrus kivuensis* Compere (Hymenoptera: Encyrtidae) and *Anagyrus beneficians* Compere (Hymenoptera: Encyrtidae) (Hill, 1975). The parasites attack the CCM on coffee and indigenous plants. However, the coffee tree has to be kept free from unwanted sucker growth (upright shoots growing low on the trunk of the tree) because these suckers are attacked by the green scale *Coccus alpinus* De Lotto (Hemiptera: Coccidae) and *A. kuviensis* is less effective in the presence of green scale. Ants, *Pheidole punctulata* Mayr (Hymenoptera: Formicidae), also aid the flourishing of the CCM by attacking its parasites (Abasa, 1981). To prevent this, trees were sprayed with a band of dieldrin (100 ml of 18 percent dieldrin in 20 ml water) mixed with methylene blue for identification of banded trees. This prevented ants from reaching the parasites. Only badly infested trees were sprayed to runoff with 60 percent diazinon. A number of coccid pests of coffee were controlled with this integrated pest control method. This included the white waxy scale *Gascardia brevicauda* Hall (Hemiptera: Coccidae), the green scale (*C. alpinus*), the star scale or yellow-fringed scale *Asterolecanium coffeae* Newstead (Hemiptera: Asterolecaniidae), and the root mealybug *Planococcus citri* Risso (Homoptera: Pseudococcidae) (Abasa, 1981). However, dieldrin has now been banned for use in Kenya and ethion is currently used to control the scales and mealy bugs in coffee and the ants that attend to them.

The cassava mealybug *Phenacoccus manihoti* Metile-Ferrero (Homoptera: Pseudo-coccidae) is a major pest of cassava *Manihoti esculenta* Crantz (Euphorbiaceae), a major source of carbohydrates in Kenya. A parasitic wasp *Epidinocarsis lopezi* De Santis (Hymenoptera: Encytidae) was released and has been shown to control populations of the cassava mealybug (Kariuki *et al.*, 1991a). Another pest of cassava, the cassava green mite *Mononychellus tanajoa* Bodar (Acari: Tetranychdae), has been found to be affected by exotic phytoseeids *Neoseiulus ideaus* Denmark and Muma (Acari: Phytoseiidae) (Kariuki *et al.*, 1993). A newly introduced phytoseeid *Typhlondromalus aripo* De Leons (Acari: Phytoseeidae) was released in 1995 and 1996 and has established itself in the western and coastal regions of Kenya (Kariuki *et al.*, 1998). Populations of *N. ideaus* were found to have established themselves 33 months after their initial release (Mambiri *et al.*, 1994).

Another biological control program, which has registered success, is that for the larger grain borer (LGB) *Prostephanus truncatus* Horn (Coleoptera: Bostrichidae). This is a pest causing serious losses of stored maize and cassava that is spreading widely in East and West Africa (Nang'ayo *et al.*, 1994). Releases of its natural predator *Teretriosoma nigrescens* (Coleoptera: Bostrichidae) resulted in a strong negative pressure on LGB populations (Nang'ayo *et al.*, 1994). Irish potato *Solanum plantanim* L. (Solanaceae), another important source of carbohydrates in Kenya, has been attacked by the potato tuber moth *Pthorimaea opercullela* Zeller (Lepidoptera: Pyrilladae). However, releases of the parasitoid *Copidosoma koehleri* Blanchard (Hymenoptera: Encyrtidae), which parasitizes eggs of PTM, did not bring it under control (Mambiri *et al.*, 1993).

Biological control methods have been used to fight weeds. *Salvinia molesta* Mitchell (Salviniceae) is a free floating aquatic fern native to South America, which was introduced into Kenya as an ornamental. It is a fast-growing weed and was reported to double in weight every 4.5 days in Lake Naivasha, a closed basin, freshwater lake on the floor of the Rift Valley (Kariuki *et al.*, 1991b). It forms dense mats, thus interfering with fishing activities and water pumping (for both domestic and irrigation purposes), among other activities. The weevil *Cyrtobagous salviniae* Calder and Sands (Coleoptera: Curculionidae) was released in 1991. It established itself and brought the weed under control (Kariuki *et al.*, 1991b; Oduor *et al.*, 1995). However, the water hyacinth *Eichhornia crassipes* Solms (Pontederiaceae) is quickly taking over as the major aquatic weed (Oduor *et al.*, 1995). Trials with the bruchids *Neochetina* bruchi Hustache (Coleoptera: Curculionidae) and *Neochetina eichhorniae*

Warner (Coleoptera: Curculionidae) have shown that they are establishing themselves (Oduor et al., 1995) and may bring this aquatic weed under control.

Pest resistance to insecticides

Armstrong and Smith (1958) studied the effects of commonly used OC insecticides on the mosquito vector for malaria and found it was susceptible to DDT, λ-BHC, and dieldrin. In Kenya, rice is grown under irrigation in two main areas: Ahero, in Nyanza Province near Kisumu, and Mwea Tabere, in Eastern Province near Embu (Okedi, 1988). Mosha and Subra (1982) found *Anopheles gambiae* Giles (Diptera: Culicidae) was the most common malaria vector in these areas. Fields are sprayed regularly with insecticides for control of agricultural pests and applied chemicals include fenitrothion, carbofuran, and, previously, DDT. These spraying activities result in pesticides being present in *A. gambiae* breeding sites in rice paddies where its larvae are found in large numbers. Chapin and Wasserstrom (1981) suggested that direct exposure of mosquitoes to agricultural insecticides may exert a selection pressure leading to the development of resistance to those insecticides present and those with similar modes of action. Okedi (1988) found that there were *A. gambiae* larvae, which showed high resistance to fenitrothion and DDT, at Ahero and Mwea. However, little or no resistance existed for pesticides not used for pest control, e.g. dieldrin and malathion. The development of resistance to insecticides by malaria vectors has been one of the causes of the resurgence of malaria in the region (Okedi, 1988).

Location of pesticide use with respect to the marine environment

Figure 4.2 is a map of Tana River basin showing the irrigation schemes, including the Hola irrigation scheme. In the 1960s and 1970s, DDT was extensively used in the Hola irrigation scheme as the primary insecticide for cotton, maize, and horticultural crops. For cotton pests, it was applied as a 5 percent dust at 5.5 to 11.0 kg ha^{-1}. The quantity of DDT used during this period was estimated at 12 T annually with at least 80 percent being used for the control of cotton pests (Munga, 1985). Monocrotophos was also used in the early stages of the scheme for cotton pests (Munga, 1985). As a combination spray mixture, monocrotophos/DDT (10 percent/40 percent) was applied with ultra low volume (ULV) equipment at 2.5 to 3.0 L ha^{-1}. Spraying of cotton at Hola was initiated using an EIL to justify the need for pesticide application and this threshold value was the appearance of pests on >5 percent of cotton plants (Munga, 1985). The 1983 cotton season required seven aerial applications of endosulfan 25 percent (ULV of 6 to 12 g a.i. ha^{-1}), deltamethrin 0.5 percent (ULV of 12.5 g a.i. ha^{-1}), and hostathion 25 percent (ULV) over the period June to September. In addition to aerial spraying, hand spraying continued for small areas affected by pests until December, although no specific schedule was followed (Munga, 1985).

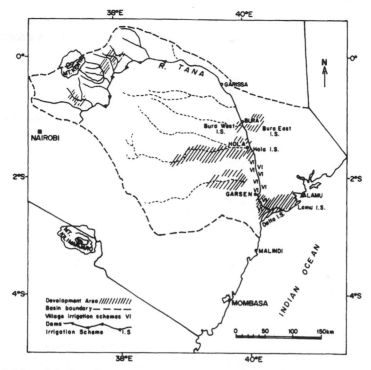

Figure 4.2 Map of the Tana River basin showing irrigation schemes and other developments (from Tana and Athi Rivers Development Authority, Nairobi, Kenya)

Starting in 1980, synthetic pyrethroids and endosulfan (as thiodan) began to replace DDT for control of cotton pests (Munga, 1985). During the 1980–81 cotton growing season, an estimated 350 kg of cypermethrin was used and, during the 1982 to 1984 cotton growing seasons, an estimated 3.9 T of endosulfan and 76 kg of deltamethrin were used per season (Munga, 1985). Endosulfan has also been widely used as an alternative to DDT and dieldrin for tsetse fly and cotton pest control in Kenya and other tropical and subtropical countries of Africa (Munga, 1985). Table 4.7 lists some important pests, the crops susceptible to attack, and the control measures used in the Tana River District (1992 to 1993) (Pest Control Products Board of Kenya).

Aerial spraying is the pesticide application technique that is most prone to spray drift. This can result in pesticide falling onto non-target areas, especially freshwater bodies located in the irrigation scheme. Soil erosion also contributes to moving pesticide residues into water bodies. Athi River (see Figure 4.3) was found to contain a number of pesticide residues arising from the extensive use of agricultural pesticides in the Kiambu District (UNEP, 1982).

Approximately 97 percent of Kenya's rice crop is produced under irrigation schemes covering 9000 ha (Anonymous, 1985). Notable among these are the Ahero

Table 4.7 Important pests and control measures taken in Tana River district (1992 and 1993)

Pest	Crop attacked	Chemical control
Maize stalk borer	Maize, sorghum, millet	Cypemethrin, trichlorfon, carbofuran
American bollworm	Cowpeas, green grams, tomatoes	Cypermethrin, carbosulfan, profenofos/cypermethrin
Diamondback moth	Kales	Cypermethrin
Aphids	Cowpeas, green grams, beans	Cypermethrin
Stainer pink bollworm, apican bollworm, spun bollworm, spider mites	Cotton	Cypermethrin, carbonsulfan, λ-cyhalothrin, profenofos/cypermethrin, fenvalerate
Army worms	Cereals, pastures	Cypermethrin 2.5% ULV[a]
Scales	Citrus, robusta coffee	Carbosulfan, chlorpyrifos, omethoate
Mango weevil	Mangoes	Carbosulfan, fenvalerate

Note:
a ULV indicates that pesticide was applied with ultra low volume application technology.

Rice Research Station in western Kenya and the Mwea Irrigation Scheme in the upper reaches of the Tana River. Lalah (1993) reported that carbofuran (as 5 percent technical furadan granules) is applied in the seed furrow at the rate of 0.54 kg a.i. ha^{-1} to control soil-dwelling or foliar-feeding insects and mites. Lalah (1993) found that carbofuran dissipated faster from flooded soil than from non-flooded soil, with levels approaching 40 percent in less than 25 days and falling below 20 percent after 111 days. Carbofuran is highly soluble in water, and thus tends to move into the water column above flooded soil. While most of the pesticide was found in the top 10 cm layer of soil, its movement into the water column poses a risk of contamination of nearby streams and canals. These waters flow into the Nyando River and, ultimately into Lake Victoria, the world's largest freshwater lake (Figure 4.3). The potential for contamination of freshwater streams and lakes was highest in the first three weeks following pesticide application.

PESTICIDE CONTAMINATION OF THE ENVIRONMENT

Sediments

Everaarts *et al.* (1997) examined pesticide residues in sediments and macro-invertebrate organisms along the Kenyan coast. They found PCBs and pesticide residues in sediment samples from two shallow coastal stations at the mouth of Sabaki River (Figure 4.3). PCB congeners 28, 52, 101, 153, and 138 were detected at the two sites in a concentration range of 7.1 to 62.2 ng g^{-1} of organic carbon.

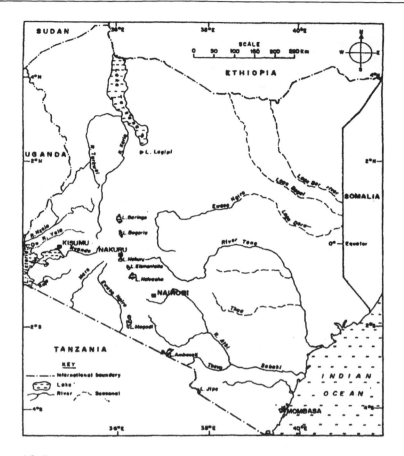

Figure 4.3 Drainage map of Kenya

Similarly p,p'-DDE was measured at concentrations ranging from 32.1 to 508.8 ng g^{-1} organic carbon. Furthermore, α-HCH was detected in increasing amounts across the continental shelf at both shallow and deep water stations along the Kenyan coast and γ-HCH was found at only six stations (concentration range 7.3 to 53.2 ng g^{-1} organic carbon). Shallow sediment samples near the Sabaki River mouth contained high levels of dieldrin (37 ng g^{-1} of organic carbon) and p,p'-DDE (510 ng g^{-1} of organic carbon).

Macroinvertebrate organisms

Samples of benthic organisms from the Kenyan coast have been analyzed for the presence of PCBs and cyclic pesticides (Everaarts *et al.*, 1997). They found that concentrations of PCB congeners and cyclic pesticides were higher at the mouth of the Sabaki River than at the mouth of the Tana River. Bivalve molluscs from the mouth of the Sabaki River and Kiwaya Bay had the highest levels of PCBs (30 and 65 ng g^{-1} of lipid for congener 153) and 40 ng g^{-1} of lipid for congener 153. They found p,p'-DDE, was present in all samples at levels ranging from 15 to 48 ng g^{-1} of lipid in both bivalve and gastropod molluscs. Based on the presence of only seven PCB congeners and the presence of p,p'-DDE at just a few sample sites around the outflow of the Sabaki River, they concluded that there was a low degree of contamination of surface estuarine sediments and shallow coastal regions of Kenya. They observed a 'wash-out' effect from river flow as evidenced by the concentration gradient (increasing) across the continental slope toward deep water. All animal groups analyzed showed the presence of PCBs and p,p'-DDE. Gastropod molluscs and edible penacid prawns had the highest levels of PCBs and p,p'-DDE.

Freshwater and estuarine ecosystems

A study conducted in the Hola irrigation scheme demonstrated a strong correlation between DDT and endosulfan tissue residues and the level of fat in fish (Munga, 1985), Tables 4.8 (a) and (b). Munga examined pesticide residues of DDT and endosulfan in four species, *Clarias gariepinus* Burchell (Siluriformes: Claridae)(syn. *C. mossambicus*), *Labeo gregorii* Boulenger (Cypriniformes: Cyprinidae), *Oreochromis mossambicus* Peters (Perciformes: Cichlidae), and *Tilapia zilli* Gervais (Perciformes: Cichlidae). He also studied various factors that might affect pesticide residue levels including, species differences, fat content, tissue type, and sampling site distance from the application site. Of the four species, *C. gariepinus* had the highest pesticide residue levels. He suggested this occurred because *C. gariepinus* is a bottom feeder while the other species are essentially surface feeders. Pesticide concentrations from lateral muscle and liver tissue and eggs were measured. Liver had the highest concentrations of total ($\sum$) DDT and endosulfan (based on ww), followed by eggs and muscle tissue, Table 4.8 (a–d). The mean concentration of $\sum$ DDT in liver was approximately 7.1 times and 2.4 times higher than in muscle and eggs, respectively. The concentration of endosulfan in liver was 12.5- and 5-fold higher than in muscle and eggs, respectively. The relative concentrations of $\sum$ DDT and endosulfan in liver, egg and muscle tissue (based on ww) from *C. gariepinus* showed a pattern different from that of *L. gregorii* (Table 4.8 (c)).

Munga (1985) found in *L. gregorii* the liver had the highest fat content followed by eggs and lateral muscle tissue, respectively. *C. gariepinus* liver samples had the highest fat content, but, unlike *L. gregorii*, egg and lateral muscle tissue had similar fat content. The residue concentrations of $\sum$ DDT in liver tissue and eggs of *C. gariepinus* were relatively higher compared to muscle tissue, and the residue concentrations in the liver were relatively lower than in eggs. Endosulfan residues in tissues of *C. gariepinus* showed a different pattern to that of *L. gregorii* in that the lowest residue concentration was in the liver, (Table 4.8 (d)).

The primary metabolites of p,p'-DDT are p,p'-DDE and p,p'-DDD (Wedemeyer, 1968). The metabolite p,p'-DDE is more stable than either p,p'-DDD or p,p'-DDT and tends to accumulate in adipose tissue (Wedemeyer, 1968; Cherrington *et al.*,

Table 4.8 Pesticide residues found in fish tissues[a]

(a) DDT residues in Labeo gregorii tissues

Tissue	Mean fat content (%)	Sum ($\sum$) DDT	
		mg kg^{-1} fat	mg kg^{-1} ww
Muscle	0.27	17.14	0.13
Liver	8.59	10.68	0.92
Eggs	1.99	19.63	0.38

(b) Endosulfan residues in Labeo gregorii tissues

Tissue	Mean fat content (%)	$\sum$ Endosulfan	
		mg kg^{-1} fat	mg kg^{-1} ww
Muscle	0.22	1.81	0.004
Liver	8.59	0.61	0.050
Eggs	1.99	0.68	0.010

(c) DDT residues in Clarias gariepinus tissues

Tissue	Mean fat content (%)	$\sum$ DDT (mg kg^{-1} ww)
Muscle	0.48	0.19
Liver	3.50	2.47
Eggs	0.49	6.01

(d) Endosulfan residues in C. gariepinus tissues

Tissue	Mean fat content (%)	$\sum$ Endosulfan (mg kg^{-1} ww)
Muscle	0.32	0.09
Liver	3.05	0.01
Eggs	0.49	0.07

Note:
a After Munga, 1985.

1969). *C. gariepinus* had slightly higher proportions of *p,p'*-DDE, and lower proportions of *p,p'*-DDD and *p,p'*-DDT in the lateral muscle than *L. gregorii*, *C. gariepinus*, and *T. zilli* (Munga, 1985). The relatively higher concentrations may be the result of the breakdown of *p,p'*-DDT to *p,p'*-DDE in the muscle tissues. It may also be due to chronic or long-term exposure to *p,p'*-DDE as a result of the breakdown of *p,p'*-DDT to *p,p'*-DDE in soil (Munga, 1985).

α-endosulfan, β-endosulfan and endosulfan sulfate (the oxidized fat soluble metabolite of the two isomers of endosulfan) residues were found in the lateral muscle, eggs, and liver of the four fish species studied (Munga, 1985). Based on the proportion of α-endosulfan and β-endosulfan in muscle tissues of *C. mossanbicus* and *L. gregorii*, α-endosulfan was metabolized faster than β-endosulfan. The higher proportion of β-endosulfan found in the *Clarias* muscle tissue may be due to a more ready availability of residues of the isomer adsorbed onto bottom sediment and organic matter that the fish take up through feeding at the bottom (Munga, 1985).

In fat and muscle tissue of Nile perch from Lake Victoria, Mitema and Gitau (1990) detected low levels of α-BHC (2.88×10^{-3} and 5.12×10^{-3} ppm, respectively),

β-BHC (0.22 and 0.26 ppm), aldrin (not detected and 0.02 ppm), dieldrin (0.2 and 0.07 ppm), and lindane (1.19 × 10^{-3} and 7.74 × 10^{-3} ppm). Mean DDT levels were 0.99 and 0.45 ppm in fat and fillets of Nile perch, respectively, and ranged from 0.002 to 4.51 mg kg^{-1} lipid and 0.004 to 0.19 mg kg^{-1}. $\sum$ HCH residues ranged from 0.001 to 0.11 mg kg^{-1} in Nile perch. DDT and its metabolites formed the largest proportion of OC residues in fish samples, a finding consistent with previous studies. These insecticides had previously been used extensively in agriculture and aerial control of mosquitoes in the Lake Victoria region.

Mugachia *et al.* (1992b) investigated OC residue levels in fish from the Athi River estuary. Eight OC pesticide residues were detected in tissues from six species of fish and they were in order of decreasing frequency: *p,p'*-DDE, *p,p'*-DDT, *o,p'*-DDT, *p,p'*-DDD, β-HCH, α-HCH, heptachlor, and *o,p'*-DDD. Seventy-three percent of samples were positive for one or more of the residues. OC residues were detected more frequently and at higher levels in liver and egg samples than in the fillet. Sharks, at the top of the food chain, had the widest range of pesticide residues and significantly higher mean $\sum$ DDT levels (0.702 mg kg^{-1}) compared to breams and catfish (0.213 and 0.145 mg kg^{-1}, respectively).

Inland lakes ecosystems

$\sum$ DDT residue levels found in fish from inland lakes are given in Table 4.9. Lincer *et al.* (1981) found a bottom feeding fish *Labeo cylindricus* Peters (Cypriniformes: Cyprinidae) from Lake Baringo had a concentration of 0.4 mg kg^{-1} ww of *p,p'*-DDE in muscle tissue. Munga (1985) also found a maximum mean concentration of DDT in *C. gariepinus*, another bottom feeding species, muscle samples from the Hola irrigation scheme was 0.4 mg kg^{-1} ww. Apart from the isolated sample of *L. cylindricus* from Lake Baringo, $\sum$ DDT residue levels in fish were higher in samples from the Hola irrigation scheme than elsewhere in Kenya.

Koeman *et al.* (1972) found $\sum$ DDT residue levels in fish of 1.0 to 7.0 × 10^{-3} mg kg^{-1} ww, while Lincer *et al.* (1981) found DDE levels in fish of 7.4 × 10^{-2} mg kg^{-1} ww and DDE levels in biota of 4 × 10^{-2} mg kg^{-1} ww from Lake Nakuru. Mugachia *et al.* (1992a) measured OC pesticide residues in 208 samples representing five species of fish collected from Lake Naivasha, Masinga Dam on the Tana River and the lower Tana River at Garsen and Tarasaa between October 1988 and 1989. They found no residues in any of the samples from Lake Naivasha or the lower Tana River and detectable residues in only 36.8 percent of the samples from Masinga Dam (Table 4.10). These levels are much lower than concentrations found in marine species (Everaarts *et al.*, 1997; Barasa, 1998). The differences may be attributed to the drainage areas covered by rivers emptying into the lakes and the agricultural activities upstream. The Sabaki River drains a larger area with varied agricultural and industrial activities into Indian Ocean.

Everaarts *et al.* (1996) found PCB residues in samples from the Kenyan coastal ecosystem, yet the compounds have never been widely used in Kenya. Their source

Table 4.9 Reported level of $\sum$ DDT residues in fish from Kenyan lakes

Location and species	Residue concentration (mg kg⁻¹ ww)	Reference
Lake Baringo		
Tilapia nilotica	0.009	Lincer et al., 1981
Clarias mossambicus	0.019	
Burbus gregorii	0.028	
Labeo cylindricus	0.400	
Lake Naivasha		
Tilapia spirulus nigrax	0.001	Lincer et al., 1981
Micropterus salmoids	0.003	Lincer et al., 1981
Lake Nakuru		
Tilapia grahami	0.015	Lincer et al., 1981
Lake Victoria		
Lates nilotica	0.004	Foxall, 1983

Table 4.10 OC pesticide residues in fish from Masinga Dam on the Tana River

Species	Tissue	Mean residue concentration (mg kg⁻¹)				
		p,p´-DDE	p,p´-DDT	$\sum$ DDT	Lindane	α-HCH
Common carp	Fillet	0.030	0.223	0.234	0.14	–ᵃ
Catfish	Fillet	0.102	0.052	0.113	0.009	0.013
	Liver	0.138	0.052	0.163	0.010	0.0009
Tilapia	Fillet	–	–	–	0.011	–
	Eggs	0.068	–	0.075	0.0009	0.021

Note:
a En dash (–) indicates below detection limit. Adapted from Mugachia et al., 1992a.

can only be speculated, but it is likely to result from disposal of PCB wastes or leaks from power transformers in use around Kenya. The low level of pesticide residues found in the Everaarts *et al.* (1997) study is consistent with earlier investigations. For instance, Koeman *et al.* (1972) found very low levels of $\sum$ DDT (<0.001 to 0.064 mg kg⁻¹) in Lake Nakuru birds and fish. Later, Greichus *et al.* (1978) found slightly higher residue levels of DDE, DDD and dieldrin in the same lake. Wandiga and Mutere (1988) detected very low levels of lindane, 9×10^{-6} to 1.0 ppm, in human milk from nursing mothers in a Nairobi hospital. Similarly Kanja (1988) found low levels of DDT in human milk, chicken eggs, and other food sources. Moderate concentrations of OCs and OPs, mainly dioxathion, in cow's milk and meat products in the Athi River and Ngong areas, have been reported (Munga, 1985). Munga (1985) found Tana River fish contained DDT, DDE, DDD, and endosulfan residues. Bottom feeding species (*C. gariepinus*) had higher levels than surface feeding species (*L. gregorii*, *O. mossambicus*, and *T. zilli*).

Soils

The persistence of OC pesticides in Kenyan soils has been extensively studied (Sleischer and Hopcraft, 1984; Wandiga and Natwaluma, 1984; Wandiga and Mghenyi, 1988; Lalah et al., 1994; Ng'ang'a, 1994). Table 4.11 shows the rate of loss of DDT, DDE, and lindane in Kenyan soils. The accumulated evidence for the tropics indicates that OC pesticide persistence is lower than is found in temperate climates (Wandiga, 1996). This conclusion is consistent with levels described above. For instance, based on the quantity of OC pesticides used in the Hola irrigation scheme in the Tana District, one would predict higher concentrations in marine species where the Tana River enters the Indian Ocean. Given the high organic content of sediments along the Kenyan coast as a result of heavy upstream soil erosion, one would expect that the movement of OC pesticides to other areas as a result of ocean action would be minimal. However, the distribution of γ-HCH along the continental slope observed by Everaarts et al. (1996) confirms that pesticide movement from shallow coastal areas to deep ocean exists. The effect of low levels of DDT and its metabolites on marine life, wildlife, and the surrounding ecosystems have drawn the attention of the authors. Recently the authors initiated studies to examine the distribution and effect of DDT in the marine environment, using a laboratory-based marine ecosystem.

ECOTOXICOLOGICAL RISK ASSOCIATED WITH PESTICIDE RESIDUES

Fish and other aquatic organisms

In a study of the lethal and sublethal effects of DDT, carbofuran, trifenmorph, and niclosamide on *Oreochromis niger* Gunther (Perciformes: Cichlidae), Wangia (1989) determined 24 hour LC_{50}s of 0.042, 0.225, 0.118, and 0.103 mg L^{-1}, respectively. To achieve snail control in flowing waters, e.g. irrigation canals, a concentration of niclosamide at 0.3 to 1 mg L^{-1} for 24 hours is recommended. This concentration would be toxic to fish in the same waters. DDT and trifenmorph can accumulate in fish tissues (Munga, 1985), which poses a risk to the human population who consume the fish. This is one reason the use of these two pesticides at the Mwea Tabere settlement scheme was discontinued.

Matthiessen et al. (1982) reported that residues of endosulfan and endosulfan sulphate accumulated in several fish species and their predators during aerial spraying for tsetse fly control in the Okavango Delta of Botswana. However, the residues were rapidly metabolized after cessation of spraying. For example, the highest concentration of endosulfan residues found in *C. gariepinus* and *C. ngamensis* Castelnau (Silviformes: Claridae) muscle samples was 0.19 mg kg^{-1} ww during the spraying period; it returned to less than 0.03 mg kg^{-1} ww within three months, and less than 0.005 mg kg^{-1} ww after a year. Apart from the mortality of fish to endosulfan, adverse effects from sublethal doses of endosulfan have been recorded for some species. The aerial spraying of endosulfan for tsetse fly control in the

Table 4.11 Dissipation of DDT, DDE, and lindane in Kenyan soils

Kenyan soil site	Pesticide	Half-life (d)	Metabolites identified	Reference
Nairobi soil				
	DDT	117 ± 10	DDT, DDE, 4 PCB	Wandiga and Natwaluma (1984)
	DDT	118 ± 13	DDT, DDE, 4 PCB	
	DDT	98	–a	Wandiga and Mghenyi (1988)
	DDT	64.6	DDT, DDE, DDD	Lalah et al. (1994)
	DDE	145	DDE	Lalah et al. (1994)
	Lindane			
	1st phase	4	–	Wandiga and Mghenyi (1988)
	2nd phase	48	–	
Mombasa soil				
	DDT	88	–	Wandiga and Mghenyi (1988)
	DDT	270	DDT, DDE	Ng'ang'a (1994)
	Lindane			
	1st phase	5	–	Wandiga and Mghenyi (1988)
	2nd phase	6	–	
Lake Nakuru soil	DDT	110	–	Sleischer and Hopcraft (1984)

Note:
a En dash (–) indicates metabolites were not detected.

Okavango Delta at 6 to 12 g a.i. ha⁻¹, caused some fish mortality and, additionally, behavior changes in the surviving fish, short-term hematological changes, and non-lethal damage to the brain and liver (Matthiessen and Roberts, 1982). Male *O. mossambicus* (syn. *Sarotherodon mossambicus*) exposed to 0.5 mg L⁻¹ endosulfan in water experienced delayed breeding behavior, some females tended to abandon their unfertilized clutches, and newly hatched fry died (Matthiessen and Logan, 1984). The effect in the field was a reduction in fish recruitment in sprayed areas.

Pollution of streams and rivers by agricultural wastes and chemicals has led to habitat destruction and exposure to pollution for species that live in brackish water (notably *Macrobrachium rude* Heller (Palaemonunae), penacid prawns, mullet, and oysters) on the island of Mauritius in the Indian Ocean (UNEP, 1984). Mbuvi (1996) found that oysters have very high bioconcentration factors for the pesticide, DDT.

Animals, birds, and humans

Pesticides have been found in the milk and tissues of animals, both domestic and wild (Maitho, 1978; Lincer *et al.*, 1981; Kahunyo *et al.*, 1986; Kituyi *et al.*, 1997). Maitho (1978) found low levels of *p,p′*-DDT, *p,p′*- DDE, lindane, aldrin, and dieldrin in the fat of cattle. Kituyi *et al.* (1997) measured chlorfenvinphos residues in cows' milk in Kenya. Chlorfenvinphos is an acaricide used to control ticks, especially in western Kenya. He found that the concentration of chlorfenvinphos in milk samples varied between 0.52 and 3.90 mg kg^{-1} in the dry season and from 1.58 to 10.69 mg kg^{-1} during the wet season. Milk collected from plunge-dipped cows (cows forced to jump into and swim across a deep pit filled with the diluted insecticide) had significantly higher ($P < 0.05$) concentrations than milk obtained from hand-sprayed animals. The exposure levels were below permitted limits for adults (8 mg kg^{-1})) but exceeded by 7 to 15 times the acceptable daily intake for infants (*Codex Alimentarius*, 1993). Kituyi *et al.* (1997) concluded that breast-feeding mothers among the women involved in the hand spraying were at risk of contamination.

Kahunyo *et al.* (1986) found high levels of DDT and dieldrin in eggs from the Embu District. Pesticide accumulation in this case may have been favored by the practice of allowing chickens to freely forage for food, thus exposing them to pesticide residues around the farm. Cowpea plants from two to twelve weeks old were found to have residue levels ranging from 0.945 ± 0.040 mg kg^{-1} to 7.765 ± 0.211 mg kg^{-1} in Mombasa in a study on the uptake of ^{14}C *p,p′*-DDT (Kiflom *et al.*, 1999). Koeman *et al.* (1972) reported low tissue concentrations of $\sum$ DDT in birds collected from Lake Nakuru. They found levels ranged from 3.4 to 6.4 × 10^{-2} mg kg^{-1} ww in the birds. Lincer *et al.* (1981) measured DDE residue levels in two bird species, African cormorants *Phalacrocorac africanus* Gmelin (Pelicaniformes: Phalacrocoracidae) and single white-necked cormorants *P. carbo lucidus* L. (Pelicaniformes: Phalacrocoracidae) collected in 1970 from Lake Nakuru. They found that the former had 15 times as much DDE as the latter species. They suggested the cause was a difference in the diets of the two bird species. The African cormorant feeds mainly on fish, frogs, aquatic insects, crustaceans, and small birds while the single white-necked cormorant feeds on fish, frogs, crustaceans, and molluscs (Brown *et al.*, 1982). They contend that the diet of small birds, which consume grain contaminated by pesticides, causes the higher level of pesticide in the African cormorant. Lincer *et al.* (1981) found levels of DDE in the white pelican *Pelecanus onocrotalus* L. (Pelecaniformes: Pelecanidae) had doubled since 1970, indicating an increase in contamination of the lake system by DDT and its metabolites. They also found bioaccumulation of DDE in various food chains in Lakes Naivasha, Baringo, and Elementaita. Greichus *et al.* (1978) found DDE levels of 4.0 × 10^{-2}, 7.4 × 10^{-2}, and 0.4 mg kg^{-1} ww in biota, fish, and birds, respectively, in Lake Nakuru. Wild animals have also been affected by pesticide use. Alsopp (1978) found low levels of dieldrin and its photo-isomer in game animals after aerial spraying of dieldrin for tsetse fly control in the Lambwe Valley.

DDT, β-HCH, dieldrin, and heptachlor epoxide have been detected in the adipose tissue of humans in Kenya, with DDT being the main contaminant (Wasserman *et al.*, 1972). The people sampled had no occupational exposure to the pesticides, indicating other sources were responsible. Pesticide residues have also been detected in the milk of Kenyan women (Wandiga and Mutere, 1988). Human milk procured from a Nairobi hospital had levels of 9 × 10^{-6} to 1.0 mg kg^{-1} of

γ-HCH. Kanja (1988) found thirteen OC pesticides in human milk collected from eight areas of Kenya. The pesticides occurred in the following frequencies: p,p'-DDT(100 percent), p,p'-DDE (100 percent), HCB (60 percent), aldrin (35 percent), lindane (30 percent), β-HCH (27 percent), dieldrin (20 percent), α-HCH (8 percent), transnonachlor (6 percent), heptachlor (4 percent), endrin (4 percent), and heptachlor epoxide (0.4 percent). There were differences in the level of these compounds from one region to another, i.e. the mean level of DDT ranged from 1.69 mg kg^{-1} in human milk fat of nomads from Loitokitok to 18.73 mg kg^{-1} human milk fat in women from Rusinga Island (Kanja, 1988). The mean ratio of p,p'-DDT to p,p'-DDE also varied with the area, i.e. 0.7 in Karatina to 4.4 in Turkana. The highest mean level of α-HCH, 10.3 mg kg^{-1}, was found in milk samples from Embu with corresponding high levels of β-HCH and γ-HCH at 11.1 mg kg^{-1} and 22.1 mg kg^{-1}, respectively (Kanja, 1988). The main route of exposure to DDT and other OC pesticides in humans is through oral intake, e.g. vegetables, beef, and dairy milk containing these residues (Matsumura, 1972; Kanja, 1988). Inhalation of pesticide vapors and consumption of contaminated water are other potential sources, especially in agricultural areas (Kanja, 1988). In one instance, people complained of stomach pains after eating fish sold by farmers near the Sagana Bridge in Muranga (Kanja, 1988). Contamination by pesticides coming from surrounding coffee plantations was suspected. Chronic toxicity may occur with ingestion of DDT over long time periods (Kanja, 1988) and may affect development of fetuses and infants (Hayes, 1982). DDT and other OC pesticides have been shown to affect the reproduction of various test animals, but effects varied with species making extrapolation to humans difficult (Hayes, 1982).

Mbuvi (1996) monitored the distribution of p,p'-DDT in aquaria set up to simulate a tropical marine ecosystem and containing oysters, seawater, and sediments. Concentrations of ^{14}C-DDT residues were found to decline very rapidly (within 24 hours) in the water to 70 percent of the initial level and further declined to less than 10 percent after three days. Oysters accumulated ^{14}C-DDT at a very fast rate reaching a maximum level in 24 hours, with a bioconcentration factor as high as 19,273 before gradually declining. There was a steady buildup of ^{14}C-DDT residue in sediment, concentrated in the top 1 cm layer, during the first 24 hours, with the unbound form accounting for more than 95 percent of the total at any given time. Residue analysis showed the presence of p,p'-DDE in all samples. Thus, Mbuvi concluded that DDT distributes itself fairly widely in the marine environment with most of it being absorbed by marine organisms and sediments. A similar study on the depuration of DDT and chlorpyrifos contaminated oysters and sediments in uncontaminated seawater was conducted by Ongeri et al. (1998). They found that DDT and chlorpyrifos were rapidly distributed in the water. The oysters accumulated DDT residues (giving a BCF value of 19,266) and chlorpyrifos residues (giving a BCF value of 4,334) rapidly after 24 hours of exposure to sublethal concentrations of the pesticides. The sediment absorbed 9 to 16 percent of DDT residues and 18 to 28 percent of chlorpyrifos residues of the applied pesticide within 24 hours. Most of the residues in seawater were lost through volatility, i.e.

70.8 to 87.8 percent of DDT and 48.6 to 78.6 percent of chlorpyrifos over a period of 24 hours.

PESTICIDE MANAGEMENT

Regulating bodies

Apart from the Pest Products Control Board (PCPB) mentioned at length previously, there are three other bodies involved in pesticide regulation in Kenya. The Kenya Environmental Secretariat is the coordinating body for all matters pertaining to the protection of the environment. This body provides linkages between Kenya and various international organizations, e.g. UNEP, FAO, and WHO, through which important information and policy guidelines are formulated. In this connection, the agricultural industries are required to implement the FAO code of conduct on the distribution and use of pesticides.

Formerly the Pesticides Chemicals Association of Kenya (PCAK), the Agro-chemicals Association of Kenya (AAK) incorporates most of the manufacturers and distributors of agrochemicals and related products and services throughout Kenya. The main objective of the association is to ensure that members ascribe to its ethical objectives as they relate to safety, packaging, labeling, and use of these products.

The Kenya Safe Use Project is a non-governmental organization supported by GIFAP, now the Global Crop Protection Federation (GCPF). Through various task forces, the Kenya Safe Use Project is pursuing the following objectives: to improve standards in formulation plants, to improve pesticide registration procedures, to train transporters of pesticides, to train farmers, to train pesticide retailers and stockists, to establish poisoning information and treatment centers, to eliminate stocks of waste agrochemicals in an environmentally safe manner, to promote protective clothing use, to improve pesticide labels and promote the use of pictograms, and to educate school children on the hazards of pesticides and what precautions must be taken when using them. Since its inception in 1991, this project has trained approximately 500,000 farmers, stockists, distributors, and health professionals in Kenya on the safe and effective handling of pesticides (Rocco, 1999).

Pesticide handling and use by farmers in Kenya

Most farmers in Kenya buy their pesticides from stockists, distributors, or through their cooperative societies and unions. Credit facilities are available in the latter, especially if the farmer is a member. In such a case, their produce acts as the guarantee to cover the loan. Various factors dictate the choice of pesticide a farmer makes from the many products available on the market. This may be the price (cheapest) or advice from agricultural extension officers or other successful and more knowledgeable farmers. They are also attracted by the container (if it is reusable). Pesticides are normally transported to the farm by public means. Most farmers store the pesticides along with household goods or farm produce. Kimani and Mwanthi (1995) observed that some community members rinsed empty pesticide containers at community water sites, which could lead to contamination of drinking water sources. They also showed that in Githunguri location, Kiambu District OCs, OPs, and carbamates were being used all year round. Mwanthi (1998) found that of 72 households in Kiaria – 50 km NW of Nairobi – who all conceded to have been using eight types of pesticides over the 1994 to 1995 period, 25 percent had detectable levels of DDTs, carbofuran, or carbaryl in their water. The Kenya Government has not yet established maximum contaminant limits (MCL) for pesticides in drinking water. While mean concentrations for each pesticide did not exceed World Health Organization guidelines, individual sites exceeded the WHO MCL for carbofuran (5 ppb) by 20-fold.

The most commonly used applicators are tractor-mounted sprayers and knapsack sprayers. Aircraft-mounted sprayers are used especially when controlling outbreaks of army worms, tsetse-flies, or locusts. Cattle dipping is also widely practiced. Unorthodox applicators include fly whisks and specially-bound leaves or brooms (Nyaga, 1988). Some of the problems associated with pesticide use in Kenya include the emergence of resistant pests; a lack of protective clothing for individuals performing spraying, mixing, or cleanup duties due in large part to its expense; and that, often, symptoms of pesticide poisoning are not diagnosed, or if they are, proper therapy or prophylaxis is not instituted (Kanja, 1988).

A lack of personnel and funds exists to test pesticides in the Kenyan environment and to ensure that private companies are adhering to the regulations as stated in the Pest Control Products Act. Additionally a lack of proper knowledge about pesticides by farmers leads to indiscriminate use or an inability to identify banned or restricted pesticides. Despite government efforts, some of these chemicals find their way back to store shelves and become available to farmers.

REFERENCES

Abasa, R.O. 1981. Management of coffee pests. International Centre for Insect Physiology and Ecology (ICIPE)/UNEP Group Training Course. 19 July–7 August 1981. Nairobi, Kenya: ICIPE.

Alsopp, T. 1978. The effect of dieldrin, sprayed by aerial application for tsetse control, on game animals. *J Appl Ecol.* 15:117–27.

[Anonymous]. 1961. Annual report, Department of Veterinary Services, Ministry of Agriculture. Nairobi, Kenya: Department of Veterinary Services, Ministry of Agriculture.

[Anonymous]. 1962. Annual report, Department of Veterinary Services, Ministry of Agriculture. Nairobi, Kenya: Department of Veterinary Services, Ministry of Agriculture.

[Anonymous]. 1974. Pest control in produce stores: Standard recommendations. National Agricultural Laboratories, Ministry of Agriculture. Nairobi, Kenya: National Agricultural Laboratories, Ministry of Agriculture.

[Anonymous]. 1985. Annual reports 1983–85, National Irrigation Board of Kenya, Ministry of Agriculture. Nairobi, Kenya: National Irrigation Board of Kenya, Ministry of Agriculture.

Armstrong, J.A. and Smith, A. 1958. Insecticide susceptibility and resistance. In: *Report of Pare Taveta malaria scheme 1954–1959*. Joint publication of the East African Institute of Malaria and Vector-borne Diseases and the Colonial Pesticides Research Unit. Dar-es-Salaam, Tanzania: EAIMVBD/CPRU.

Asman, F. 1966. An assessment of the value of dilute dust insecticides for the protection of stored maize in Kenya. *J Appl Ecol.* 3:169–79.

Ayertey, J.N. 1978. An unusual mortality of *Sitophilus zeamais* Motsch. caused by *Sitotroga cerealella* Olivier in mixed laboratory cultures. In: Proc 2nd Int Conf on Stored Products Entomol. 10–16 Sept 1978. Ibadan, Nigeria: Int Conf on Stored Products Entomol, pp. 328–38.

Barasa, M.W. 1998. Studies on distribution of organochlorine pesticide residues in a tropical marine environment along the Kenyan Coast [MSc thesis]. Department of Chemistry, University of Nairobi. Nairobi, Kenya.

Brown, L.H., Urban, E.K. and Newman, K. 1982. *The Birds of Africa*. Volume 1. London: Academic Press, pp. 108–10, 117–18, 122.

Casida, J.E. 1973. *Pyrethrum: The Natural Insecticide*. London: Academic Press.

Champ, B.R. and Dyte, C.E. 1976. Report of the FAO global survey of pesticide susceptibility of stored grain pests. FAO Plant Product Protection Serial nr 5. Rome: FAO.

Chapin, G. and Wasserstrom R. 1981. Agricultural products and malaria resurgence in Central America and India. *Nature.* 293:181–4.

Cherrington, A.D., Paim, U. and Page, O.T. 1969. In-vitro degradation of DDT by intestinal contents of Atlantic salmon (*Salmon salar*). *J Fish Res Biol Can.* 26:47.

Codex Alimentarius. 1993. Pesticide Residues in Food, Suppl 1, 2nd edn. Joint FAO/WHO Food Standards Program. Rome: Codex Alimentarius Commission.

Delima, C.P.F. 1973. A technical report on 22 grain storage projects at the subsistence farmer level in Kenya. PROJ/RES/AG 21, Nos 1–8. Nairobi, Kenya: National Agriculture Laboratories, Ministry of Agriculture.

Everaarts, J.M., van Weerlee, E.M., Fischer, C.V. and Hillebrand, M.Th.J. 1997. Polychlorinated biphenyls and cyclic pesticides in sediments and macroinvertebrates from the coastal regions of different climatological zones. In: Environmental behavior of crop protection chemicals. Proc Int Symp on Use of Nuclear Related Techniques for Studying Environmental Behavior of Crop Protection Chemicals. 1–5 July 1996. IAEA-SM-343/45. Vienna: IAEA/FAO, pp. 407–31.

Floyd, E.H. 1971. Relationship between maize weevil infestation in corn at harvest and progressive infestation during storage. *J Econ Entomol.* 64:408–11.

Foxall, C.D. 1983. Pesticides in Kenya. *Swara.* 6:25.

Graham, W.M. 1966. Annual report of the senior entomologist. Nairobi, Kenya: National Agriculture Laboratories, Ministry of Agriculture, p. 22.

Graham, W.M. 1970a. Warehouse ecology studies to bagged maize in Kenya. I. The distribution of adult *Ephestia (Cadra) cautella* (Walker) (Lepidoptera: Phycitidae). *J Stored Prod Res.* 6:147–55.

Graham, W.M. 1970b. Warehouse ecology studies of bagged maize in Kenya. II. Ecological observations of an infestation by *Ephestia (Cadra) cautella* (Walker) (Lepidoptera: Phycitidae). *J Stored Prod Res.* 6:157–69.

Greichus, U.A., Greichus, A., Amman, B.D. and Hopecraft, J. 1978. Insecticides, poly-chlorinated biphenyls and metals in African lake ecosystems. III. Lake Nakuru, Kenya. *Bull Environ Toxicol.* 19:455–61.

Hall, D.W. 1970. Handling and storage of food grains in tropical and subtropical areas. FAO Agricultural Development Paper nr 90. Vienna: FAO, p. 350.

Hayes Jr, W.J. 1982. *Pesticide Studies in Man.* Baltimore: Lippincott, Williams and Wilkins, pp. 172–283.

Hill, D.S. 1975. *Agricultural Insects of the Tropics and Their Control.* London: Cambridge University Press, p. 31.

Kahunyo, J.M., Maitai, C.K. and Froslie, A. 1986. Organochlorine pesticide residues in chicken fat: A survey. *Poult Sci.* 65:1084–9.

Kaine, K.W.S. 1976. The control of tickborne diseases in cattle in Kenya. *Bulletin de L'Office International des Epizooties.* 86:71–5.

Kanja, L.W. 1988. Organochlorine pesticides in Kenyan mothers milk: levels and sources [PhD dissertation]. Department of Public Health, Pharmacology and Toxicology, University of Nairobi. Nairobi, Kenya.

Kariuki, C.W., Ngare, B.M., Osodoloo, R., Mweki, D., Mutisya, D. and Kairu, A. 1991a. Biological control of cassava mealybug (*Phenacoccus manihoti*) in Kenya. Annual report of the National Agricultural Research Centre. Muguga, Kenya: National Agricultural Research Centre, pp. 80–6.

Kariuki C.W., Thiga, S.M., Majisu, A.L. and Mutisya, D. 1991b. Biological control of the water fern *Salvinia molesta* in Lake Naivasha (Kenya) by *Cyrtobagous salviniae* (Coleoptera: Curculionidae). Annual report of the National Agricultural Research Centre. Muguga, Kenya: National Agricultural Research Centre, pp. 88–93.

Kariuki, C.W., Ngari, B.M., Osodoloo, R., Mutisya, D. and Mweki, P. 1993. Biological control of cassava green mite *Mononychellus tanajoa* in Kenya. Annual report of the National Agricultural Research Centre. Muguga, Kenya: National Agricultural Research Centre, pp. 90–2.

Kariuki, C.W., Ngari, B.M. and Kusewa, T.M. 1998. The current situation on the establishment of *Typhlodromelas aripo* de Leon in Kenya. In: Proc 7th Triennial Symp Int Soc Trop Root Crops–Africa Branch held 11–17 October 1998 in Cotonou, Benin.

Keating, M.I. 1983. Tick control by chemical ixocides in Kenya: A review 1912 to 1981. *Trop Anim Health Prod.* 15:1–6.

Kenya. 1976. Legal Notice nr 25: The Cattle Cleansing Act (Cap 358). Nairobi, Kenya: Government Printers.

Kiflom, W.G., Wandiga, S.O., Ng'ang'a, P.K. and Kamau, G.N. 1999. Variation of plant p,p'-DDT uptake with age and soil type and dependence of dissipation on temperature. *Environ Int.* 25(4):479–87.

Kimani, V.N. and Mwanthi, M.A. 1995. Agrochemicals exposure and health implication in Githunguri location, Kenya. *East Afr Med J.* 72:531–5.

Kituyi, E., Jumba, I.O. and Wandiga, S.O. 1997. Occurrence of chlorfenvinphos residues in cow milk in Kenya. *Bull Environ Contam Toxicol.* 58(6):969–75.

Koeman, J.H., Pennings, J.H., de Goeijj, J.J.M., Tjioe, P.S., Olindo, P.M. and Hopcraft, J. 1972. A preliminary survey of possible contamination of Lake Nakuru in Kenya with some metal and chlorinated hydrocarbon pesticides. *J Appl Ecol.* 9:411.

Lalah, J.O. 1993. Studies on the dissipation and metabolism of a variety of insecticides under Kenyan environmental conditions [PhD dissertation]. Department of Chemistry, University of Nairobi. Nairobi, Kenya.

Lalah, J.O., Acholla, F.V. and Wandiga, S.O. 1994. The fate of ^{14}C p,p'-DDT in Kenyan tropical soil. *J Environ Sci Health B.* 29(1):57–64.

Lalah, J.O. and Wandiga, S.O. 1996. The persistence and fate of malathion residues in stored beans (*Phaseolus vulgaris*) and maize (*Zea mays*). *Pest Sci.* 46(3):215–20.

Lincer, J.L., Zalkind, D., Brown, L.H. and Hopcraft, J. 1981. Organochlorine residues in Kenya's Rift Valley Lakes. *J Appl Ecol.* 18:157.

Maitho, T. 1978. Organochlorine and organophosphorus insecticides in milk and body fat [MSc thesis]. Department of Public Health, Pharmacology and Toxicology, University of Nairobi. Nairobi, Kenya.

Mambiri, A.M., Lwili, S., Majisu,A., Mugii, P. and Odhiambo, F. 1993. Control of potato tuber moth *Pthorimaea opercullela* (Zeller). Annual report of the National Agricultural Research Centre. Muguga, Kenya: National Agricultural Research Institute, pp. 92–8.

Mambiri, M.M., Ngari, B., Kusewa, M., Osodoloo, R., Mutisya, D. and Mugii, P. 1994. Biological control of cassava pests. Annual report of the National Agricultural Research Centre. Muguga, Kenya: National Agricultural Research Institute, pp. 35–6.

Matsumura, F. 1972. Current pesticide situation in United States. In: Matsumura, F., Boush, G.M. and Misaito, T. eds. Environmental toxicology of pesticides. New York: Academic Press, p. 44.

Matthiessen, P. and Roberts, R.J. 1982. Histological changes in the liver and brain of fish exposed to endosulfan insecticide during tsetse fly control operations in Botswana. *J Fish Dis.* 5:153.

Matthiessen, P. and Logan, J.W. 1984. Low concentration effects of endosulfan insecticide on reproductive behavior in the tropical cichlid fish *Sarotherodom mossambicus. Bull Environ Contam Toxicol.* 33:575.

Matthiessen, P., Fox, P.J., Douthwaite, R.J. and Wood, A.B. 1982. Accumulation of endosulfan residues in fish and their predators after aerial spraying for the control of tsetse flies in Botswana. *Pest Sci.* 13:39.

Mbatha, J.K. 1988. Distribution systems of pesticides in Kenya. Paper presented to a workshop on responsible and effective crop protection. PCAK (now the Agrochemical Association of Kenya)/GIFAP(now the GCPF–Global Crop Protection Federation). 11–12 February 1988. Nairobi, Kenya.

Mbuvi, L. 1996. Distribution and fate of ^{14}C-DDT on model ecosystem simulating tropical marine environment [MSc thesis]. Department of Chemistry, University of Nairobi. Nairobi, Kenya.

McFarlane, J.A. 1969. Stored products insect control in Kenya. *J Stored Prod Inf.* 18:13–23.

McFarlane, J.A. and Sylvester, N.K. 1969. A practical trial of pyrethrins-in-oil surface sprays for the protection of bagged grain against infestation by *E. (C.) cautella* (Wlk.) in Kenya. *J Stored Prod Res.* 4:285–93.

Mitema, E.S. and Gitau, F.K. 1990. Organochlorine residues in fish from Lake Victoria, Kenya. *Afr J Ecol.* 28(3):234–9.

Mosha, F.W. and Subra, R. 1982. Ecological studies on *A. gambiae* complex sibling species in Kenya: preliminary observations on their geographical distribution and chromosomal polymorphic inversions. WHO report VBC/82.867. Geneva: WHO.

Mugachia, J.C., Kanja, L. and Gitau, F. 1992a. Organochlorine pesticide residues in fish from Lake Naivasha and Tana River, Kenya. *Bull Environ Contam Toxicol.* 49(2):207–10.

Mugachia, J.C., Kanja, L. and Maitho, T.E. 1992b. Organochlorine pesticide residues in estuarine fish from the Athi River, Kenya. *Bull Environ Contam Toxicol.* 49(2):199–206.

Muhihu, S.K. 1996. Ecological studies on the tropical warehouse moth *Ephesta cautella* Walker (Lepidoptera: Pyrilidae) and its predator mite *Blattisocius tarsalis* Berlese (Acari: Ascidae) in relation to effective warehouse integrated pest management programmes [PhD dissertation]. Department of Zoology, University of Nairobi. Nairobi, Kenya.

Munga, D. 1985. DDT and endosulfan residues in fish from Hola Irrigation Scheme, Tana River, Kenya [MSc thesis]. Department of Chemistry, University of Nairobi. Nairobi, Kenya.

Mwaisaka, P. 1999. Government applauds efforts. *Agrochem News*, Nairobi, Kenya. 5(11):1.

Mwanthi, M.A. and Kimani, V.N. 1993. Patterns of agrochemical handling and community response in Central Kenya. *J Environ Health*. 55:11–6.

Mwanthi, M.A. 1998. Occurrence of three pesticides in community water supplies, Kenya. *Bull Environ Contam Toxicol*. 60(4):601–8.

Nang'ayo, F.L.O., Hill, M.G., Chandi, E.A., Chiro, C.T., Nzeve, D. and Wadenje, J.O. 1994. Ecological studies of the larger grain borer, *Prostephanus truncatus* Horn (Coleoptera: Bostrichidae) in the natural vegetation in Kenya. Annual report of the National Agricultural Research Centre. Muguga, Kenya: National Agricultural Research Centre, pp. 43–8.

NDP. 1997. National development plan 1997–2001, Republic of Kenya. Nairobi, Kenya: Government Printers.

NEAP. 1994. Report – The Kenya national environment action plan. Ministry of Environment and Natural Resources. Nairobi, Kenya: Ministry of Environment and Natural Resources.

Ng'ang'a, P.K. 1994. Studies of the dissipation of p,p'-DDT in Mtwapa, Kilifi District, of the Coast Province, Kenya [MSc thesis]. Department of Chemistry, University of Nairobi. Nairobi, Kenya.

Nyaga, J.M. 1988. Farmer perspective on pesticide handling and use. Paper presented at a workshop on responsible and effective crop protection. PCAK/GIFAP. 11–12 February 1988. Nairobi, Kenya.

Oduor, G.I., Mambiri, A.M., Kusewa, T.M. and Lwili ,S. 1995. Biological control of the water hyacinth (*Eichhornia crassipes*) in Lake Naivasha. Annual report of the National Agricultural Research Centre. Muguga, Kenya: National Agricultural Research Centre, pp. 65–7.

Ohayo-Mitoko, G.J.A. 1997. Occupational pesticide exposure among Kenyan agricultural workers: an epidemiological and public health perspective. [PhD Dissertation]. Kenyan Medical Research Institute and Department of Epidemiology and Public Health, Wageningen Agricultural University. Wageningen, The Netherlands.

Okedi, L.M.A. 1988. Evaluation of resistance to insecticides in the mosquito larvae of *Anopheles gambiae* Giles and its control by two microbial bacterial pathogens [MSc thesis]. Department of Zoology, University of Nairobi. Nairobi, Kenya.

Ongeri, D., Wandiga, S.O. and Lalah, J.O. 1998. The distribution, metabolism and persistence of ^{14}C-DDT and ^{14}C-chlorpyrifos residues in sediment and fish in a model ecosystem. In: Proc. 7th Int Conf in Africa and 34th Convention of the South African Chemical Institute held 6–10 July 1998 at University of Natal in Durban, South Africa, p. 72.

Pest Control Products Board (PCPB).1994. Pesticide imports into Kenya, 1987–90. Nairobi, Kenya: PCPB.

Rocco, D.M. 1999. The history of the Agorchemicals Association of Kenya. *Agrochem News* Nairobi, Kenya. 5(11):3.

Sleischer, C.A. and Hopcraft, J. 1984. Persistence of pesticides in surface soil and relation to sublimation. *Environ Sci Technol.* 18:514–8.

UNEP. 1982. Pollution of the marine environment in the Indian Ocean. UNEP Regional Seas Reports and Studies, nr 13. Geneva: Regional Seas Programme Activity Centre, p. 123.

UNEP. 1984. Marine and coastal conservation in the East Africa region. UNEP Regional Seas Reports and Studies nr 39. Geneva: Regional Seas Programme Activity Centre, p. 39.

Vickery, M.L. and Vickery, B. 1979. Plant products of tropical Africa. In: *Macmillan tropical agriculture, horticulture and applied ecology series.* London: Macmillan.

Wandiga, S.O. 1996. Organochlorine pesticides: Curse or blessing of tropical environment. In: Environment and development: A public lecture series. Nairobi, Kenya: Kenya National Academic Sciience Press, pp. 64–92.

Wandiga, S.O. and Natwaluma, H.C.B. 1984. Degradation and dissipation of persistent pesticides in the tropics. *Kenya J Sci Technol.* 5:31–44.

Wandiga, S.O. and Mghenyi, J.M. 1988. Persistence of ^{14}C-DDT in the tropical soils of Kenya. In: *Isotope techniques for studying the fate of persistent pesticides in the tropics.* IAEA TECDOC-476. Vienna: IAEA, pp. 19–26.

Wandiga, S.O. and Mutere, A. 1988. Determination of gamma-BHC in breast milk of Kenyan women. *Bull Chem Soc Ethiopia.* 2:39–44.

Wangia, B.M. 1989. Lethal and sublethal effects of DDT, carbofuran, trifenmorph and niclosamide on *Oreochromis niger* Gunther [MSc thesis]. Department of Zoology, University of Nairobi. Nairobi, Kenya.

Wasserman, M., Rogoff, M.G., Tomatis, L., Day, N.E., Wasserman, D., Djavahierian, M. and Guttel, G. 1972. Storage of organochlorine insecticides in the adipose tissue of people in Kenya. *Societe Belge de Medecine Tropicale.* 52:509–14.

Wedemeyer, G. 1968. Role of intestinal microflora in the degradation of DDT by rainbow trout. *Life Sci.* 7:219.

Chapter 5

Distribution, fate, and effects of pesticides in the tropical coastal zones of India

K. Raghu, G.G. Pandit, N.B.K. Murthy, A.M. Mohan Rao, P.K. Mukherjee, S.P. Kale, and J. Mitra

INTRODUCTION

India has a long coastline with an exclusive economic zone stretching out to 200 nautical miles offshore. Pesticides are used for agricultural and public health purposes and for nearly four decades after their introduction in India, OC pesticides dominated the scene. Large portions of applied pesticide are dissipated at the site of application through chemical and biological degradation processes. However, a fraction of the OCs reach the marine environment through canals, estuaries, and through a network of river systems. Because OCs are known for their persistence, toxicity, and bio-accumulation characteristics, there is concern about their impact on the marine environment. Recently efforts have been made to substitute other classes of pesticides for these environmentally damaging compounds. Although pesticide consumption is low in India compared to other developed countries, the indiscriminate use of these pesticides has resulted in sporadic observations of residues in food and the environment. After presenting an overview of Indian topography, geology, climate, land use, and principal crops grown, this review chapter examines pesticide use and distribution, the environmental impact of pesticides, and the legislative management of pesticides in India.

OVERVIEW

Topography

The Indian subcontinent is the middle of three irregularly shaped peninsulas, which jut out southward from mainland Asia. India lies to the north of the equator between Lat. 8°4'N and 37°69'N and Long. 68°7'E and 97°25'E. It is bounded on the southwest by the Arabian Sea and on the southeast by the Bay of Bengal and the southern tip is washed by the Indian Ocean. India measures 3,214 km from north to south and 2,933 km from east to west with a total land area of 3,287,263 km^2. Its land frontier stretches 15,200 km and its coast line 7,516 km. The entire territory is rich in a variety of scenery and climate, from high mountains in the north to the extensive sandy wastes of the Thar Desert and vast river deltas.

India has seven major physiographic regions: the northern mountains, including the Himalayas and mountain ranges in the northeast; the Indo-Gangetic Plain; the central highlands; the peninsular plateau; the East Coast; the West Coast; and the bordering seas and islands. All the major landforms including hills, mountains, plateaus, and plains are well represented in India, and much of the land surface of India has developed a plateau character. There are extensive plains, either flat or rolling, at elevations ranging from 300 to 900 m, dotted with conical or rounded hills or traversed by flat-topped ridges. These are mostly in the central highlands and the peninsular Deccan Plateau. India has seven major mountainous regions that include the Himalayas, which is both the highest mountain system and the youngest mountain range in the world (*Manorama Year Book*, 1994).

Geology

Geologically the Indian subcontinent consists of three elements: the Himalayas, the Indo-Gangetic Plain and the Deccan Plateau plate. The structure of the Himalayas, with huge overfolds and nappes, has all the main horizons from Precambrian to Recent. It is certain that during Mesozoic times the Himalayan area was occupied by the great geosyncline, which coincided with the Tethys Sea or Ocean basin. The sediments laid down during the Paleozoic and Mesozoic eras in the Tibetan section of this great basin constitute the Tibetan zone in which the fossiliferous beds differ entirely in their mineralogy. The second or Himalayan section, which comprises the great and lesser Himalayas, is composed chiefly of metamorphic rocks and sediments that are generally unfossiliferous. Thus, the Himalayas have arisen as a result of the collision between the drifting Indian Plate and the Tibetan Plate of south Asia, occurring about 50 M years ago (*Encyclopaedia Britannica*, 1967).

The Indo-Gangetic Plain depression is the foredeep of the Himalayas lying between the folded bed, which forms the mountain range and the rigid foreland constituted by the Indian peninsula. The Indian peninsula has three major rock groups. The Archean include crystalline and schistose rocks of Precambrian age while the rocks of the Purana group (comprising the Vindhyan and Cuddappah systems) consist of normal sedimentary deposits, which rest unconformably on the underlying Archean rocks, but which, from the total absence of fossils, are also believed to be Precambrian. The Gondana system (the Aryan group) is the most important and interesting set of beds in India. They consist chiefly of sandstones, shales, and clays with seams of coal in the lower division. A major interest in marine Cretaceous deposits lies in the evidence that in Upper Cretaceous times the Narmada area lay in the northern sea, which stretched into Europe, while Asam and the southeast border of India lay in the southern sea, which extended up to South Africa.

The Deccan lavas are an extensive series of basaltic lava flows with a total thickness on the order of a few kilometers (typically approximately 3 km) on the Western Ghats. They thin out as one goes east to as little as a few tens of meters in thickness. The Deccan lavas cover the northwestern quadrant of the peninsula, are absent in the Indo-Gangetic Plain, but are present in the Himalayas. They were laid down from the Paleozoic to the Upper Carboniferous periods. The most interesting of recent geologic deposits in India is laterite, which caps many of the hills and plateaus of the Deccan lava area.

Climate and rainfall

The Indian year may be divided into four seasons: the cold season during the months of January and February, the hot season from March to May, the southwest monsoon season from June to October, and the retreating monsoon period (northeast monsoon), which includes November and December. The mean temperature is nearly constant in southern India (27° to 30°C) during the entire year, but in northern India, the temperature variation is very large because in the cold season the mean temperature averages around 17°C and then rises to 31° to 33°C during the hot season.

The southwest monsoon atmospheric currents usually set in during the first fortnight of June on the west and east coasts and give more or less general rain in every part of India during the next three months, but distribution of the rainfall is very uneven. The rains are heavy in the Western Ghats and in the Assam region. In September, the force of the southwest monsoon begins to decline rapidly and, in its place, the gentle, steady northeast monsoon springs up, generally extending over the Bay of Bengal. On the basis of total annual rainfall, four zones are recognized: the wet zone (more than 2000 mm annual precipitation), an intermediate zone having 1,000 to 2,000 mm annual rainfall, the dry zone (500 to 1,000 mm), and a semidesert and desert zone (<500 mm). In general, rice is the main crop in the wet zone, while mixed crops are raised in the intermediate zone, and crops like millet and cotton are grown in the dry zone.

Watersheds and land use

There are three main watersheds in India: the Himalayan range with its Karakoram branch in the north, the Vindhya and Satpura ranges in central India, and the Sahayadri in the Western Ghats on the West Coast. All the major rivers of India originate in one or the other of these watersheds (Figure 5.1).

The primary rivers of the Himalayan group are the Indus, the Ganga (Ganges), and the Brahmaputra. The Indus river arises on Mount Kailas in Tibet and traverses many miles through the Himalayas before being joined by its major tributaries in the Punjab. It passes into the Sind in Pakistan to empty into the Arabian Sea. The Ganga, which rises near the glacier Gangotri in the Himalayas, and its tributaries

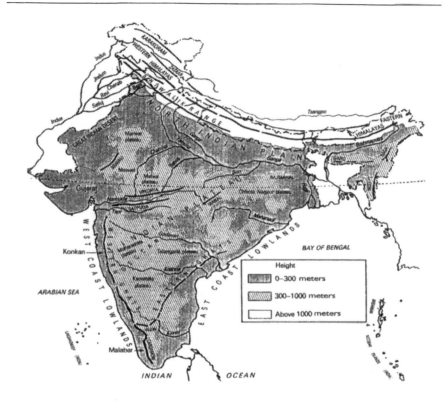

Figure 5.1 A map of India showing topographical details relevant for pesticide movement

spread out like a fan in the North India Plain, thus forming the largest river basin in India with an area equal to one quarter of the total area of India. The Brahmaputra, after rising in western Tibet, flows through the Himalayas, turns southwest, and then south, joining a branch of the Ganga and emptying together with the Ganga into the Bay of Bengal.

The rivers of the Deccan Plateau are entirely rain-fed. The Godavari, the Krishna, the Cauvery, and the Pennar all rise in Western Ghats, traverse the plateau, the Eastern Ghats, and the East Coast lowlands to fall into the Bay of Bengal. The Mahanadi and the Damodar rise in the northwest of the plateau and flow east into the Bay of Bengal. The Narmada and Tapti arise in the northern most extremity of the plateau and fall into Gulf of Cambay in the Arabian Sea.

Of a total geographical area of 328.73 M ha, the land use for 305.02 M ha has been detailed (Directorate of Economics and Statistics, 1994). Of this reported area, net crop area occupies 142.24 M ha (46.6 percent of the total); forests, 67.99 M ha (23.3 percent); permanent pastures and other grazing lands, 11.80 M ha (3.9 percent); tree crops and groves, 3.7 M ha (1.2 percent); and the remainder are fallows, barren areas, and areas under non-agricultural use (Figure 5.2). The total irrigated area is 61.78 M ha and it represents 33 percent of the total cropped area.

Principal economic crops

With the availability of flat land, fertile soil, good climate, and an adequate water supply to help in raising crops, farming has been the traditional occupation since ancient times. Soil characteristics vary in different parts of India and various type of crops are grown according to the soil characteristics and the amount of rainfall. The crops grown from June to September when India receives southwest monsoon rains are called kharif crops. Rice, red gram (pigeon-pea *Cajanus cajan* (L.) Millsp. (Fabaceae)), and cotton are examples of kharif crops. Following the monsoon, the soil continues to retain moisture, helped by the gradual lowering of temperatures during the ensuing three months. Winter crops, including wheat and mustard, are called the rabi crops. Some crops are grown during both seasons and, because rainfall is irregular, sometimes it is necessary to supply water to both kharif and rabi crops through irrigation. However, some crops are totally dependent on irrigation, e.g. sugarcane, bananas, and vegetables.

Cereals, pulses (edible leguminous crops, e.g. peas, beans, or lentils), and oilseeds constitute the principal crops grown over an area of 150.2 M ha (Directorate of

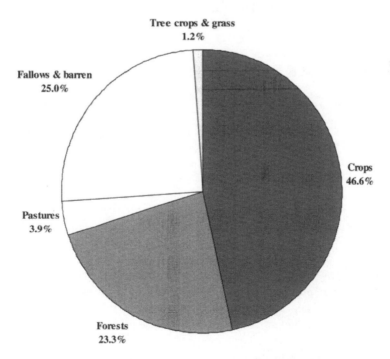

Figure 5.2 The percentage distribution of accounted land area in India by use

Economics and Statistics, 1994). During the 1992–93 season, a total of 166.4 M T of cereals was produced with rice and wheat the major crops, followed by the major millets bajra or spiked millet *Pennisetum glaucum* L. (Panicoideae: Panicodae) and jowar or great millet *Sorghum bicolor* (L.) Moench (Panicoideae: Andropogonodae) and the minor millets, e.g. ragi or finger millet *Eleusine coracana* L., foxtail millet *Setaria italica* L., kodo millet *Paspalum scrobiculatum* L., little millet *Panicum sumatrense* Roth ex Roemer & Schultes, and barnyard or sawa millet *Echinochloa crus-galli* (L.) P.B. and *Echinochloa colona* (L.) Link. Among the pulses red gram and bengal gram *Cicer arietinum* L. are the major ones with a total 1992 to 1993 production of 13.6 M T. Peanut, rapeseed, mustard, soybean, sesame, sunflower, linseed, safflower, castor, and niger *Guizotia abyssinica* (L. f.) Cass. (Compositae) are the principal oilseeds with 20.3 M T produced.

Cotton (11.6 M bales of 170 kg each), sugarcane (280.8 M T), potato (15.7 M T), onion (3.6 M T), jute and mesta (fibre from the stems of roselle *Hibiscus sabdariffa* L. var. *altissima Webster* and kenaf *H. cannabinus* L. (Malvaceae)) (9.0 M bales of 180 kg each) are also important economic crops grown in India. In addition, spices (including black pepper *Piper nigrum* L. (Piperaceae), dry chillies *Capsicum* spp. (Solanaceae), dry ginger *Zingiber officinale* Rosc. (Zingiberaceae), turmeric *Curcuma domestica* Valet. (Zingiberaceae), and cardamom *Elettaria cardamomum* White & Mason (Zingiberaceae)) and plantation crops (including tea, coffee, and rubber) are also considered important crops because of their considerable export value. Among horticultural crops, mango, banana, coconut, apples, citrus, grapes, guava, and pineapple are grown in significant quantities.

PESTICIDE USE AND DISTRIBUTION

Pesticide sources

India has become self-sufficient in food production and, during 1995 to 1996, a total of 180 M T was produced. The sharp growth in agricultural production in India in the last three decades has taken place due to the increasing use of pesticides, fertilizers, and high yielding varieties of seeds. Pesticides are also used for control of diseases as part of the vital public health programs for malaria, filariasis (a non-hemorrhagic fever without a rash caused by one of several nematodes including *Wuchereria bancrofti*, *Brugia malayi*, and *B. timori* and vectored by mosquitoes from the genera *Culex, Aedes, Anopheles,* and *Mansonia*), kala-azar (visceral leishmaniasis, a non hemorrhagic fever without a rash caused by the protozoan *Leishmania donovani* and vectored by sandflies (Diptera: Psychodidae) from the genera *Phlebotomus* and *Lutzomyia*), Kyasanur Forest disease (from a tick-borne *Flavivirus* of the Russian spring-summer encephalitis virus complex), and others. Pesticide consumption in 1996 was over Rs.1800 crores (one crore is equal to ten million rupees) and the Indian pesticide market ranks as the twelfth largest in the world. India is virtually self-sufficient with regards to its pesticide requirements, with a growing export market that stood at around Rs.312 crores in 1994 and 1995 (Bami, 1996).

Until 1949 plant protection was meager in India; so also was food grain production. The small scale introduction of DDT for malaria control just after World War II defines the start of the use of modern synthetic pesticides in India. DDT was followed by the use of HCH (BHC) for locust control in 1948. With the introduction of the 'Grow More Food Campaign' and 'National Malaria Control' programs, pesticides became an essential part of the programs and plans were made for local production of pesticides. This started in 1952 with the installation of a plant at Rishra near Calcutta for production of HCH (BHC) technical.

Pesticide production during 1978 and 1979 was 49,849 T of which HCH and DDT accounted for 76.2 percent. By 1992 to 1993, production was 84,446 T (50 percent of which was HCH and DDT) and, in 1994 to 1995, production was 96,688 T (HCH and DDT constituted 31 percent of total production). Forecast production for pesticides by the year 2000 is expected to be 118,000 T. However the announced planned phasing out of HCH by the government of India during 1997 would mean a decrease in HCH production to 25,000 T (Bami, 1996). Today India produces around 144 technical grade pesticides and their formulations. Fifteen companies, including multinational as well as large Indian companies, share 70 percent of the technical pesticide market. Pesticide formulators consist of two categories, multinationals and large-scale companies and the small-scale pesticide formulators. Producers of technical grade pesticides and their formulations involve about 60 sizeable companies, while small independent formulators number well over 500.

Past and current usage patterns

Crop protection in India was revolutionized with the introduction of HCH (lindane) and DDT during 1947 and 1948. This was followed by the introduction of OPs such as methyl parathion and TEPP (tetraethyl pyrophosphate), the OCs endrin and endosulfan, and the carbamate carbaryl and, later, systemic OPs. The next phase in the evolution of pesticides came in the 1980s and was marked by the introduction of pyrethroids and the chitin inhibitor diflubenzuron. Many acaricides, e.g. dicofol, tetradifon, ethion, aramite, and chlorobenzilate, were introduced on tea plantations. Early fungicides were mostly based on copper compounds, but, in recent years, antibiotics, organic compounds, and systemic fungicides have been used for crop disease control. For plant parasitic nematodes, fumigants like DD (1,2-dichloropropane + 1,3-dichloropropene), EDB (ethylene dibromide), and Nemagon (dibromochloropropane) are used. Herbicides have gained acceptance recently, including 2,4-D, butachlor, paraquat, isoproturon, anilofos, and a few others (David, 1993).

Pesticide consumption has risen from 434 T technical grade in 1954 to about 30,000 T in 1996. The average consumption of pesticides per ha in India has risen from 15.4 g ha^{-1} in 1960 and 1961 to 440 g ha^{-1} in 1990 (Paroda, 1997). Presently, pesticide use in India is 570 g ha^{-1} compared to use rates as high as 12,000 g ha^{-1} in Japan (Bami, 1996). Compared with other Southeast Asian countries, Indian farmers' pesticide expenditures for rice (24.9 US$ ha^{-1}) were similar to those of farmers in the Philippines (26.1 US$ ha^{-1}), the Peoples' Republic of China (25.6), and northern Vietnam (22.3) based on 1990 to 1992 data (Anonymous, 1992). Expenditures in southern Vietnam were considerably higher (39.3 US$ ha^{-1}) probably due to the lack of IPM implementation (farmers' field schools to teach IPM were begun later in the decade). Expenditures in Bangladesh were considerably lower (7.7 US$ ha^{-1}) because so many of Bangladesh's farmers could not afford synthetic pesticides. Despite the significant historical increase in pesticide consumption, India, with almost 4 percent of the world's cropped area, consumes only around 3.75 percent of the world's pesticide production.

Pest control in India is still primarily based on insecticides though the use of fungicides and herbicides has gradually increased (Figure 5.3). The differences in use patterns become more significant when one considers that 62 percent of total pesticide use is on cotton (50.0 percent) and rice (12.0 percent) with a further nearly 8 percent used on vegetable crops (Bami, 1996; Singbal, 1997) (Figure 5.4). These are used primarily in the five states of Andhra Pradesh, Karnataka, Gujarat, Punjab, and Maharashtra. Plantation crops like rubber, tapioca, coconut, and spices require little or no pesticides. In India, the use of pesticides falls drastically in a drought year and increases significantly in a year of good monsoons.

Nearly 95 percent of the total herbicides used in India today are for wheat, rice, and tea. In crops like sugarcane, barley, mustard, tobacco, and gram, herbicide use accounts for less than 1 percent. Herbicides are used less in India than in other countries because weeding is done by hand due to the availability of inexpensive labor.

Research and development

In India research and development work is confined mostly to developing process technology rather than to discovery of new pesticide compounds. Government institutes like the National Chemical Laboratory, regional Research Laboratories, and the Indian Institute of Chemical Technology plus a few Indian companies develop technologies for pesticides, e.g. monocrotophos, carbendazim, pyrethroids, isoproturon, anilofos, etc. They have also created R&D facilities for process improvement, formulation technology, and large scale production of critical intermediates.

India is also embarking on a program of IPM with a 10–15 percent growth rate for biopesticides envisioned. The essence of IPM is to reduce the use of synthetic chemical pesticides by promoting the application of nonchemical pest control measures. In India, considerable work has been done on the management of insect pests and plant pathogens through the use of non-conventional pest management practices (Jayaraj, 1992; Mukhopadhyay et al., 1992; Rananvare et al., 1996). Further, management of weeds through nonconventional methods (e.g. bioherbicides) has recently been initiated by some laboratories.

The first stage in IPM research is to make available various pest management options that are effective, economical, and environmentally safe. Good progress

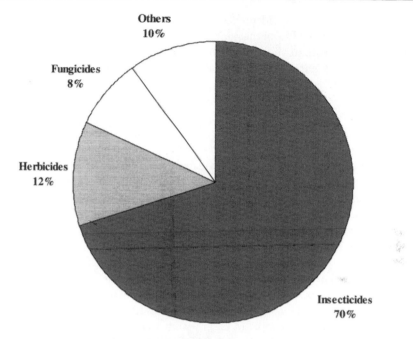

Figure 5.3 Percentage consumption of pesticides by use category In India

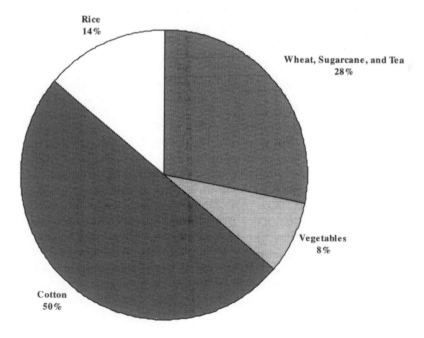

Figure 5.4 Percentage of pesticide use in various crops in India

has been made in India in identifying and developing a range of nonchemical pest management options. Use of the sterile insect technique (radiation-induced steriliza-tion of male insects) has been successfully demonstrated under field conditions for control of red palm weevil *Rhynchophorus ferrugineus* Oliver (Coleoptera: Curculionidae) and potato tuber moth *Phthorimaea operculella* Zeller (Lepidoptera: Gelechiidae) (Rananvare *et al.*, 1996). Much work has focused on the use of predators and parasites for controlling agricultural pests (Jayaraj, 1992). Various species of *Trichogramma* have been identified for the control of insect pests and some are commercially available in India. Pheromones (hormones used by insects for communication between opposite sexes) could be an effective component in IPM. These chemicals have been used successfully in traps for monitoring population levels at the early stage of infestation as well as for disorienting the insects resulting in mating disruption. A synthetic pheromone against pink boll worm *Pectinophora gossypiella* Saunders (Lepidoptera: Gelechiidae) developed at the Bhabha Atomic Research Centre (BARC) in Mumbai, India has shown great commercial promise for pest management in field trials (Tamhankar *et al.*, 1995). By far, microbial biopesticides have proved to be the most important substitute for chemical pesticides (Mukhopadhyay and Mukherjee, 1995). Among the various microbial products, *Bacillus thuringiensis* has received the most attention, resulting in the availability of a number of commercial formulations. An indigenous strain of *B. thuringiensis* var *kenyae* isolated at BARC is active against various lepidopteran pests and is under active consideration for technology transfer to commercial firms (Amonkar *et al.*, 1979; 1985).

Exhaustive research over the past two decades has focused on development of microbial fungicides as a substitute for broad spectrum seed-dressing and soil fungicides. Among the various microorganisms examined, fungi belonging to the genera *Trichoderma* and *Gliocladium* and the bacterial genera *Bacillus* and *Pseudomonas* have received widespread attention as potential biofungicides. *Trichoderma* spp. are hyperparasites (an organism that is parasitic on another parasite) of a number of plant pathogenic fungi. Several strains are being actively researched in India for development as commercial biofungicides (Mukhopadhyay and Mukherjee, 1996). At least two *Trichoderma*-based formulations (developed at Tamil Nadu Agricultural University, Coimbatore and Punjabrao Krishi Vidhyapeeth, Nagpur) are available to farmers for seed treatment and soil application against various soil-borne plant pathogens. There are also reports on the development of *Trichoderma*-based formulations by many private companies (Mukhopadhyay *et al.*, 1992). Recently a strain of *T. pseudokoningii* has been isolated from a sclerotium of *Sclerotium rolfsii* that has shown great promise for the control of this pathogen in ginger rhizomes and several vegetables (Mukherjee *et al.*, 1995; Mukherjee and Raghu, 1997).

The development of pesticides from plant origins (e.g. neem oil) has received significant attention. Apart from direct neem-based formulations, other pesticide formulations enriched in terms of azadirachtin (the primary a.i. in neem) are being prepared.

Location of pesticide applications with respect to the marine environment

India has a long coastline encompassing large land masses. There are 14 major, 44 medium, and 162 small rivers with a mean annual runoff of 1,645 km³. Some of these rivers discharge into the sea after traveling long distances and draining both agricultural and urban areas (Figure 5.1). Many cities and small towns are located in coastal areas and nearly 25 percent of India's population live in these places. There is considerable runoff into river systems from agricultural activities. In coastal areas where rice is grown, excess water often drains off through drainage canals and into the sea. Much of India's rice cultivation is under submerged (flooded) conditions.

In urban areas and swamps, vector control insecticides are spayed. Thus, considerable quantities of pesticides are discharged into the sea through canals, rivers, and estuaries. A total of 81,000 T of pesticides is used for agriculture (80 percent) and public health purposes (20 percent). It is estimated that nearly 0.1 percent of the pesticides used over land is carried to the sea by atmospheric deposition and river outflow (Qasim and Sen Gupta, 1988).

ENVIRONMENTAL IMPACT OF PESTICIDES

Fate and behavior of pesticides

Most of the studies on persistence of pesticides in India were conducted with the OC group of pesticides as these have dominated the Indian market for nearly four decades because of their cost benefits and bioefficacy potential. The persistence of HCH, DDT, aldrin, and dieldrin have been studied in detail by many workers. A number of reviews on the persistence of pesticides in the Indian environment have appeared (Agnihotrudu and Mithyanta, 1978; Krishnamurti and Dikshith, 1982; Sethunathan et al., 1982; Mehrotra, 1985; Agnihotri, 1989; Kathpal et al., 1996; Raghu et al., 1996). Because the main emphasis of the present review is on the marine environment, only a few references are cited that focus on the fate and persistence of pesticides in soil.

The persistence of pesticides in tropical soils is influenced by climatic and cultural practices used for growing specific crops. Thus lindane (γ-HCH) is known to persist for several years in aerobic upland soils but is rapidly degraded in flooded soils, the most widely used cultural practice for India's major food crop, namely rice (Raghu and MacRae, 1966). Similar observations were made with other pesticides (MacRae et al., 1967; Sethunathan et al., 1983). Agnihotri et al. (1974; 1977) found that lindane dissipated at a lower rate under fallow conditions than under crop cover. In India, commercial HCH consists of alpha, beta, and delta isomers besides the active insecticidal γ isomer. The persistence of β-HCH in food products, in human milk, and in adipose tissues is well documented (Sakurai et al., 1979; Currie *et al.*, 1979). Rice straw (Siddaramappa and Sethunathan, 1975; Ferreira and Raghu, 1978) and green manure amendment (Ferreira and Raghu, 1981) decreased the persistence of HCH isomers including β-HCH. Studies using a continuous flow system, permitting a ^{14}C mass balance, showed that in both γ- and β-HCH treated soils there was considerable formation of volatiles and less of the parent compound in green manure amended soils. There was considerable mineralization in green manure amended soil and one of the degradation products was benzene (Drego *et al.*, 1990). Raghu and Drego (1986) demonstrated that with lindane bound residues form and can be released. Lindane dissipates rapidly from soil, primarily due to volatilization, which is promoted by high temperatures, humidity, and intense solar radiation (Samuel and Pillai, 1988a). They also found that pre-treatment of tropical soils with lindane enhanced the rate of dissipation of freshly applied lindane and significantly reduced the formation of soil-bound residues (Samuel and Pillai, 1991).

Microbial degradation of HCH isomers has been demonstrated (MacRae *et al.*, 1969; Sahu *et al.*, 1990). An anaerobic bacterium *Clostridium spinoides* capable of degrading α- and γ-HCH (MacRae *et al.*, 1969) and an anaerobic bacterium *Sphingomonas paucimoblis*, which can degrade all HCH isomers (Sahu *et al.*, 1990) have been isolated. Thus, it appears that in tropical conditions volatilization, photodegradation, and microbial degradation may all be playing key roles in the degradation of HCH isomers. Kathpal *et al.* (1996) in their review concluded that lindane dissipates at a much faster rate ($t_{1/2}$ varying from nine to 107 days) under Indian soil and climatic conditions than reported under temperate conditions ($t_{1/2}$ of 1.2 years by Edwards (1973)).

The persistence DDT decreased in flooded as compared to unflooded soil, DDD being the major degradation product upon flooding (Mitra and Raghu, 1978). Rice straw and green manure decontaminated the DDT-polluted soil to a considerable extent (Mitra and Raghu, 1986; 1988). Flooding the soil decreased mineralization but increased volatilization, which accounted for the bulk of the DDT lost (Samuel and Pillai, 1988b). Estimates of DDT half-life in soil for different seasons ranged from 234 to 317 d.

Biodegradation of OPs has been demonstrated with phosphorothioate insecticides (viz. parathion, methyl parathion, fenitrothion, and diazinon) in flooded soils after repeated applications of the respective insecticides or through acquisition of degradative abilities to a related group of pesticides or other chemicals (a phenomenon known as cross-adaptability) (Barik and Sethunathan, 1978; Misra *et al.*, 1992; 1993). The hydrogen sulfide, which was formed in an acid sulfate soil, interacted with aminoparathion (formed from parathion) to form the dealkylation product desethylparathion (Wahid and Sethunathan, 1979). Bacterial isolates of *Pseudomonas* spp. and *Flavobacterium* isolated from the soils in the above studies were shown to degrade many of the phosphorothioate insecticides (Sethunathan and Yoshida, 1973; Siddaramappa *et al.*, 1973).

The carbamate insecticides carbaryl and carbofuran are used as effective insecticides against brown plant hoppers. Rajagopal *et al.* (1984) reviewed in detail studies on the fate and persistence of carbamate pesticides in soil and water. Studies on the fate of carbaryl, under moist and flooded conditions and soil types with varying pH, showed that with an increase in soil pH there was a decrease in extractable residues and an increase in bound residues (Murthy and Raghu, 1991). Metabolic studies using ^{14}C labeled carbaryl and 1-napthol in moist and flooded soils in a continuous flow system showed that there was considerable mineralization of carbaryl in moist soils. Carbaryl was extensively metabolized and the extractable ^{14}C residues were higher in flooded than in moist soils (Murthy and Raghu, 1989). With 1-napthol there was negligible mineralization and formation of bound residues. Kale et al. (1996) studied the uptake and distribution of ^{14}C labeled carbofuran and HCH in the catfish *Heteropneustes fossilis* Bloch (Siluriformes: Heteropneustidae) – a popular variety of fish grown in rice paddy fields – and concluded that there is no bioaccumulation of either of these insecticides in catfish at recommended field application rates.

Carbofuran undergoes rapid chemical hydrolysis under alkaline conditions both in water and soil (Venkateswarlu et al., 1977). It degraded faster under flooded conditions than under non-flooded conditions (Venkateswarlu et al., 1977; Rajukannu and Ramulu, 1981). Under flooded soils, the hydrolysis product carbofuran phenol accumulated and the addition of rice straw accelerated this hydrolysis (Venkateswarlu and Sethunathan, 1979). A study with ^{14}C carbofuran showed that, in sterilized and unsterilized and green manure amended clay soil under moist and flooded conditions, soil bound residues were formed (Kale and Raghu, 1996). In moist soils 48 percent was found as soil bound residues compared to 23 percent in flooded soils. Green manure enhanced the formation of bound residues in moist soil whereas in flooded soils it decreased the extractable residues and the overall persistence of the insecticide (Kale and Raghu, 1996). The notable degradation products formed under flooded soil conditions were 3-keto carbofuran and 3-hydroxy carbofuran. The fate of ^{14}C carbofuran in soils using a continuous flow system showed that mineralization to CO_2 and formation of bound residues accounted for almost all the ^{14}C residues.

There is very little information on pesticide residue levels in aquatic ecosystems. Seasonal variation in the concentration of pesticides was observed with higher values in the rainy seasons (Lakshmi, 1993). Agarwal et al. (1986) observed an increasing concentration gradient of DDT in bottom sediments. They also found faster degradation rates and, as a result, lower persistence time for OC insecticides. Bakre et al. (1990) observed a wide variation in monthly total OC values (1.1 to 81.2 μg L^{-1}) in a reservoir and attributed this to the runoff or subsoil water movements in the catchment areas. According to Ramesh et al. (1991), when compared to soils the residue levels in sediments are low and the seasonal variation less pronounced, indicating that in tropical watersheds, the relative flux of residues into the aquatic environment is smaller than the amount volatilized into the atmosphere. An assessment of the work done on the fate and behavior of pesticides shows dissipation of OC pesticides takes place in soil and water, although complete degradation was not seen. Because of their persistence they are likely to reach water bodies like canals, estuaries, rivers, and eventually the sea.

Pesticide residues in river systems

Pesticide residues resulting from agricultural and industrial applications that are carried off by rivers eventually drain into the sea. Sometimes, these residues reach the sea through estuaries. Two major reports, one on pesticide residues monitored all along the river systems of the Ganga, Yamuna and Hooghly (Ray, 1990) and another on pesticide residues from agricultural fields and their transport to the Ganga (Agnihotri, 1993) are of particular interest.

The levels of OC pesticide residues at 48 points in the Ganga basin were monitored (Ray, 1990). In the upper reaches (the Himalayas) of these river systems, only low pesticide concentrations (<200 ng L^{-1}) were found, presumably due to low use of pesticides and rapid water currents. However, in the Gangetic Plain, around the main potato belt near Kanpur and Allahabad, an increase in pesticide content (200 to 600 ng L^{-1}) of Ganga waters was observed (Ray, 1990). Higher levels were detected in Yamuna and in the Agra-Etawah River regions. After the confluence of these two rivers, pesticide concentration rose again at Varanasi, which incidently is another agricultural zone, to 200 to 400 ng L^{-1}. The levels then decrease (<100 ng L^{-1}) along the Hooghly River toward the Bay of Bengal. These studies indicate that pesticide residues are high in regions of heavy agriculture or industrialization.

To study the effect of agricultural application of pesticides on water quality of streams and ground water, pesticide residues were monitored in Ganga River water over a stretch of 8 km and in soil and ground water samples from an agricultural field (about 6 km^2) situated on the Ganga River bank at Farrukabad (Agnihotri, 1993). The average concentration in river water of $\sum$ HCH, $\sum$ DTT, $\sum$ aldrin, $\sum$ endosulfan, and total OPs was 230, 290, 105, 34, and 44 ng L^{-1}, respectively. The quantity of pesticides transported from the agricultural field to the river at Farrukabad was determined on the basis of runoff, sedimentary transport, and ground water movement. Agnihotri (1993) found that, following field application of 300 g ha^{-1} HCH, 620 g ha^{-1} DDT, 46 g ha^{-1} aldrin, 378 g ha^{-1} endosulfan, and 594 g ha^{-1} OPs, the contribution from each to the total of pesticides transported from the agricultural field to the river was 0.708, 0.682, 0.200, 0.374 and 0.926 g ha^{-1}, respectively. The above measurements were calculated based on losses that occurred through surface runoff and sediment transport from the field and movement of ground water from aquifer to the river. Of a total 1938 g ha^{-1} of field-applied pesticide, only 2.89 g ha ^{1} were transported to the river.

Joshi and Ghosh (1982) analyzed water samples from the Hooghly Estuary (a mouth of the Ganga River in West Bengal) and found no OC pesticide residues. However, Haldar et al. (1989) found the concentration of DDT in water samples of the Ganga varied from 0.0006 to 4.0 μg L^{-1} in 12 of 35 samples analyzed. Ahmad et al. (1996) monitored the presence of OC pesticides in sediments collected along 236 km of the Ganga River from Narora to Kannuaj. Of the various OCs detected, γ-HCH, aldrin, dieldrin, heptachlor, and heptachlor oxide were the most frequently detected, being identified in 56 percent (0.002 to 0.014 μg g^{-1}), 56 percent

(0.0012 to 0.12 μg g^{-1}), 50 percent (0.002 to 0.014 μg g^{-1}), 43 percent (0.0014 to 0.008 μg g^{-1}), and 53 percent (0.002 to 0.018 μg g^{-1}), respectively of the samples analyzed. In their studies of Ganga River water, Agnihotri *et al.* (1994) observed that α-HCH, *p,p'* -DDT and α-endosulfan were found as the dominant isomers. However, an enhanced percentage of β-HCH over the expected value suggests accumulation of this isomer in the aquatic environment. Total DDT concentrations were less than previously reported values for further downstream in the Ganga River (Haldar *et al.*, 1989) and in the Jamuna River, a tributary of the Ganga (Agarwal *et al.*, 1986). However, these concentrations were much higher than the concentrations found in many other Indian rivers (Ramesh *et al.*, 1990b; Pathak *et al.*, 1992). Nayak *et al.* (1995) monitored OC residues in the Ganga River from western bank to eastern bank over a span of 10 km in the city of Varanasi. They found higher levels of HCH, DDT, and endosulfan near the urbanized western bank than near the cultivated eastern bank, and attributed this to increased municipal public heath activities in the monsoon period as compared to agricultural pest management activities.

Thus, numerous studies indicate that river waters are polluted by pesticides. It would be interesting to determine the amount of pesticide discharged from the mouth of various rivers into the sea to assist in the proper evaluation of pesticide loading into estuarine ecosystems.

Effect of pesticides on biota

Pesticides entering the marine environment may affect biota and this can lead to ecological damage. Lakshmi (1993) reviewed the risk of pesticide exposure in aquatic systems and Nair *et al.* (1985) reviewed the effect of marine pollution on aquaculture. Kime (1995) reviewed the effects of pollutants on the reproduction of fish, especially on species of commercial importance in India. Acute toxicity (LC$_{50}$) data for a number of organisms from Indian aquatic ecosystems show that invertebrates are more sensitive to pesticides. Rajendran and Venugopalan (1988) found that DDT followed by lindane and endosulfan were highly toxic to zooplankton (*Acartia* sp., *Eucalamus* sp., *Lucifer* sp., *Sagitta* sp.) collected from the Vellar Estuary (located at Portonovo-Tamilnadu (Lat. 11°29'N, Long. 79°46'E). *Sagitta* sp. was more tolerant to DDT and endosulfan than the other zooplankton. Seawater temperature determines the distribution and abundance of fish to a large extent (Gopal *et al.*, 1990). Rao and Nagabhushanam (1986) reported that the toxicity of Dimecron (phosphamidon) to a marine edible crab showed a seasonal variation with an LC$_{50}$ value for 30.5°C in April (the hot season) that was almost half the LC$_{50}$ value at 19.5 °C in November (the cold season).

The sublethal effects of pesticides are of concern because of their long-term implications. They seem to have indirect effects on ecosystems through depression of primary productivity, disturbance of population dynamics, change in food habits, and reproductive behavior (Nasar, 1976; Rao and Venugopalan, 1984; Somasekhar and Sreenath, 1984; Jindal and Singh, 1989; Piska and Waghray, 1991). Increased respiration in phytoplankton was attributed to metabolized food reserves. Rotifer population *Brachionus patulus* (Rotifera) declined drastically at low food (*Chlorella*) concentrations caused by DDT (Rao and Sarma, 1990). Chlorpyrifos appears to affect zooplankton and chironomid larvae, and at or above 0.005 ppm it may damage fisheries (Mani and Konar, 1988).

Pesticides at sublethal concentrations affect the surfacing and opercular movement of fish (Anbu and Ramaswamy, 1991; Gaikwad, 1989), their reproductive system (Pawar and Katdare, 1983), and their growth (Bhaktavatsalam and Reddy, 1981). Fish production in Chilkha Lake (a brackish water lagoon near the Bay of Bengal) decreased owing to the discharge of pesticides (Misra, 1991).

Rajendran and Venugopalan (1991) measured the bioconcentration of endosulfan in different tissues of estuarine organisms including the fishes striped or gray mullet *Mughil cephalus* L. (Perciformes: Mugilidae) and long-whiskered catfish *Mystus gulio* Hamilton (Siluriformes: Bagridae), the oyster *Crassostrea madrasensis*, and the clam *Katelysia opima* in a continuous flow system for a period of 10 d. Endosulfan was bioconcentrated rapidly and showed the following trend among the four species tested: *M. gulio* > *C. madrasensis* > *M. cephalus* > *K. opima*. Jacob and Menon (1991) found that the organophosphate insecticides Ekalux (quinalphos) and Dimecron impaired the filtering mechanism of the black clam *Villorita cyprinoides* var. *cochinensis* (Harvey) in the Cochin backwater system.

Mullick and Konar (1994), in their 26-month study on the effects of effluents, including pesticides, discharged into the Hooghly Estuary, found reduced phytoplankton populations. Ghatak and Konar (1994) reported that industrial effluents discharged into the Hooghly ecosystem at the port of Haldia in West Bengal affected plankton and benthic organisms (chironomids and molluscs). It is evident that pesticides affect biota but more studies with invertebrates are needed to conduct an adequate risk assessment.

FATE AND DISTRIBUTION OF PESTICIDE RESIDUES IN THE INDIAN MARINE ENVIRONMENT

Primarily three groups of researchers have made major contributions to the study of the fate and distribution of pesticide residues in the Indian marine environment. For over a decade, Sen Gupta and his associates at the National Institute of Oceanography at Goa, India have monitored pesticide residues in sediment and biota from the Arabian Sea and the Bay of Bengal. Tanabe and his associates at Ehime University in Tarumi, Masuyama, Japan have studied the geographical distribution of pesticide residues in the seas of southeastern Asia and Oceania. Venugopalan and his associates at the Center for Advanced Study in the Department of Marine Biology of Annamalai University in Parangipettai, India have studied marine samples of the Vellar Estuary near the Bay of Bengal. There are a number of reviews covering aspects of pesticide residues and their contribution to marine pollution (Nair et al., 1985; Bhaktavatsalam and Reddy, 1981; Qasim et al., 1988; Sen Gupta and Kureishy, 1989; Madhyastha, 1990; Sen Gupta, 1991; Tanabe, 1991; Desai, 1992; Sarkar and Sen Gupta, 1992; Shailaja and Sarkar, 1992; Sarkar, 1994; Glasby and Roonwal, 1995).

The results obtained by various authors on the fate and distribution of pesticide residues in various components of marine ecosystems, namely water, plankton, sediments, bivalves, fish, and dolphins, are discussed and presented below. Some of the important results of the published work are summarized and presented in Tables 5.1 to 5.8 (Table 5.1: Water, Table 5.2: Plankton, Table 5.3: Sediment, Table 5.4: Bivalves, Tables 5.5 and 5.6: Fish, Table 5.7: Other Organisms, and Table 5.8: Birds).

Marine waters

There is scarce literature on pesticide residues in seawater around India. Venugopalan and Rajendran (1984) reported pesticide concentration ranges in Vellar estuarine water for total $\sum$ DDT of 1.63 to 14.09 ng L^{-1}, for lindane of 0.09 to 2.79 ng L^{-1}, and for endosulfan of 0.02 to 1.37 ng L^{-1}. Their observations showed no definite seasonal variations for any of the pesticides monitored. The authors attributed the low residue concentrations in water to high surface water temperatures, which resulted in a high vaporization rate for the pesticides. Sarkar and Sen Gupta (1989) using a moored in-situ sampler to determine the concentration of residues in seawater off the central West Coast of India. Their results have been discussed in a review by Shailaja. and Sarkar (1992). The order of distribution of different chlorinated compounds along the central West Coast of the Arabian Sea was as follows: o,p'-DDD < o,p'-DDT < p,p'-DDT < o,p'-DDE < dieldrin < p,p'-DDE < aldrin < γ-HCH. The $\sum$ DDT concentrations ranged from 15.8 to 444.0 ng L^{-1}. Among DDT isomers, p,p'-DDT was found to be more abundant than others in the southern part of the region whereas o,p'-DDT was present in significant concentrations off the Ratnagiri Coast. Gamma-HCH ranged from 0.26 to 9.4 ng L^{-1}, whereas aldrin and dieldrin were found at concentrations ranging from 1.4 to 9.8 ng L^{-1} and 2.1 to 50.9 ng L^{-1}, respectively. In general, γ-HCH and the two cyclodiene compounds aldrin and dieldrin were found more consistently in seawater samples than compounds of the DDT family.

Sujatha et al. (1994a) assessed the distribution of DDT and its metabolites in the Cochin estuarine system in the state of Kerala. Total DDT concentration was as high as 54.4 μg L^{-1} and the predominant metabolite was p,p'-DDE. Total HCH concentration was as high as 1.125 μg L^{-1} on the Cochin backwaters due to a pre-monsoonal accumulation of the pesticide (Sujatha et al., 1993). However, during the monsoon $\sum$-HCH concentrations ranged from not detectable to 0.175 μg L^{-1} through the estuary followed by an increase in the post-monsoon period to 0.242 to 0.518 μg L^{-1} (Sujatha et al., 1993). Sujatha et al. (1999) found a similar monsoon seasonal pattern of pre-monsoon loading for endosulfan and malathion (Table 5.1) for water samples from the Cochin Estuary. Endosulfan's distribution throughout

Table 5.1 Pesticide residues in marine, estuarine, and fresh waters of India

	Concentration range and (mean) in ng L^{-1}							
	DDT	HCH	Aldrin	Dieldrin	Endosulfan	Chlorpyrifos	Malathion	References
Vellar Estuary	1.63–14.09	0.09–2.79	–[a]	–	0.02–1.37	–	–	Venugopalan and Rajendran, 1984
Arabian Sea	ND–444.0 (115.2)	0.26–9.4 (2.7)	ND–9.8 (3.2)	ND–51.0 (10.4)	–	–	–	Sarkar and Sen Gupta, 1989
Cochin Estuary	–	ND–1,125	–	–	206–2,035[b]	–	1,373–13,013[b]	Sujatha et al., 1993 Sujatha et al., 1999
Vellar River (wet season)	0.12–0.23 (0.17)	170–2,000 (1,026.7)	–	–	–	–	–	Ramesh et al., 1990a
Pichavaram mangroves (wet season)	0.16–0.63 (0.32)	110–630 (255)	–	–	–	–	–	
Vellar River (dry season)	0.06–0.38 (0.18)	14–90 (33)	–	–	–	–	–	
Pichavaram mangroves (dry season)	0.06–0.07 (0.07)	21–24 (22.8)	–	–	–	–	–	
Kolleru Lake	198	544	46[c]	86	2,396	415	699	Amaraneni Rao and Pillala, 2000
Rivers Kaveri and Coleroon (premonsoon)	0.8–4.17 (2.4)	4.7–182 (102.2)	–	–	–	–	–	Rajendran and Subramanian, 1997
Rivers Kaveri and Coleroon (monsoon)	0.75–3.27 (2.0)	3.2–131 (68.2)	–	–	–	–	–	
Rivers Kaveri and Coleroon (post monsoon)	0.87–3.48 (2.05)	17.5–106 (45.2)	–	–	–	–	–	
Rivers Kaveri and Coleroon (summer)	0.94–3.05 (1.96)	7.35–41.3 (19.5)	–	–	–	–	–	

Notes:
a En dash (–) indicates no data.
b Maximum concentrations of endosulfan and malathion were primarily the result of the close proximity of a pesticide manufacturing facility.
c Aldrin's endo-endo isomer, isodrin.

the estuary varied seasonally with pre-monsoon loading always being higher than post-monsoon loading (generally a factor of two) and the pesticide was undetectable throughout the monsoon. The distribution of malathion followed a very similar pattern. However, in the upper reaches of the Cochin estuary, which is largely riverine in nature and slightly acidic, OP pesticides are not subject to removal by sorption onto sediments (Sujatha *et al.*, 1994b) and malathion concentrations reached 13.05 μg L^{-1} during pre-monsoon loading from agricultural use. Comparatively lower levels of malathion were observed in the lower reaches of the estuary where higher pH would promote bio-degradation and low levels would be further diluted with seawater moving into the estuary. The riverine nature and acidic pH of the upper estuary combined with the influence of industrial effluents from a pesticide manufacturing plant accounted for the very high concentrations of both pesticides found there. Levels in the mid-estuarine region reflected the prominent influence of agricultural runoff that characterized this area. As expected, levels of pesticides in water from Lake Kolleru, India's largest natural freshwater lake, were higher than seawater and estuarine samples (Amaraneni Rao, 1997) but similar to the levels found by Sujatha *et al.* (1999) from the upper reaches of the Cochin Estuary.

Studies by Ramesh *et al.* (1990a) and Rajendran and Subramanian (1997) measured DDT and HCH residues in several rivers of South India. Neither study found significant changes in DDT residue concentrations in waters of the rivers Vellar, Kaveri, and Coleroon or in the Pichavaram mangroves based on seasonal changes, wet or dry season or summer, pre-, post, or monsoon season. However, there was a significant increase in mean $\sum$ HCH levels during the wet season for the Vellar River and the Pichavaram mangroves (Ramesh *et al.*, 1990a) and in the pre-monsoon season for the rivers Kaveri and Coleroon (Rajendran and Subramanian, 1997). These increases in $\sum$ HCH concentrations corresponded with the time of increased agricultural use of the pesticide and the absence of a similar pattern for $\sum$ DDT strongly suggests that DDT is not being employed by farmers for pest control nor is it being excessively employed in public health programs in South India. Rao Amaraneni and Pillala (2000) measured the concentration of several OCs and OPs in water from Kolleru Lake in southeast India. Endosulfan was present at 2,396 ng L^{-1} and the OPs chlorpyrifos and malathion were present at 415 and 699 ng L^{-1}, respectively.

Marine plankton

Kureishy *et al.* (1978) studied the samples of zooplankton in the eastern Arabian Sea and found residues of DDT, HCH, and other unidentified compounds (Table 5.2). In their study area south of Bombay (Lat. 16°02'N by Long. 71°49'E to Lat. 17°34'N by Long. 71°32'E), $\sum$ DDT was found in the range of 0.50 to 3.21 μg g^{-1} ww. They could detect *o,p'*-DDT, *p,p'*-DDT, *o,p'*-DDE, and *p,p'*-DDE but no DDD. Subsequently Kannan and Sen Gupta (1987) collected zooplankton samples off the Saurashtra Coast (north of Bombay) in the eastern Arabian Sea (Lat.

Table 5.2 OC pesticide residues in plankton samples from Indian waters

Location	Year of sample	Concentration range and (mean) in ng g^{-1} ww				References
		DDT	HCH	Aldrin	Endosulfan	
Vellar Estuary	1979	1.2–47.3	0.3–8.5	–[a]	0.1–0.4	Venugopalan and Rajendran, 1984
Northeast Arabian Sea	1978	50–3,210 380–1,630	– –	– –	– –	Kureishy et al., 1978 Kannan and Sen Gupta, 1987
Southeast Bay of Bengal Kaveri and Coleroon River mouths	1990	4.0–6.2 (5.2)	–	0.36–0.78 (0.6)	–	Shailaja and Singbal, 1994
Northern Bay of Bengal Mahanadhi and Hoogly River mouths	1991	310.2–1,587.8 (844.6)				

Note:
a En dash (–) indicates no data.

22°30'N and Long. 67°36'E) and found that $\sum$ DDT ranged from 0.38 to 1.63 μg g^{-1} ww and from 17.5 to 379.46 μg g^{-1} lipid. DDD was the major metabolite detected in most of the samples. Phytoplankton and copepods showed higher $\sum$ DDT levels. Shailaja and Singbal (1994) determined residues in zooplankton samples collected from two oceanographic cruises, viz. October 1990 during the southwest monsoon covering the southeast coast and in December 1991 during the northeast monsoon covering the northern coast of the Bay of Bengal. In samples from the southeast coast, $\sum$ DDT concentrations ranged from 4.0 to 6.2 ng g^{-1} ww while aldrin levels varied between 0.36 and 0.78 ng g^{-1} ww. However, in samples from the northen Bay of Bengal, DDT ranged from 310.2 to 1,587.8 ng g^{-1} ww and aldrin was not detected. Shailaja and Sen Gupta (1990) had confirmed the presence of DDT and its metabolites, DDD and DDE, in their study on samples obtained from three stations in a transect off Bombay. In this study, DDD was the major product formed in zooplankton, indicating DDT metabolism in zooplankton. They observed that $\sum$ DDT concentrations showed a declining gradient from near-shore to offshore while DDD concentrations increased from the coast to offshore. The concentration of $\sum$ DDT in zooplankton was higher than in both plankton-feeding and carnivorous fish from coastal as well as open ocean regions of the Arabian Sea. This indicates that zooplankton do not serve as an important link in the transfer of organic material from the primary producers to plankton-feeding fish. In zooplankton samples taken from coastal regions of the Bay of Bengal, $\sum$ DDT was 5.2 ng g^{-1} and aldrin 0.6 ng g^{-1} ww. Venugopalan and Rajendran (1984) detected $\sum$ DDT in the range of 1.2 to 47.3 ng g^{-1} ww, HCH from 0.3 to 8.5 ng g^{-1} ww, and endosulfan from 0.1 to 0.4 ng g^{-1} ww. Toxicity studies with zooplankton indicated that DDT was more toxic than either lindane or endosulfan (Venugopalan and Rajendran, 1984; Rajendran and Venugopalan, 1988). These same authors ranked zooplankton sensitivity in the order *Acartia* sp. > *Eucalanus* sp. > *Lucifer* sp. > *Sugitta* sp. Shailaja and Sarkar (1993) summarized their analysis of zooplankton pesticide residues, with reference to the influence of the monsoon, from coastal and offshore sampling in the Arabian Sea and the coastal region of the Bay of Bengal. For pre- southwest monsoon samples from the Arabian Sea, $\sum$ DDT of zooplankton varied from 3.36 to 38.78 ng g^{-1} ww. However, for the post-monsoon period $\sum$ DDT ranged from 31.78 to 614.81 ng g^{-1} ww. DDT accounted for nearly 83 percent of $\sum$ DDT in the post-monsoon samples but only about 70 percent of $\sum$ DDT in the pre-monsoon samples. The increase in DDT's contribution to $\sum$ DDT for the post-monsoon samples may be attributed to land surface runoff carrying DDT to coastal waters. DDD, which contributed 29 percent of $\sum$ DDT in the pre-monsoon samples contributed only 16 percent to the post-monsoon samples. They hypothesized that because DDD could be further meta-bolized to DDA, there was no food-chain magnification in samples tested from this region (Shailaja and Sen Gupta, 1990). Venugopalan and Rajendran (1984) in their studies of the Vellar Estuary found that residue levels were low during the monsoon season and that it is likely that residues swiftly reach the Bay of Bengal. In fact, Shailaja and Nair (1997) observed higher residues in ocean plankton during the post-monsoon season.

Marine sediments

Data on the accumulation of pesticides in marine sediments is limited, especially considering the length of India's coastline (see Table 5.3). Sarkar and Sen Gupta (1987) measured pesticide residues in sediments from 14 locations along the west-central coast of India in the Arabian Sea. Sediment samples taken during September 1986 showed OC pesticides in the following order: HCH > aldrin > p,p'-DDE > p,p'-DDT > o,p'-DDE > p,p'-DDD > dieldrin. HCH was detected in almost all sediment samples at concentrations ranging from 0.44 to 17.9 ng g^{-1} ww. Although the paper does not state which HCH isomer was detected, it is likely that only γ-HCH was estimated. Aldrin was found at concentrations ranging from 0.95 to 35.7 ng g^{-1} ww. Residues of p,p'-DDT, p,p'-DDE, and o,p'-DDE were found in some sediment samples. Total DDT detected ranged from 7.01 to 179.10 ng g^{-1} ww. It is interesting to note that HCH was found in 12 of 14 samples while DDT and its metabolites were found in only 5 of 14 sediment samples.

Sarkar and Sen Gupta (1991) collected sediment samples off the West Coast of India in the Arabian Sea during the period January to February 1988 – some 1986 sample sites were repeated. Residue levels were in the order, DDT > HCH > aldrin > dieldrin. The concentration of $\sum$ DDT was high in most of the samples. However, DDTs could not be detected in a few sediment samples that contained the maximum percentage of silt and sand. Aldrin was much more prevalent than dieldrin and γ-HCH was dominant over its other isomers.

A more recent survey of sediments from within and without estuaries on the Arabian Sea by Sarkar et al. (1997) found higher mean concentrations of $\sum$ DDT and dieldrin (by factors of 1.72 and 2.44, respectively) in estuarine sediments compared to offshore sediments while mean concentrations of $\sum$ HCH, aldrin, and endrin were similar from both offshore and estuarine samples. In both offshore and estuarine sediments, α-HCH was the most dominant isomer and for the DDTs, p,p'-DDE was the most predominant metabolite except near Cannanore and Murmugao where o,p'-DDE and p,p'-DDT were significant in offshore samples. Their overall assessment of which estuaries were most susceptible to $\sum$ DDT contamination grouped the Zuari and Kali estuaries followed by the other estuaries.

Sarkar and Sen Gupta (1988a; 1988b) determined residues of OC pesticides in sediments from the Bay of Bengal. The compounds and concentrations detected were aldrin at 0.02 to 0.53 μg g^{-1} (all sample weights expressed as ww), γ-HCH at 0.01 to 0.21 μg g^{-1}, dieldrin 0.05 to 0.51 μg g^{-1} (Sarkar and Sen Gupta, 1988b), and $\sum$ DDT at 0.02 to 0.79 μg g^{-1} (Sarkar and Sen Gupta, 1988a). The variability in pesticide residue concentrations was attributed to the presence of numerous rivers along the East Coast of India including the Hooghly, Mahanadi, Vamsodhara, Godavari, Krishna, Pennar, and Palar rivers, and the fact that most sediment samples were collected from the mouth of these rivers. The highest concentration of aldrin (0.53 ppm) was present in a sample collected from the mouth of the Palar River, whereas the highest concentrations of HCH (0.21 ppm), dieldrin (0.51 ppm) and $\sum$ DDT (0.79 ppm) were found in samples from the mouth of

Mahanadi, Hooghly and Palar rivers, respectively. Among metabolites of DDT, both p,p'-DDE and o,p'-DDE were consistently found along the East Coast of India, attributable to the degradation of DDT to DDE in the coastal sediments. Sarkar and Sen Gupta (1988a) suggested that the following phenomena, either singly or in combination, are responsible for the degradation of DDT to DDE: the ability of marine benthic organisms to biodegrade DDT; the alkaline nature of the marine system; the presence of chemical constituents and their characteristics, e.g. salinity, clay mineral concentration, major elements, and the presence and concentration of organic matter; and stored thermal energy in ocean waters. DDT isomers and their metabolites were detected in the sediments of coastal Bay of Bengal in the following order, o,p'-DDT < p,p'-DDD < o,p'-DDD < p,p'-DDT < p,p'-DDE < o,p'-DDE. Venugopalan and Rajendran (1984) showed DDT was in the range of 1.8 to 25.8 ng g^{-1} ww and HCH in the range of 0.4 to 7.1 ng g^{-1} ww in sediment samples of Vellar Estuary.

Shailaja and Sarkar (1992) attributed the high concentration of residues, especially $\sum$ DDT, in the Bay of Bengal to the rapid transport by rivers of their high suspended sediment load (approximately 8.1 mg L^{-1}) compared to rivers emptying into the Arabian Sea, where suspended particulate loads are 4.6 mg L^{-1}. Several studies have measured sediment concentrations of pesticides in Indian rivers, including the Vellar, Ganges, Kaveri, and Mahanadi rivers, and Lake Kolleru. Ramesh et al. (1991) found higher sediment loads of DDTs and HCHs in the Vellar River during the wet season (3.4 and 12.4 ng g^{-1} dw, respectively) compared to the dry season (1.0 and 2.3 ng g^{-1} dw, respectively). Senthilkumar et al. (1999) found mean sediment concentrations in the Ganges of 5.6 and 2.6 ng g^{-1} dw DDT and HCH, respectively. Rajendran and Subramanian (1999) measured chlorinated pesticide residues in surface sediments from the Kaveri River on the southeastern coast of India. Total DDT was low and showed no wide variations among sampling sites, reflecting its ban from agriculture and use for mosquito control. The concentrations of HCHs (39.4–158.4 ng g^{-1} dw) at the various collection sites reflected its agricultural use, with peak values reflecting the application of HCH for rice crops during the monsoon and subsequently to grams during the post-monsoon season. Pandit et al. (2001) found residues of DDTs, HCHs, aldrin, and endosulfan in sediments from the Mahanadi River (and its delta). Both former and current agricultural use patterns for pesticides were reflected in residue levels measured in Kolleru Lake, Krishna District, Andhra Pradesh State, India (Amaraneni Rao and Pillala, 2001). Residues of isodrin, dieldrin, and p,p'-DDT remain despite their no longer being used in agriculture in India and the impact of pesticides on Kolleru Lake is relatively high.

Iwata et al. (1994) analyzed air, river water, and sediment samples from eastern and southern Asia (India, Thailand, Vietnam, Malaysia, and Indonesia) and Oceania (Papua, New Guinea and the Solomon Islands) for the presence of OC pesticides to elucidate their geographical distribution. The distribution patterns in sediments showed smaller spatial variations on global terms, indicating that OCs released in the tropical environment are dissipated rapidly through air and water

Table 5.3 Pesticide residues in marine, estuarine and freshwater sediments of India

Location	Year of sample	Concentration range and (mean) in ng g⁻¹ (dw or ww as indicated by location)							References
		∑DDT	Aldrin	Dieldrin	∑HCH	Endosulfan	Malathion	Chlorpyrifos	
Arabian Sea (ww)	1986	ND[a]–179.1 (43.3)	ND–35.7 (8.7)	ND–0.88 (0.06)	ND–17.9 (5.0)	–[b]	–	–	Sarkar and Sen Gupta, 1987
Bay of Bengal (ww)	1984–85	ND–790 (138.6)	20–530 (136)	50–510 (185.5)	10–210[c] (54)	–	–	–	Sarkar and Sen Gupta, 1988a,b
Arabian Sea (ww)	1988	ND–43 (11.8)	ND–5.3 (1.8)	ND–2.5 (1.0)	1.2–8.7 (4.9)	–	–	–	Sarkar and Sen Gupta, 1991
Vellar Estuary (East Coast) (ww)		1.8–25.8	–	–	0.4–7.1	–	–	–	Venugopalan and Rajendran, 1984
Vellar River dry season (dw)	1988–89	0.78–1.4 (1.0)	–	–	1.9–2.6 (2.3)	–	–	–	Ramesh et al., 1991
Vellar River wet season (dw)		1.6–8.6 (3.4)	–	–	5.4–27 (12.4)	–	–	–	
Pichavaram mangroves dry season (dw)		0.25–1.6 (0.7)	–	–	0.95–9.3 (4.1)	–	–	–	
Pichavaram mangroves wet season (dw)		0.56–2.0 (1.0)	–	–	3.3–17 (8.7)	–	–	–	
West Coast Estuaries from Veraval to Cochin (dw)	~early 1990s	1.47–25.17	0.10–0.27	0.7–3.33	0.85–7.87	–	–	–	Sarkar et al., 1997
Arabian Sea	~early 1990s	1.41–17.55	0.10–6.20	0.09–0.26	0.10–6.20	–	–	–	
Veraval to Cochin (dw)									
Bassim, north of Mumbai (formerly Bombay) (ww)	1995	29.3–464.6 (100)	–	–	18–1,053.6 (289.3)	–	–	–	Pandit et al., 2001

Concentration range and (mean) in ng g^{-1} (dw or ww as indicated by location)

Location	Year of sample	$\sum$DDT	Aldrin	Dieldrin	$\sum$HCH	Endosulfan	Malathion	Chlorpyrifos	References
Alibagh, south of Mumbai (ww)	1996–97	BDL–109.5 (16.5)	BDL–3.0 (0.6)	–	2.3–108.1 (25.5)	BDL–1.1 (0.34)	–	–	Rajendran and Subramanian, 1999
Paradeep, East Coast (ww)	1996	3.3–12 (7.65)	BDL–0.3 (0.15)	–	1.5–10.4 (5.9)	0.2 (0.2)	–	–	
Mahanadi River and Delta, East Coast (ww)	1996	1.3–5.1 (3.79)	BDL–0.1 (0.03)	–	10.3–25.4 (16.7)	0.06–1.4 (0.65)	–	–	
Kaveri River (on Bay of Bengal) (dw)	~mid-1990s	0.69–4.85 (1.87)	–	–	39.4–158.4 (84.8)	–	–	–	
Kolleru Lake (dw)	~late 1990s	BDL–191 (87.1)	BDL–38.6[e] (9.7)	BDL–128 (35.9)	107.2–512.4 (250.6)	BDL–206 (117.4)	BDL–186 (58.5)	BDL–292 (118.4)	Amaraneni Rao and Pillala, 2001
Ganges River[f] (dw)	1989–97	0.1–36 (5.6)	–	–	<0.1–8.1 (2.6)	–	–	–	Senthilkumar et al., 1999
Cochin, West Coast (dw)	1998	3.8	–	–	4.8	–	–	–	Senthilkumar et al., 2001
Chennai (formerly Madras), East Coast (dw)		<0.1–35	–	–	0.1–2.1	–	–	–	
Visakapattanam, East Coast (dw)		<0.1	–	–	0.21	–	–	–	

Notes:
a ND indicates compound not detected and BDL indicates below detection limit.
b En dash (–) indicates no data.
c Study measured only γ-HCH.
d Study only analyzed for p,p'-DDT.
e Aldrin's endo-endo isomer, isodrin.
f Chlordanes (range < 0.1–49; mean 9.6 ng g^{-1} dw) and HCB (range <0.1–0.2; mean 0.2 ng g^{-1} dw) were also measured.

and retained less in sediments. They noted that atmospheric and hydrospheric concentrations of HCHs (hexachlorocyclohexanes) and DDTs (DDT and its metabolites) in tropical developing countries were higher than those observed in developed nations.

There are only a few studies on the fate and behavior of OP pesticides in marine sediments (Sarkar and Sen Gupta, 1985; Sarkar and Sen Gupta, 1986; Sujatha and Chacko, 1991). They found that the stability and fate of the pesticides monocrotophos, phosphamidon, and DDVP in sediment samples were influenced by pH, salinity, and exchangeable cations. The low stability of these pesticides in sea sediments indicates that major element cations have reduced the toxic effects of these pesticides in marine organisms (Sarkar and Banerjee, 1987).

Sujatha and Chacko (1991) investigated the sorptional behavior of malathion through an experiment designed to assess the effects of varying pesticide concentrations and environmental influences on the sorptional behavior of sediment samples from a tropical estuary. They concluded that malathion adsorption was essentially a heterogenous multilayer interaction.

A review of published data of pesticide residues in sediments indicates that isomers of DDT and its metabolites, HCHs, aldrin, dieldrin, and endrin are present but that, while all may not be found in every sample, the most common and abundant pesticide residues were aldrin, HCHs, and DDTs. Furthermore, there is a striking contrast in pesticide residue concentrations in sediments of the east and west coasts of India. East Coast sediment samples contain pesticide residue concentrations an order of magnitude higher than samples from off the West Coast of India. This circumstance may be attributed to several factors. All major Indian rivers flow east and drain through fertile agricultural lands, which receive a considerable input of pesticides, and carry their sediment loads to marine waters on the coast. Also, the use pattern of pesticides shows a preponderance of pesticides are applied to lands bordering the eastern coast.

Marine bivalves

Bivalves have been widely accepted and used as sentinel organisms to monitor the concentration of pollutants in coastal marine environments (see Table 5.4). Different species of oysters, mussels, and clams have been used as bioindicators for Mussel Watch programs in different countries to monitor marine pollution from pesticide residues. Also, Ramesh et al. (1990b) suggest the bioaccumulation of OCs by mussels might have implications for human exposure, because mussels are marketed immediately following collection and without depuration.

Ramesh et al. (1990b) measured the concentrations of OC residues in green-lipped mussels Perna viridis L. (Mollusca: Bivalvia) collected from nine locations along the South Indian coast, which includes the east and west coasts covering the Bay of Bengal and the Arabian Sea, respectively. They found that HCH isomer (α, β, and γ) concentrations ranged from 4.3 to 16 ng g^{-1} whereas $\sum$ DDT (the sum of pp'-DDT, pp'-DDE, pp'-DDD and op'-DDT) varied from 2.8 to 39 ng g^{-1}.

Table 5.4 OC pesticide and PCB residues in bivalves from Indian waters

Location	Concentration range and (mean) in ng g⁻¹ ww				
	HCH	DDT	PCB	Endosulfan	References
Vellar Estuary	0.1–3.1 (0.8)[a]	0.3–7.3 (3.4)	5.0	2.1–0.8 (0.4)	Venugopalan and Rajendran, 1984
East Coast (Bay of Bengal)	4.3–16.0 (7.9)	2.8–33 (26.6)	1.0–7.1 (3.3)	–[b]	Ramesh et al., 1990b
West Coast (Arabian Sea)	4.9–9.7 (6.7)	6.0–39 (27.3)	1.03	4.2–5.7 (4.7)	Venugopalan and Rajendran, 1984
Porto Novo and Cuddalore	6.6	15	1	–	Senthilkumar et al., 2001

Notes
a Figures in parentheses indicate mean values.
b En dash (–) indicates no data.

PCB levels were between 1.0 and 7.1 ng g⁻¹. Mussels collected from the West Coast had higher levels of DDT, suggesting DDT used for vector control in urban locales is in storm water runoff from coastal cities. However, in Porto Novo and Pondicherry harbors on the East Coast and Suratkal on the West Coast HCH levels were slightly higher than DDT, which is indicative of the use of HCH for agricultural purposes in nearby areas. Studies by Ramesh et al. (1989; 1991) of air and water pollution showed considerable pesticide contamination, especially in agricultural areas. A comparison of OC residue levels reported by the Mussel Watch Programs of other countries with the Ramesh et al. (1990b) study shows PCB levels are lower in India than elsewhere. The highest levels of HCH in Indian mussels are comparable to those reported for Hong Kong and Taiwan and this is probably attributable to the large-scale use of HCH in mainland China and its eventual transport to the South China Sea.

Although OPs are generally considered as a non-persistent group of pesticides, Caravalho et al. (1992) found that they remain in sediments for long periods and, thus, may be harmful to marine species like clams, which remain buried in the sediment most of the time. Sathe (1995) exposed the bivalve Mytilus hiantina Lamarck (Mollusca: Bivalvia) to three concentrations of fenthion under laboratory conditions. When exposed to 0.02, 0.12, and 0.22 mg L⁻¹ fenthion, the viscera of the clams accumulated fenthion to the extent of 1.33, 16.73, 47.15 μg g⁻¹, respectively, after 15 d of exposure. The tissue concentration factor was 214.3 for clams subjected to 0.22 mg L⁻¹ fenthion. In further experiments with clams exposed to the high concentration of fenthion (0.22 mg L⁻¹), Sathe (1995) found that, although they showed higher assimilation compared to the lowest concentration (0.02 mg L⁻¹), more than 94 percent of the fenthion was lost after a 15 d depuration period. However, in clams exposed to the lowest concentration (0.02 mg L⁻¹), he found comparatively lower residue levels, and that more than 75 percent of the accumulated residues were eliminated after the depuration period (Sathe, 1995).

Venugopalan and Rajendran (1984) detected pesticide residues in three molluscs (the oyster *Crassostrea madrasensis* and the clams *Meretrix casta* and *Katalysia opima*) collected from Vellar Estuary. The mean pesticide residues in these three species were 3.44 ng g^{-1} ww for DDT, 0.82 ng g^{-1} ww for lindane, and 0.42 ng g^{-1} ww for endosulfan. There was no correlation between the pesticide concentration in the mollusks and the concentration in solution or in the particulate fraction of the water. Venugopalan and Rajendran (1984) studied the toxicity of DDT, lindane, and endosulfan using the same three species of mollusks and found the order of toxicity was DDD > endosulfan > lindane and the sensitivity of the bivalves was in the order *C. madrasensis* > *K. opima* > *M. casta*. Rajendran and Venugopalan (1991) studied the uptake of endosulfan in all tissues of the oyster *C. madrasensis* and the clam *K. opima* exposed to 0 to 14 g L^{-1} for 10 d under laboratory conditions.

As part of the Asia–Pacific Mussel Watch Program, Senthilkumar et al. (2001) measured the concentrations of butyltin, PCBs, DDTs, HCHs, chlordanes, and HCB in green mussel tissues collected from coastal regions of India, Thailand, and the Philippines during 1994 through 1997. Mussels collected in India contained low but still detectable butyltin concentrations. Butyltin residues were highest in samples from areas with intense boating suggesting the use of tributyltin as an antifouling agent in marine paints. However, they were not significant in green mussels from coastal aquaculture areas in India unlike samples from aquaculture areas in Thailand. After comparing measured PCB levels to earlier reports (Ramesh et al., 1991; Kannan et al., 1995), they concluded that inputs of PCBs are still occurring in Indian coastal waters. They found pesticide concentrations were highest for DDTs, followed by HCHs, chlordane, and HCB, respectively and were comparable to an earlier study by Ramesh et al. (1991).

Marine fish

Venugopalan and Rajendran (1984) studied the pesticide residues in fish samples obtained from Vellar Estuary adjoining the Bay of Bengal including the grey or striped mullet *Mugil cephalus* L. (Perciformes: Mugilidae), therapon or target-fish *Therapon jarbua* (Forsskål) Klunzinger (Perciformes: Terapontidae), spine foot *Siganus javus* L. (Perciformes: Siganidae), and the catfishes *Mystus gulio* Hamilton (Siluriformes: Bagrinae) and *Arius (Tachysurus)* sp. (Siluriformes: Ariidae). Residue concentrations were in the following ranges (ng g^{-1} ww): $\sum$ DDT, 0.4 to 89.27; lindane, 0.05 to 4.64; and endosulfan, 0.02 to 2.47. Between DDT and its metabolites, *p,p'*-DDE was the dominant degradation product. There was no significant variation in the concentration of various pesticides between fish species and they observed no obvious differences in pesticide concentrations between herbivorous and carnivorous fish species. They studied the toxicity of these pesticides under laboratory conditions and found the order of toxicity was DDT > lindane > endosulfan. *M. cephalus* was more sensitive than *Mys. gulio* to all three pesticides.

Radhakrishnan and Antony (1989) measured pesticide residues from Indian mackeral, *Rastrelliger kanagurta* Cuvier (Perciformes: Scombridae), marine ulva *Chirocentrus* sp. (Clupeiformes: Chirocentridae), black pomfret *Parasromateus niger* Bloch (Perciformes: Carangidae), and eastern little tuna *Euthynnus affinis* Cantor (Perciformes: Scombridae) caught near Cochin, a town on the West Coast of India on the Arabian Sea. They reported concentration ranges of α-HCH (2 to 200 ng g^{-1} ww), p,p'-DDE (1 to 3 ng g^{-1} ww), p,p'-DDD (2 to 9 ng g^{-1} ww), o,p'-DDT (0 to 7 ng g^{-1} ww) and p,p'-DDT (0 to 42 ng g^{-1} ww).

Shailaja and Sen Gupta (1989) studied the occurrence of pesticide residues in 48 fish specimens obtained from coastal and open sea samples from the eastern Arabian Sea. Metabolites of DDT were ubiquitous in the muscle samples analyzed, with p,p'-DDE frequently detected in both coastal and open ocean fish samples. The concentration of p,p'-DDE ranged from 14.89 to 36.62 ng g^{-1} (ww) in coastal samples and from traces to 50.42 ng g^{-1} (ww) in open ocean samples. In coastal samples, besides metabolites of DDT, other chlorinated hydrocarbons, e.g. hexa-chlorobenzene, HCH isomers, and 1-hydroxychlordane were also found. Fish included representatives from different feeding habits (herbivores, carnivores, and omnivores) but no remarkable food-chain magnification of the accumulated pesticide was evident. They suggested that migratory species, because of their high metabolic rate, could perhaps rapidly convert DDT into its metabolites. They found p,p'-DDD in the livers of the slender bambooshark *Chiloscyllium indicum* Gunther (Orectolobiformes: Hemiscyllidae) and silver pomfret *Pampus argenteus* Euphrasen (Perciformes: Stromateidae) but not in muscle tissue, and suggested this was due to extensive (rapid) metabolism in the liver.

Shailaja and Sarkar (1993) found interesting trends from a review of earlier findings from 35 fish taken from coastal waters of the Arabian Sea, the Bay of Bengal, and the Goa Estuary. There were variations in levels of DDT and its metabolites during the pre- and post-monsoon seasons for samples collected from the Arabian Sea. In edible portions of fish collected from the coast, $\sum$ DDT concentrations ranged from 0.43 to 40.41 ng g^{-1} ww during the pre-monsoon season and increased to 57.9 to 430.55 ng g^{-1} during the post-monsoon season. Runoff from public health pesticide use and atmospheric deposition and sediment runoff from agricultural use would have entered coastal regions through rivers and estuaries. In offshore samples, all the DDT residues were in the form of p,p'-DDE and concentrations ranged from non-detectable to 41.4 ng g^{-1} ww in edible parts of the fish. Fish samples collected from the Goa Estuary after the southwest monsoon had residue levels of $\sum$ DDT from 20 to 3,780 ng g^{-1} ww. This also suggested runoff of pesticides during the monsoon. Shailaja and Nair (1997) found that fish samples taken from different locations in the northern Arabian Sea showed $\sum$ DDT and aldrin concentrations that were 10 to 30 times and 3 to 40 times higher in post-monsoon samples than in pre-monsoon samples. Interestingly, of the 23 species of fish analyzed from the Arabian Sea, the estuarine species had the highest DDT residue burden (see Tables 5.5 and 5.6) followed by the coastal species (Shailaja and Sarkar, 1993).

Table 5.5 OC pesticide residues in fish from Indian waters

Location	Concentration range and (mean) in ng g⁻¹ ww						References
	ΣDDT	ΣHCH	Aldrin	Endosulfan	ΣChlordane	HCB	
Arabian Sea	1.4–430	–[a]	0.09–6.7	–	–	–	Shailaja and Sarkar, 1993
Mumbai (fish market samples)	6.1–140 (44)	2.2–49 (23)	0.5–2.0[b] (1.2)	–	0.47–2.2 (1.2)	–	Kannan et al., 1995
Alibagh, south of Mumbai	0.4–16.9 (12)	0.5–6.8 (4.8)	BDL[c]–0.3	BDL	–	–	Pandit et al., 2001
Goa Estuary near Panaji	17–2,200	–	–	–	–	–	Shailaja and Sarkar, 1993
South India (prey items of birds)	12–43 (21.75)	6.4–2,000 (528.8)	–	–	0.2–1.31 (0.53)	<0.01–0.1 (<0.06)	Senthilkumar et al., 2001
Porto Novo (fish market samples)	0.86–75 (7.6)	0.48–150 (40)	<0.1–2.0[b] (1.2)	–	<0.01–0.34 (0.19)	–	Kannan et al., 1995
Vellar Estuary near Cuddalore	0.4–89.27	0.05–4.64	–	0.02–2.47	–	–	Venugopalan and Rajendran, 1984
Vellar River and coastal seaboard of Parangipettai, South India	4.3–75[d] (31)	0.48–4.4 (1.7)	–	–	–	0.02–0.07 (0.04)	Ramesh et al., 1992
	1.4–45 (12)	1.5–14 (4.8)	–	–	–	0.02–0.07 (0.04)	
	1.9–16 (3.8)	1.8–13 (4.8)	–	–	–	<0.01–0.05 (0.07)	
	2.2–7.4 (4.6)	32–93 (61)	–	–	–	<0.01–0.03 (0.02)	
	0.9–8.1 (3.8)	8.9–150 (63)	–	–	–	0.04–0.09 (0.07)	
	0.86–2.6 (1.8)	61–130 (84)	–	–	–	0.01–0.20 (0.08)	
Porto Novo and Cuddalore, South India	1.5–43 (17.6) (18)[e]	6.4–68 (30.5) (2,000)	–	–	ND–0.31 (0.2) (1.31)	<0.01–0.1 (<0.1) (<0.03)	Senthilkumar et al., 2001

Location	Concentration range and (mean) in ng g⁻¹ ww						
	ΣDDT	ΣHCH	Aldrin	Endosulfan	ΣChlordane	HCB	References
Bay of Bengal	1.3–116.0	0.14–8.98	–	–	–	–	Shailaja and Sarkar, 1993
	0.03–2.38	–	–	–	–	–	Rajendran et al., 1992
Calcutta (fish market samples)	4.2–62 (18)	4.6–380 (73)	0.37–3.0[b] (1.8)	–	<0.01–0.27 (0.21)	–	Kannan et al., 1995
Kolleru Lake[f]	89,500[g]	149,400	400[h]	51,900	–	–	Rao Amaraneni, 1997
Ganges River[i]	160	77	2.7	–	30	0.24	Kannan et al., 1994
Ganges River	60–3,700 (1,100)	28–110 (76)	–	–	0.8–18 (5.6)	0.3–0.5 (0.38)	Senthilkumar et al., 1999

Notes:
a En dash (–) indicates no data.
b Aldrin and dieldrin combined.
c BDL indicates below detection limit.
d Fish species by row were seer fish *Scomberomorus commerson* Lacpède (Perciformes: Scombridae), sole fish *Cynoglossus paraplagusia* (Pleuronectiformes: Cynoglossidae), catfish *Arius arius* Hamilton (Siluriformes: Ariidae), catfish *A. maculatus* Thunberg (Siluriformes: Ariidae), striped mullet *Mugil cephalus* L. (Perciformes: Mugilidae), and pearl spot *Etroplus suratensis* Bloch (Perciformes: Cichlidae), respectively.
e Concentration found in Java tilapia *Oreochromis mossambica* Peters (Perciformes: Cichlidae).
f Concentrations of OCs found in the gills of fish. Malathion (1.8 μg g⁻¹ ww), chlorpyrifos (45.5 μg g⁻¹ ww), and dieldrin (1.2 μg g⁻¹ ww) were also detected.
g Value for *p,p′*-DDT.
h Aldrin's *endo-endo* isomer, isodrin.
i Pooled sample of fish in the gut of Ganges river dolphins.

Table 5.6 DDT residues in ng g⁻¹ ww from different types of fish from Indian waters

Type of fish	DDT residues
Estuarine	17.0–577.1
	(344.7)[a]
Free swimming	ND[b]–618.3
	(170.4)
Sessile	125.0–206.1
	(165.5)
Open ocean	1.04–41.4
	(25.4)

Source: Adapted from Shailaja and Sarkar, 1992.

Notes:
a Figures in parentheses indicate mean value.
b ND indicates DDT not detected.

In a recent study, Shailaja and Singbal (1994) measured residue levels in bottom-feeding fish species, e.g. *Upeneus* sp. (Perciformes: Mullidae or goatfishes), *Nemipterus* (Perciformes: Nemipteridae or threadfin bream), *Nemipterus (Synagris)* sp., and *Sillago* sp. (Perciformes: Sillaginidae or sillagos), collected in the vicinity of the Coleroon River mouth in the Bay of Bengal. The $\sum$ DDT concentration was 1.31 to 115.9 ng g⁻¹ and the concentration of aldrin was 0.32 to 4.23 ng g⁻¹ (ww). The observed mean values for $\sum$ DDT and aldrin (9.95 and 0.35 ng g⁻¹, respectively) were comparable to levels found in pelagic fish (10.42 and 0.39 ng g⁻¹, respectively) from the northeastern Arabian Sea. *Upeneus* sp. had the lowest concentration of $\sum$ DDT.

Rajendran *et al.* (1992) determined residue levels for $\sum$ DDT and HCH isomers in 14 species of marine fish collected along the coast of the Bay of Bengal. In general, $\sum$ HCH was greater than $\sum$ DDT concentration, barring a few exceptions. Among the 14 species examined *Dasyatis* sp. (Rajiformes: Dasyatidae or rays), spotted Spanish mackeral *Scomberomorous guttatus* Bloch and Schneider (Perciformes: Scombridae), Barramundi *Lates calcarifer* Bloch (Perciformes: Centropomidae), *Nemipterus japonicus* Bloch, *Mugil cephalus* L., and *Carangoides* spp. (Perciformes: Carangidae, jacks and pompanos) showed high $\sum$ HCH and $\sum$ DDT residue levels compared to other species. The concentration of $\sum$ HCH ranged between 0.14 to 8.98 ng g⁻¹ and $\sum$ DDT between 0.03 to 2.38 ng g⁻¹ on a ww basis.

Takeoka *et al.* (1991) opined that post application volatilization is a major route of transport for insecticides in tropical areas based on their studies of HCH applied in paddy areas of the Vellar watershed in South India. They observed that 99.6 percent of applied HCH was removed to the atmosphere and only 0.1 percent reached the sea through water and sediment transport. Rajendran *et al.* (1992) attributed the low detection of HCH residue levels in their studies to this factor. Shailaja and Sarkar (1993) found that post-monsoon fish samples from coastal and estuarine environments had higher pesticide residues than pre-monsoon samples. Kannan *et al.* (1993) argued that HCH – with a higher vapor pressure than DDT – would rapidly dissipate in the tropics leaving less residue in the biosphere. However, a clear picture will emerge only after comparing residue levels from pre- and post-monsoon biota samples from both the eastern and western coasts of India (especially river and estuarine samples).

In the final analysis, results obtained so far from fish samples show pesticide residues less than MRLs for fish set by the FDA of the USA. However, a systematic and continuous monitoring program based on large numbers of samples is necessary to protect public health in India.

Marine dolphins and other marine, estuarine, and freshwater organisms

Blubber samples were taken and examined for OC pesticide residues from 12 dolphins that were accidentally netted in the coastal waters of Porto Novo (Tanabe *et al.*, 1993). The dolphins had been living in the Bay of Bengal and belonged to three species from three genera, the spinner dolphin *Stenella longirostris* Gray (Cetacea: Delphinidae), the bottlenose dolphin *Turisiops truncatus* Montague (Cetacea: Delphinidae), and the Indo-Pacific humpbacked dolphin *Sousa chinensis* Osbeck (Cetacea: Delphinidae). Residues of HCH isomers (α, β, γ and Δ), DDT (*p,p'*-DDT, and *o,p'*-DDT), *p,p'*-DDE, PCBs, and HCB were detected. DDT concentrations in blubber were 4,500 to 35,000 ng g^{-1} ww, PCBs from 240 to 1,800 ng g^{-1} ww, HCH from 60 to 1,100 ng g^{-1} ww, and HCB residues were low. These toxicant residues stored in the lipid rich blubber may pass into later generations of dolphins through lactation.

The relatively high concentration of DDT compared to other OCs in dolphin blubber is an interesting finding when one considers that HCH is used in large amounts compared to DDT in the areas adjoining coastal zones. Tanabe *et al.* (1993) attribute this to several factors including that DDT is more lipophilic, less biodegradable, and less likely to undergo atmospheric transport, owing to its relatively low vapor pressure. The concentration of *p,p'*-DDE (1,700 to 23,000 ng g^{-1} ww) was the highest among the DDT family, while β-HCH (39 to 960 ng g^{-1}) and α-HCH (6 to 410 ng g^{-1}) were highest among HCH isomers. Tanabe (1991) reported that the DDT concentrations are within the range of concentrations found in dolphin blubber from other tropical waters (southern Indian and eastern Pacific oceans). DDE and β-HCH are known persistent environmental contaminants. That cetaceans have a unique ability to transfer substantial quantities of lipophilic pollutants from one generation to the next through lactation implies long-term accumulation in animals from tropical waters. Despite the DDT ban by many developed nations, the levels of DDT in the striped dolphin *Stenella coeruleoalba* Meyen (Cetacea: Delphinidae) from the western North Pacific remained unchanged for nearly ten years (Loganathan *et al.*, 1990).

It is interesting to compare the results discussed above with river dolphins. Kannan *et al.* (1993; 1994) measured pesticide residues in the Ganges river dolphin *Platanista gangetica* (Cetacea: Platanistidae), see Table 5.7. OC residues were in the

Table 5.7 OC pesticide residues in marine, estuarine, and freshwater organisms of India

Organism	Location	Sample date	Concentration range (mean) in ng g⁻¹ ww				Reference
			ΣDDT	ΣHCH	ΣChlordane	HCB	
Benthic invertebrates	Ganges River	1992–97	250–740 (500)	26–96 (61)	3.0–13 (8)	1.0–21 (11)	Senthilkumar et al., 1999
Dwarf prawn *Macrobrachium equidens* Dana (Decapoda: Palaemonidae)	Porto Novo and Cuddalore, East Coast	1998	(0.9)	(20)	<0.03	<0.03	Senthilkumar et al., 2001
Tiger frog *Rana tigrina* Cantor (Anura: Ranidae)			(0.8)	(17)	ND[a]	ND	
Brick red box crab *Calappa philargius* L. (Decapoda: Calappidae)			(42)	(300)	(3.7)	(0.21)	
Black rice crab *Somanniathelpusa* sp.			(6.3)	(4.8)	(0.41)	(0.24)	
Cone snail *Augur territella*			(1.1)	(25)	(0.5)	ND	
Bell (apple) snail *Pila ampullacea* L. (Mesogastropoda: Ampullariidae)			(1.1)	(4,100)	(1.51)	(0.22)	
Freshwater field crab *Barytelphusa guerini* Milne Edwards (Decapoda: Brachyura)	Vellar River watershed	1988–89	6.9[b]	62	–[c]	0.03	Ramesh et al., 1992
Mangrove crab *Scylla serrata* Forskål (Decapoda: Portunidae)	Vellar River watershed	1987	5.8–59 (20)	9.5–130 (45)	—	<0.01–0.02 (0.02)	

Organism	Location	Sample date	Concentration range (mean) in ng g⁻¹ ww				Reference
			ΣDDT	ΣHCH	$\Sigma Chlordane$	HCB	
Indian flap-shell turtle *Lissemys punctata punctata* Lacépède (Testudines: Trionychidae)	Vellar River watershed	1991	0.52–1.4 (0.83)	5.5–15 (9)	–	0.01–0.02 (0.02)	
Ganges river dolphin *Plantanista gangetica* (Cetacea: Plantanistidae)	Ganges River	Blubber 1989–92	4,700–13,000 (9,700)	190–610 (425)	9.1–60 (36.5)	2.8–7.2 (5.6)	Kannan et al., 1994
		Muscle 1989–92	100–5,100 (2,775)	8.7–300 (150)	1.4–76 (30.4)	0.06–4.6 (1.9)	
		Kidney 1989–92	88–460 (219.5)	8.8–28 (21.5)	1.1–5.8 (3.3)	0.11–2.4 (0.83)	
		Liver 1989–92	77–520 (289)	23–50 (36.3)	4.9–14 (8.5)	0.08–0.66 (0.37)	
Ganges river dolphin	Ganges River	Blubber 1993–96	21,000–64,000 (41,800)	860–1,900 (1,400)	45–240 (160)	7.7–19 (13)	Senthilkumar et al., 1999
		Liver 1994–96	750–1,700 1,200	83–150 (117)	1.9–4.5 (3)	0.4–1.1 (1)	
		Milk[d] 1992	48,000	400	33	4	

Notes:
a ND indicated compound not detected.
b Pooled sample from 18 crabs.
c En dash (–) indicates no data.
d Sample size of one.

order $\sum$ DDT > $\sum$ PCB > $\sum$ HCH > dieldrin and aldrin > $\sum$ chlordane > heptachlor, heptachlor oxide, and HCB, a pattern similar to that found in dolphins from the Bay of Bengal. River dolphins also had higher $\sum$ DDT than $\sum$ HCH and p,p'-DDE constituted 46 percent of $\sum$ DDT. However, α-HCH constituted the major HCH isomer unlike in marine dolphins where β-HCH was dominant. It is possible that during the course of residue transport from river to ocean, only persistent contaminants like β-HCH and p,p'-DDE reach the oceans, the others being dissipated by degradation or volatilization.

Senthilkumar et al. (1999) found a similar pattern of accumulation for $\sum$ DDT > $\sum$ PCB > $\sum$ HCH > $\sum$ chlordane > HCB; however, the concentrations of DDTs, HCHs, and chlordanes were 2- to 4-fold higher – following conversion to a lipid weight basis – in blubber and liver samples than the levels found several years earlier by Kannan et al. (1993; 1994). They attributed the increase in DDT levels to the widespread use of DDT for malaria vector control and Kala-azar disease vecton – the sandflies, Phlebotomus argintepes and P. papatasi Scopoli (Diptera: Psychodidae) – control in India. Again, α-HCH was the predominant isomer in river dolphin blubber, although they found β-HCH accounted for 48 and 65 percent of the total HCH concentrations in milk and liver, respectively. Residue patterns for DDT, HCH, and chlordanes suggested recent inputs into the Ganga of each of these pesticides.

Benthic invertebrates from the Ganges showed a similar pattern of pesticide accumulation (Senthilkumar et al., 1999) with $\sum$ DDT (500 ng g^{-1} ww) much greater than $\sum$ HCH (61 ng g^{-1} ww). Ramesh et al. (1992) found much lower levels of $\sum$ DDT (6.9 and 20 ng g^{-1} ww) in two species of crab from the Vellar River watershed, although $\sum$ HCH residues (62 and 45 ng g^{-1} ww) were roughly similar to those found in Ganges River benthic invertebrates. Senthilkumar et al. (2001) also recorded low levels of $\sum$ DDT, $\sum$ chlordane, and HCB in a variety of organisms, which are prey items for resident and migratory birds of South India, near Porto Novo and Cuddalore. However, they found that $\sum$ HCH varied from 4.8 to 4,100 ng g^{-1} ww in the same prey items with the highest concentrations found in the brick red box crab Calappa philargius L. (Decapoda: Calappidae) and bell (apple) snail Pila ampullacea L. (Mesogastropoda: Ampullariidae), at 300 and 4,100 ng g^{-1} ww, respectively.

Recent studies by Tanabe et al. (1998) and Senthilkumar et al. (2001) have measured OC pesticide and PCB residues in tissues of resident and migratory birds – including local migrants and both short- and long-range migratory species – and there exists one earlier study by Ramesh et al. (1992) with data for resident and local migrant species (Table 5.8). After grouping Indian wild birds according to feeding habits and predominant prey, higher DDT and HCH residues were found in piscivores followed by insectivores, omnivores, and granivores. Inland piscivore resident species had higher levels of both OCs than their coastal counterparts. There was no discernible pattern to DDT and HCH accumulation based on species' migratory pattern. Tanabe et al. (1998) proposed that migratory birds wintering in India acquire considerable amounts of HCHs and DDTs based on a global comparison of OC concentrations in birds, which indicated that resident birds in India had the highest residues of HCHs and moderate to high residues of DDTs.

Model ecosystems

The phenomenon of bioconcentration is considered to be the most important among the chronic effects of pesticides on non-target organisms from the viewpoint of protecting the aquatic environment and preventing protein resources for mankind from becoming contaminated. Biomagnification of residues in succeeding tropic levels is another phenomenon of importance and concern.

Model ecosystems using aquarium tanks with ecosystem components like soil or sand, fish, snails, and algae have proven useful in predicting the fate and behavior of pesticides in terrestrial ecosystems, especially their distribution among compartments (Metcalf, 1974). A model ecosystem using seawater and sediments was envisaged and developed by Raghu *et al.* (1994). The fate of ^{14}C-DDT and ^{14}C-chlorpyrifos in seawater, clams, and sediment was followed using a model ecosystem under an IAEA research co-ordinated project (nr 7936) to examine the fate and effects of pesticides in the tropical environment. Experiments with ^{14}C-DDT showed that, within 72 h, negligible ^{14}C-residues (only 1 percent of the applied radioactivity) could be detected in the seawater (Figure 5.5). The uptake of ^{14}C-residues in clams peaked after 3 d and decreased considerably at the end of 30 d (Kale *et al.*, 1995; Raghu *et al.*, 1996; 1997). To illustrate in terms of quantity of extractable residues, of 700 ng g^{-1} ^{14}C-residues recovered after 3 d, the parent compound, DDT, was 109 ng g^{-1} and the primary metabolite, DDE, was 522 ng g^{-1}. After 30 d, of the total ^{14}C-residues recovered (52 ng g^{-1}), only a negligible amount was in the form of DDT. In sediments the degradation product distribution showed that there was no DDE formation but DDD was formed to some extent. Bound activity in sediments was negligible. Algae, like clams, also showed the highest residue levels after 3 d and, thereafter, residues decreased to almost negligible levels after 30 d. Recently Kale *et al.* (1999) studied the degradation of DDT in marine sediments under moist and flooded conditions using a continuous flow system for a period of 130 d. DDT underwent degradation and about 22 percent of the applied ^{14}C-activity was recovered as volatiles under both conditions. In sediments, extractable ^{14}C-residues accounted for about 30 and 19 percent under moist and flooded conditions, respectively, and DDT was the major compound found.

Pandit *et al.* (1996) conducted a similar model ecosystem experiment using ^{14}C-chlorpyrifos. They observed that more than 50 percent of the insecticide moved out of the water phase within 24 h and only 7 percent remained after five days. Sediments accounted for 6 percent of the applied activity after five hours. Interestingly there was no further build up of ^{14}C-residues in sediments. A major portion of ^{14}C-residues was found in clams. After three days, there was a sharp decline in the extent of ^{14}C-residues and bound residues were formed to only a small extent. Thin layer chromatography of clam extracts indicated the presence

Table 5.8 OC pesticide and PCB residues in resident and migratory birds of India

Species	Location	Resident status	Feeding habit	Sample date	Concentration range (mean) in ng g⁻¹ ww				Reference
					$\sum DDT$	$\sum HCH$	HCB	$\sum PCB$	
Pond heron *Ardeola grayii* Sykes (Ciconiiformes: Ardeidae)	Vellar River watershed	Resident	Inland piscivore (fish, frogs, crabs, insects)	1989	98–1,800 (550)	210–4,000 (1,400)	0.06–0.49 (0.20)	2.0–35 (17)	Ramesh et al., 1992
	Porto Novo and Pudukottai			1995	3,100–3,600 (3,400)	1,100 (1,100)	0.9–1.1 (1.0)	22–65 (44)	Tanabe et al., 1998; Senthilkumar et al., 2001
White-throated kingfisher *Halcyon smyrnensis* Braunliest (Coraciiformes: Alcedinidae)	Vellar River watershed	Resident	Coastal piscivore (fish, tadpoles, lizards, insects)	1991	46–340 (150)	490–1,400 (950)	0.19–3.3 (1.0)	5.7–16 (11)	Ramesh et al., 1992
	Porto Novo			1995	410	420	0.3	40	Tanabe et al., 1998; Senthilkumar et al., 2001
Kentish plover *Charadrius alexandrinus* L. (Charadriiformes: Charadriidae)	Vellar River watershed	Local migrant	Insectivore (insects, crustaceans, molluscs)	1991	130–270 (220)	41–360 (130)	0.01–0.05 (0.03)	14–64 (41)	Ramesh et al., 1992
	Porto Novo			1995	67–330 (210)	280–590 (450)	<0.1–1.4 (0.8)	69–300 (150)	Tanabe et al., 1998; Senthilkumar et al., 2001

Species	Location	Feeding habit	Resident status	Sample date	Concentration range (mean) in ng g⁻¹ ww				Reference
					$\sum DDT$	$\sum HCH$	HCB	$\sum PCB$	
Jungle crow Corvus macrorhynchos Wagler (Passeriformes: Corvidae)	Vellar River watershed	Omnivore	Resident	1987	180–370 (280)	250–1,100 (680)	0.01–0.01 (0.01)	19–35 (27)	Ramesh et al., 1992
White-breasted waterhen Amaurornis phoenicurus Pennant (Gruiformes: Rallidae)	Vellar River watershed	Granivore (grains, molluscs, snails, worms)	Resident	1991	3.9–4.0 (4.0)	26–65 (46)	0.05–0.09 (0.07)	9.6–12 (11)	Ramesh et al., 1992
	Pudukottai			1995	170	840	0.2	23	Senthilkumar et al., 2001
Long-billed Mongolian plover Ch. mongolus Pallas sub sp. atrifrons	Porto Novo, South India	Insects, crabs, fish, bivalves	Short-range migrant	1995	120–620 (260)	62–480 (310)	1.3–4.6 (3.4)	130–420 (250)	Tanabe et al., 1998; Senthilkumar et al., 2001
Short-billed Mongolian plover Ch. ongolus Pallas sub sp. schaeferi	Porto Novo, South India	Insects, crabs, fish, bivalves	Short-range migrant	1995	17–370 (110)	180–470 (320)	4.4–23 (11)	71–300 (160)	Tanabe et al., 1998
Red-capped lark Calandrella cinerea Gmelin (Passeriformes: Alaudidae)	Porto Novo and Pudukottai	Insects, grains	Short-range migrant	1995	19–200 (110)	120–3,000 (1,100)	0.3–2.0 (0.2)	20 (20)	Senthilkumar et al., 2001

continued…

Table 5.8 continued

Species	Location	Feeding habit	Resident status	Sample date	Concentration range (mean) in ng g⁻¹ ww					Reference
					$\sum DDT$	$\sum HCH$	HCB	$\sum PCB$		
Common redshank *Tringa totanus* L. (Charadriiformes: Scolopacidae)	Pudukottai Southeast India	Insects, grains, molluscs, worms	Short-range migrant	1995	160–1,100 (600)	19–89 (54)	0.9–3.4 (1.5)	40–210 (90)		Senthilkumar et al., 2001
Common sandpiper *Actitis hypoleucos* L. (Charadriiformes: Scolopacidae)	Porto Novo, Southeast India	Omni-vorous (molluscs, insects, worms)	Long-range migrant	1995	140–1,900 (620)	82–380 (230)	0.3–1.6 (0.6)	70–170 (120)		Tanabe et al., 1998; Senthilkumar et al., 2001
Curlew sandpiper *Calidris ferruginea* (Charadriiformes: Scolopacidae)	Pudukottai Southeast India	Molluscs, shrimps, insects, crustaceans	Long-range migrant	1995	9.2–16 (11)	40–82 (54)	0.2–0.6 (0.4)	27–48 (36)		Senthilkumar et al., 2001
White-winged black tern *Chlidonias leucopterus* Temminck (Charadriiformes: Sternidae)	Porto Novo, Southeast India	Piscivore (fish, crabs, tadpoles, insects)	Long-range migrant	1995	850–1,700 (1,300)	36–710 (360)	1.1–1.8 (1.4)	210–850 (550)		Tanabe et al., 1998; Senthilkumar et al., 2001

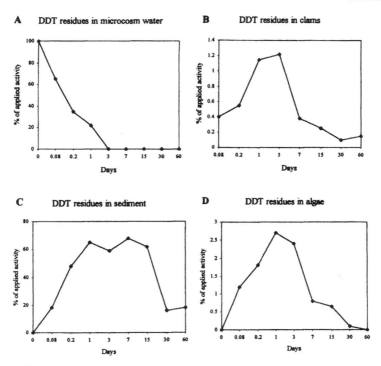

Figure 5.5 ¹⁴C-DDT distribution in different compartments of a model ecosystem

of TCP (3,5,6-trichloro-2-pyridinol) as the major metabolite. Recently Kale *et al.* (1999) studied the degradation of ^{14}C-chlorpyrifos in marine sediment under moist and flooded conditions using a continuous flow system. Volatilization accounted for 0.8 to 1 percent loss during the first 10 d post application. The amounts of extractable ^{14}C-activity were higher in flooded sediments than in moist sediments. More bound residues were formed under moist conditions. TCP was the major degradation product formed under both moist and flooded conditions, its formation being higher under flooded conditions. These studies underline the role of clams in the degradation of chlorpyrifos and the lack of microbial degradation. In the absence of clams, chlorpyrifos underwent abiotic degradation in marine sediment with formation of bound residues. Kale *et al.* (1997) followed the fate of ^{14}C-chlorpyrifos in marine sediment using a continuous flow system, under moist and flooded conditions. There was no mineralization of the insecticide under either moisture condition. Most of the ^{14}C-activity was recovered in the extractable and bound fractions. This was in sharp contrast to the model ecosystem experiment where there was no substantial build-up of ^{14}C-residues in sediment. Thus, it appears that clams were responsible for rapid degradation of the ^{14}C-chlorpyrifos in the model ecosystem experiment and so model ecosystems are indeed helpful in predicting the behavior of pesticides.

LEGISLATIVE MANAGEMENT OF PESTICIDES IN INDIA

Legislative acts

Details regarding legislative regulation of pesticides have been summarized (Narayanaswamy, 1993). At first, pesticide use was regulated under the Indian Poison Act of 1909; the State Poison Rules covering nine toxic pesticides; and the Indian Drugs and Cosmetic Act of 1940, which included four pesticides used in the public health sector. The Prevention of Food Adulteration Act of 1954 and its Rules, 1955, prescribed tolerance limits for pesticide residues in common food items. These piecemeal provisions had a limited impact as reports of food poisoning by OPs occurred in the states of Kerala and Madras. In 1962, 450 people were reported to have been crippled with paralysis in the Malda and Dinajpur districts because wheat had been contaminated by tri-ortho cresyl phosphate. A number of other poisoning incidents caused the Government of India to appoint an Enquiry Commission under the chairmanship of Justice J.C. Shah. Based on the recommendations of a subcommittee headed by Professor M.S. Thaker, the Government of India passed comprehensive pesticide legislation titled the Indian Insecticides Act, 1968. Implementation of the Act and the Rules framed under it began on 1 August 1971. Enforcement responsibility is divided between the Central and State governments. Registration of pesticides, their formulations, and all matters connected with them are the responsibility of the Central Government, while the burden of field enforcement of provisions of the Act rests with the State governments and the Union territories. Two statutory bodies were constituted under the Act: the Central Insecticides Board (CIB) and the Registration Committee.

The Central Insecticide Board consists of 29 members who are experts in various fields, e.g. health, plant protection, factory advisory services, agriculture, medical research, standardization, transport and shipping, pharmacology and medical toxicology, industrial health, and occupational hazards. The CIB advises the Central and State governments on matters related to: 1) risks to human beings or animals involved in the use of insecticides and the safety measures needed to ameliorate such risks, and 2) the manufacture, sale, storage, transport, and distribution of insecticides with a focus on ensuring safety.

The Registration Committee, consisting of a Chairman and five members, including the Drug Controller and Plant Protection Adviser to the Government of India, has the responsibility to register insecticides – with such specific conditions as may be necessary – after scrutinizing their formulations and verifying the claims made by importers or manufacturers as regards their efficacy and safety to human beings, animals, non-target plants, and non-target insects. Implementation of the Act and its enforcement is carried out by several levels of official bureaucratic machinery such as Licensing Officers, Insecticide Inspectors and Insecticide Analysts, representing either the Central or State governments. Each of these officers has specific responsibilities.

Under the schedule forming part of the Insecticides Act, 438 substances are listed, out of which 12 substances are banned from manufacture and use, 13 are registered for restricted uses, 17 substances are under review, and nine new chemicals are allowed to begin production (David, 1993).

Environmental guidelines

The United Nations Conference on the Environment held in Stockholm, Sweden in 1972 probably paved the way for India to enact comprehensive legislation of its own in 1974 in the form of the Water (Prevention and Control of Pollution) Act. This was followed by the Air Pollution Control Act in 1981 and the Environmental Protection from Hazardous Chemicals and Microorganisms Act in 1986. Goel and Sharma (1996) provide detailed information covering various aspects of environmental guidelines and standards in India.

Development of environmental standards is addressed by two agencies, the Bureau of Standards and the Central Pollution Control Board (CPCB). In addition to the CPCB, State Pollution Control Boards (SPCB) play roles in preventing pollution in various parts of the country. The standards developed by the CPCB are included in the Schedules of the Environment (Protection) Rules, 1986. The CPCB has also developed industry specific Minimal National Standards (MINAS) taking into account the technology and economic acceptability of the suggested treatment, particularly factors such as the ratio of total annual cost of the pollution control to the annual turnover of the industrial unit. Drinking water standards have been provided by the Indian Council of Medical Research (ICMR) and the Ministry of Works and Housing. Biological standards are by the ICMR.

The CPCB has a defined infrastructure to deal with pollution assessment and control. The CPCB survey and assessment for pesticides comprises inventorying storage, handling of hazardous wastes, and application of pesticides. Pesticide monitoring activities keep a check on the quality of inland and coastal waters, air quality, industrial emissions, and effluents. The CPCB also conducts data interpretation and quality assurance. A few salient points with regard to pesticide residues are discussed below.

The Drinking Water Standard as prescribed by the Bureau of Indian Standards (BIS) stipulates that the drinking water should not contain any pesticide residues. For marine coastal areas, BIS prescribes tolerance limits for water quality at the point where sewage and other wastes are drained into the sea. Effluent discharge is not permitted if the Tolerance Limits are exceeded. The Tolerance Limit for Chlorinated Pesticides in water used for bathing, recreation, shell fish, commercial fish culture, and salt manufacture is set at a maximum of 0.002 mg (as Cl) L^{-1}. The CPCB has prescribed effluent standards for industrial waste discharge from many industries including the pesticide formulation industry. Tolerance Limits for the various pesticides apply to discharge levels from the treatment plant before any dilution.

The significance of releases from industrial accidents was brought into focus through the tragedy at Bhopal, India in 1983, where a leak at the (then) Union Carbide pesticide plant led to the release of large quantities of methyl isocyanate (MIC). Rules have been framed under the India Environment (Prevention) Act, 1986 to establish procedures and provide guidelines for safe handling and storage of hazardous substances. The rules are applicable to all the manufacturers of pesticides in India. The general MINAS for discharge of effluents into marine coastal areas stipulates the absence of pesticides.

Formulation of these standards was based on information provided in environmental quality criteria and also data derived from biological toxicity studies. The standards may be legally enforceable or issued as guidelines only, depending on the agency stipulating the standards. For example, the standards suggested by the Bureau of Standards are taken as guidelines, while the Discharge Standards set by the CPRB and SPCBs are legally enforceable (Goel and Sharma, 1996).

National Standards are prescribed through Technical Documents formulated (by consensus) among all interest groups concerned and for the benefit of the consumer, industry, and the trades. They lay down optimum product quality parameters based on available resources, technology, and local consumer needs. The BIS has completed more than 250 publications on Indian Standards for various pesticides, both technical and formulation grade. BIS also stipulates tests for determination of an active ingredient, physico-chemical characteristics, stability, sampling, and packing. It has about 27 standards for various types of sprayers, dusters, and related equipment. BIS has also adopted the ISO 9000 series of standards of the International Organization for Standardization, published them as IS 14000 series, and has launched Quality Systems Certification against these standards.

Safety considerations

Under the Insecticides Act, 1968 administered by the Ministry of Agriculture, the Registration Committee may register a pesticide or its formulation for use on agricultural crops after determining that use of the pesticide will not result in hazardous residues. Data on the range of residues from supervised trials carried out at manufacturing companies' fields or from sponsored studies through agricultural universities or institutes are provided to the committee by those companies wishing to register their products.

The Factory Act, 1948 and Industries Development and Regulation Act, 1951 control safety aspects of pesticide production. Conversion of a 'letter of intent' into an industrial license requires a site approval and environmental pollution clearance by the CPCB and the applicable SPCB tasked with enforcing the Water Act, 1974, the Air Act, 1981, and the MINAS for water, air pollution, and effluent discharges.

The Pesticide Registration Committee thoroughly examines aspects of consumer safety before allowing product registration. The Indian Council of Agricultural

Research sponsored the 'All India Coordinated Research Project' on pesticide residues in crops from supervised trials with recommended pesticides. The project determined safe time limits between pesticide application and consumption of produce. These safe waiting period values have been determined for many different types of crops grown in different areas of India.

Another statutory provision regulating residues on food is the Prevention of Food Adulteration Act (PFA), 1954. Rules were promulgated in 1955 and are administered through the Ministry of Health. The MRL for tolerance of pesticides is prescribed under the PFA rules. The use of insecticides directly on foods is prohibited. MRL values for around 50 pesticides have been prescribed and most of these are based on FAO values – barring a few exceptions like the HCH isomers for which MRL values have been specifically determined for India. However, MRL values for other food commodities are not being determined regularly. Kannan *et al.* (1992) determined OC pesticide and PCB residues in foodstuffs collected from different locations throughout India (Table 5.9). Concentrations of aldrin, dieldrin, and HCB, in all foodstuffs tested except butter, were comparable to levels found in fish and prey organisms (of wildlife) in India (see Tables 5.5 and 5.7) but were much higher, and quite variable, in butter samples, especially for dieldrin and aldrin. Mean $\sum$ HCH concentrations were below FAO MRLs but individual samples of pulses, butter, and meat and animal fat exceeded FAO standards. DDTs were low except for butter, fish and prawn, and meat and animal fat samples, although among these samples concentrations were highly variable. Absence of a permanent national monitoring system for pesticide residues in agriculture and the environment has resulted in an unclear national picture and a failure to initiate suitable measures based on field data (Bami, 1996).

SUMMARY AND FUTURE PERSPECTIVES

Measurements of marine waters from the Arabian Sea showed low OC residues. Concentrations of DDT in the zooplankton showed a decreasing gradient from the near-shore to offshore, while the DDD concentrations showed the reverse trend. Aldrin, HCHs, and DDTs are the most abundant and most commonly found pesticides in the Arabian Sea. Bay of Bengal sediments contain an order of magnitude higher residues than sediments in the Arabian Sea and this was attributed to residues carried by the major rivers, which primarily flow east through heavily agricultural lands. Green mussels (bivalves) collected from the East Coast had high levels of HCH while West Coast samples had high levels of DDT indicative of the different pesticide usage for agriculture (HCH) and public health purposes (DDT). The distribution of pesticide residues in fish shows DDT is ubiquitous, while p,p'-DDE is frequently encountered.

Food chain magnification was not evident, but estuarine fish carried more residues than fish from the open sea. Higher concentrations of DDT compared to other OCs were found in dolphin blubber samples. It appears that, in general,

Table 5.9 OC pesticide and PCB residues in foodstuffs collected from different locations in India

Foodstuff	Concentration range (mean) in ng g⁻¹ ww						
	Σ DDT	Σ BHC	Heptachlor	Aldrin	Dieldrin	HCB	Σ PCB
Cereals	1.7–5.4 (3.5)	27–39 2(35)	0.04–0.11 (0.08)	0.17–2.3 (1.3)	0.28–1.4 (0.75)	0.01–0.04 (0.03)	0.21–0.64 (0.45)
Pulses	1.1–40 (20)	5.4–1,600 (420)	<0.01–0.27 (0.09)	0.80–1.6 (1.3)	0.82–3.3 (2.1)	0.02–0.16 (0.07)	0.11–2.9 (1.2)
Oils	1.8–57 (21)	6.9–480 (220)	0.08–1.6 (0.45)	<0.1–47 (19)	<0.1–47 (24)	0.09–2.8 (1.5)	2.6–15 (11)
Milk	0.17–5.2 (1.4)	82–490 (180)	<0.01	<0.01–0.03 (0.02)	0.01–0.17 (0.09)	0.01–0.10 (0.03)	0.03–1.7 (0.52)
Butter	780–3,000 (1,400)	2,100–3,800 (2,800)	<0.01–0.07 (0.04)	7.7–140 (42)	8.9–2,900 (740)	0.86–2.4 (1.7)	2.4–9.3 (6.0)
Fish and prawn	0.86–140 (15)	0.48–380 (28)	<0.01–0.50 (0.06)	<0.1–6.1 (1.3)	<0.1–8.5 (1.8)	<0.01–0.55 (0.07)	0.38–110 (3.5)
Meat and animal fat	0.13–820 (100)	3.3–5,500 (480)	<0.01–0.91 (0.09)	<0.1–23 (2.4)	<0.1–37 (8.3)	0.02–4.8 (0.61)	0.43–33 (3.6)

Source: Adapted from Kannan et al., 1992.

pesticide residues are low in the Indian marine environment. Sen Gupta (1991) suggests that the impact of semi-diurnal tides with an amplitude range of one to eight meters, coupled with the influence of the biannual reversal of the direction of monsoon winds ensures adequate distribution of pollutants in the Arabian Sea, the Bay of Bengal, and the northern arms of the Indian Ocean.

Although attempts have been made to determine residues in the marine environment and a large amount of data is available, a planned and systematic survey is lacking. Such a study would help in arriving at residue distribution among different compartments of marine ecosystems and interactions between compartments. For example, the simultaneous determination of residues in water, sediments, and biota in pre- and post-monsoon samples from both the Arabian Sea and the Bay of Bengal would help determine the actual distribution of these residues.

The impact of pesticide residues contributed from agricultural activities could be discerned if river mouths of major rivers passing through the agricultural fields' drainage areas are monitored. Furthermore, in addition to agricultural activities, there is a significant environmental contribution from waste discharges from pesticide manufacturers. Characterization of these waste products is important for understanding their effects on ecosystems.

Legislative efforts are underway to restrict and eliminate the future use of DDT and HCH isomers. It will be quite some time before we see the impact of these measures in the marine ecosystem; hence, the monitoring program needs to be continued. The toxicity and effects of many of the currently used pesticides and their metabolites to India's aquatic organisms, especially invertebrates, need to be studied. Such a data base will greatly assist our efforts to conduct risk assessments. A systematic study with well-defined short and long-range objectives would be of great value in evaluating and sustaining the health of India's marine environment.

REFERENCES

Agarwal, H.C., Mittal, T.K., Menon, K.B. and Pillai, M.K.K. 1986. DDT residues in the river Jamuna in India. *Water A S P.* 28:89–104.

Agnihotri, N.P. 1989. Loss of organochlorine insecticides from soils under semitropical conditions. In: Proc. Nat. Symp. on Nuclear Techniques in the Study of Pesticides in Food, Agriculture, and Environment. Bombay, India: Department of Atomic Energy (DAE), pp. 93–5.

Agnihotri, N.P. 1993. *Contribution of Agricultural Application of Pesticides on Quality of Ground and River Water.* Ganga Project Directorate Final Report. New Delhi: IARI.

Agnihotri, N.P., Gajbhiye, V.T., Kumar, M. and Mohapatra, S.P. 1994. Organochlorine insecticide residues in Ganga River water near Farrukhabad, India. *Environmental Monitoring and Assessment.* 30:105–12.

Agnihotri, N.P., Pandey, S.Y. and Jain, H.K. 1974. Persistence of BHC and aldrin in soil and translocations in mung (*Phaseolus aureus* L) and Lobia (*Vigna sinensis* siva). *Indian J Ent.* 36:261–7.

Agnihotri, N.P., Pandey, S.Y., Jain, H.K. and Srivastava, K.P. 1977. Persistence of aldrin, dieldrin, lindane, heptachlor and *p,p'*-DDT in soil. *J Ent Res.* 1:89–91.

Agnihotrudu, V. and Mithyanta, M.S. 1978. *Pesticide Residues: A Review of Indian Work.* Bangalore, India: Rallis India Ltd, RD Laboratories, Fertilizers and Pesticides Division.

Ahmad, S., Ajmal, M. and Nomani, A.A. 1996. Organochlorines and polycyclic aromatic hydrocarbons in the sediments of Ganges River (India). *Bull Environ Contam Toxicol.* 57(5):794–802.

Amaraneni Rao, S. 1997. Studies on the pollution problems of Kolleru Lake with special reference to pesticides, polycyclic aromatic hydrocarbons and heavy metals. [PhD dissertation]. Andhra University. Visakhapatnam, India.

Amaraneni Rao, S. and Pillala, R.R. 2000. Kolleru Lake water pollution pesticides. *Ind J Environ Health.* 42:169–75.

Amaraneni Rao, S. and Pillala, R.R. 2001. The concentration of pesticides in sediments from Kolleru Lake in India. *Pest Manag Sci.* 57:620–4.

Amonkar, S.V., Pal, A.K., Vijayalakshmi, L. and Amaraneni Rao, S. 1979. Microbial control of potato tubermoth (*Phthorimea operculella* Zell.). *Indian J Exp Biol.* 17:1127–33.

Amonkar, S.V., Urmila Kulkarni and Amardeep Anand 1985. Comparative toxicity of *Bacillus thuringiensis* subspecies to *Spodoptera littura* (F). *Curr Science.* 54:475–8.

Anbu, R.B. and Ramaswamy, M. 1991. Adaptive changes in respiratory movements of an air breathing fish *Channa straiatus* (Berker) exposed to carbamate pesticide Sevin. *J Ecobiol.* 3:11–6.

Anonymous. 1992. *Intercountry Program for Integrated Pest Control in Rice in South and Southeast Asia.* Rome: FAO.

Bakre, P.P., Misra, V. and Bhatnagar, P. 1990. Organochlorine residues from the Mahala Reservoir, Jaipur, India. *Environ Poll.* 63:275–81.

Bami, H.L. 1996. Pesticide use in India: Ten questions. *Pesticide Information.* 22(3):19–27.

Barik, S. and Sethunathan, N. 1978. Biological hydrolysis of parathion in natural ecosystems. *J Environ Qual.* 7:346–8.

Bhaktavatsalam, R. and Reddy, Y.S. 1981. Survival of the fish *Anabas testudineus* (Bloch) exposed to lindane. *Indian J Environ Health.* 33:126–7.

Caravalho, F.P., Fowler, S.W., Readman, J.W. and Mu, L.D. 1992. Pesticide residues in tropical coastal lagoons: application of isotopes and radiation in conservation of the environments. In: Proc. Int. Symp. on Applications of Isotopes and Radiation in Conservation of the Environment held 9–13 March 1992 in Karlsruhe, Baden-Württemberg, Germany. Vienna: IAEA, pp. 637–53.

Currie, R.A., Kadis, V.W., Breitkreitz, W.E., Cunningham, G.B. and Bruns, G.W. 1979. Pesticide residues in human milk, Alberta, Canada, 1966–70; 1977–78. *Pest Monit J.* 13:52–5.

David, B.V. 1993. Indian pesticides industry: An overview. In: Vasanyharaj, D.B. (ed.) *Kothari's Desk Book Services: The Pesticide Industry.* Madras, India: HC Kothari Group, Pub. Div., pp. 16–35.

Desai, B.N. 1992. *Oceanography of Indian Ocean.* New Delhi, India: Oxford and IBH Pub.

Directorate of Economics and Statistics. 1994. *Indian Agriculture in Brief,* 25th edn. New Delhi: Department of Agriculture and Co-operation, Ministry of Agriculture, Government of India.

Drego, J., Murthy, N.B.K. and Raghu, K. 1990. ^{14}C-gamma hexachlorocyclohexane in flooded soil with green manuring. *J Agri Food Chem.* 38:266–8.

Edwards, C.A. 1973. Persistent pesticides in the environment, 2nd edn. Cleveland, Ohio: CRC Press.

Encyclopaedia Britannica 1967. Volume 12. Chicago: Encyclopedia Britannica, pp. 122–3.

Ferreira, J. and Raghu, K. 1978. Persistence of hexachlorocyclohexane in soils. In: *Nuclear Techniques in Studies of Metabolism, Effects and Degradation of Pesticides*. Bombay: DAE, pp. 126–33.

Ferreira, J. and Raghu, K. 1981. Decontamination of hexachlorocyclohexane in soils by green manure applications. *Environ Toxicol Lett*. 2:357–64.

Gaikwad, S.R. 1989. Effects of Thiodan on the oxygen consumption of *Etroplus matulatus*. *Indian J Environ Health*. 31:267–9.

Ghatak, D.B. and Konar, S.K. 1994. Field survey on the status of pollution by various industrial effluents on Hooghly Estuary ecosystem at Haldia, West Bengal. *Environ Ecol*. 12:128–52.

Glasby, G.P. and Roonwal, G.S. 1995. Marine pollution in India: an emerging problem. *Curr Sci*. 68:495–7.

Goel, P.K. and Sharma, K.P. 1996. *Environmental guidelines and standards in India*. Jaipur, India: Technoscience Publishers, pp. 1–318.

Gopal, K., Ram, M.D., Gupta, G.S.D. and Anand, M. 1990. Effect of elevated temperature and pesticide stress to teleosts. *Indian J Environ Protect*. 10:161–9.

Haldar, P., Raha, P., Bhattacharya, P., Choudhury, A. and Aditya Chaudhury, N. 1989. Studies on the residues of DDT and endosulfan occurring in Ganges water. *Indian J Environ Health*. 31:156–61.

Iwata, H., Tanabe, S., Sakai, N., Nishiraura, A. and Tatsakawa, R. 1994. Geographical distribution of persistent organochlorines in air water and sediments from Asia and Oceania and their implications for global redistribution from lower latitudes. *Environ Poll*. 85:15–33.

Jacob, J.P. and Menon, N.R. 1991. Effects of organophosphates on the life and activity of the black clam *Villorita cyprinoides* var. *cochinensis* (Hanley). In: Nair, N.B. (ed.) Proc. Nat. Seminar on Estuarine Management held in Trvandrum, India 4–5 June 1987. Kerala, India: State Committee on Science, Technology, and Environment, pp. 472–6.

Jayaraj, S. 1992. Integrated pest management. In: David, B.V. (ed.) *Pest Management and Pesticides: Indian Scenario*. Madras, India: Namrata Publishers, pp. 7–16.

Jindal, R. and Singh, J. 1989. Toxicity of pesticides to the primary productivity of a freshwater pond. *Indian J Environ Health*. 31:257–61.

Joshi, H.C. and Ghosh, B.B. 1982. Studies on the distribution of organochlorine pesticide residues in the Hooghly Estuary. In: Proc. First Nat. Environ. Conf. New Delhi: Indian Council of Agricultural Research (ICAR), p. 27.

Kale, S.P. and Raghu, K. 1996. Fate of ^{14}C carbofuran in soils. *Z pflanzenernahr Bodenk*. 159:519–23.

Kale, S.P., Carvalho, F.P., Raghu, K., Sherkane, P.D., Pandit, C.G., Mohan Rao, A., Mukherjee, P.K. and Murthy, N.B.K. 1999. Studies on degradation of ^{14}C-chlorpyrifos in the marine environment. *Chemosphere*. 39:969–76.

Kale, S.P., Murthy, N.B.K., Raghu, K., Sherkane, P.D. and Carvalho, F.P. 1999. Studies on degradation of ^{14}C-DDT in the marine environment. *Chemosphere*. 39(6):959–68.

Kale, S.P., Raghu. K., Mohan Rao, A.M., Murthy, N.B.K. and Pandit, G.G. 1995. Fate of ^{14}C DDT in the Indian marine environment using microcosm experiments. Report presented at 2nd Research Coordination Meeting (RCM) of IAEA Cooperative Research Program (CRP) held 12–16 June 1995 in Kuala Lumpur, Malaysia on Distribution, Fate and Effects of Pesticides on Biota in Tropical Environment.

Kale, S.P., Sarma, G., Goswami, U.C. and Raghu, K. 1996. Uptake and distribution of ^{14}C-Carbofuran and ^{14}C-HCH in catfish. *Chemosphere*. 33(3):449–51.

Kale, S.P., Sherkhane, P.D., Mukherjee, P.K. and Murthy, N.B.K. 1997. Studies on degradation pattern of ¹⁴C Chlorpyrifos in marine environment using a continuous flow system. Report presented at 4th RCM of IAEA CRP held 16–20 June 1997 in Nairobi, Kenya on Distribution, Fate and Effects of Pesticides on Biota in Tropical Environment.

Kannan, K., Tanabe, S. and Tatsukawa, R. 1994. Biodegradation capacity and residue pattern of organochlorines in Ganges river dolphins from India. *Toxicol Environ Chem.* 42:249–61.

Kannan, K., Tanabe, S. and Tatsukawa. R. 1995. Geographical distribution and accumulation features of organochlorine residues in fish in tropical Asia and Oceania. *Environ Sci Technol.* 29(10):2673–83.

Kannan, K., Sinha, R.K., Tanabe, S., Ichihashi, H. and Tatsukawa, R. 1993. Heavy metals and organochlorine residues in Ganges river dolphins from India. *Mar Poll Bull.* 26:159–62.

Kannan, K., Tanabe, S., Ramesh, A., Subramanian, A. and Tatsukawa, R. 1992. Persistent organochlorine residues in foodstuffs from India and their implications on human dietary exposure. *J Agric Food Chem.* 40(3):518–24.

Kannan, S.T. and Sen Gupta, R. 1987. Organochlorine residues in zooplankton of the Saurashtra Coast, India. *Mar Poll Bull.* 18:92–4.

Kathpal, T.S., Madan, V.K., Arora, K. and Arora, S.S. 1996. Environmental fate and role of lindane in soil under Indian conditions. *Pesticide Information.* 1:8.

Kime, D.E. 1995. The effects of pollution on reproduction in fish. *Rev Fish Biol.* 5:52–96.

Krishnamurti, C.R. and Dikshith, T.S.S. 1982. Application of biodegradable pesticides in India. In: Matsumura, F. and Krishnamurthi, C.R. (eds) *Biodegradation of pesticides.* London: Plenum, pp. 257–305.

Kureishy, T.W., George, M.D. and Sen Gupta, R. 1978. Concentration in zooplankton from the Arabian Sea. *Indian J Mar Sci.* 7:54–5.

Lakshmi, A. 1993. Pesticides in India: risk assessment to aquatic ecosystems. *Sci. Total Environ.* Suppl Part 1, pp. 243–53. In: Sloof, W. and de Kruijf, H. (eds) Proc. 2nd European Conference on Ecotoxicology held 11–15 March 1992 in Amsterdam, The Netherlands.

Loganathan, B.G., Tanabe, S., Tanaka, H., Watanabe, S., Miyazaki, N.M., Amano, M. and Tatsukawa, R. 1990. Comparison of organochlorine residue levels in the striped dolphin from western North Pacific 1978–79 and 1986. *Mar Poll Bull.* 21:435–9.

MacRae. I.C., Raghu, K. and Castro, T.F. 1967. Persistence and biodegradation of four common isomers of benzene hexachloride in submerged soils. *J Agric Food Chem.* 15:911–14.

MacRae, I.C., Raghu, K. and Bautista, E.M. 1969. Anaerobic degradation of insecticide lindane by *Clostridium* sp. *Nature.* 221:839–60.

Madhyastha, M.N. 1990. Aquaculture and aquatic pollution. In: Manna, G.K. and Jana, B.B. (eds) *Impacts of Environment on Animals and Aquaculture.* Proc. Nat. Symp. held 14–16 May 1988 at University of Kalyani, Kalyani, India, pp. 67–8.

Mani, V.G.T. and Konar, S.K. 1988. Pollution hazards of the pesticide chlorpyrifos on aquatic ecosystem. *Environ Ecol.* 6:460–2.

Manorama Year Book. 1994. Malayala Manorama, Kottayam. Kerala, India.

Mehrotra, K.N. 1985. Use of DDT and its environmental effects in India. *Proc Indian Nat Sci Acad (B).* 51:169–84.

Metcalf, R.L. 1974. A laboratory model ecosystem for evaluating the chemical and biological behaviour of radiolabeled micropollutants. In: *Comparative Studies of Food and*

Environmental Contamination. Proc. Symp. Otaniemi. IAEA. Vienna, Austria. 1973, pp. 49–62.

Misra, D., Bhuyan, S., Adhya, T.K. and Sethunathan, N. 1992. Accelerated degradation of methyl parathion, parathion and fenitrothion by suspensions from methyl parathion and *p*-nitrophenol treated soil. *Soil Biol Biochem.* 24:1035–42.

Misra, D., Sreedharan, B., Bhuyan, S. and Sethunathan, N. 1993. Accelerated degradation of parathion in fenitrothion on 3-methyl, 4-nitro phenol acclimatized soil suspensions. *Chemosphere.* 27:1529–38.

Misra, P.M. 1991. Improvement of Chilka lake as an index to coastal development with special reference to fisheries. *Seafood Export J.* 23:14–19.

Mitra, J. and Raghu, K. 1978. Persistence of DDT in soil. In: *Nuclear Techniques in Studies of Metabolism, Effects and Degradation of Pesticides.* Bombay, India: DAE, pp. 134–42.

Mitra, J. and Raghu, K. 1986. Rice straw amendment and degradation of DDT in soils. *J Toxicol Environ Chem.* 11:171–81.

Mitra, J. and Raghu, K. 1988. Influence of green manuring on persistence of DDT in soil. *Environ Soil Lett.* 9:847–52.

Mullick, S. and Konar, S.K. 1994. Field studies on the effect of effluents of detergent, petroleum oil refinery, pesticides and fertilizer factories on biotic and abiotic parameters in the Hooghly Estuary. *Environ Ecol.* 12:802–9.

Mukherjee, P.K. and Raghu, K. 1997. Effect of temperature on antagonistic and biocontrol potential of *Trichoderma* sp. on *Sclerotium rolfsii. Mycopathologia.* 139:151–7.

Mukherjee, P.K., Thomas, P. and Raghu, K. 1995. Shelf-life enhancement of fresh ginger rhizomes at ambient temperatures by combination of gamma-irradiation bio-control and closed polyethylene storage. *Ann App Biol.* 127:375–84.

Mukhopadhyay, A.N. and Mukherjee, P.K. 1995. Biopesticides. In: Prakash, J. (ed.) *Biotechnology Research and Industry Survey.* Coimbatore, India: Vadamalai Media, pp. 179–83.

Mukhopadhyay, A.N. and Mukherjee, P.K. 1996. Fungi as fungicides. *Int J Trop Plant Dis.* 14:1–17.

Mukhopadhyay, A.N., Shrestha, S.M. and Mukherjee, P.K. 1992. Biological seed treatment for control of soil-borne plant pathogens. *FAO Plant Prot Bull.* 40:21–30.

Murthy, N.B.K. and Raghu, K. 1991. Metabolism of ^{14}C carbaryl and ^{14}C 1-napthol in moist and flooded soils. *J Environ Sci Health B.* 24:479–91.

Murthy, N.B.K. and Raghu, K. 1989. Fate of ^{14}C carbaryl in soils as a function of pH. *Bull Environ Contam Toxicol.* 46:374–9.

Nair, P.V.R, Rajagopalan, M.J., Pillai, V.K., Gopinathan, C.P., Chandrika, V. and Vincent, D. 1985. Marine pollution, its effects on living resources with special reference to aquaculture. In: Proc. Symp. on Coastal Aquaculture held in Cochin, India 12–18 January 1980. Cochin, India: Marine Biology Association of India, pp. 1352–8.

Narayanaswamy, M. 1993. The Act '68 and The Rules '71. In: David, B.V. (ed.) *Kothari's Desk Book Series: The Pesticides Industry.* Madras, India: HC Kothari Group, Pub Div., pp. 102–10.

Nasar, S.A.K. 1976. The effect of endrin on primary production in pond ecosystem. *Phykos.* 15:47–8.

Nayak, B.K., Raha, R. and Das, A.K. 1995. Organochlorine pesticide residues in middle stream of Ganga River. *Indian Bull Environ Contam Toxicol.* 54:68–75.

Pandit, G.G., Kale, S.P., Mohan Rao, A.M., Murthy, N.B.K., Mukherjee, P.K. and Raghu, K. 1996. Fate of ^{14}C chlorpyrifos in Indian marine environment using model ecosystem.

Report presented at 3rd RCM of IAEA CRP held 9–13 September 1996 in San Jose, Costa Rica on Distribution, Fate and Effects of Pesticides on Biota in Tropical Environment.

Pandit, G.G., Mohan Rao, A.M., Jha, S.K., Krishnamoorthy, T.M., Kale, S.P., Raghu, K. and Murthy. N.B.K. 2001. Monitoring of organochlorine pesticide residues in the Indian marine environment. *Chemosphere.* 44(2):301–5.

Paroda, R.S. 1997. Integrated pest management: need and relevance. Mumbai, India: Asia Pacific Crop Protection Conf. pp. 1–9.

Pathak, S.P., Kumar, S., Kamteke, P.W., Murthy, R.C., Singh, K.P., Bhattachajee, J.W. and Ray, P.K. 1992. Riverine pollution in some northern and north eastern states of India. *Environmental Monitoring and Assessment* 22:227–36.

Pawar, K.R. and Katdare, M. 1983. Effect of sumithion on the ovaries of fresh water fish *Garrya mullya* (Sykes). *Curr Sci.* 52:784–5.

Piska, R.S. and Waghray, S. 1991. Toxic effects of dimethoate on primary production of lake ecosystem. *Indian J Environ Health* 33:126–7.

Qasim, S.Z. and Sen Gupta, R. 1988. Some problems of coastal pollution in India. *Mar Poll Bull.* 19:100–6.

Qasim, S.Z., Sen Gupta, R. and Kureishy, T.W. 1988. Pollution of the seas around India. *Proc Indian Acad Sci Animal Sci.* 97:117–31.

Radhakrishnan, A.G. and Antony, P.D. 1989. Pesticide residues in marine fishes. *Fish Technol Soc Fish Technol Cochin.* 26:60–1.

Raghu, K. and MacRae, I.C. 1966. Biodegradation of the gamma isomers of benzene hexachloride in submerged soils. *Science.* 154:263–4.

Raghu, K. and Drego, J. 1986. Bound residues of lindane, magnitude, microbial release, plant uptake and effects on microbial activities. In: *Quantification, Nature, and Bioavailability of Bound ^{14}C-pesticide Residues in Soil, Plant and Food.* Vienna: IAEA, pp. 41–50.

Raghu, K., Kale, S.P., Murthy, N.B.K. and Mitra. J. 1996. Radiotracer techniques: an aid in management of pesticides for human welfare. *Sci and Cult.* 62:177–80.

Raghu, K., Murthy, N.B.K., Kale, S.P. and Kulkarni, M.G. 1997. Model ecosystems for predicting the behavior of pesticides. In: *Environmental Behavior of Crop Protection Chemicals.* Proc. Int. Symp. on Use of Nuclear and Related Techniques for Studying the Environmental Behavior of Crop Protection Chemicals, 1–5 July 1996. Vienna: IAEA/FAO, pp. 205–14.

Raghu, K., Kale, S.P., Murthy, N.B.K., Pandit, G.G. and Mohan Rao, A.M. 1994. Organochlorine pesticides in Indian marine environment. Report presented at 1st RCM of IAEA CRP held 20–24 June 1994 in Monaco on Distribution, Fate and Effects of Pesticides on Biota in the Tropical Environment.

Rajagopal, B.S., Brahmaprakash, G.P., Reddy, B.R., Singh, U.D. and Sethunathan, N. 1984. Effect and persistence of selected carbamate pesticides in soils. *Residue Rev.* 93:1–199.

Rajendran, N. and Venugopalan, V.K. 1988. Toxicity of organochlorine pesticides to zooplankton of Vellar Estuary. *Indian J Mar Sci.* 17:168–9.

Rajendran, N. and Venugopalan, V.K. 1991. Bioconcentration of endosulfan in different body tissues of estuarine organisms under sublethal exposure. *Bull Environ Contam Toxicol.* 46:151–8.

Rajendran, R.B. and Subramanian, A.N. 1997. Pesticide residues in water from the River Kaveri, South India. *Chemistry and Ecology.* 13:223–36.

Rajendran, R.B. and Subramanian, A.N. 1999. Chlorinated pesticide residues in surface sediments from the River Kaveri, South India. *J Environ Sci Health, B.* 34(2):269–88.

Rajendran, R.B., Karunagaran, V.M., Balu, S. and Subramanian, A.N. 1992. Levels of chlorinated insecticides in fishes from the Bay of Bengal. *Mar Poll Bull.* 24:567–70.

Rajukannu, K. and Ramulu, V.S. 1981. Degradation and persistence of carbofuran in soil. *Madras Agri J.* 68:668–71.

Ramesh, A., Tanabe, S., Iwata, H., Tatsukawa, R., Subramanian, A.N., Mohan, D. and Venugopalan, V.K. 1990a. Seasonal variations of persistent organochlorine insecticide residues in Vellar River waters Tamil Nadu, South India. *Environ Poll.* 67(4):289–304.

Ramesh, A., Tanabe, S., Subramanian, A.N., Mohan, D., Venugopalan, V.K. and Tatsukawa, R. 1990b. Persistent organochlorine residues in green mussels from coastal waters of South India. *Mar Poll Bull.* 12:587–90.

Ramesh, A., Tanabe, S., Tatsukawa, R., Subramanian, A.N., Palanichamy, S., Mohan, D. and Venugopalan, V.K. 1989. Seasonal variations of organochlorine insecticide residues in air from Porto Novo, South India. *Environ Poll.* 62:213–22.

Ramesh, A., Tanabe, S., Morase, H., Subramanian, A. and Tatsukawa, R. 1991. Distribution and behaviour of persistent organochlorine insecticides in paddy soil and sediments in tropical environment: a case study in South India. *Environ Poll.* 74:293–307.

Ramesh, A., Tanabe, S., Kannan, K., Subramanian, A., Kumaran, P.L. and Tatsukawa, R. 1992. Characteristic trend of persistent organochlorine contamination in wildlife from a tropical agricultural watershed, South India. *Arch Environ Contam Toxicol.* 23(1):26–36.

Rananvare, H.D., Tamhankar, A.J. and Dongre, T.K. 1996. Biotechnological approach to pest management. *Sci and Cult.* 62:173–6.

Rao, K.S. and Nagabhushanam, R. 1986. Temperature dependent toxicity of Dimecron to marine edible crab, *Scylla serrata. Environ Ecology.* 4:597–9.

Rao, P.S.B. and Venugopalan, V.K. 1984. Lindane induced respiratory changes in juveniles of estuarine fish *Mugil cephalus* (L). *Indian J Mar Sci.* 13:196–8.

Rao, T.R. and Sarma, S.S.S. 1990. Interaction of chlorella density on the population dynamics of the rotifer *Brachionus patulus* (Rotifera). *Indian J Environ Health.* 32:157–60.

Ray, P.K. 1990. *Toxicology Atlas of India: Pesticides.* Lucknow, India: Industrial Toxicology Research Centre, pp. 64–7.

Sahu, S.K., Patnaik, K.K., Sharmila, M. and Sethunathan, N. 1990. Degradation of alpha, beta and gamma hexachlorocyclohexane by a soil bacterium under aerobic conditions. *Appl Environ Microbiol.* 56:3620–2.

Sakurai, K., Mori, S., Sudo, T., Fujio, A., Ishisu, T. and Ho, F. 1979. Contamination of milk and human milk by organochlorine pesticides. *Mieken Eisei Kenkyusho Nenpo.* 25: 33–9.

Samuel, T. and Pillai, M.K. 1988a. Persistence and fate of ^{14}C gamma HCH in an Indian sandy loam soil under field and laboratory conditions. In: *Isotopic Techniques for Studying the Fate of Persistent Pesticides in the Tropics.* Vienna: IAEA. IAEA TECHDOC-476, pp. 93–102.

Samuel, T. and Pillai, M.K. 1988b. Persistence and fate of ^{14}C DDT in an Indian sandy loam soil under field and laboratory conditions. In: *Isotopic Techniques for Studying the Fate of Persistent Pesticides in the Tropics.* Vienna: IAEA. IAEA TECHDOC-476, pp. 27–40.

Samuel, T. and Pillai, M.K. 1991. Impact of repeated application on the binding and persistence of ^{14}C DDT and ^{14}C-HCH in a tropical soil. *Environ Poll.* 74:205–16.

Sarkar, A. 1994. Occurrence and distribution of persistent chlorinated hydrocarbons in the seas around India, Chapter 28. In: Majumdar, S.K., Miller, G.S., Miller, E.R., Schmalz, R.F., Forbes, G.S. and Panah, A.A. (eds) *The Oceans: Physical-chemical Dynamics and Human Impact.* Easton, PA: Pennsylvania Academy of Science, pp. 444–58.

Sarkar, A. and Banerjee, G. 1987. Component analysis of some chemical parameters influencing the stability of DDVP in sediments along the East Coast of India. *Int J Environ Studies.* 29:171–4.

Sarkar, A. and Sen Gupta, R. 1985. Persistence and fate of some organophosphorous pesticides in sea sediments. In: Proc. 2nd Annual Conf. Nat. Environ. Sci. Acad. Faizabad, India: NESA, pp. 5–8.

Sarkar, A. and Sen Gupta, R. 1986. Persistence and fate of some organophosphorous pesticides in the sea sediments along the East Coast of India. *Indian J Mar Sci.* 15:72–4.

Sarkar, A. and Sen Gupta, R. 1987. Chlorinated pesticide residues in sediments from the Arabian Sea along the central West Coast of India. *Bull Environ Contam Toxicol.* 39:1049–54.

Sarkar, A. and Sen Gupta, R. 1988a. DDT residues in sediments from the Bay of Bengal. *Bull Environ Contam Toxicol.* 41:664–9.

Sarkar, A. and Sen Gupta, R. 1988b. Chlorinated pesticide residues in marine sediments. *Mar Poll Bull.* 19:35–7.

Sarkar, A. and Sen Gupta, R. 1989. Determination of organochlorine pesticides in Indian coastal waters using a moored in situ sample. *Water Res.* 23:975–8.

Sarkar, A. and Sen Gupta, R. 1991. Pesticide residues in sediments from the West Coast of India. *Mar Poll Bull.* 22(1):42–5.

Sarkar A and Sen Gupta R. 1992. On chlorinated hydrocarbons in Indian Oceans. In: Desai, B.N. ed. *Oceanography of the Indian Ocean.* New Delhi: Oxford, IBH Publ. pp. 385–95.

Sarkar, A., Nagarajan, R., Chaphadkar, S., Pal, S. and Singbal, S.Y.S. 1997. Contamination of organochlorine pesticides in sediments from the Arabian Sea along the West Coast of India. *Water Research.* 31(2):195–200.

Sathe, M.C. 1995. Observations on the toxicity of insecticide fenthion to the marine edible clam *Marcia hiantina* (Lamarck) [PhD dissertation]. University of Bombay. Mumbai, India. 234 p.

Sen Gupta, R. 1991. Health of the seas around India. In: Rao, B.L.S.H. and Pandian, T.J. (eds) *Ocean Science and Technology.* Proc. Seminar Ocean Sci Tech held 2 January 1990 in Trivandrem, Kerala, India. New Delhi: Indian National Science Academy, pp. 91–115.

Sen Gupta, R. and Kureishy, T.W. 1989. Marine pollution levels and potential threat to the Indian marine environment: State of the art. In: Sinha, A.K., Viswanathan, P.N. and Boojah, R. (eds) *Water Pollution Conservation Management.* Nainital, India: Gnanodaya Prakashan, pp. 165–81.

Senthilkumar, K., Kannan, K., Sinha, R.K., Tanabe, S. and Giesy, J.P. 1999. Bioaccumulation profiles of polychlorinated biphenyl congeners and organochlorine pesticides in Ganges river dolphins. *Environ Toxicol Chem.* 18(7):1511–20.

Senthilkumar, K., Kannan, K., Subramanian, A. and Tanabe, S. 2001. Accumulation of organochlorine pesticides and polychlorinated biphenyls in sediments, aquatic organisms, birds, bird eggs and bats collected from South India. *Environ Sci Poll Res.* 8(1):35–47.

Sethunathan, N. and Yoshida, T. 1973. A *Flavobacterium* sp that degrades diazinon and parathion as sole carbon sources. *Can J Microbiol.* 19:873–5.

Sethunathan, N., Adhya, T.K. and Raghu, K. 1982. Microbial degradation of pesticides in tropical soils. In: Matsumura, F. and Krishnamoorthy, C.R. (eds) *Biodegradation of Pesticides.* New York: Plenum, pp. 91–115.

Sethunathan, N., Rao, V.R., Adhya, T.K. and Raghu, K. 1983. Microbiology of rice soils. In: *CRC Critical Reviews in Microbiology*. Boca Raton: CRC Press, pp. 125–72.

Shailaja, M.S. and Nair, M. 1997. Seasonal differences in organochlorine pesticide concentrations of zooplankton and fish in the Arabian Sea. *Mar Environ Res*. 44:264–74.

Shailaja, M.S. and Sen Gupta, R. 1989. DDT residues in fishes from the eastern Arabian Sea. *Mar Poll Bull*. 20:620–30.

Shailaja, M.S. and Sen Gupta, R. 1990. Residues of dichlorodiphenyltrichloroethane and metabolites in zooplankton from the Arabian Sea. *Curr Sci*. 59:929–31.

Shailaja, M.S. and Sarkar, A. 1992. Chlorinated hydrocarbon pesticides in the Northern Indian Ocean. In: Desai, B.N. (ed.) *Oceanography of the Indian Ocean*. New Delhi: Oxford, IBH Publishers, pp. 379–83.

Shailaja, M.S. and Sarkar, A. 1993. Organochlorine pesticide residues in the seas around India. Internal Report for Chemical Research and Environmental Needs (CREN) – joint program of Council of Scientific and Industrial Research (CSIR) – New Delhi and Commonwealth Science Council (CSC), London. Goa, India: National Institute of Oceanography (NIO).

Shailaja, M.S. and Singbal, S.Y.S. 1994. Organochlorine pesticide compounds in organisms from the Bay of Bengal. *Estuarine Coastal Shelf Sci*. 39(3):219–21.

Siddaramappa, R. and Sethunathan, N. 1975. Persistence of gamma-BHC and beta-BHC in Indian rice soils under flooded conditions. *Pest Sci*. 6:395–403.

Siddaramappa, R., Rajaram, K.P. and Sethunathan, N. 1973. Degradation of parathion by bacteria isolated from flooded soil. *Appl Microbiol*. 26:846–9.

Singbal, S. 1997. Market potential of agrochemicals in India. In: Proc. Asia Pacific Crop Protection Conference. Mumbai, India, pp. 251–60.

Somasekhar, R.K. and Sreenath, K.P. 1984. Effect of fungicides on primary productivity. *Indian J Environ Health*. 26:355–7.

Sujatha, C.H. and Chacko, J. 1991. Malathion sorption by sediments from tropical estuary. *Chemosphere*. 23:167–80.

Sujatha, C.H., Nair, S.M., Kumar, N.C. and Chacko, J. 1993. Distribution of organochlorine pesticides in a tropical waterway: HCH isomers. *Toxicol Environ Chem*. 39(1–2): 103–11.

Sujatha, C.H., Nair, S.M., Kumar, N.C. and Chacko, J. 1994a. Distribution of dichlorodiphenyltrichloroethane (DDT) and its metabolites in an Indian waterway. *Environ Toxicol Water Qual*. 9:155–60.

Sujatha, C.H., Nair, S.M. and Chacko, J. 1994b. Sorption of malathion and methylparathion by tropical aquatic sediments: influence of pH. *Toxicol Environ Chem*. 41: 47–55.

Sujatha, C.H., Nair, S.M. and Chacko, J. 1999. Determination and distribution of endosulfan and malathion in an Indian Estuary. *Water Research*. 33(1):109–14.

Takeoka, H., Ramesh, A., Iwata, H., Tanabe, S., Subramanian, A.N., Mohan, D., Mahendran, A. and Tatsukawa, R. 1991. Fate of the insecticide HCH in the tropical coast area of South India. *Mar Poll Bull*. 22:290–7.

Tamhankar, A.J., Rajendran, T.P. and Mamdapur, V.R. 1995. Novel synthesis of cotton pest pheromones and their field evaluation studies. In: Krishnaiahand, K. and Diwakar, B.J. (eds) *Pheromones in Integrated Pest Management*. Hyderabad, India: Plant Protection Association of India, pp. 43–4.

Tanabe, S. 1991. Fate of persistent chemicals in the tropics. *Mar Poll Bull*. 22:259–60.

Tanabe, S., Senthilkumar, K., Kannan, K. and Subramanian, A.N. 1998. Accumulation features of polychlorinated biphenyls and organochlorine pesticides in resident and migratory birds from South India. *Arch Environ Contam Toxicol.* 34(4):387–97.

Tanabe, S., Subramanian, A.N., Ramesh, A., Kumaran, P.L., Miyazaki, N. and Tatsukawa, R. 1993. Persistent organochlorine residues in dolphins from the Bay of Bengal, South India. *Mar Poll Bull.* 26:311–16.

Venkateswarlu, K., Gowada, T.K.S. and Sethunathan, N. 1977. Persistence and biodegradation of carbofuran in flooded soils. *J Agric Food Chem.* 25:533–6.

Venkateswarlu, K. and Sethunathan, N. 1979. Metabolism of carbofuran in rice straw amended and unamended rice soils. *J Environ Qual.* 8:365–8.

Venugopalan, V.K. and Rajendran, N. 1984. Pesticide pollution effects on marine and estuarine resources. DAE Research Project Report. Parangippettai, India: Centre for Advanced Study in Marine Biology, Annamalai University, pp. 1–316.

Wahid, P.A. and Sethunathan, N. 1979. Involvement of hydrogen sulphide in the degradation of parathion in flooded acid sulphate soil. *Nature.* 282:401–2.

Chapter 6

Pesticides in Bangladesh

M. A. Matin

INTRODUCTION

Bangladesh is a south Asian country of 144,000 km² with 120 million inhabitants and a population growth rate of 2.2 percent per annum. Topographically it is a vast riverine delta, situated at the apex of the Bay of Bengal with a coastal plain of >3,400 km² and a coastline of 710 km (Matin, 1995).

Bangladesh's economy is agriculture-based with 10 M ha under cultivation. Arable land per capita (<0.13 ha) is the world's lowest with a cropping intensity of 159 percent (i.e. land produces 1.59 crops per year), equivalent to >15 M ha cultivated. Agriculture contributes 35 percent to GDP and is growing at an annual rate of 1.9 percent. Obviously fallow land that can be brought under cultivation is minimal in Bangladesh (less than 2 percent), and that small amount is in jeopardy due to pressure from population growth.

Rice *Oryza sativa* L. (Gramineae) is the principal food crop in Bangladesh as it is in other South and Southeast Asian countries (Abdullah *et al.*, 1997). Rice accounts for about 75 percent of the total cropped area, with about 4 M ha of HYV and about 6.7 M ha of local varieties. The country is almost self-sufficient in production of cereals but has a shortfall in many other food crops. Priority at the state level is given to enhancing food production, especially of cereals and pulses (seed-bearing leguminous crops, e.g. beans, peas, lentils), by utilizing all available means including HYV of rice and intensive fanning practices like irrigation, improved seed, chemical fertilizer, and pesticide application. These agricultural practices have had a positive impact on farm productivity. Bangladesh's food grain production rose to 19.5 M T in 1993 compared with 9.7 M T in 1967 (Rahman *et al.*, 1995). The country is striving for further boosts in agricultural production based on improved management of the entire agricultural sector.

With respect to environmental management and agrochemical control practices and policies in Bangladesh, the current status is disappointing with inadequate and ineffective legislation and enforcement mechanisms. There is a lack of coordination among the various agencies involved in the protection of the environment; however, several national and international agencies have conducted evaluations and put forward recommendations to the government for improvement

(Coastal Environment Management, 1985; Environmental profile: Bangladesh, 1989; Environment Strategy Review, 1991; National Environment Management Plan, 1991).

PESTICIDE REGULATION

History of early pesticide legislation

Synthetic pesticides, which play an essential role as crop protection agents in modern agricultural systems, began to be used in Bangladesh (then, East Pakistan) in the early 1950s. Marketing of commercial pesticides, mainly OC and OP insecticides, was administered through government agencies and departments. Under the 'Pesticide Management Rules of Pakistan' crop protection chemicals were distributed free of cost (fully subsidized), with spraying services and equipment, during the years 1950 through 1971. The Pesticide Control System in Pakistan charged the Agriculture Extension Department with responsibility for pesticide registration and issuing licences. Guidelines for registration and the requisite enforcement mechanisms were not well developed.

Current pesticide regulations

Bangladesh was founded in 1971 when East Pakistan split away from Pakistan. The Pesticide Ordinance, 1971, of Bangladesh was its first law regulating uses of pesticides (Government of the People's Republic of Bangladesh, Ministry of Law and Justice, 1984). In its original form the ordinance is a virtual copy of the old Pakistan Pesticide Rules. The Pesticide Ordinance was modified at various times through 1984 to accommodate new pesticides. Subsequently the Pesticide Rules, 1985, was enacted and published in the *Bangladesh Gazette*-Extraordinary on 16 November 1985, in exercise of powers conferred under section 29 of the Pesticide Ordinance, 1971 (*Bangladesh Gazette*, 1985). These laws form the legal framework for regulating pesticide uses and associated affairs in Bangladesh.

Under the provisions of the Pesticide Ordinance, 1971 and the Pesticide Rules, 1985, a technical advisory committee advises the government on technical matters arising out of the administration of these laws and performs any other functions assigned to it by law. The committee consists of a chairman and a number of members including government officers and representatives of the pesticide industry. The Pesticide Rules Director, in the Plant Protection wing of the Department of Agriculture Extension, or any person authorized by him in writing, constitutes the Pesticide Registration Authority, and also functions as the Pesticide Licensing Authority. An analytical laboratory, still poorly organized and staffed, is also located in the Plant Protection wing of the Department of Agriculture Extension and provides technical services for testing commercial formulations and technical pesticide products. Field tests and efficiency trials may be required in certain cases.

Government subsidization, provided during the East Pakistan period, continued during the early years of the Bangladesh government. Changing circumstances and economic considerations led the Bangladesh government to withdraw subsidies beginning in 1974, and ending them in 1979. At that time pesticide marketing was turned over to the private sector. As a consequence of the subsidy withdrawal, there was a sharp decline in consumption of pesticides. However, pesticide uses continued to be controlled by the government. Then in the early 1980s, agricultural use of pesticides gradually increased.

Regulatory provisions describe pesticide registration licensing formalities, although how closely the regulatory mandate is adhered to is a matter of debate. Regulatory provisions are clearly inadequate and need amendment in the areas of use, labeling, and residues. The system of regulation places emphasis on control mechanisms, which are regrettably very poorly enforced. Ideally to ensure safe and effective use of pesticides, use application and application instructions should be evaluated by an independent scientific panel. They should consider data collected by the manufactures' own research laboratories and other available data concerning toxicity, persistence behavior, and the sensitivity of analytical techniques with respect to formulated products and residues. Bangladesh's regulatory mechanisms are inadequate in this area. Additionally some control must be exercised over residue levels permitted on foodstuffs. Because it is impossible to test all farm produce, this approach requires establishment of regulations setting residue limits. If exceeded in marketed foods, legal action can be taken against the offending farmer and the condemned produce destroyed. Such regulations are meaningless unless the government employs trustworthy inspectors and analysts with an adequate laboratory infrastructure. Neither the present regulatory framework nor the enforcement mechanisms are currently exercised. This situation must improve to ensure safe use of crop protection chemicals.

Under the regulatory provisions, the registration of a pesticide remains valid for a period of three years (until 30 June of the third year following the year of registration). However, the government may cancel the registration following a hearing if it believes the registration was secured fraudulently; the pesticide is ineffective; or the pesticide is hazardous to vegetation (other than weeds), humans, or animal life. A license may be issued by the licensing authority to any person or business intending to import, manufacture, formulate, repack, sell, offer for sale, hold in stock for sale, engage in a pest control operation on a commercial basis, or advertise any brand of registered pesticide. A license, unless suspended or canceled, remains valid for a period of two years from the date of issue and, on payment of such fees as may be prescribed, may be renewed for a like period. Regulations regarding adulteration of pesticides are provided in the legal mandate. Any pesticide found to be adulterated; incorrectly or misleadingly tagged, labeled, or named; or its sale contravenes any provision of the Rules or Ordinance may be prohibited from further importation by publication of a notice to that effect in the official *Gazette* (*Bangladesh Gazette*, 1985).

PAST AND CURRENT PESTICIDE USAGE PATTERNS

Until recently, rice accounted for more than 80 percent of the quantity of pesticides consumed in Bangladesh. Tea, sugarcane, and potato were next in importance, but with limited use. The use of pesticides has now been extended to vegetables, oilseeds, fruits, tobacco, and other crops.

Synthetic pesticides were distributed free of cost along with spraying services and equipment by the Agriculture Department during the 1950s and 1960s. Figure 6.1 shows pesticide consumption for granular and conventional products since 1972 and 1973. With the lifting of subsidies beginning in 1974 and ending in 1979, there was a sharp decline in the consumption of pesticides. Subsequently use of pesticides increased gradually. During subsidies the consumption of liquid and other conventional products were three times greater than granular products. Withdrawal of the subsidy resulted in reduced consumption of liquid products and higher usage of granular formulations. High consumption of liquid pesticide formulations in the early years of plant protection activities was primarily because there was free distribution of products, lending of spray equipment, and sharing of responsibility for application by the Plant Protection wing of the Department of Agriculture Extension. During epidemics of specific pests, e.g. the rice ear-cutting caterpillar *Mythimna separata* Walk. (Lepidoptera: Noctuidae) or rice hispa *Dicladispa armigera* Olivier (Coleoptera: Chrysomelidae), government applicators conducted aerial spraying. With the introduction of HYV, use of pesticides increased

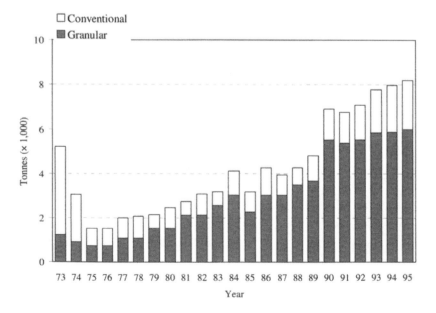

Figure 6.1 Pesticide consumption in Bangladesh from 1973–95

beginning in the early 1980s. The 'Grow More Food' campaign, beginning in the early 1980s, also promoted increased consumption of pesticides.

Currently there is increased use of granular insecticides. The reasons include ensured efficacy, a longer protection period, the scarcity of workable spray equipment, and 'ready to use' formulations. About 6,000 T of granular products were used in 1994 compared with 1,220 T liquid (EC or emulsifiable concentrate) formulations and less than 100 T soluble (SP) or wettable powder (WP) formulations. A review of the consumption of insecticides, herbicides, and fungicides finds that Bangladesh is predominantly an insecticide market (Figure 6.2). However, fungicides and herbicides are steadily gaining market share. By 1994, 500 T of fungicides and herbicides were used annually in Bangladesh. Fungicides are primarily used for cash crops such as potato, but fungicide use has also expanded in rice production. During 1990 to 1995, the average increase in consumption of 1 percent, 22 percent and 3 percent for insecticides, fungicides, and herbicides, respectively, was mostly due to increased awareness of losses from diseases in field crops. From Figure 6.2, it is evident that OP and carbamate pesticides constitute about 90 percent of the market, while OCs and other types of crop protection chemicals make up the remainder.

Pesticides are used on <10 percent of cropped land with approximately 100 gm a.i. being applied per ha. Pesticides are considered crucial for improving agricultural productivity and to prevent crop loss both pre- and post-harvest. The annual consumption of formulated pesticides for agriculture in Bangladesh is gradually increasing (Matin, 1995; Rahman *et al.*, 1995). Consumption has risen from 2,510 T in 1982 to 1983 and 5,150 T in 1988 to 1989 (Rahman *et al.*, 1995) to 7,800 T in 1993 and just over 8,000 T in 1995 (Matin, 1995). The value of these agrochemicals was 1 billion taka (40 taka = US$1) or US$25 M in 1993.

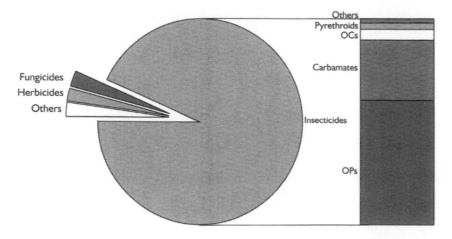

Figure 6.2 Distribution of pesticides classified by function and insecticides by chemical group

Crop protection chemicals have had a positive impact on the increased production of rice (Rahman *et al.*, 1995). However, consumption of pesticides is still low compared to other countries in the region. The real problem with pesticide use may be weak control mechanisms, which may result in widespread misuse and excess application. Thus, there is considerable concern about the misuse of pesticides in Bangladesh. Rice paddies, especially those planted with HYV, require strict pest management through the application of pesticides. HYV rice is vulnerable to various insect pests (Table 6.1) including rice hispa, stem borers *Scirpophaga* sp. and *Chilo* sp. (Lepidoptera: Pyralidae), plant hoppers *Pyrilla perilla perpusilla* Wlk. (Homoptera: Fulgoridae) and leafhoppers (Hemiptera: Cicadellidae), the rice ear-cutting caterpillar, rice case worm *Nymphula depuunctalis* Guen. (Lepidoptera: Pyralidae), and the rice army worm *Spodoptera mauritia* Boisd. (Lepidoptera: Noctuidae) among others. Consequently with such a broad array of potential insect pests to attack the HYV rice, pesticides are required for adequate control (Howlader and Matin, 1988). Per hectare consumption of crop protectants (as formulated products) has gradually increased from 0.35 kg ha^{-1} in 1984 to 0.60 kg ha^{-1} in 1989 and an estimated 0.8 kg ha^{-1} in 1995. Generally insecticide is applied once during the growing season (primarily during the winter season from December to March), but repeated applications may be required to control pest outbreaks and to increase the growing season of the crop.

The population of Bangladesh is estimated to reach 223 million by 2030 with little additional land available to boost food production. Presently over 10 M ha of agricultural land are used for rice cultivation, mostly by small farmers. Rice production, which was 9.7 M T in 1967, increased to about 20 M T in 1994 through intensive farming techniques and the use of HYV rice (Matin, 1995).

Table 6.2 lists the active ingredients of pesticides registered for use in Bangladesh. Of 39 listed insecticides, 24 are labeled for rice. Table 6.3 sets forth the pesticides that are used for rice cultivation in Bangladesh and their application rates. It is apparent from the table that most rice pesticides, none of which are OCs, are insecticides with only a few being fungicides. While not registered, some persistent OCs are illegally applied because of weak regulatory controls and inadequate surveillance programs. Heptachlor, however, is registered in Bangladesh for other agricultural purposes (sugarcane cultivation). OPs constitute the greater part of insecticides (60 percent); carbamates (28 percent); and pyrethroids, OCs, and others (12 percent). Of the commonly used rice insecticides, diazinon, carbofuran, malathion, chlorpyrifos, fenthion, fenitrothion, carbaryl, dimethoate, dichlorvos, and fenvelarate find wider application in rice field ecosystems. Herbicides are not generally used for the rice paddy field although they are used for other crops such as tea. Of the 21 fungicides and 10 herbicides registered for use in Bangladesh only three fungicides find application in the rice paddy field.

Geographically Bangladesh is divided into five regions, Dhaka, Chittagong, Rajshahi, Khulna, and Barisal. Pesticide consumption for each region from four time periods is shown in Table 6.4. For the country, per ha consumption of pesticides gradually increased from 0.35 kg ha^{-1} in 1984 to 0.60 kg ha^{-1} in 1989 and was estimated as 0.80 kg ha^{-1} in 1995. Annual consumption (kg ha^{-1} yr^{-1}) of pesticides

Table 6.1 Principal pests of important crops in Bangladesh

Name of the crop	Common name of the pest	Scientific name
Rice	Rice hispa	*Dicladispa armigera* Olivier
	Gold-fringed borer	*Chilo auricilius* Dudgeon
	White rice borer	*Tryporyza innotata* Walker
	Pink borer	*Seasamia inferens* Walker
	Stem borer	*Chilo partellus* Swinhoe
	Yellow stem borer	*Scirpophaga incertulas* Walker
	Pale headed stripped rice borer	*Chilo suppresalis* Walker
	Rice case worm	*Nymphula depuunctalis* Guenée
	Rice ear-cutting caterpillar	*Mythimna separata* Walker
	Rice army worm	*Spodoptera mauritia* Boisduval
Wheat	Termite	*Microtermes anandi* Holmgr.
	Wheat aphid	*Rhopalosiphumn rufiabdominali* Sasaki
Tea	Red spider mite	*Oligonychus coffeae* Nietner
	Shoot-hole borer	*Xyleborus* sp.
	Trank borer	*Heterobostrychus aequalis* Waterhouse
	Tea mosquito bug	*Helopeltis* sp.
Jute	Jute hairy caterpillar	*Diacrsia obliqua* Walker
	Jute semi-looper	*Anomis sabulifera* Guenée
	Jute stem weevil	*Apion corchori* Marshall
Sugar cane	Top shoot borer	*Scirpophaga monostigmata* Zeller
	Stem borer	*Scirpophaga exerptralis* Walker
	Sugar cane top shoot borer	*Scirpophaga auriflua* Zeller
	Sugar cane stem borer	*Chilo tumidicostalis* Hampson
	Leafhopper	*Pyrilla perilla perpusilla* Walker
Pulses	Pod weevil	*Bruchus pisorum* L.
	Pod beetle	*Pachymerus chinensis* L.
	Leaf caterpillar	*Chaetochema cohcicpennis* Bally
	Gram weevil	*Alcides colloris* P.
	Azuki bean weevil	*Callosobruchus chinensis* L.
	Pulse beetle	*Gonocephalum elongatum* Fabricius
Stored grains (rice and wheat)	Rice weevil	*Sitophilus oryzae* L.
	Angoumois grain moth	*Sitotroga cerealella* Olivier
	Red flour beetle	*Tribolium castaneum* Herbst
	Grain weevil	*Sitophilus granarius* L.
	Depressed flour beetle	*Palorus subdepressus* Wollaston

was highest in Chittagong (having a higher population density and more extensive farming practices) and lowest in Dhaka (land much more devoted to industrial enterprises). Because there is little additional land to boost food production for an expanding population, multiple cropping or intensive farming practices, involving HYV use, are necessary, thus increasing pesticide consumption from multiple applications of crop protection chemicals.

Table 6.2 List of registered pesticides (a.i.) in Bangladesh

Acaricides	Bromopropylate	Sulphur
	Dicofol	Propargite
	Ethion	Tetradifon
	Fenbutatin oxide (Hexakis)	
Fungicides	Chinomethionate (Oxythioquinox)	Triadimefon
	Carbendazim	Triadimenol
	Carboxin + Thiram	Thiophanate-methyl
	Edifenphos	Thiabendazole
	Copper oxychloride	Aluminum phosphide
	Iprodione	Methyl bromide
	Mancozeb	Cufraneb
	Methacrifos	Propineb
	Metiram	Propiconazole
	Primiphos-methyl	Metalaxyl
	Tridemorph	
Rodenticides	Flocoumafen	Coumatetralyl
	Brodifacoum	Zinc phosphide
	Bromadiolone	
Herbicides	Glyphosate	Dazomet
	Terbuthylazine	Diuron
	Glufosinate-ammonium	Paraquat
	2,4-D	Propanil
	Dalapon Na	Oxadiazon
Insecticides	BPMC (fenobucarb)	Fenitrothion
	Carbaryl	Fenvalerate
	Carbofuran	Esfenvalerate
	Carbosulfan	Formothion
	Cartap	Isazofos
	Chlordane	Heptachlor
	Chlorpyrifos	Malathion
	Chlorpyrifos methyl	Monocrotophos
	Cypermethrin	Methamidophos
	Cyfluthrin	Phenthoate
	Fenpropathrin	Pirimicarb
	Deltamethrin	Quinalphos
	Diazinon	Tetrachlorvinphos
	Dichlorvos	Trichlorfon
	Dimethoate	Isoprocarb
	Endosulfan	Methyl demeton
	Etofenprox	Phosalone
	Fenthion	Phosphamidon

PESTICIDE RESIDUES IN BIOTA AND ABIOTIC MATRICES

Pesticide residues originate from application of the formulated a.i.(s) to crop fields and may be transported offsite through spray drift and runoff aided by rain, floods, tidal surge etc. (Hassall, 1990). Consequently contamination of canals, ponds,

Table 6.3 Pesticides recommended (registered) for rice fields in Bangladesh with application rates[a]

Common name	Concentration/formulation[b]	Rate (ha⁻¹)
Insecticides		
BPMC (fenobucarb)	50 EC	1.0 L
Carbaryl	85 WP	1.4 kg
Carbofuran	3 G	16.7 kg
Carbosulfan	20 EC	1.5 L
Cartap	10 G	16.8 kg
Chlorpyrifos	20 EC	1 L
Diazinon	10 G	16.8 kg
	60 EC	1.5 L
	14 G	13.5 kg
Dichlorvos	100 EC	560 ml
	50 EC	1 L
Dimethoate	40 EC	1.2 L
Etofenprox	10 EC	0.5 L
Fenthion	50 EC	1.5 L
Fenitrothion	50 EC	1.12 L
Fenvalarate	20 EC	250 ml
Isazofos	3 G	16.8 kg
Formothion	25 EC	1.12 L
Malathion	57 EC	1.12 L
Monocrotophos	40 WSC	1.5 L
Isoprocarb	75 WP	1.5 L
Phosalone	35 EC	1.0 L
Phosphamidon	100 SL	0.5 L
Quinalphos	5 G	16.3 kg
	25 EC	1.5 L
Tetrachlorvinphos	75 WP	1.12 L
Primiphos methyl	50 EC	1.0 L
Fungicides		
Edifenphos	50 EC	840 ml
Pyroquilon	50 WP	600 g
Thiophanate methyl	70 WP	2.4 kg

Notes:
a Adapted from Plant Protection Wing, Department of Agriculture Extension, Khamar Bari, Farmgate, Dhaka, Bangladesh.
b Formulation abbreviations are as follows: G, granular; WSC, water soluble concentrate; SL, slurry.

rivers, and other waterways with residues of the parent compounds or their degradation products is possible (Rahman, 1995). Furthermore, food harvested from the application sites and elsewhere may be contaminated with residues (Rahman, 1995). Fish cultured or living in the fields or nearby and fish in downstream waterways can be affected by toxic residues (Abdullah *et al.*, 1997). Fish are the major non-target species adversely affected by application of hazardous pesticides (Abdullah *et al.*, 1997). Populations of both flora and fauna have been reported to be experiencing significant declines due to application of pesticides in Bangladesh and elsewhere (Hassall, 1990). Many other undesirable side-effects may appear in the aftermath of repeated pesticide application (Hassall, 1990; Rahman, 1995). Organochlorine and pyrethroid insecticides, in general, may cause damage to fish and other non target species (Hassall, 1990).

In Bangladesh a complete risk assessment of the use of pesticides, especially in the rice paddy ecosystem, has yet to be made. Scattered published reports coupled with some preliminary findings (since 1992) by the Institute of Food and Radiation Biology of the Bangladesh Atomic Energy Commission suggest misuses of crop protection chemicals in Bangladesh. This is due to a lack of or inadequate enforcement of regulatory measures. Unregistered compounds find application in agriculture due to poor enforcement of the relevant law by authorities (Matin *et al.*, 1995). Unlawful use of DDT to treat pest infestations in dried fish has been detected and residue levels were found to be high (IAEA, 1995; Matin *et al.*, 1995). In subsequent studies using dried fish treated with ^{14}C- DDT, Matin *et al.* (1996) found that, luckily for consumers, most of the applied DDT remains on the surface of the sun-dried fish and also that most of the residue could be eliminated during pre-cooking processing using traditional household preparation techniques (Figure 6.3).

Table 6.4 shows the range of pesticide residues (including toxic metabolites) measured in food and environmental samples from Bangladesh. Although OC insecticides are not registered, residues are found in different components of the

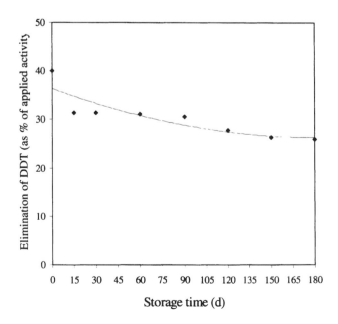

Figure 6.3 Removal of DDT by water washing of dried fish treated with ^{14}C-DDT

Table 6.4 Region-wise crop area and pesticide (formulation) consumption in Bangladesh

Region	1984–85		1985–86		1988–89		1994–96	
	Crop area (×1000 ha)	Annual pesticide consumption (kg ha^{-1})	Crop area (×1000 ha)	Annual pesticide consumption (kg ha^{-1})	Crop area (×1000 ha)	Annual pesticide consumption (kg ha^{-1})	Crop area (×1000 ha)	Annual pesticide consumption (kg ha^{-1})
Dhaka	2,227	0.27	2,239	0.33	2,259	0.40	2,497	0.60
Chittagong	1,900	0.50	1,936	0.63	1,985	0.78	2,289	0.92
Rajshahi	2,565	0.32	2,600	0.39	2,751	0.54	2,964	0.74
Khulna-Barishal	1,957	0.34	1,995	0.39	2,178	0.58	2,375	0.80
Bangladesh	8,649	0.35	8,770	0.42	9,173	0.58	10,125	0.77

ecosystems indicating continued illegal use of persistent OC insecticides. In addition to heptachlor and lindane, DDT and its metabolites were detected in water, sediment, and fish muscle samples from various sources. Dieldrin and endosulfan were also found in a small number of samples. However, OC residues were found at low and, in general, within acceptable levels (WHO, 1993).

More than 200 samples from natural waters throughout the country including rice-paddy depressions, rivers, canals, and ponds in addition to sources of irrigation and drinking water were analyzed. Slight to moderate pollution levels from OC residues were found (Matin *et al.*, 1998), but no OP or carbamate residues could be detected, despite their predominance among chemicals registered for use around these water sources.

Fish, sediment, and cleaned rice samples were analyzed to develop baseline data for pesticide residue levels. However, no report of residues from the predominantly used pesticides (OPs and carbamates) was found despite detection of persistent OCs in many samples. From Table 6.5, among the DDT family, p,p'-DDT, p,p'-DDE, and p,p'-DDD were most often observed in various food and environmental samples and, in some cases, lindane in sediment samples, and endosulfan and dieldrin in fish samples were detected (Flood Action Plan, 1992).

ECOTOXICOLOGY OF PESTICIDES

Agricultural practices for pesticides may lead to various effects in plant and animal species under cultivation and in the wild. The aquatic environment including marine coastal and estuarine waters support the growth of fish and other biota that may be affected by pesticide residues transported into both inland waterways and coastal waters through flooding, rainfall, and runoff. Fish migrate to crop fields during flooding and may be exposed to pesticide residues. Submergence of rice plants (with pesticide application) in water and runoff from rice paddies to adjacent ponds and waterways may have adverse effects on aquatic biota, including fishes. This may result in contamination of both terrestrial and aquatic environments. Two to three crops of rice including both high yield and local varieties would increase the toxicological burden in the rice cultivation ecosystem.

Fish are regarded as good indicator species for evaluating pesticide toxicity in aquatic ecosystems. The susceptibility of different fish species to pesticides varies widely. The toxicological impact of pesticides is a reduction in species richness. Pesticides may cause detrimental effects in other biota. Of particular concern is pesticide toxicity to the natural predators of pest species. Many pest insect predators, such as coleopteran, hymenopteran, and arachnid species can be adversely affected by pesticide toxicity (Abdullah *et al.*, 1997). Also, snakes, other reptiles, and amphibians may be adversely affected by pesticide residues. A noticeable population reduction among different fish species coupled with declines in predator populations are matters of grave concern.

OCs represent the major class of synthetic insecticides used in Bangladesh from the 1950s to the early 1990s. DDT, lindane, aldrin, dieldrin, endrin, heptachlor,

Table 6.5 Pesticide residues in food and the environment in Bangladesh

Location	Sample/ components	Year	Compounds detected	Concentrations (range/mean value)
Tangail/Sirajganj FAP embankment	Rice-field fish	1992	p,p´-DDT	1.10–24.9 mg kg⁻¹
			p,p´-DDE	0.75–12.0 mg kg⁻¹
			p,p´-DDD	0.05 mg kg⁻¹
			Endosulfan	Trace–281.88 mg kg⁻¹
River, ponds, major rice growing fields rice field runoff	Fish	1993	p,p´-DDT	7.86–142.66 mg kg⁻¹
			o,p´-DDT	1.76–38.00 mg kg⁻¹
			p,p´-DDE	9.60–101.12 mg kg⁻¹
Dhaka (Narshindi) Chittagong	Rice	1993	p,p´-DDT	4.12–276.29 mg kg⁻¹
			o,p´-DDT	0.45–1.80 mg kg⁻¹
			p,p´-DDE	2.19–7.62 mg kg⁻¹
			p,p´-DDD	0.08 mg kg⁻¹
Mymensingh Narsinghdi (paddy field)	Sediment	1993	p,p´-DDT	0.71–10.05 mg kg⁻¹
			o,p´-DDT	0.38–2.43 mg kg⁻¹
			p,p´-DDE	1.48–2.26 mg kg⁻¹
			Lindane	22.90–30.42 mg kg⁻¹
Major rice growing regions	Water (paddy field, irrigation canal, adjoining river and ponds)	1994–95	Dieldrin	0.64 mg L⁻¹
			p,p´-DDT	0.06–19.60 mg L⁻¹
			p,p´-DDE	0.01–2.51 mg L⁻¹
			p,p´-DDD	0.01–0.37 mg L⁻¹
			o,p´-DDT	0.01–0.26 mg L⁻¹
			Lindane	0.23–0.55 mg L⁻¹
			Heptachlor	0.025–1.020 mg L⁻¹
Dhaka (Savar, Manikganj)	Vegetables (leafy vegetables), cabbage, beans, brinjal, peas	1993	p,p´-DDT	0.231–4.75 μg kg⁻¹
			p,p´-DDE	0.555–2.74 μg kg⁻¹
			o,p´-DDT	0.294–0.788 μg kg⁻¹
	Fruits	1993	o,p´-DDT	0.065–0.31 μg kg⁻¹
			p,p´-DDT	2.52–7.74 μg kg⁻¹
			p,p´-DDE	1.101–28.20 μg kg⁻¹
	Pulses	1993	p,p´-DDT	5.87–38.43 μg kg⁻¹
			o,p´-DDT	0.326–1.19 μg kg⁻¹
			p,p´-DDE	1.233–3.854 μg kg⁻¹

endosulfan, chlordane, and toxaphene were among those used in the agricultural and public health sectors. The contribution of DDT and lindane to malaria control through eradication of vector insects during the 1980s was significant to improved public health. Although such persistent compounds are no longer registered, their past use may already have caused ecological problems. Fish populations are reduced, many species of birds are affected, and species richness has declined in Bangladesh's major ecosystems (Matin, 1995). Although the exact causative factors are not yet known, the use of pesticides may be related to recent occurrences of ulcerative diseases in fresh water fish species (Flood Action Plan, 1992; Matin *et al.*, 1997b). The interaction of pesticides and aquatic habitats has not been studied in Bangladesh. Published information on the concentrations of pesticide residues and their effects on components of the country's major ecosystems is very limited.

Marine microcosm experiments conducted with ^{14}C-DDT showed no effect on snails or algal species that were present in the ecosystem. DDT applied to sea water in aquaria was found to distribute between sediment and biota. More than 30 percent of the applied DDT volatilized from the system (Matin *et al.*, 1997b).

OPs do not accumulate in the food chain or in mammalian tissues, and chronic effects are minimal (Cremlyn, 1980). Many of this class of compounds function as systemic insecticides enabling applicators to utilize less a.i. and thereby reducing the harmful effects on natural insect predators (Hassall, 1990). OPs, if applied carefully and correctly, are unlikely to cause serious harm to non-target organisms and are unlikely to bioaccumulate or biomagnify and, thus, may not harm the environment (Cremlyn, 1980; Hassall, 1990). Malathion labeled with ^{14}C was applied to a rice fish model ecosystem and was found to cause minimal effects in non-target species. Fish were unaffected, and residues on the rice grain and abiotic components of the ecosystem were virtually absent (Islam, 1996). Controlled experiments in marine microcosms were conducted with ^{14}C-chlorpyrifos and showed no adverse effects on marine organisms (mussels and algae) but low-level accumulation of ^{14}C-activity in organisms was noted (Matin *et al.*, 1997c).

Carbamate compounds are similar to OPs in their effects on beneficial organisms (Hassall, 1990). Carbofuran applied to a rice fish model ecosystem was found to degrade to non-toxic compounds within five days of application (Hoque, 1994). Indian catfish *Heteropneustes fossilis* Bloch introduced into the ecosystem five days after application of the carbofuran grew without showing any untoward effects, residue in the rice grain was negligible, and no effects were observed in microflora (Hoque, 1994). Aquatic organisms appear to be more vulnerable to pyrethroid insecticides, but fortunately very little of pyrethroid compounds are used in Bangladesh at the present time.

DISTRIBUTION AND FATE OF PESTICIDES IN MODEL MICRO-ECOSYSTEMS

Recently there have been attempts to develop aquatic, terrestrial, and mixed model ecosystems to assess the dispersal and degradation of pesticides using radio-labeled compounds. These ecosystems were designed to show how pesticides behave in the environment and to predict ecotoxicological effects (International Union of Pure and Applied Chemistry, 1985). The concept of model ecosystems has been developed primarily for aquatic systems. For experiments with flora and fauna in their abiotic environment such model ecosystems are regarded as very useful. Unlike a natural ecosystem, a model ecosystem is generally closed by boundaries and composed of more than two compartments, with at least two of the compartments being biotic and from different trophic levels. These models can be used for controlled laboratory experiments as well as less-controlled outdoor systems. Sediment and water are necessary abiotic components of model ecosystems used for studying distribution and fate of pesticides (Guth, 1991). Microcosms, which generally consist of a variety of microbial and other low-trophic level organisms, are a special group of model ecosystems.

Microcosms are used to investigate the fate (including mobility, transformation, and degradation) of pesticides in the environment and employ a number of different compartments and defined components under various boundary conditions. By careful manipulation of variables in the model system, experimental attempts are made to achieve comparable results on the environmental behavior of a particular pesticide, including metabolic pathways involved in its degradation (IAEA, 1993).

Pesticides can enter the aquatic environment through direct application, spray drift, atmospheric deposition, leaching and runoff from agricultural land, or by the indirect routes of equipment washing and disposal (Howlader and Matin, 1988). In the aquatic environment, pesticides tend to absorb to suspended solids or become bound to sediment, although a small fraction remains in the aqueous phase due to continuous exchange between the sediment and water (Matin et al., 1997b). Pesticides in the aqueous phase can be taken up and stored in aquatic organisms. They are also subject to metabolic transformation in biota and to various physicochemical reactions with abiotic components (Guth, 1991). To evaluate the translocation and interaction between pesticides and biota continuously exposed to the compounds, model ecosystems have proven to be useful (Guth, 1991). Pesticides that enter into the aquatic environment, including rivers, ultimately reach the sea. Published information on the distribution and environmental fate of pesticides in Bangladesh's ecosystems is limited. Hence, model ecosystem experiments using ^{14}C labeled pesticides can provide essential information on the distribution and behavior of pesticides in various components of these ecosystems. Marine microcosm experiments were recently conducted to study the distribution and fate of ^{14}C-DDT in water, sediment, algae, and mussels collected from the Bay of Bengal coast at St Martin Island in southern Bangladesh (Matin et al., 1997b). Samples exposed to ^{14}C-DDT in the model ecosystem were analyzed at 0, 2, 4, and 24 h, 6 d, 14 d and 30 d to determine the fate and distribution of ^{14}C-DDT residues. Techniques utilizing a liquid scintillation counter (LSC) and biological oxidizer (BO) were used to determine the radioactivity representing DDT (or its degradation products) in components of the ecosystem. Thin-layer chromatographic techniques were used to investigate the metabolic transformation of DDT. Matin et al. (1997a) found that ^{14}C-DDT applied in sea water translocated from water to sediment, algae, and mussels at varying rates (Figure 6.4). Exchange of DDT from one component of the ecosystem to another was observed. Algae were found to accumulate DDT more rapidly than mussels and sediment was found to contain substantial activity throughout during the experimental period. The parent DDT was metabolized, in part, to DDE and DDD by both biota and abiotic components. Substantial volatilization of the applied DDT was observed.

Model micro-ecosystem studies were conducted to evaluate behavior of ^{14}C-carbofuran, a common insecticide for rice-paddies, in a rice fish mixed agricultural system (Hoque, 1994). Because fish is regarded as major source of animal protein in Bangladesh, fish aquaculture in rice-paddy fields has been practiced recently to

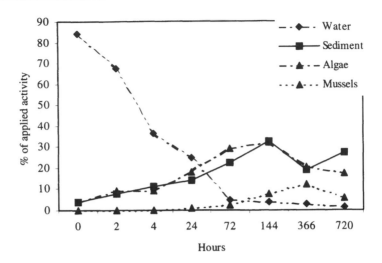

Figure 6.4 Distribution of DDT residues in a marine microcosm

produce more freshwater fish (Hoque, 1994). The study examined the specific conditions under which carbofuran can be used in a rice fish ecosystem without risk of toxicity to the fish being raised in the paddy. It also evaluated the distribution behavior of the pesticide, the occurrence of any acute toxic effects from the pesticide, and whether bioaccumulation of the pesticide occurs in fish and other components in the mixed agroecosystem. He found that 3 μg g^{-1} carbofuran in water remained acutely toxic to catfish (*H. fossilis*) until the fifth day. Thereafter, it was no longer lethal to catfish under the experimental conditions. The highest carbofuran residue concentrations were found in the emergent portion of rice plants; concentrations in the submerged portion were low. In catfish, residues were higher in the viscera than in muscle and no chronic toxicity effects were found. In paddy water and soil, residues were low (0.33 μg g^{-1} and 1.28 μg g^{-1}, respectively).

In a similar investigation with ^{14}C-malathion in a rice fish ecosystem, radioactivity in water, soil, rice plants (submerged and emergent portions), fish, and grain samples was measured (Islam, 1996). Extractable and bound activity in soil was determined using LSC and BO. The highest residues were found in soil on day16 after pesticide application. In catfish, residues were again higher in the viscera (1.66 μg g^{-1}) than that in muscle but in rice the highest residue concentrations were found in the submerged portion. Controlled experiments in marine micro-ecosystems were conducted with ^{14}C chlorpyrifos and showed no adverse effects on marine organisms (mussels and algae) though low-level accumulation of ^{14}C-activity in the marine organisms was observed (Matin *et al.*, 1997c).

Recently, the many oxbow lakes in Bangladesh – a total of 5,488 ha of oxbow lakes exist – have gained importance as a potential fishery resource (Chowdhury and Yakupitiyage, 2000). Consideration is being given to cage culture so that resource-poor fishing villages may more fully utilize the potential of this resource.

This would complement existing stock enhancement programs but would require a unified management system to replace existing dispersed systems under different management bodies. Because the use of agricultural pesticides in the lake catchment is potentially harmful to fish, an integrated pest management program based on the rice fish ecosystem rearing system currently in use (Chowdhury and Yakupitiyage, 2000).

SUSTAINABLE PEST CONTROL THROUGH IPM

IPM calls for limited pesticide use and the use of botanical, biological, mechanical, and cultural means to control insect and disease attacks on food crops. Examples of biological control agents are the pest insect predators such as spiders, parasitoid organisms, and plants with insect repellent or insecticidal properties such as the neem tree (*Azadirachta indica* A. Juss.).

Among Asian counties, Indonesia has been the most successful in implementing IPM, followed by Thailand, Taiwan, and the Philippines. China, India, Bangladesh, and Vietnam are emerging as significant players in the attempt to integrate IPM into their respective agricultural systems. Farmers in at least 17 countries are practicing IPM to some extent, as a way to maintain production and profit levels and to safeguard the environment. Specialists believe that the use of synthetic chemical pesticides, farming costs, accidental fatalities from pesticide exposure, and undesirable impacts on the environment can be significantly reduced through effective implementation of IPM practices.

Under an IPM program, specialists are trained in season-long (equivalent to one cropping season) activities in farmers' field schools, which produce rice or vegetables. IPM specialists, in turn, train individual farmers to become aware of pests and their characteristics, identify beneficial insects, learn plant physiology, perform agro-ecosystem analysis, and learn farm management techniques. These skills enable them to make informed decisions when faced with farming problems, especially pests and diseases.

IPM implementation for rice and vegetable crops in Asia has resulted in dramatic drops in the use of insecticides and fungicides in recent years, lowering use by more than 35 percent in the Philippines, 49 percent in Indonesia, 15 percent in China, 14 percent in Vietnam, and 29 percent in India (Ramaswamy, 1995). Labor costs have likewise dropped in all countries practicing IPM because the average number of pesticide applications for one rice cropping season fell from a high of 20 applications to as low as two (Ramaswamy, 1995). Indonesia has banned 35 pesticides, while four have been banned in the Philippines (Ramaswamy, 1995). India has imposed an excise duty on chemical companies (Ramaswamy, 1995) and the Bangladesh Government's Technical Advisory Committee on pesticides has recently moved to cancel the registration of the more harmful compounds still in use, while restricting the retail sale of all aluminum phosphides and temporarily suspending pyrethroid pesticides (Matin *et al.*, 1998). How well these publicized bans, suspensions, and restrictions will work in the prevalent socioeconomic context remains to be seen. However, it should be pointed out that many Asian countries still use and misuse many deadly and environmentally harmful pesticides.

The brightest contribution of IPM programs has been the reduction in human and environmental poisoning. Further intensification of IPM training activities and implementation and strengthening of pesticide regulations in Asian countries is needed. Research on IPM needs to be accelerated and tailored to meet farmers' needs. Researchers, agricultural extension agents, and IPM training personnel must learn from and with farmers. The IPM program in Asia is being spearheaded by FAO with funding from the Asian Development Bank, the US Agency for International Development, and the national governments of countries involved in the program. IPM activities are required to be coordinated to address all relevant issues of pest control and safe use of agrochemicals. The issues in Bangladesh include the provisions for funding research on biological and botanical pesticides, tax levies on chemical companies based on the 'polluter pays' principle (proceeds from which may go to promotion of IPM), formulation of national policies to reduce or eliminate pesticide use and instead put priority on human health and environmental safety issues, the banning of OC and some OP pesticides, and the review of maximum permissible limits for pesticide residues.

IPM activities for rice in Bangladesh needs coordination and planning for expansion. Certainly a good start has been made by Bangladesh. There is a huge area under rice cultivation (approximately 10 M ha) and a very large number of farmers (>2.0 M) involved in rice cultivation in Bangladesh. The concepts and components of IPM have been successfully assembled and packaged in a practical way by Dr Peter E. Cenmore and his staff from FAO's Inter-country Program for IPM in rice in South and Southeast Asia. The core of IPM activities involves teaching the principles and practices of IPM to farmers through practical field training in such a way that poor and illiterate farmers do not merely *learn* IPM but are able to successfully practice it on their farms. The Department of Agriculture Extension, Khamar Bari, Dhaka, Bangladesh and the FAO Inter-country IPM Program endeavored to introduce this practical IPM training to the rice farmers of Bangladesh beginning in late 1988.

For coordination and effective implementation of IPM activities, eight IPM regions were selected in Bangladesh based on the quantity of pesticides used and on the intensive nature of rice cultivation. The regions are Dhaka, Comilla, Chittagong, Jessore, Bogra, Mymensingh, Patuakhali, and Thakurgaon. In each IPM region, three to 12 farmers' field schools, one school per Thana (sub-district), were established. An IPM field school consists of about 20 ha of rice field called the 'IPM Plot' and a 2 ha rice field called the 'farmers' practice plot'. Each school trains 50 selected local rice farmers. Each school is run by a local (grass-roots level) extension officer such as the Block Supervisor (BS), Plant Protection Inspector (PPI), or Subject Matter Officer for Plant Protection (SMOPP), and is supervised by the Thana Agriculture Officer (TAO) at the Thana (sub-district) level and the

Subject Matter Specialist at the district level. Technical management and supervisory assistance for each school are provided by IPM Master Trainers (who have received in-depth IPM training both in Bangladesh and abroad) from the Department of Agriculture Extension in Dhaka and the FAO Inter-country IPM Project. So far, 91 farmers' IPM field schools have been established in 74 Thanas in 39 districts. Using the field schools as the training ground, senior extension officers in the eight regions and from all the 64 districts were given IPM exposure and training. In addition, 309 field officers from Non-Governmental Organizations (NGOs) were given a crash course on IPM. Bangladesh now has 4,025 core (trained) rice farmers and an additional 20,750 farmers have been exposed to some IPM training for rice cultivation.

Each IPM field school has conducted a benchmark survey to obtain information on plant protection practices that farmers generally followed, the amount of money they spent on pesticides, the yield of rice they obtained, etc. The results were assembled as a baseline data set to use for comparison between IPM and non-IPM plots. This shows the impact of IPM training and practices during the growing season. Available data shows that 3,000 farmers from 60 field schools spent an average of 887 taka per ha for pesticides during one cropping season before IPM training (Ramaswamy, 1995). The same 3,000 farmers, after IPM field training, managed pests with their acquired IPM skills (allowing the naturally occurring parasites and predators to suppress pest populations; adopting improved cultivation practices such as optimal spacing, optimal fertilizer application, etc.; and controlling pests by mechanical means) while spending only 74 taka per ha on pesticides. Before receiving IPM training and adopting IPM practices, farmers at the 60 IPM field schools, on average, produced 3.70 T of rice ha^{-1} and the same farmers, after IPM training, produced 4.61 T ha^{-1}, or 24 percent more rice from their fields (Ramaswamy, 1995). Even farmers using high production systems (involving HYVs and mechanized cultivation with irrigation and high agrochemical inputs) were able to reduce pesticide use. Without IPM training, farmers are overcautious when under high input, high production systems and tend to use more pesticides under the mistaken idea that their crops can only be protected with heavy applications of pesticides. Under high input systems, untrained farmers apply pesticides on a calendar basis before they even see pests (prophylactic treatment). IPM trained farmers avoid applying pesticides until an economic threshold level of a pest population develops. Comparing rice yields, IPM farmers obtained 4.61 T ha^{-1} while their neighbors, without employing IPM practices, produced only 4.17 T ha^{-1} (Ramaswamy, 1995).

The government of Bangladesh is giving priority to the successful practice of IPM especially for control of the most troublesome rice paddy pests. Parliament members have been advised to support IPM activities, especially for rice. To protect dwindling environmental quality, it is very worthwhile to make IPM training and knowledge available throughout the country, and especially among the 2.0 million rice farmers in Bangladesh (Ramaswamy, 1995).

APPROACHES FOR PESTICIDE MANAGEMENT

The Bangladesh authorities responsible for regulatory control of pesticides are becoming adequately conversant with control systems already in effect in other countries so that they may effectively streamline the guidelines and standards currently in effect to ensure the future safe use of pesticide chemicals. Effective schemes for minimizing the risks associated with the use of pesticides exist already in many countries, including the USA and the European community. United Nations agencies and their established networks are extending cooperation, collaboration, and expert guidance in devising practical steps for the control of pesticides. They are assisting in maximizing pesticides' beneficial role while minimizing risks associated with undesirable levels of residues in food chains and untoward effects on non-target organisms in the environment (Ambrush, 1997).

Two new and vital control systems are expected to be implemented as far as possible in Bangladesh. If a request for registration of a pesticide product is to be approved, the manufacturer must deliver to a panel of independent, conversant scientists a comprehensive data set collected by their research laboratory that covers a wide range of toxicity data, persistence data, and details of the nature and sensitivity of analytical techniques used to collect the data. Also, a level of control will be exercised over residue levels present in food at the time it is offered for sale. Because it is impossible to test all farm produce, this approach requires the establishment of regulations concerning maximum permissible residue limits that must not be exceeded in marketed food. Exceeding these limits will lead to legal action against the offending farmer or trader and destruction of the condemned produce. Obviously these regulations will be meaningless unless the Bangladesh Government establishes a well-equipped laboratory of international caliber and reputation and staffs it with a team of trustworthy analysts and inspectors to oversee the correct use of pest control chemicals.

No pesticide product, or active ingredient, should be registered without limits placed on its use. A pesticide product should be registered for a specific purpose on a particular crop with guidelines to describe the proper manner of application. It is dangerous to have a list of registered products with no statement as to the purpose for which they have been registered. This can become a recipe for disaster if poorly educated farmers can use a product indiscriminately. The choice of pesticide for use for a particular purpose is a highly skilled task and cannot be left to the whim of the man in the field; it must be legislated and properly controlled.

It is essential that laboratories that produce data on pesticide interactions with environmental compartments and residues have quality assurance (QA) and control (QC) procedures that meet the standard criteria of ISO-25. Good Laboratory Practices (GLP) and laboratory Standard Operating Procedures (SOPs) are necessary for reliable and dependable analytical systems and include standardization of facilities for analysis. The reliability of data generated by these laboratories must be assured and internationally accepted.

Supervized field trials must be arranged to supplement a manufacturer's data and to ensure that local climatic and environmental factors are accounted for in registration deliberations. Safety in the use of pesticides is a dynamic challenge and locally generated data must cover formulations in use, use patterns, and cropping systems. Ecotoxicological aspects of pesticide use under a given ecological scenario are an essential requirement for safe use of pesticide chemicals.

Registration authorities must also address issues of impurities in commercial pesticide products to ensure user and environmental safety.

REFERENCES

Abdullah, A.R., Bajet, C.M., Matin, M.A., Nhan, D.D. and Sulaiman, A.H. 1997. Ecotoxicology of pesticides in the tropical paddy-field ecosystems. *Environ Toxicol Chem.* 16(1):59–70.

Ambrush, A. 1997. Main provisions of the International Code of Conduct of the distribution and use of pesticides. In: *Environmental Behavior of Crop Protection Chemicals.* Proc. Int. Symp. on Use of Nuclear and Related Techniques for Studying the Environmental Behavior of Crop Protection Chemicals, 1–5 July 1996. Vienna: IAEA/FAO, pp. 11–34.

Bangladesh Gazette. 1985. Ministry of Agriculture, Government of the People's Republic of Bangladesh. 16 November 1985.

Chowdhury, M.A.K. and Yakupitiyage, A. 2000. Efficiency of oxbow lake management systems in Bangladesh to introduce cage culture for resource-poor fisheries. *Fisheries Management and Ecology.* 7(1–2):5–74.

Coastal Environment Management, ESCAP Study. 1985. Dhaka, Bangladesh.

Cremlyn, R. 1980. *Pesticides: Preparation and Mode of Action.* Singapore: J. Wiley, pp. 210–21.

Environment Strategy Review, World Bank. 1991. Dhaka, Bangladesh: World Bank.

Environmental Profile: Bangladesh, DANIDA Report. 1989. Dhaka, Bangladesh.

Flood Action Plan. 1992. Studies on pesticide residues in floodplain fish species. Report for Government of Bangladesh, FAP-17. Dhaka, Bangladesh: Government Printers.

Government of the People's Republic of Bangladesh, Ministry of Law and Justice. 1984. The Pesticide Ordinance, 1971 (Ordinance nr 11 of 1971). Dhaka, Bangladesh: Government Printers.

Guth, J.A. 1991. Experimental approaches to studying the fate of pesticides in soil. In: Houston, D.H. and Roberts, T.R. (eds) *Progress in pesticide biochemistry*, Volume 1. Singapore: J, Wiley, pp. 85–114.

Hassall, K.A. 1990. *The Biochemistry and Uses of Pesticides*, 2nd edn: *Structure, Metabolism, Mode of Action and Uses in Crop Protection.* London: English Language Book Society and New York: VCH Publishers.

Hoque, E. 1994. Studies on the fate of ^{14}C-carbofuran in rice fish model ecosystem [MSc thesis]. Jahangirnagar University. Savar, Bangladesh. 81 p.

Howlader, A.J. and Matin, M.A. 1988. Observation on the pre-harvest infestation of paddy by stored grain pests in Bangladesh. *J Stored Prod Res.* 24:229–31.

International Atomic Energy Agency (IAEA). 1993. Use of isotopes in studies of pesticides in rice fish eco-systems. IAEA TECDOC-695. Vienna: IAEA.

IAEA. 1995. Inside technical co-operation: Bangladesh-more dried fish not DDT. *IAEA Bull.* 37(3):2.

International Union of Pure and Applied Chemistry. 1985. IUPAC report on pesticides (20): Critical evaluation of model ecosystem. *Pure Appl Chem.* 57(10):1523–36.

Islam, M.N. 1996. Studies on the fate of ^{14}C-malathion in a model paddy-fish ecosystem [MSc thesis]. Jahangirnagar University. Savar, Bangladesh.

Matin, M.A. 1995. Environmental pollution and its control in Bangladesh. *Trends Anal Chem.* 4:468–73.

Matin, M.A., Malek, M.A., Amin, M.R., Khatoon, J., Rahman, M.S. and Rahman, M. 1995. DDT residues in dried fish of Bangladesh. *Nucl Sci Appl.* 4(1):61.

Matin, M.A., Khatoon, J., Rahman, M. and Mian, A.J. 1996. Influence of pre-cooking processing and storage on reduction of DDT residues from dried fish treated with ^{14}C-DDT. *Nucl Sci Appl.* (1–2):17–26.

Matin, M.A, Amin, M.R., Rahman, S., Khatoon, J., Malek, M.A., Rahman, M. and Mian, A.J. 1997a. Studies on pesticide residues in flood plain fish species. *Nucl Sci Appl.* 5(1–2): 55–60.

Matin, M.A., Hoque, E., Khatoon, J., Khan, Y.S.A., Hossain, M.M. and Mian, A.J. 1997b. Distribution and fate of ^{14}C-DDT in microcosm experiments simulating the tropical marine environment of the Bay of Bengal. In: *Environmental Behavior of Crop Protection Chemicals.* Proc. Int. Symp. on Use of Nuclear and Related Techniques for Studying the Environmental Behavior of Crop Protection Chemicals, 1–5 July 1996. IAEA-SM-343/15. Vienna: IAEA/FAO, pp. 279–87.

Matin, M.A., Hoque, E., Khatoon, J., Rahman, M.S., Malek, M.A., Khan, Y.S.A., Hossain, M.M. and Aminuddin, M. 1997c. *Pesticide Residues in the Bangladesh Marine Environment.* Research report presented at the IAEA/IAEA-MEL sponsored 4th Research Coordination Meeting (RCM) of IAEA Cooperative Research Program (CRP) held 16–20 June 1997 in Nairobi, Kenya.

Matin, M.A., Malek, M.A., Amin, M.R., Rahman, S., Khatoon, J., Rahman, M., Aminuddin, M. and Mian, A.J. 1998. Organochlorine insecticide residues in surface and underground water from different regions of Bangladesh. *Agric Ecosystems Environ.* 69:11–5.

National Environment Management Plan. 1991. GOB/UNDP Report. Dhaka, Bangladesh.

Rahman MS, Malek MA and Matin MA. 1995. Trends of pesticide usage in Bangladesh. *Sci Total Environ.* 59:33–9.

Ramaswamy, S. 1995. A summary of Integrated Pest Management activities and their impacts in Bangladesh. Project GCP/RAS/145/NET. Dhaka, Bangladesh: FAO.

WHO. 1993. *Guidelines for Drinking Water Quality,* 2nd edn. Volume 1, (Recommendations). Geneva, Switzerland: WHO.

Chapter 7

Pesticide use in Malaysia
Trends and impacts

Abdul Rani Abdullah

INTRODUCTION

Agriculture has always been an important sector of the Malaysian economy. Although its contribution to Malaysia's GDP has been steadily decreasing over the last few decades (down from 33 percent in 1960 to 12.7 percent in 1996 due predominantly to an increasing emphasis on industrialization), the agricultural sector continues to grow in absolute terms (MADI, 1996; MACA, 1997). In 1985, agriculture accounted for 20.8 percent of GDP with a value of US$4.6 billion. However, in 1996, the contribution to GDP had decreased to 12.7 percent while the absolute value had increased to US$6.6 billion (MADI, 1997).

Malaysia is currently one of the world's primary exporters of natural rubber and the world's primary exporter of palm oil. These together with cocoa, pepper, pineapple, and tobacco comprise the main crops responsible for the growth of this sector. Agriculture has also been an important base for the development of other sectors of the Malaysian economy, particularly the manufacturing sector as exemplified by the food and beverage industry.

The pesticide industry is one of the most important support industries in agriculture. The economic benefits of pesticide use in producing high crop yields and the role of pesticides in the control of disease-borne pests are undeniable. Equally the adverse effects of elevated pesticide residues in water, soil, and crops to man, domestic animals, wildlife, and the environment are well recognized and documented.

In tropical countries like Malaysia, crops such as rice and vegetables are particularly susceptible to the negative impacts of pesticide use (ADB, 1987). This is attributed to the often indiscriminate and intensive use of pesticides associated with these crops. The problem is exaggerated by the inadvertent destruction of the pest's natural enemies, and the emergence of resistant pest strains, the consequence of which is the application of increasingly larger amounts of pesticides. Other crops, including palm oil and rubber, also require intensive use of pesticides, particularly herbicides.

In addition to their use in agriculture, pesticides have also contributed to the control of insect-borne diseases. Pest control programs to improve public health in Malaysia have been primarily directed toward the eradication of mosquitoes. Under the Malaria Eradication Program initiated in 1967, wall surfaces inside homes of malaria-infected areas were sprayed with DDT. Dengue fever was similarly brought under control by large-scale spraying programs using insecticides such as pyrethrins, malathion, and temephos (Abate). Other diseases such as typhus (carried by body lice) and dysentery (carried by flies), once rampant and greatly feared, have been either curtailed or practically eradicated by the application of pesticides in addition to other public-health related strategies. Recently, research has focused on evaluating the efficacy of alternatives to DDT and other OCs – specifically OPs, pyrazoles, and pyrethroids – or controlling disease vectors (Yap *et al.*, 1996; Sulaiman *et al.*, 1999; 2000).

The objective of this chapter is to examine various aspects of pesticide use in Malaysia, including current trends, levels of contamination in the aquatic environment, as well as impacts of pesticide use. In addition, recommendations and suggestions are put forward with respect to both mitigation measures, and essential areas of additional research.

AGRICULTURE AND PESTICIDES

In a relatively short period of time, Malaysia became a major producer of primary commodities and assumed a dominant world position in rubber, palm oil, and cocoa. The location of the major agricultural areas on Peninsular Malaysia is given in Figure 7.1.

Currently oil palm remains the favored crop, while rubber and cocoa have undergone a decline in acreage in recent years (see Figure 7.2) (MACA 1997). In 1996, the number of hectares devoted to oil palm increased by 2.4 percent from the previous year to 2.6 M ha. Indeed, Malaysia is currently the world's leading producer of palm oil at 53 percent of total world palm oil production in 1993 (MADI, 1996).

The decrease in acreage for rubber and cocoa has been attributed to the shortage of labor as well as the conversion of land to other crops, particularly oil palm, and for commercial and residential uses. The area under rice paddy culture has also been on the decline (Figure 7.2). In order to achieve a targeted 65 percent self-sufficiency in rice, there has been an emphasis on increased crop intensity, mechanization, and varietal yield (HYV) improvements. It should also be noted that IPM has been widely promoted for rice paddies and has resulted in a reduction of incidences of pest population explosions as was frequently reported in the 1970s and early 1980s. These incidences particularly related to severe outbreaks of the brown planthopper *Nilaparvata lugens* Stål (Homóptera: Delphacidae) and the white-backed planthopper *Sogatella furcifera* Horváth (Homóptera: Delphacidae) (MACA, 1997).

Although the importance of agriculture is declining, it continues to play an important role in the development strategy of Malaysia. Agriculture's continued

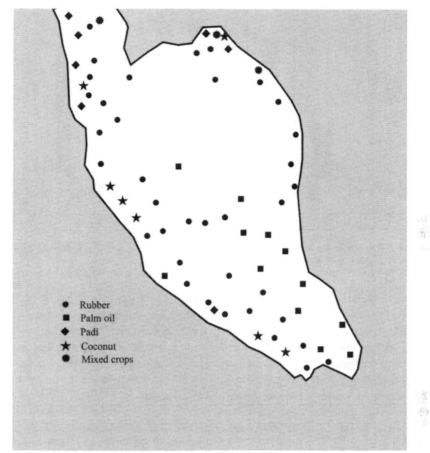

Figure 7.1 The location of major agricultural areas by crop in Peninsular Malaysia
(Yeop, *et al.* 1982)

importance lies in its contribution to the rural economy and to its link with other
sectors of the economy by providing the raw materials for manufacturing and
agronomic-based industries.

A National Agricultural Policy (NAP, 1984) was promulgated in 1984 to serve
as a guideline for Malaysia's agricultural development up to the year 2000. A few
years later, the policy was redefined as the NAP, 1992 to 2010 to emphasize various
strategies such as the optimization of resource use, the development of related
agro-based industries, and the enhancement of research and development activities.

The success of the agriculture sector in Malaysia was achieved by the introduction of sound and effective agricultural policies, coupled with the application of
modern technologies. NAP (1984) was aimed toward greater agricultural productivity, emphasizing higher-value crops such as oil palm, cocoa, vegetables, fruits,
and flowers. Modern practices of large-scale continuous cropping of individual

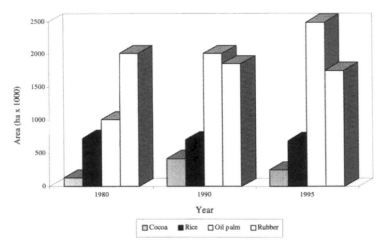

Figure 7.2 Change in the area planted in oil palm, rubber, cocoa, and rice in Malyasia for the years 1980, 1990, and 1995 (MACA, 1997)

crops and the use of high-yielding strains were also encouraged. Many of the benefits obtained from the adoption of monoculture technology and the use of HYV(s) are dependent on effective plant protection strategies. Monocultures tend to encourage rampant population growth of pest species by providing ideal conditions for their development and reproduction. Crop losses have also resulted due to disease and pests encouraged by large-scale planting and genetic uniformity of the crop, a consequence of using a limited range of HYV(s). Large-scale continuous cropping of individual crops also tends to encourage the growth of pest species by providing the necessary conditions for their development. Hence, improved management practices, including more cost-effective pest control strategies, have been introduced.

The majority of the pesticides used in Malaysia are applied in the rubber, oil palm, and rice sectors of agriculture. As can be seen in Table 7.1, herbicides account for 75 percent of the total pesticide market, followed by insecticides (16 percent), fungicides (5 percent), and rodenticides (4 percent) (MACA, 1997). Table 7.2 lists

Table 7.1 Estimate of the pesticide market[a] in Malaysia (US$ M)[b]

Pesticide class	1990	1991	1992	1993	1994	1995	1996
Herbicides	104.5	92	84	80	80.4	88	90.8
Insecticides	17.1	16	16.4	15.6	16.4	17.2	18.8
Fungicides	5.8	5.2	5.2	5.2	5.6	6	6.4
Rodenticides	4.2	4	4.8	4	4.4	4.4	4.4
Total	131.6	117.2	110.4	104.8	106.8	115.6	120.4

Notes:
a End-user value at a constant exchange rate of US$1 = RM2.5.
b Adapted from MACA, 1996; 1997.

some examples of commonly applied pesticides in oil palm, rubber, cocoa, and rice. Herbicides used in Malaysia are predominantly in the form of aqueous concentrates while the majority of insecticides and fungicides used are in the forms of emulsifiable concentrates and wettable powders respectively (Abdullah, 1993).

The use of herbicides has been and will continue to be an important aspect of the crop protection strategy in Malaysia as long as a labor shortage makes manual

Table 7.2 Commonly used pesticides in oil palm, rubber, cocoa, and rice in Malaysia[a]

Crop	Pesticide class		
	Herbicide	*Insecticide*	*Fungicide*
Oil palm	2,4-D dimethylamine Diuron DSMA (disodium methylarsonate) + diuron + dicamba Fluazifop-butyl Glufosinate ammonium Glyphosate Metsulfuron methyl Paraquat Quinclorac, (a quinolinecarboxylic acid herbicide)	Carbofuran Chlorpyrifos Cypermethrin Endosulfan Methamidophos Monocrotophos	Captan Chlorothalonil Maneb Thiram
Rubber	Same as for oil palm	Chlorpyrifos Cypermethrin Dicofol Dimethoate	Hexaconazole, (a conazole fungicide) Propineb Tridemorph Chlorothalonil
Cocoa	Fluroxypyr methyl heptyl ester, (a pyridyloxyacetic acid herbicide) Glufosinate ammonium Glyphosate Oxyfluorfen Sodium chlorate	α-cypermethrin Cypermethrin + chlorpyrifos Deltamethrin Lindane Methamidophos	Captafol Copper oxychloride Hexaconazole Triadimenol
Rice	Propanil Quinclorac 2,4-D butyl ester Bentazone Metsulfuron methyl Oxadiazon	Acephate α-cypermethrin BPMC [2-(1-methyl- propyl)phenyl methylcarbamate] Carbaryl Carbofuran Endosulfan Lindane Diazinon	Benomyl Carbendazim Thiram Flutolanil, (a benzanilide fungicide) Mancozeb

Note:
a Derived from MADI, 1996.

weeding uneconomical. Most herbicide is applied on rubber and palm oil plantations. While older herbicides, e.g. paraquat, glyphosate, glufosinate-ammonium, and 2,4-D, represent the bulk of herbicides used in Malaysia, there has also been an increase in popularity for the use of newer chemicals, e.g. metsulfuron methyl, which require only low concentrations to be effective (MADI, 1996). It is also noteworthy that Malaysia exports a considerable quantity of pesticides, particularly herbicides. In 1993, herbicides worth US$10.8 million, e.g. paraquat, 2,4-D amine, sodium chlorate, glyphosate, diuron, monuron, and linuron, were exported primarily to other countries in the region (MADI, 1997).

The use of insecticides, predominantly in vegetable production, is characterized by the wide variety of available chemicals. As with herbicides, newer, more biologically active and environmentally friendly chemicals are increasingly popular. These chemicals tend to be more costly but are effective at lower concentrations. Fungicides are for the most part imported because their consumption is low, being used mainly in vegetable, fruit, and flower production.

The most common method of applying pesticides in Malaysia is spraying the pesticide solution onto crops with a knapsack sprayer. This application technique has been proven to be inefficient as only about 20 percent of the spray reaches the plants, and less than 1 percent of the chemical contributes to pest control, resulting in wastage and contamination of the environment (MADI 1996). This is attributable to the variable size of the droplets, which tend to coalesce and run off the leaf surface, produced by such sprayers. The fine mist generated by these sprayers also tends to evaporate before reaching the plants. Controlled droplet application (CDA) technology has been introduced to increase the efficacy of pesticides by restricting the droplet size to an optimum range. In Malaysia the use of CDA has not been widespread due mainly to the higher cost of the sprayer but also to higher operational and maintenance costs. However, the use of CDA has become quite common on the larger oil palm plantations (MADI, 1996).

In recent years there has been a gradual decrease in the growth of the pesticide market in Malaysia from an annual increase of 9 percent in 1992 to 3 percent in 1996 (estimated at end-user level) (MADI, 1997). The decreasing trend in pesticide use has been attributed to the introduction of improved products with greater efficacy and selectivity, more judicious application of the pesticides, and the development of biological control and integrated pest management strategies. Some examples of recently introduced pesticides include cyhalofop butyl (an aryloxyphenoxypropionic herbicide), tralomethrin (a pyrethroid ester insecticide), and acetamiprid (a pyridine insecticide). These chemicals exhibit substantial reduction in dosage rates and therefore smaller amounts are applied to treat the same area of cultivated land. Improved pest management practices have also contributed to this trend, including such specialized biocontrol techniques as arthropod-dependent (ant) protection from pests in cashew nut *Anacardium occidentale* (Anacardiaceae) production (Rickson and Rickson, 1998) and parasitoid control of the diamondback moth *Plutella xylotella* L. (Lepidoptera: Plutellidae) in vegetable production (Verkerk and Wright, 1997). New pesticide chemistry, biocontrol techniques, and IPM procedures all contribute to the management of developed/developing resistance in pest species because of improperly or excessively used pesticides. Pesticide resistance studies in Malaysia have most often implicated phase I microsomal monoxygenases and (or) esterases in the development of resistance to such diverse pesticides as the carbamates propoxur and bendiocarb, the OP chlorpyrifos, the pyrethroids cypermethrin and permethrin, the antibiotic pesticide abamectin, the chitin synthesis inhibitor teflubenzuron, and the biological insecticide Bt (Lee *et al.*, 1996; Iqbal and Wright, 1997; Verkerk and Wright 1997). However, Iqbal and Wright (1997) did find some evidence for the involvement of phase II glutathione-S-transferases in decreasing the toxicity of abamectin to *P. xylostella.* Resistance management strategies must account for the development of cross-resistance, the use and timing of synergists, and the management of parasitoids and other bio-control species.

The pesticide industry in Malaysia is made up of about 140 companies – both multinational and local companies – that are involved in manufacturing, formulating, or trading activities (MACA, 1997). The majority of pesticides are imported as technical materials, which are then blended, diluted, or formulated. However, in recent years an increasing variety of pesticides are being manufactured in Malaysia. These include herbicides such as paraquat, sodium chlorate, dalapon, and glyphosate. At present, the quantity of wastes generated by these industries is rather small and generally manageable. Current waste treatment systems include those based on chemical degradation (alkaline hydrolysis, particularly for OPs) and those using oxidation ponds (Samad, 1991).

In addition to chemical pesticides, biological pesticides – best exemplified by the bacterium *Bacillus thuringiensis* (Berliner) (BT) – have been introduced into the Malaysian pesticide market. BT's specificity and versatility, in that different variants can be developed for different pest species, are an important contribution in the promotion of environmentally friendly and sustainable agricultural production.

The pesticide industry is expected to continue to be an important component of the agriculture sector in Malaysia. Pesticides will continue to provide a reliable and cost-effective solution to pest and diseases problems. As the number of hectares of arable land is not expected to increase in the future – indeed, land conversions to other non-agricultural uses have been taking place – the focus in agriculture has been on increasing crop intensity and on yield improvements. Hence, to meet the increasing demands of a growing population, pesticides will continue to make an important contribution to increasing yields and to the reduction of post-harvest losses.

PESTICIDE REGULATORY POLICIES

For many years, pesticide use in Malaysia was controlled by the Ministry of Health under the Poisons Ordinance, 1952 and the Poison List Order, 1970 under which certain highly toxic chemicals are banned from import or manufacture. A Voluntary

Registration Scheme for pesticides was introduced in the 1960s and 1970s but was deemed to be unsatisfactory due to poor and inadequate response by the pesticide industry. This led to the introduction of specific legislative acts governing pesticide use in Malaysia.

At present, the use of pesticides in Malaysia is governed by the Pesticides Act of 1974, which regulates the importation, manufacture, distribution, sale, and use of pesticides in Malaysia. The Act came into force on 1 October 1976 and is administered by the Pesticide Board of the Department of Agriculture. The Act requires that all pesticides be registered. Registration is implemented under the Pesticide (Registration) Regulations, 1976 (Pesticides Board Malaysia, 1991). Data requirements are essentially in accordance with FAO guidelines. The information required to be submitted with the application for registration of new pesticides includes the physicochemical properties of the chemical, its efficacy, storage stability, toxicological data, residue data, known environmental impacts, and a declaration that the pesticide has been approved for use in other countries that practice an acceptable registration procedure. A proposal for a label for the pesticide is also required based largely on FAO Guidelines on Good Labeling Practices for Pesticides (FAO, 1995).

The registration is valid for five years after which each registered pesticide is reassessed for its continued use. In this way, several pesticides have either been deregistered, e.g. aldrin and dieldrin, or their use restricted following reassessment, e.g. HCH and endosulfan. This procedure takes into consideration safer alternatives and reported abuses among other factors.

The use of pesticides deemed to be highly toxic is further governed by the Pesticide (Highly Toxic Pesticides) Regulations of 1996 – particularly with respect to specific handling restrictions – ensuring that workers handling these chemicals do so with the utmost care. Employers are required to provide adequate training to workers, who must also be medically fit. Workers are only permitted to work a maximum eight hours per day. Employers are further required to maintain strict records detailing the number of hours worked, the type and amount of pesticide used, and the method of application. Their workers are required to wear proper protective clothing and complete an annual medical examination. Furthermore, these regulations dictate that, in the case of female workers, only those who are not pregnant or lactating are permitted to handle pesticides. Other requirements include safe and proper storage of the chemicals and safe disposal of empty containers (Pesticides Board Malaysia, 1996). Handling pesticides such as paraquat, monocrotophos, and calcium cyanide is subject to these regulations. The handling of other pesticides not included in this particular list is not subject to the rigorous set of conditions stipulated in the Pesticide (Highly Toxic Pesticides) Regulations, 1996 but is, of course, required to comply with the Pesticide Act, 1974.

Pesticide residues in food are controlled by The Food Regulations Act of 1985, which is enforced by the Ministry of Health. MRL(s) for pesticide residues are stipulated in this Act. The Act also provides for punitive action against those who misuse pesticides and by their actions cause unacceptable residue levels in food.

Pesticide use in Malaysia is also subject to the Environmental Quality (Scheduled Wastes) Regulations, 1987. The objectives of these regulations are to control and manage the generation, storage, transportation, recycling, treatment or destruction, and disposal of toxic and hazardous wastes.

Malaysia began active environmental management with enactment of the Environmental Quality Act, 1974 and the creation of the Department of the Environment in 1976 (Abdullah, 1995). The Environmental Quality Act was amended in 1985 to include submission of Environmental Impact Assessment (EIA) reports to the Department of the Environment for proposed development projects, thus moving the country toward a proactive, preventive strategy of environmental management. EIAs, which became mandatory in 1988, are used to predict potential environmental impacts from the proposed development. Thereafter, mitigation measures can be identified and prescribed to minimize the predicted impacts. The Department of the Environment also regularly monitors air and water quality throughout the country. To monitor river water quality, samples are collected from 116 major rivers at 892 monitoring stations and, to assess marine water quality, there are an additional 229 sampling stations in coastal and estuarine areas (Abdullah, 1995). The majority of analyses for river and marine samples are conducted in laboratories of the Department of Chemistry under the Ministry of Science, Technology, and the Environment (Abdullah, 1995). Currently, there is no national or regional monitoring program designed to identify and measure pesticide residues in the environment beyond those tasked with evaluating possible risk to humans.

PESTICIDE RESIDUES IN THE AQUATIC ENVIRONMENT

Contamination of the environment by pesticides arises primarily from their application. Surface water contamination can occur as a result of spray drift from aerial spraying or runoff from agricultural areas as a consequence of rain, and to a lesser extent, leaching from the soil. Hence, runoff water contains dissolved pesticides as well as chemicals sorbed onto particulate matter. Pesticide residues can also be transported sorbed on airborne particles and then washed into the aquatic environment by rainfall. Pesticides applied on fields have been known to volatilize and be deposited in areas far removed from the point of application. Volatile pesticides have been observed to be more rapidly lost in tropical agro-ecosystems because of the high temperatures associated with this region. Several studies involving volatile OC insecticides such as DDT and HCH have shown volatilization to be a major route of dissipation of these chemicals from tropical agro-ecosystems and other tropical environments (Abdullah et al., 1997). Ultimately the ocean acts as the final reservoir for these chemicals.

In addition to surface water, groundwater can also be contaminated by pesticide residues as a result of leaching from the soil and the inherent interaction between groundwater and contaminated surface water. Contamination of the environment by pesticides arises not only from their application but also from accidental or intentional discharges of pesticides and pesticide wastes and rinses from mixing areas (on the farm) and from manufacturing plants.

There is at present no national or regional monitoring program designed to investigate pesticide residues in the environment, apart from those intended to evaluate possible risk to humans. However, over the years some data have been accumulated from studies conducted by various groups and researchers. Table 7.3 provides an indication of the extent of contamination by OC pesticides in both biotic and abiotic components of the freshwater environment of Malaysia. OCs are of particular concern due to their persistent nature, and their bio-accumulative and toxic properties. OC pesticide residues are commonly detected in the aquatic environment. Pesticides, which have been banned or whose use has been restricted, e.g. dieldrin, endrin, and DDT, continue to be detected in the environment due to their persistent character. In a recent study by Tan *et al.* (1991), DDE, DDT, and heptachlor were found in samples of water from almost every river surveyed in Peninsular Malaysia. Endosulfan, an insecticide with a well-documented piscidal activity, was also commonly detected. As to be expected, higher levels of pesticide residues were observed in the vicinity of agricultural land. HCH, heptachlor, aldrin, and endosulfan were detected in sediments in a recent survey conducted in the vicinity of a rice growing area (Tan and Vijayaletchumy, 1994). Significant levels of both HCH and endosulfan were due to current usage of these chemicals in rice fields.

In general, there appears to be a decreasing trend in the levels of OC pesticides detected in the Malaysian aquatic environment. This can be seen when a comparison is made between earlier studies and more recent studies (see Table 7.3) of Malaysia's freshwater environment. Additionally, this trend is apparent in relation to similar studies conducted in India where OC insecticides, in particular HCH and DDT, have been the major pesticides for years. Surveys in 1982 in Tanjong Karang (Table 7.3) for example, showed levels of 600 ng L^{-1} of α-HCH and γ-HCH in water. These levels are comparable to similar studies conducted in India (Ramesh *et al.*, 1990). However, a recent survey by Tan *et al.* (1991) showed substantially less contamination. This encouraging trend can be attributed to several factors, including the increasingly popular use of less persistent OP and carbamate pesticides in favor of the OC class. Educational programs in the safe and effective use of these chemicals conducted by the pesticide industry and the Department of Agriculture and directed toward end users and suppliers of pesticides have also contributed to the observed trend.

The marine environment, in particular near-shore coastal waters, has also been observed to be contaminated by OC pesticide residues. The majority of Malaysia's agricultural land is located in the vicinity of the western coastline of Peninsular Malaysia (Figure 7.1). Hence, agricultural runoff and spray drift allow the deposition of applied pesticides into near-shore coastal waters. In Malaysia, surveys conducted to measure pesticide residues in the marine environment have been

Table 7.3 OC pesticide residues in the aquatic environment of Malaysia

Location	Survey year	Matrix	Pesticide	Concentration (ng mL⁻¹or ng g⁻¹)	References
Krian River Basin, Perak	1981	Water	Dieldrin	0.2–0.5	Meier et al., 1983
			β-HCH	0.1–0.9	
			γ-HCH	0.1–0.6	
			Aldrin	0.1–1.8	
		Sediment	Dieldrin	0.8–4.7	
			β-HCH	0.6–8.0	
			γ-HCH	0.4–0.8	
			Aldrin	0.1	
		Rice-field fish	Dieldrin	6.6–2.49	
			α-Chlordane	2.8–17.1	
			β-HCH	3.3–8.2	
			Aldrin	0.3–1.1	
Tanjong Karang, Selangor	1982	Water	α-HCH	0.5	Soon and Hock, 1987
			γ-HCH	0.1	
		Rice-field fish	α-HCH	18–58	
			γ-HCH	10–100	
			α-endosulfan	5130	
			β-endosulfan	1700	
Penang	1984–87	Rice field + marine fish	α-HCH	2.3	Jothy et al. 1987
			Dieldrin	0.2	
			DDT	0.8	
			α-endosulfan	3.4	
			β-endosulfan	2	
Sabah (East Malaysia)	1988	Sediment	Lindane	<0.1–1.11	Heng et al. 1989
			Heptachlor	<0.1–0.51	
			∑ DDT	<0.1–34.7	
West Malaysia major river systems	1989–90	Water	Dieldrin	NDᵃ–0.00025	Tan et al. 1991
			Endrin	ND–0.00323	
			DDT	ND–0.0687	
			Heptachlor	ND–0.00338	
			α-endosulfan	ND–0.044	
			β-endosulfan	ND–0.01	
Bernam River	1992–93	Sediment	HCH	3.52	Tan and Vijayaletchumy, 1994
			Heptachlor	1.275	
			Endosulfan	0.96	
			Aldrin	0.045	

Note:
a ND indicates compound not detected.

directed primarily toward assessing possible hazards to human health by determining pesticide residues in seafood. The results of some of these surveys are given in Table 7.4. OC pesticides in marine biota have been detected since the first surveys, conducted in the mid-1970s by the Fisheries Research Institute. The study on OC residues in fish and shellfish from the coastal waters off the Straits of Malacca by Jothy *et al.* (1983) is probably the earliest report on OC pesticide levels in marine species from Malaysia (Table 7.4). OC residue levels were found to be low in all samples analyzed with the exception of cockles (*Anadara granosa*) L. (Bivalvia: Arcidae) collected from Penang and Perak. Lindane levels in fish ranged

Table 7.4 OC pesticide residues in marine biota from Malaysian waters

Location	Sample type	Survey year	Pesticide	Concentration (ppb) (range/mean)	References
Coastal waters off the Straits of Malacca	Cockles (Anadara granosa)	1977	DDT	50	Jothy et al., 1983
			Lindane	1–12	
			Dieldrin	<1–4	
	Fish		DDT	0.4–0.8	
Jeram, Selangor	Shrimp	1985	γ-HCH	3	Everaarts et al., 1991
	Crab		γ-HCH	4	
	Polychaetae worm		γ-HCH	8	
	Bivalve mollusc		γ-HCH	17	
	Shrimp		Dieldrin	94	
	Crab		Dieldrin	232	
	Polychaetae worm		Dieldrin	57	
	Bivalve mollusc		Dieldrin	52	
Various locations in the coastal waters off the Malay Peninsular	Various species of marine fish	1984 1986 and 1987	α-HCH	0.02–5.3	Goethe et al., 1987
			β-HCH	0.025–3.1	
			γ-HCH	0.02–3.6	
			Dieldrin	0.04–4.1	
			DDD	0.02–2.4	
			DDE	0.02–4.7	
			DDT	0.04–3.9	
			α & β-endosulfan	ND[a]	
			Endosulfan sulfate	ND	
Penang	Mussels (Perna vendis)		Lindane	180.9	Rohani et al., 1992
	Cockles (Anadara granosa)		Lindane	0.222–3.01	
			DDT	1.23	
Muar, Johor	Oysters (Cronsostrea belcherei)		Lindane	27.50–66.46	
			DDT	1.46–7.41	
Batu Lindang, Kedah	Mussels		α-endosulfan	0.05	
Lekir, Perak	Cockles		Aldrin	0.24	

Note: a ND indicates pesticide not detected.

from 0.001 to 0.012 mg kg^{-1} ww, dieldrin from below the detection limit (<0.001 mg kg^{-1}) to 0.004 mg kg^{-1} ww and total ($\sum$) DDT (DDT, DDE, and DDD) from 0.0004 to 0.0008 mg kg^{-1} ww. The $\sum$ DDT concentrations in cockles were found to be significantly higher than in fish with a mean value of 0.05 mg kg^{-1} ww. In another survey conducted in Jeram off the west coast of Peninsular Malaysia, concentrations of lindane ranged from 0.003 to 0.017 mg kg^{-1} ww in various marine organisms while those of dieldrin were found to be in the range 0.052 to 0.232 mg kg^{-1} ww (Table 7.4). Low concentrations of OC residues in a variety of marine fish were observed in a survey conducted by Jothy *et al.* (1987) (see Table 7.4) who also included freshwater fish in their survey (see Table 7.3). Endosulfan and its metabolite endosulfan sulfate were detected in freshwater fish but not detectable (ND) in marine fish. This is evidence of the relatively non-persistent nature of this OC insecticide.

Between 1987 and 1991, Rohani *et al.* (1992) conducted a study of OC pesticide residues in molluscs, collected mainly from the west coast of Peninsular Malaysia (see Table 7.4). They found that OC residue levels were generally low except for lindane, which they found in high concentrations in some samples, e.g. mussels *Perna viridis* L. (Bivalvia: Mytilidae), collected in 1990 from Gertak Sanggul and Pulau Jerajak in Penang. Samples from these sites contained 180.9 μg kg^{-1} and 123.7 μg kg^{-1} ww lindane, respectively. Samples of oysters *Crassostrea belcherei* (Bivalvia: Ostreidae) taken from Muar, Johor in 1990 had lindane residue concentrations ranging from 27.50 to 66.46 μg kg^{-1} ww. Lindane levels in cockles in Juru, Penang ranged from 0.22 to 3.01 μg kg^{-1} ww. Alpha-endosulfan was detected only in mussel samples from Batu Lintang, Kedah and they contained 0.05 μg kg^{-1} ww. None of the samples analyzed contained β-endosulfan. Aldrin was not detectable in most samples; the highest concentration (0.24 μg kg^{-1} ww) was found in cockles from Lekir, Perak. Cockle samples collected in 1987 were found to have higher levels of dieldrin (ND to 41.09 μg kg^{-1} ww) compared to other years (0.03 to 5.75 μg kg^{-1} ww) (Rohani *et al.*, 1992). DDT was not detected in most samples except for slipper oysters *C. iredalei* (Bivalvia: Ostreidae) from Muar, Johor, whose DDT content ranged from 1.46 to 7.41 μg kg^{-1} ww. Samples of cockles analyzed by Rohani *et al.* (1992) from Penang and Perak had non-detectable levels of DDT (except for one sample in Sg. Belanak, Penang containing 1.23 μg kg^{-1} ww). This differs markedly from the study by Jothy *et al.* (1983) in which cockles collected from Penang and Perak had DDT levels averaging 50 μg kg^{-1} ww. This difference is probably due to the phasing out of the use of DDT for the Malaria Eradication Program in Malaysia.

OC pesticide levels in tiger shrimp *Penaeus monodon* Fabricius (Decapoda: Penaeoidae) in Ban Merbok, Kedah were reported by Liong (1993). He found OC pesticide residues were low except for lindane (1 to 3.41 μg kg^{-1} ww). An absence of DDT was noted and attributed to the recent cessation of the use of the chemical by the Ministry of Health (Liong, 1993). The slightly higher levels of lindane were attributed to the current use of this pesticide in nearby rice fields. On the whole, levels of OC pesticide residues in the aquatic environment in Malaysia reflect the banned status of the chemicals including DDT, aldrin, and dieldrin. For HCH, only the γ isomer is permitted, which should result in a decline of other isomers that are less insecticidal but more persistent. Previous surveys have shown the α and β isomers to be present at higher concentrations than the γ isomer as exemplified by observations made by Jothy *et al.* (1987) (Tables 7.3 and 7.4). Furthermore, endosulfan, which is currently in use, is expected to be present primarily in freshwater ecosystems near the point of application due to its rather short half-life.

IMPACT OF PESTICIDES

The primary concern for the presence of pesticide residues in the environment arises from their toxicity to living organisms. Hence, in addition to the intended target pests, non-target organisms including man are exposed to the toxic effects of pesticides. Other effects, particularly bioaccumulation in the food chain, are mainly associated with the OC class of chemicals and contribute to the negative impacts of pesticides. Most of the currently used pesticides have relatively low toxicity to mammals, a primary consideration in the registration approval process. As the aquatic environment is a major focus for the evaluation of pesticides discharged into the general environment, various aquatic species have been used as the test species of choice in toxicity tests.

There is a great variation in the way different classes of pesticides affect aquatic organisms, in terms of both acute and of chronic toxicities, just as there are variations in the toxicities of individual pesticides within each class. The toxic effects of these chemicals may also manifest themselves in many forms, e.g. physiological, morphological, and behavioral. Although these effects may not be immediately fatal, they may affect the ability of the organisms to search for food or flee from predators (fitness). The susceptibility of a particular aquatic organism to a pesticide is also subject to many variables, notably the stage of development of the organism concerned, ambient temperature, rainfall amounts, water chemistry and dissolved organic matter, the presence of suspended sediments and cations, and others.

There is a great wealth of toxicity data, particularly acute toxicity information, for a number of commonly used aquatic test species. However, much of this data has been derived from temperate countries. In Malaysia some acute toxicity data has been generated using local species of fish (Table 7.5). From Table 7.5 and other more comprehensive compilations, it can be seen that insecticides have moderate to high acute toxicities and the same is generally true for fungicides. Herbicides, on the other hand, are generally less toxic. However herbicides, as with other pesticides, are more toxic to aquatic invertebrates, which constitute an important source of food for many species of fish.

In addition to their toxic effects, pesticides, in particular the OC chemicals, also tend to accumulate in the fatty tissues of aquatic organisms. In this way they are transported through the food chain and can be biomagnified from lower trophic

Table 7.5 Some acute toxicity data on local Malaysian aquatic species

Pesticide	Test species	96h LC_{50} (ppm)	References
Diquat	Common carp	50	Yew and
Methamidophos	(*Cyprinus carpio*)	68	Sudderuddin,
Carbaryl		1.7	1979
Lindane		0.21	
Endosulfan	Catfish(*Clarius*	0.002	Gill, 1982
Lindane	*batrachus*)	0.13	
Carbofuran		5	
Carbaryl		20	
Malathion		1.3	
Fenitrothion		117	
Malathion	Tilapia Sepat Siam	68	Mohsin et al.,
	(*Trichogaster*	0.98	1984
	pectoralis)		
Agrocide	Tilapia	62	Liong et al., 1988
Furadan 2G (carbofuran)		23	
Gramaxone (paraquat)		67	
Thiodan 35% (endosulfan)		0.01	
Bensulfuron (a	Tilapia Sepat Siam	>1,000	Ooi and Lo, 1990
sulfonylurea herbicide)	(*Trichogaster pectoralis*)		
2,4-D		153	
Metsulfuron		>1,000	
Quinclorac		50	
Butachlor		3	
Molinate (a			
thiocarbamate herbicide)		5	
Oxadiazon		1	
Propanil		3.5	
Fenoxaprop		0.2	

levels to higher levels of the food chain. Hence, high concentrations may be found in aquatic organisms compared to the levels in the water column, which are usually a few orders of magnitudes lower. Table 7.6 illustrates the bioaccumulation of HCH and endosulfan in paddy-field fish in Malaysia. Bioconcentration factors for α-endosulfan and lindane were 424 and 239, respectively (Soon and Hock, 1987).

Pesticides have been implicated in the decline of fish production on agricultural land, in particular paddy-fields (Yunus and Lim, 1971; Alabaster, 1986). In Indonesia where agricultural practices and climatic conditions are similar to those in Malaysia, the majority of the documented pesticide poisonings of aquatic organisms have been attributed to agricultural runoff (Chua *et al.*, 1989). It should also be noted that fish kills have also been attributed to other factors including suspended solids, the presence of other pollutants, and disease outbreaks.

Sources of pollution in coastal waters arise from both marine activities such as shipping and land-based activities. The latter represent non-point sources and

Table 7.6 Bioaccumulation of insecticides in paddy field fish in Malaysia

Insecticide	Concentration (ppb)		Bioconcentration factor	Reference
	Fish	Water		
α-HCH	38	0.05	85	Soon and Hock,
β-HCH	31	0.1	239	1987
α-Endosulfan	4,660	11.0	424	
β-Endosulfan	1,540	6.0	275	

include pesticide residues and fertilizers from agricultural runoff. Coastal aquaculture as an economic activity has been greatly affected by the discharge of domestic, agricultural, and industrial effluents, which results in an increasing deterioration of coastal water quality (Chua *et al.*, 1989).

Aquaculture activities in Malaysia, primarily located in coastal areas, comprise freshwater and marine fish, prawn, cockles, mussels, and oysters (Table 7.7) (MADI, 1996). Although aquaculture is a relatively young industry in Malaysia – comprising approximately 10 percent of the net income from marine fishing activities in 1993 – it is of increasing importance because of its rapid growth. Aquaculture production increased by 100 percent in 1993 compared to 1990 (MADI, 1996). In 1993, cockles were the largest component in terms of production (77,755 T) valued at US$10.7 million, while freshwater fish and prawn culture was second in volume with 15,468 T, although with a higher commercial value of US$45 million. The highest value component (US$61.3 million) was brackish water and marine aquaculture of prawn and fresh market quality fish (MADI, 1996).

Aquaculture has been identified as being an essential component in the agricultural production of Malaysia in the NAP, 1992 to 2000. Production is envisaged to expand from 52,000 T in 1990 to 200,000 T by the year 2010. Development of the industry will include identification of suitable land and water sites for aquaculture activities. Suitable aquaculture sites will be concentrated in areas designated as Aquaculture Development Area with the necessary infrastructure and technical support services.

Although there has been a general paucity of information correlating negative impacts on aquaculture activities to pesticide residues, pesticide residues certainly contribute to the land-based sources of pollutants, which result in the deterioration of coastal water quality (Abdullah *et al.*, 1999). Indeed, persistent pesticide residues get transported to the marine environment particularly near-shore coastal waters as evidenced by the detection of OC residues in a variety of marine biota, many of which are consumed as seafood (Table 7.4).

FUTURE CONSIDERATIONS

IPM, involving a combination of chemical, biological, and cultural methods of pest control, is a realistic and viable means of decreasing the negative impacts of

Table 7.7 Area used for aquaculture in Malaysia in 1993[a]

Type of aquaculture	Area (ha)
Fresh water fish culture in pond and disused mining pools	5,754
Cockle culture in mudflats	5,041
Penaeid prawn culture in brackish water ponds	1,878
Marine finfish in floating net cages in coastal water	671[b]
Freshwater finfish in floating net cages in ponds	5[b]
Culture of mussel and oyster on rafts/racks	17[c]
Cement tank	3
Total	12,765

Notes:
a Adapted from MADI, 1996.
b Area of net cage.
c Area of rafts/racks.

the excessive use of pesticides (Babu and Hallam, 1989). The development, promotion, and implementation of IPM in Malaysia have had some success, particularly in rice paddy-fields (Majid *et al.*, 1984). There is clearly an increasing need to develop and adopt IPM strategies for other crops in Malaysia. This would require extensive research in various approaches to pest control in specific agroecosystems, including the introduction of multi pest resistant cultivars, biological control methods (see Verkerk and Wright, 1997; Rickson and Rickson 1998), and effective training of farmers in the implementation of IPM strategies and techniques.

A major concern for the continued success and expansion of IPM is the belief by farmers that pesticides are the only viable solution to their farming problems. In a 1992 survey conducted in the Cameron Highlands, this was by far the most often mentioned solution (32.1 percent) and, furthermore, 97 percent of all respondents regarded pesticides as a necessary input (Midmore *et al.*, 1996). About half of respondents had heard about IPM, but the Malaysian Agricultural Research and Development Institute's (MARDI) IPM package for vegetable farming was used by less than 10 percent of vegetable farmers (Midmore *et al.*, 1996). More farmers did claim to be using smaller quantities of insecticides than in the past (42 percent) compared to those using more pesticides (34 percent) than previously, one hopeful sign. Vegetable farmers in the Cameron Highlands spent M$912 ha^{-1} for pesticides of M$33,783 ha^{-1} in gross sales during the 1991 growing season (Midmore *et al.*, 1996). They also reported a change in the type of pesticide used (by 77 percent of farmers) in recent years but this change was almost exclusively because of increased resistance of insect pests to previously used chemicals (Midmore *et al.*, 1996). It is clear that much work remains to be accomplished by MARDI to develop and promote IPM packages throughout Malaysian agriculture and to educate farmers on the benefits to be derived from IPM, biocontrol, and other alternatives to synthetic chemical control of pests.

There is really little information available on the fate and behavior of pesticides in the Malaysian tropical environment. Predicted impacts are at present largely based on data derived from temperate countries. Although much can be learned from studies conducted in these countries, there is clearly a need to conduct similar studies to elucidate the distribution, behavior, fate, and bioavailability of pesticides in tropical ecosystems to assess the potential impacts of these chemicals. Working with Malaysian agricultural soils, e.g. a sandy loam from a vegetable-growing area in the Cameron Highlands and a muck soil from a rice-growing area in Tanjong Karang, Cheah *et al.* (1997) derived Freundlich adsorption distribution coefficients for paraquat, glyphosate, 2,4-D and lindane. Adsorption of the pesticides was not affected by temperature, pH, or addition of the pesticides as a mixture. Only the herbicide 2,4-D was mobile in both the muck soil and the sandy loam. The adsorption-desorption characteristics and leaching behavior of these pesticides in Malaysian soils showed little difference from results obtained in other parts of the world (Cheah *et al.*, 1997). There is also a critical need to compile toxicity data including sublethal and chronic exposure information using local test species.

The limitations of acute toxicity data are well recognized. In order to properly assess impacts, information on pesticide residue effects in whole ecosystems is essential. As a means to achieve this, micro- and mesocosm studies are considered a bridge between simple LC_{50} data and comprehensive ecosystem assessments. Finally, field validation is required to match predictions derived from laboratory, micro-, and mesocosm tests to observations of responses in complex natural ecosystems (Cairns, 1992).

Compilation of the relevant toxicity data will allow determination of the types and levels of pesticides causing significant impacts on aquatic organisms at the individual, population, and community levels. This information will contribute to the establishment of both freshwater and marine water quality criteria by applying appropriate threshold concentrations. Currently used criteria in Malaysia are of an interim nature (Goh *et al.*, 1986; Yap, 1988). In addition, knowledge on the fate, distribution, and bioavailability of pesticides is essential for risk assessment, prudent management decision-making, and the improvement of aquatic and coastal management policies.

While the widespread use of pesticides continues, there is a need for extensive monitoring of their residues in the environment. Such monitoring programs must be supported by the necessary regulatory capacities, coupled with effective enforcement mechanisms to prevent contamination levels from exceeding locally established limits as stipulated by the appropriate legislatures. Furthermore, because contamination is due predominantly to application in the field, an extensive program of education and public awareness on the proper uses of pesticides needs to be continued, improved, and reinforced to minimize the indiscriminate and irresponsible use of pesticides and to reflect advances in pesticide science.

As far as OC insecticides are concerned, lindane and endosulfan are the only two remaining OC(s) in widespread use in Malaysia. However, they are of primary concern with respect to the aquatic environment. Both these compounds have proven highly toxic to aquatic life forms. While these compounds may in the near future be restricted, the implementation of buffer zones in sensitive areas may help to minimize their entry into waterways. In areas where such an approach may not be practical, alternative pesticides with minimal toxic effects to aquatic organisms, while still maintaining field efficacy, should replace those being currently used.

REFERENCES

Abdullah, A.R. 1993. Pesticide formulations – Present trends and recent developments. *The Planter.* 69:3–13.

Abdullah, A.R. 1995. Environmental pollution in Malaysia: trends and prospects. *Trends Anal Chem.* 14(5):191–8.

Abdullah, A.R., Bajet, C.M., Matin, M.A., Nhan, D.D. and Sulaiman, A.H. 1997. Ecotoxicology of pesticides in the tropical paddy field ecosystem. *Environ Toxicol Chem.* 16:59–70.

Abdullah, A.R., Mohd Tahir, N., Tong, S.L., Mohd Hoque, T. and Sulaiman, A.H. 1999. The GEF/UNDP/IMO Malacca Straits Demonstration Project: Sources of pollution. *Mar Poll Bull.* 39:229–33.

Asian Development Bank (ADB). 1987. *Handbook on the Use of Pesticides in the Asia-Pacific Region.* Manila, Philippines: ADB, p. 14.

Alabaster, J.S. 1986. Review of the state of water pollution affecting inland fisheries in Southeast Asia. FAO Fisheries Technical Paper Nr 260. Rome, Italy: FAO.

Babu, S.C. and Hallam, A. 1989. Environmental pollution from pest control, integrated pest management and pesticide regulation policies. *J Environ Manag.* 29:377–89.

Cairns, J. 1992. The threshold problem in ecotoxicology. *Ecotoxicology.* 1:3–16.

Cheah, U.B., Kirkwood, R.C. and Lum, K.Y. 1997. Adsorption, desorption and mobility of four commonly used pesticides in Malaysian agricultural soils. *Pestic Sci.* 50(1):53–63.

Chua, T.E., Paw, J.N. and Guarin, F.Y. 1989. The environmental impact of aquaculture and the effects of pollution on coastal aquaculture development in Southeast Asia. *Mar Poll Bull.* 20:335–43.

Everaarts, J.M., Bano, N., Swennen, C. and Hillebrand, M.T.J. 1991. Cyclic chlorinated hydrocarbons in benthic invertebrates from three coastal areas in Thailand and Malaysia. *J Sci Soc Thailand.* 17:31–49.

FAO. 1995. *Guidelines on Good Labeling Practices for Pesticides.* Rome: FAO.

Gill, S.S. 1982. Pesticides and the environment. In: Proc. Symp. on the Malaysian Environment in Crisis held 18–19 April 1981 in Pulau Pinang, Malaysia. Penang, Malaysia: Consumer Association of Penang, pp. 37–42.

Goh, S.H., Lim, R.P. and Yap, S.Y. 1986. *Water Quality Criteria and Standards for Malaysia,* Volume 4. Kuala Lumpur, Malaysia: Institute of Advanced Studies, University of Malaya, pp. 59–148.

Heng, L.Y., Mohamed, M. and Lee, O.K. 1989. Paper presented to the Intensification of Research in Priority Areas Seminar, September 1989 in Malacca, Malaysia.

Iqbal, M. and Wright, D.J. 1997. Evaluation of resistance, cross-resistance and synergism of abamectin and teflubenzuron in a multi-resistant field population of *Plutella xylostella* (Lepidoptera: Plutellidae). *Bull Entomol Res.* 87(5):481–6.

Jothy, A.A., Huschenbeth, E. and Harms, U. 1983. On the detection of heavy metals, organochlorine pesticides and polychlorinated biphenyls in fish and shellfish from the coastal waters of Peninsular Malaysia. *Arch Fish Disease.* 33:161–206.

Jothy, A.A., Kruse, G.H. and Macht-Hansmann, M. 1987. Proc. Int. Conf. on Pesticides in Tropical Agriculture held 23–25 September 1987 in Kuala Lumpur, Malaysia. Malaysian Agricultural Research and Development Institute and the Malaysian Plant Protection Society, pp. 189–200.

Lee, C.Y., Yap, H.H., Chong, N.L. and Lee, R.S.T. 1996. Insecticide resistance and synergism in field collected German cockroaches (Dictyoptera: Blattellidae) in Peninsular Malaysia. *Bull Entomol Res.* 86(6):675–82.

Liong, P.C. 1993. On the detection of organochlorine pesticides and polychlorinated biphenyls in pond-raised shrimp (*Penaeus monodon*). Fisheries Bulletin Nr 85. Kuala Lumpur, Malaysia: Ministry of Agriculture.

Liong, P.C., Hamzah, W.P. and Murugan, V. 1988. Toxicity of some pesticides towards freshwater fishes. *Malaysian J Agri.* 54:147–56.

MACA. 1996. *Annual report and directory 1994/95.* Petaling jaya, Malaysia: Malaysian Agricultural Chemicals Association, p. 2.

MACA. 1997. *Annual Report and Directory 1996/97.* Petaling jaya, Malaysia: Malaysian Agricultural Chemicals Association, pp. 41–3.

MADI. 1996. *Malaysian Agriculture Directory and Index 1995/96.* Petaling jaya, Malaysia: Agriquest Sdn Bhd (pte or Private Ltd), pp. 33–64.

MADI. 1997. *Malaysian Agriculture Directory and Index 1997/98.* Petaling jaya, Malaysia: Agriquest Sdn Bhd., pp. 59–71.

Majid, T., Lim, B.K. and Booty, A. 1984. Implementation of the IPM programme for rice in Malaysia. In: Lee, B.S., Loke, W.H. and Heong, K.L. (eds) *Integrated Pest Management in Malaysia.* Kuala Lumpur, Malaysia: Nan Yang Muda Publishers, pp. 319–23.

Meier, P.G., Fook, D.C. and Lagler, K.F. 1983. Organochlorine pesticide residues in rice paddies in Malaysia 1981. *Bull Environ Contam Toxicol.* 30:351–7.

Midmore, D.J., Jansen, H.G.P. and Dumsday, R.G. 1996. Soil erosion and environmental impact of vegetable production in the Cameron Highlands, Malaysia. *Agric Ecosyst and Environ.* 60(1):29–46.

Mohsin, A.K.M., Ong, S.L. and Ambak, M.A. 1984. Effect of malathion on Sepat Siam and Tilapia. *Pertanika.* 7:57–60.

NAP (National Agricultural Policy). 1984. Cabinet Committee for Agriculture Policy. Kuala Lumpur, Malaysia: Government Printers.

Ooi, G.G. and Lo, N.P. 1990. Toxicity of herbicides to Malaysian rice field fish. In: Proc. 3rd Int. Conf. Plant Protection in the Tropics held 20–23 March 1990 in Kuala Lumpur, Malaysia. Malaysian Agricultural Research and Development Institute and the Malaysian Plant Protection Society, pp. 71–4.

Pesticides Board Malaysia. 1991. *Guidelines on Registration, Labeling and Classification of Pesticides.* Kuala Lumpur, Malaysia: Department of Agriculture.

Pesticides Board Malaysia. 1996. *Guidelines on Pesticide (Highly Toxic Pesticides) Regulations.* Kuala Lumpur, Malaysia: Department of Agriculture.

Ramesh, A., Tanabe, S., Iwata, H. and Tatsukawa, R. 1990. Seasonal variation of persistent organochlorine insecticide residues in Vellar River waters in Tamil Nadu, South India. *Environ Poll.* 67:289–304.

Rickson, F.R. and Rickson, M.M. 1998. The cashew nut, Anacardium occidentale (Anacardiaceae), and its perennial association with ants: extrafloral nectary location and the potential for ant defense. *Am J Bot.* 85(6):835–49.

Rohani, I., Chan, S.M. and Ismail, I. 1992. Organochlorine pesticide and PCBs residues in some Malaysian shellfish. In: Proc. Nat. Seminar on Pesticides in the Malaysian Environment held 27 February 1992 in Kuala Lumpur, Malaysia. Malaysia Department of Agriculture, pp. 27–33.

Samad, A.H. 1991. Handling, storage and disposal of pesticides and pesticide wastes. *J Environ Manage Res Assoc Malaysia* 2:7–12.

Soon, L.G. and Hock, Q.S. 1987. Environmental problems of pesticide usage in Malaysian rice fields: perceptions and future considerations. In: Tait, J. and Napompeth, B. (eds) *Management of Pests and Pesticides*. London: Westview Press, p. 14.

Sulaiman, S., Pawanchee, Z.A., Wahab, A., Jamal, J. and Sohadi, A.R. 1999. Field efficacy of fipronil 3G, lambda-cyhalothrin 10 percent CS, and sumithion 50 EC against the dengue vector *Aedes albopictus* in discarded tires. *J Vector Ecol.* 24(2):154–7.

Sulaiman, S., Pawanchee, Z.A., Othman, H.F., Jamal, J., Wahab, A., Sohadi, A.R. and Pandak, A. 2000. Field evaluation of deltamethrin/S-bioallethrin/piperonyl butoxide against dengue vectors in Malaysia. *J Vector Ecol.* 25(1):94–7.

Tan, G.H., Goh, S.H. and Vijayaletchumy, K. 1991. Analysis of pesticide residues in Peninsular Malaysian waterways. *Environ Monit Assess.* 19:469–79.

Tan, G.H. and Vijayaletchumy, K. 1994. Determination of organochlorine pesticide residues in river sediments by soxhlet extraction with hexane-acetone. *Pestic Sci.* 40:121–6.

Verkerk, R.H.J. and Wright, D.J. 1997. Field-based studies with the diamondback moth tritrophic system in Cameron Highlands of Malaysia: implications for pest management. *Int J Pest Manage.* 43(1):27–33.

Yap, H.H., Foo, A.E.S., Lee, C.Y., Chong, N.L., Awang, A.H., Baba, R. and Yahaya, A.M. 1996. Laboratory and field trials of fenthion and cyfluthrin against *Mansonia uniformis* larvae. *J Vector Ecol.* 21(2):146–9.

Yap, S.Y. 1988. Water quality criteria for the protection of aquatic life and its users in tropical Asian resources. In: de Silva, S.S. (ed.) *Reservoir Fishery Management and Development in Asia*. Proc. Int. Development Research Centre (IRDC) Workshop held 23–28 November 1987 in Kathmandu, Nepal. Ottawa, Canada: IRDC, pp. 74–86.

Yeop, M.T., Yusoff, A. and Tan, S.L. 1982. A special report on agricultural land use in Peninsular Malaysia. Malaysia Agricultural Research and Development Institute (MARDI) Publication. Serdang, Malaysia: MARDI.

Yew, N.C. and Sudderuddin, K.I. 1979. Effect of methamidophos on the growth rate and esterase activity of the common carp *Cyprinus carpio* L. *Environ Poll.* 18:213–21.

Yunus, A. and Lim, G.S. 1971. A problem in the use of insecticides in paddy fields in West Malaysia: a case study. *Malaysian Agri J.* 48:167–78.

Chapter 8

Distribution, fate, and impact of pesticides in the tropical marine environment of Vietnam

Dang Duc Nhan

PESTICIDE REGULATION IN VIETNAM

History of pesticides and pesticide legislation

The history of pesticide use in agriculture in Vietnam is short. Before 1950, malaria was a widespread disease among the Vietnamese people. At that time, French and native physicians trained the people to sleep under nets to protect against *Anopheles* mosquitoes, which are the vectors for malaria. During the period of 1950 to 1960, dichlorodiphenyltrichloromethane (DDT) and a technical mixture of hexachlorocyclohexane (HCH) – termed '666' by the Vietnamese – were applied over all Vietnamese territory to control the mosquitoes responsible for vectoring malaria, which at that time had become epidemic. The DDT and HCH insecticides were donated to Vietnam by either the former Soviet Union or the People's Republic of China. Since 1960, Vietnam has maintained control of malaria through the use of insecticide applications.

OP pesticides, particularly methyl parathion, were introduced into Vietnam in the early 1960s, and methyl parathion quickly became the most commonly applied insecticide. It is primarily used for insect control during vegetable crop production. In 1978, 2-sec-buthylphenylmethyl carbamate (BPMC) was introduced as an effective insecticide for controlling brown planthopper (BPH) *Nilaparvata lugens* Stål (Homoptera: Delphacidae), which is problematic in rice culture. Since 1983, the pyrethroids, cypermethrin and fenvalerate, have replaced the OPs and carbamates. Many pests and rice diseases had become resistant to both classes of older pesticides. The extensive use of pesticides in Vietnamese agriculture since 1980 has made them an important element in the cost structure of agricultural activities of Vietnamese farmers. Indeed, their regular use has been linked to the rise of the BPH as a serious rice pest, as outbreaks of this delphacid in tropical rice resulted from the destruction of effective predators by overuse of insecticides (Ooi, 1996).

As mentioned above, until 1980, pesticides used in Vietnam came free of charge from the former Soviet Union and the People's Republic of China and they were distributed to farmers as a 'gift', whether they were needed to control epidemic pest outbreaks or were not needed at all. Hence, until the termination of 'gift' or free pesticides they were used in Vietnam without any regulations. On the other hand, until then, rice farmers grew low-yield but insect and disease resistant rice cultivars that did not require significant pesticide usage.

Since the middle of the 1990s when Vietnam developed its open door policy and the FAO began to operate in Vietnam, prudent pesticide use in the country has increased. FAO experts have assisted the Ministry of Agriculture and Rural Development (MARL), formerly the Ministry of Agriculture and Foodstuffs, in the establishment of a Code of Conduct for the distribution and use of pesticides.

On 15 February 1993, Presidential Decree No 08L/CTN on Plant Protection and Control was issued. This Decree had been approved by the National Assembly of Vietnam on 4 February 1993 after a determination that plant protection and insect, disease, and weed control were important elements in the food safety strategy of the Government and that pesticides were among the goods to be subjected to common control by the Government. According to the Decree, the Agency designated to regulate the use of agrochemicals in Vietnam was the Plant Protection Department (PPD) of the MARL. The PPD was given control over the importation, manufacture, distribution, sale, transport, storage, labeling, use, and disposal of pesticides throughout Vietnam. Today, PPD is mandated to register new pesticides, including new formulations; regulate pesticide availability and use; license handlers; set residue limits on foods and feeds; supervise pesticide importation; and design handler and applicator training programs.

Current pesticide regulations and registration of new pesticides and formulations

Registration with PPD is intended to ensure that pesticides will be effective and efficient for the purposes claimed when used according to registered label directions. Registering new pesticides in Vietnam is a lengthy process. Any new pesticide, before its official introduction into use in Vietnam, must pass through the three steps of the registration procedure without regard to previously generated data on toxicology, bio-efficacy, environmental fate, and residue levels of the pesticide. The procedure includes laboratory and field tests, and, if results of these tests confirm the data submitted by the registrant, the pesticide is registered for use. Registration procedures for new formulations of an a.i. are the same as for registration of new pesticides.

The registration label must show the hazard category (imprinted in color coding), trade and common name, directions for use by crop, formulation and target pests, and pre-harvest interval (PHI). Most Vietnamese farmers surveyed were familiar with which pesticides were most effective for controlling insect or disease pests for their crops, but few were aware of the hazard category, common name, and mode of use, and even fewer knew the safe PHI.

PPD can restrict or ban the use of any pesticide or pesticide formulation if there is evidence that the pesticide is an imminent hazard or has caused or is causing widespread serious damage to crops, fish, livestock, public health, or the environment. The PPD Administrator through the Pesticide Technical Advisory Committee (PTAC) can restrict pesticide use. PTAC consists of 13 members from the MARD, the Trade Ministry, the Labor Ministry, the Ministry of Science Technology and the Environment, the Ministry of Public Healthcare, and the Ministry of Interior.

Maximum residue limits for foods and feeds

To protect the public from hazardous pesticide residues on foods and feeds, PPD is mandated to establish a system of maximum residue limits (MRLs) applicable to both domestic and imported raw agricultural products. PPD has prepared a list of proposed MRLs based on extrapolation and estimation of the allowable daily intake of pesticide residues from calculations taking into account food factors, dietary intake studies, and existing MRL Code. Due to funding constraints, the list of MRLs on foods in Vietnam was based on the FAO/UNEP MRL Code.

PESTICIDE USE AND DISTRIBUTION IN VIETNAM

The pesticide industry

Vietnam, at present, is not a pesticide producer. All pesticides used in the country are imported in technical form and formulated or repackaged into smaller portions suitable for distribution by governmental pesticide supply companies. These companies are VIPESCO I (the 1st Vietnam Pesticide Supply Company), VIPESCO II, and VIPESCO III located, respectively, in the northern, central, and the southern parts of Vietnam. All VIPESCOs belong to the PPD of Vietnam. The VIPESCOs supply necessary pesticides to their daughter companies in the provinces and manage the retail price of pesticides. Provincial companies distribute pesticides to districts or sell them to farmers through stores located in individual villages.

In 1996, a pesticide production Joint Venture (JV) was established between South Korea and Vietnam in Songbe Province, located northeast of Ho Chi Minh City (formerly Saigon). The JV currently produces carbofuran, carbaryl, isoprocarb, and validamicyn. The insecticides and fungicide produced are for domestic use as well as for export to other Southeast Asia countries. The JV production has been 2,000 T a.i. (all pesticides) each year (Pesticide Management Section, Plant Protection Department of Vietnam, 1997). At present, there are more than 100 foreign pesticide producers from Europe, North America, and Asia marketing and testing their agrochemicals in every corner of Vietnam.

Past and current pesticide usage patterns

Until 1990, when farm practice shifted away from traditional mono-cropping, the need for pesticides was limited and the quantity of pesticide used each year in all of Vietnam was estimated to be 200 to 250 T a.i.(s). This quantity of pesticide was distributed based on farm area and was equal among provinces without regard to need. After distribution to the provinces, no one was responsible for proper use of the pesticide. Since 1990, a new land management policy has been introduced for rural farmers. Specifically local government endorsed the use of tillers that significantly increased crop yield. Before 1985, rice yield in the Red (Yuan) River and Mekong River deltas was 5 T ha^{-1} per year; today that figure has doubled. Currently throughout Vietnam, increased crop yield is correlated with increased pesticide use.

In 1991, the PPD imported 7,595 T a.i. of all pesticides, while in 1992, the quantity was 9,510 T; 1993 (9,969 T); 1994 (6,213 T); 1995 (7,653T); and in 1996 about 10,000 T (Technical Section, VIPESCO I, 1997). The lesser quantities of pesticides imported for 1994 and 1995 do not reflect a decrease in pesticide use; rather it is a result of the introduction into Vietnam of many new and more effective pesticides with much lower use rates. In terms of formulated products, more than 20,000 T of pesticide are used yearly in Vietnam with nearly 80 percent in the form of insecticides (Quyen et al., 1995).

On 22 May 1996, the Ministry of Agriculture and Rural Development of Vietnam, in response to Presidential Decree No. 08L/CTN on Plant Protection and Control, issued a *List of Pesticides Permitted, Restricted, and Banned for Use in Vietnam* (The List) (Ministry of Agriculture and Rural Development, 1996). There are 228 insecticide formulations, 159 fungicide formulations, 124 herbicide formulations, six rodenticide formulations, and 16 plant growth regulator formulations permitted for use in Vietnam. All OC pesticides have been banned for agricultural use throughout the country (Table 8.1). However, DDT and HCH are still permitted

Table 8.1 List of pesticides banned from use in Vietnam[a]

Pesticide class			
Insecticides	Fungicides	Herbicide	Rodenticide
Aldrin	Arsenic compounds	2,4,5 T	Thallium
BHC, lindane	Captan		compounds
Cadmium compounds	Captafol		
Chlordane	Hexachlorobenzene		
DDT	Mercury compounds		
Dieldrin	Selenium compounds		
Endrin			
Heptachlor			
Isobenzene			
Isodrin			
Lead compounds			
Parathion ethyl			
Toxaphene			
Strobane			

Note:
a Adapted from MARD, 1996.

for public health use (including vector control) in Vietnam. Among OP pesticides, methyl parathion was banned from use beginning on 1 February 1997. Before issuing The List, about 8,000 T of restricted-use pesticide formulations were imported in 1993, which constituted almost 50 percent of the total yearly quantity of pesticides used in Vietnam. However, since 1994, a maximum importation limit of 3,000 T per year of these formulations has been permitted. These restricted-use pesticide formulations are primarily used for medical and public health purposes. Table 8.2 contains a list of the restricted-use pesticides currently permitted in Vietnam. While official data are not available, it should be noted that banned pesticides are still being smuggled into, distributed, and used in Vietnam. Thus, it should be expected that sporadic OC pesticide use still occurs in Vietnam. In fact, relatively high soil residue levels of DDT have been measured near Hanoi (Thao *et al.*, 1993a; Nhan *et al.*, 2001), in the central part of the country (Thao *et al.*, 1993a), and in soils from two southwestern provinces (Thao *et al.*, 1993b).

Although there are many new formulations of permitted pesticides that could replace older, more persistent, and more environmentally hazardous pesticides, farmers are reluctant to change from formulations familiar to them. Formulation familiarity, lack of knowledge of the harmful effects of pesticides, and the high retail price of new pesticides have kept many pesticides on the market that present considerable environmental risk. Table 8.3 lists the most used pesticides for agricultural purposes in Vietnam.

A study conducted by the National Institute of Plant Protection Sciences (Cung, 1995) reported that more than 70 percent of rice farmers used older, familiar insecticides, e.g. methyl parathion, methamidophos, cartap (S,S,-2-dimethyl-aminotrimethylene bis-(thiocarbamate) hydrochloride), chlorpyrifos, deltamethrin,

Table 8.2 List of restricted use pesticides in Vietnam[a]

| | Pesticide class | | |
Insecticides	Fungicides	Herbicide	Rodenticide
Aluminum phosphide	MAFA (CH_5AsO_3)	Paraquat	Zinc phosphide
Carbaryl 4% + Lindane 4%	Methyl iso-thio-cyanate 5% +		
Carbofuran	Quaternary ammonium		
Dichlorvos	compounds 25%		
Dicofol	Sodium pentachloro-		
Dicrotophos	phenoxide		
Endosulfan	Sodium tetraborate		
Magnesium phosphide	decahydrate		
Methamidophos	(borax) 54% +		
Methyl bromide	Boric acid 36%		
Monocrotophos	Tribromophenol		
Phosphamidon	Tributyl tin naphthenate		

Note:
a Adapted from MARD, 1996.

monocrotophos, and dichlorvos; fungicides, e.g. validamicyn, isoprothiolane, and iprobenfos; and herbicides, e.g. 2,4-D and pretilachlor (Table 8.3). Farmers use these pesticides for controlling BPH, leaf roller, stem borer, and various other pest insects, leaf blights, and blast diseases of rice. A study conducted in 1993 by Chien (1994), reported that almost 100,000 ha of spring season rice was infected by BPH in the Red River Delta and that autumn season BPH damage in the Mekong River Delta was estimated at more than 200,000 ha. During the winter to spring 1990 to 1991 season, more than 200 T of BPMC (fenobucarb) was used against BPH in the Red River Delta. In the summer to autumn 1991 season, 1,758 T of all pesticides (converted into BPMC equivalents) were used in the southern provinces of Vietnam against this insect (Duy Nghi, 1991).

Table 8.4 lists the pesticides used by vegetable farmers from the suburbs of Hanoi (Cung, 1995) and the percentage of farmers using each pesticide. Pesticide application rates normally follow guidelines developed by local plant protection experts. Usually the application rate for rice is 0.3–0.5 kg a.i. ha^{-1} and that for vegetable crops is 0.5–1.0 kg a.i. ha^{-1} per year. Application frequency varies between two and three times for each crop cycle. The frequency might be somewhat higher than three times because farmers do not apply pesticides based on an economic threshold (the minimum number of pest insects per unit area that would result in a reduction of crop yield or value). Using this as a triggering mechanism, pesticides would be applied only after the insect's or disease organism's population reached

Table 8.3 **The most used pesticides for agricultural purposes in Vietnam**[a]

Pesticide	No. of formulations	Pesticide	No. of formulations
Insecticides		*Fungicides*	
Buprofezin	6	Isoprothiolane	5
Carbaryl	5	Kasugamycin	4
Cartap	4	Mancozeb	11
Cypermethrin	22	Thiophanate methyl	14
Diazinon	13	Validamicyn	10
Fenobucarb	13	Zineb	13
Isoprocarb	1		
Dimethoate	14	*Herbicides*	
Fenitrothion	15	Atrazine	7
Chlorofos	1	Benthiocarb	1
Triazophos	1	Butachlor	18
Fenvalerate	4	2,4-D	9
		Glyphosate (IPA) salt	28
Fungicides		Pretilachlor	2
Benomyl	12		
Carbendazim	17	*Rodenticides*	
Copper hydroxide	3	Brodifacoum	2
Copper oxychloride	3	Bromadiolone	2
Edifenphos	5	Diphacinone	1

Note:
a Adapted from Cung, 1995.

Table 8.4 List of pesticides used by vegetable farmers around Hanoi and the percentage of farmers using each pesticide[a]

Pesticide common name	Farmers using the pesticide (%)
Insecticides	
Methamidofos	99.0
Methyl parathion	49.5
Dimethoate	1.0
Phenthoate	34.7
Chlorofos	30.7
Cartap	38.6
Cypermethrin	11.9
Fenvalerate	18.8
Bacillus thuringiensis Berber, (Bt)	1.0
Fungicides	
Zineb	32.7
Kasugamycin + copper oxychloride	1.0
Validamycin	15.0

Note:
a Adapted from Cung, 1995.

the threshold value. Farmers have not yet adopted this approach; hence, local pesticide retailers can only supply 60–70 percent of farmers' pesticide requirements, particularly in regions specializing in vegetable production (Cung, 1995).

Pesticide use and misuse

Self-sufficiency is characteristic of Vietnamese farmers. This character trait appears in the so-called GPS (Garden, Pond and Shed) policy, which is widely encouraged by the Vietnamese Farmers Association. It is a curious fact that every farm household in rural regions tries to own a pond in addition to their land for rice cultivation. Furthermore, in most cases, the ponds are close to their rice fields or in front of the farmer's house and utilized for fish aquaculture. Around each pond, there are gardens for growing vegetables and fruit trees, e.g. citrus, bananas, etc., and the produce grown in these areas receives at least a minimum of pesticide applications. Everything raised in these areas is for the family's consumption or advantage. Farmers in some areas use water from the pond to wash their clothes and farm tools, including pesticide application equipment and pesticide containers.

Vietnamese farmers consider land, buffaloes, and ponds to be their most valuable properties. Buffaloes are used by farmers to plough land, but ponds provide not only food but also additional income to the household. It has become a tradition that when relatives or guests visit their farmer friends the farmer exhibits his prosperous lifestyle to his guests by serving fish from his pond or poultry from his farmstead. Usually farmers raise fish and chickens as a means of generating the additional income required for child rearing. At the beginning of each new school year, farmers must spend a considerable sum of money to provide their children with new clothes and books. This money is generated from the products of the farmers' gardens, ponds, or livestock sheds. The implementation of the GPS policy has greatly improved living standards in rural areas of Vietnam. It is also worth noting that many new rice cultivars combining high yields and a shorter growing season have been introduced into rice cultivation in Vietnam. This allows farmers to utilize better crop rotation cycles and adopt improved farming methods. However, these new rice cultivars are usually only weakly resistant to insect pests and diseases, so that, at present, the continued use of pesticides is a necessity in farming.

A study conducted by N.D. Thiet (1994), an ecologist from Hanoi Medical College, reported that 100 percent of farm households in Thanh Tri, a district just south of Hanoi, use pesticides in their farming activities. All farms used methamidophos and 10 percent also used chlorofos, methyl parathion, fenobucarb, and validamycin. The author found that pesticide storage was a serious problem. He found that 58 percent of pesticide storage was in water closets (toilet facilities) or animal sheds, 27 percent was outside under garden bushes, while 12 percent was stored in farmers' kitchens, and 3 percent inside living areas or bedrooms. A similar study by an ecologist from the Center for Occupational Health (Ministry of Health Care) examined pesticide storage practices in rural regions distant from Hanoi (Hoi, 1994). He found that 20 percent of farm households stored pesticides in their living spaces, 50 percent stored them in their kitchens, and 30 percent outside in their gardens. Furthermore, 55 percent of farmers stored pesticides in unlabeled bottles. This lack of the knowledge about how to properly store and safely handle pesticides combined with a tradition of perpetuating old habits among farmers provides ample opportunity for food and environmental contamination from pesticide use in rural areas. Kannan et al. (1992) found elevated levels of PCBs, DDTs, HCHs, aldrin, and dieldrin in foodstuffs collected from different locations in Vietnam (Table 8.5). Average daily intake of some OCs by the Vietnamese people were higher than those observed in most developed nations with DDTs highest among the various chemicals studies and PCBs comparable to developed countries. Other Ministry of Health studies (Quyen and Van San, 1994; Quyen et al., 1994) found that most pesticides are applied by farmers, themselves, wearing neither protective clothing, nor eye-wear, nor shoes and fully 15 percent of rural households store pesticides inside the house. OPs were the major cause of occupational poisoning cases from 1986 through 1991 (Le Trung, 1994).

Historically, DDT was heavily used by farmers in Vietnam and some evidence suggests insecticides such as p,p'-DDT have been heavily used in the recent past (Schecter et al., 1997). Schecter et al. (1997) studied the association between blood levels of p,p'-DDT and its metabolite, p,p'-DDE, and the risk of invasive breast cancer among residents of norther Vietnam. Their results suggest that recent and past exposure to p,p'-DDT does not play an important role in the etiology of breast cancer among women living in a tropical country where DDT is still used for mosquito control.

Verle, et al. (1999) conducted a cost analysis in Hoa Binh Province – a mountainous province in northern Vietnam – comparing the cost of residual

Table 8.5 OC pesticide and PCB residues in foodstuffs collected from different locations in Vietnam

Foodstuff	Concentration range (mean) in ng g⁻¹ ww							Reference
	$\sum DDT$	$\sum BHC$	$\sum Heptachlor$	Aldrin	Dieldrin	HCB	$\sum PCB$	
Rice	0.96–3.3 (2.0)	2.7–6.4 (4.3)	0.03–0.28 (0.13)	<0.1–0.49 (0.19)	0.12–0.48 (0.26)	<0.01–0.05 (0.03)	0.95–3.4 (2.5)	Kannan et al., 1992
Pulses	0.34–3.0 (1.9)	0.13–26 (5.0)	0.01–0.18 (0.07)	<0.1	<0.1–0.14 (0.08)	<0.01–0.18 (0.04)	1.0–16 (4.0)	
Butter	7.0–7.3 (7.2)	17–81 (49)	3.9–12.24 (8.13)	<0.1	1.7–2.7 (2.2)	0.2–9.7 (5.0)	11–22 (17)	
Animal fat	61–180 (130)	8.9–130 (70)	0.45–3.79 (2.44)	1.2–1.8 (1.5)	0.5–13 (5.3)	0.29–0.65 (0.41)	21–128 (61)	
Meat	10–86 (48)	0.61–34 (17)	0.18–0.51 (0.35)	0.09–0.22 (0.16)	0.73–1.9 (1.3)	0.03–0.18 (0.11)	7.3–29 (18)	
Fish[a]	3.9–76 (26)	0.58–4.0 (1.8)	0.04–0.45 (0.12)	<0.1–0.68 (0.12)	<0.1–0.42 (0.17)	0.01–0.31 (0.05)	3.1–24 (10)	
Fish from Hanoi	13–76 (36)	0.89–4.0 (2.3)	–[b]		0.12–0.8[c] (0.36)		5.2–18 (11)	Kannan et al., 1995a
Fish from Phu Da	3.9–52 (20)	0.6–1.4 (0.92)	–		<0.01–0.53[c] (0.29)		3.1–24 (12)	
Fish from Ho Chi Minh City	1.7–78 (21)	0.63–3.8 (2.0)	–		0.12–0.51[c] (0.31)		3.7–59 (18)	
Prawn[d]	1.7	1.5	0.15	0.03	0.25	0.03	6.6	Kannan et al., 1992
Shellfish	7.2	2.8	0.12	0.09	0.40	0.04	15	
Crab	78	1.4	0.33	0.13	0.38	0.17	59	

Notes:
a Purchased from fish markets in Hanoi, Thua Thien Province, and Duyen Hai district, east of Ho Chi Minh City. b En dash (–) indicates no data. c Aldrin and dieldrin combined. d Prawn, shellfish, and crab were pooled samples collected from Duyen Hai.

spraying for the control of malaria mosquito vectors with λ-cyhalothrin and the cost of permethrin-treated bednets. The actual cost of insecticide required per person per year was lower for impregnating bednets (US$0.26) than for spraying (US$0.36), although the determining factor was the cost of the net (US$0.58 assuming a five-year life of the net). For the National Malaria Control Programme of Vietnam, the cost per person per year for impregnated bednets was only US$0.32, primarily because the majority of nets are bought by the population. Mosquitos, specifically *Aedes aegypti* L. (Diptera: Culicidae), are also the vector of dengue viruses, a major health concern in Asia (Huber *et al.*, 2000). As urban centers in Southeast Asia have rapidly expanded, so to has *A. aegypti*, thriving in new ecological and demographic settings. The dramatic increase in dengue transmission is the result of these changes and the development of resistance to insecticides by the vector species (Huber *et al.*, 2000).

The use of Agent Orange – a mixture equal amounts of the two active ingredients, 2,4-D and 2,4,5-T – as a defoliant by the USA in the Second Indochina War from January 1965 to April 1970 has posed questions about long-term health effects in the south of Vietnam, particularly in light of recent developments linking TCDD or dioxin (a trace contaminant of 2,4,5-T) to a number of malignancies and other disorders. Ha *et al.* (1996) investigated claims of an increased risk for gestational trophoblastic disease in a controlled study in Ho Chi Minh City. No significant difference was found between cases and controls based on a cumulative Agent Orange Exposure index, nor was a difference found for agricultural use of pesticides – increased risk was associated with breeding of pigs. However, Schecter *et al.* (2001) found a marked elevation of TCDD in blood samples from persons living in Bien Hoa, a city in southern Vietnam located near a former US airbase where a spill of Agent Orange occurred more than 30 years ago. Compared to blood TCDD levels of residents of Hanoi (two parts per trillion), where Agent Orange was never used, residents of Bien Hoa showed an increase of up to 135-fold. Soil and sediment samples taken nearby were found to contain TCDD. They hypothesized that a major route of current and past exposure was through movement of dioxin from soil into river sediment, then into fish, and finally into people from fish consumption (Schecter *et al.*, 2001).

Location of pesticide application with respect to the marine environment

Vietnam is a tropical country (Lat. 8° to 23°30'N, Long. 110° to 103°E) with an average precipitation and temperature range of 2,000 to 2,500 mm and 15 to 34°C, respectively. It can be divided into seven major agricultural zones as shown in Figure 8.1. There is a northern midland and mountainous region (Reg. 1), the Red River Delta region (Reg. 2), the north-central coastal region (Reg. 3), the south-central coastal region (Reg. 4), the central highland region (Reg. 5), the Nambo eastern region (Reg. 6), and the Mekong River Delta region (Reg. 7) (Chuong, 1995). The northern midland and mountainous zone specializes in tea and fruit

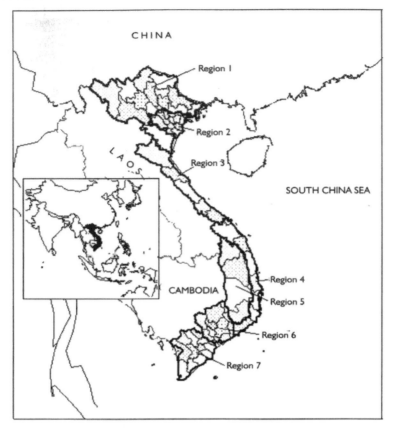

Figure 8.1 Major agricultural zones in Vietnam

production while the Red River (in the northern portion of Vietnam) and Mekong River (in the south) delta zones specialize in rice and vegetable farming. The north-central coastal zone produces legume crops and the south-central coastal zone specializes in short-lived industrial crops such as sugarcane and tobacco. The central highland zone raises long-lived industrial crops, e.g. tea and coffee, and the Nambo eastern region is the home of rubber and fruit production areas. The Red River and the Mekong River deltas, with an area of 1.29 and 3.9 million hectares, respectively, are the most important agricultural zones in Vietnam. While the Red River and Mekong River deltas occupy 15.7 percent of the country's land area, they supply food for almost 60 percent of the country's population (Chuong, 1995). Because agricultural activities in these river deltas involve pesticide-intensive rice and vegetable cultivation, much greater consumption of pesticides occurs in Regions 2 and 7 compared to other regions.

The Red River crosses through the north of Vietnam below Hanoi and flows into the Gulf of Tonkin while the Mekong River enters the southern portion of Vietnam and flows into the South China Sea below Ho Chi Minh City. They carry huge amounts of alluvial soil into their respective deltas each year. Although Vietnam has a coastline exceeding 3,260 km, pesticide and other industrial pollution of marine environments comes primarily from these two agricultural zones.

Each year the Red River transports 140 trillion T of water and 140 M T of sediment into the sea. The respective figures for the Mekong River are almost four times higher. Thus, a significant amount of pollution originating from agricultural activities could be discharged with water and sediment into the sea from these two regions. Moreover, it should be noted that Hai Phong and Ho Chi Minh City are port cities where many industrial and petrochemical plants operate. Offshore of Ho Chi Minh City, crude oil is extracted from existing oil fields and offshore drilling rigs continue exploration. These activities can also contribute pollutants into the marine environment.

PESTICIDE RESIDUES IN BIOTIC AND NONBIOTIC MATRICES IN VIETNAM

Residues in water

As pesticides enter the environment, they become subject to numerous degradation mechanisms, which ultimately determine their environmental fate. These include photolysis, hydrolysis, oxidation, isomerization, and biodegradation. OC pesticides such as DDT, the 'drins' (aldrin, dieldrin, etc.), chlordane, HCB, and poly-chlorinated biphenyls (PCBs) are some of the most resistant anthropogenic chemicals to environmental degradation. Both OCs and PCBs are hydrophobic and their solubility in water is quite limited, e.g. 2 ppb for DDT (Budavari *et al.*, 1989), but they accumulate to high levels in biota due to their strong lipophilicity. DDT and its metabolites DDD and DDE were detected in the waters of Lake Baikal (Central Siberia Russian Republic) at (87 ± 37) pg L^{-1}, where DDT insecticide had never been applied (Kucklick *et al.*, 1994). The long-term persistence and volatilization into the atmosphere of DDT and its capacity to be transported long distances from the site of application were the basis for explaining these findings. Consequently, DDT and its metabolites can be found anywhere on earth because the insecticide and its metabolites are removed from the atmosphere by deposition with rainwater.

Recently several studies in Vietnam have focused on monitoring the level of OC pollutants in freshwater from rice paddies and estuarine waters (Am *et al.*, 1995; Hoa, 1995; Thanh Tu, 1995). Hoa (1995) found OC pesticide residues during the dry season (November to April) in water from irrigation canals for rice paddies near Hanoi. Residue levels were 0.70, 0.40, and 2.10 μg L^{-1} for lindane, HCB, and $\sum$ DDT, respectively. During the rainy season (June to September), these residues were found at 0.33, 0.30, and 0.68 μg L^{-1}, respectively. Am *et al.* (1995) analyzed for OC contamination in water samples from rice paddies and the Red River Estuary in the dry season of 1994. He found that among all the OC pesticides soluble in water, only p,p'-DDE was detectable – at $0.13 \pm 0.04 \mu$g L^{-1} and independent of the site of sampling. This seems to suggest that DDT contamination in these areas was the result of long-term use of the pesticide. It is well known that, in the environment, p,p'-DDT is easily transformed to p,p'-DDE, and p,p'-DDE has higher water solubility (8 μg L^{-1}) compared to p,p'-DDT (2 μg L^{-1}) and p,p'-DDD (5 μg L^{-1}) (Budavari et al., 1989). Thus, p,p'-DDE has been found to distribute equally over a large sampling area.

Thanh Tu (1995) surveyed the level of OC contamination in waters of central Vietnam during 1990 to 1995. She found the concentration of $\sum$ DDT and lindane in grapevine irrigation water (stored in ponds) to be from 0.2 to 0.4 and 1.5 to 8.1 μg L^{-1}, respectively. All DDT and lindane concentrations measured in freshwater and estuarine water in Vietnam are lower than their water solubility.

In contrast to OC pesticides, OPs are usually polar substances with high water solubility. Thus, they are expected to exist at high concentrations in surface waters, particularly in rice paddies where they are applied. However, OP pesticides are known to be easily hydrolyzed, especially in waters with a high pH (Sethunathan and Yoshida, 1969). Methyl parathion, sumithion, and chlorpyrifos have been found in waters from rice paddies, vegetable fields, and irrigation canals in northern Vietnam (Tuan et al., 1992). Tuan et al. (1992) found concentrations of methyl parathion, sumithion, and chlorpyrifos in water from rice paddies near Hanoi ranged from 0.5 to 1.0 μg L^{-1}, while values from vegetable fields were a bit higher, at 2 to 3 μg L^{-1}. The difference could be attributed to the higher label application rate for vegetables as compared to rice.

Nhan (unpublished results) attempted to measure OP residues in estuarine waters from the coastal zone of the northeast sea of Vietnam, but residues in these areas were always below the detection limit of gas chromatography with a flame ionization detector (GC-FID detection limit 0.001 μg L^{-1}). This suggests the high ambient temperatures in the region accelerate the degradation of OP pesticides and remove them from freshwater drainage before they reach the sea.

Residues in sediments

It is well known that deposition onto suspended solids and bottom sediments is one of the distribution modes for pesticides in the environment. The distribution coefficient, K_d, defined as the ratio of the concentration of a pesticide-pollutant in sediment (μg kg^{-1}) to that in water at equilibrium (μg L^{-1}), is dependent on the water solubility, or more correctly on the K_{ow} (the octanol-water distribution coefficient) of the pesticide. As a rule, substances with high K_{ow} values appear to have high K_ds. K_d is also strongly dependent on the suspended solid content of the water column, on the particle size of the solids, and on the organic carbon content of the sediment. Gonzalez-Farias and Mee (1988) have shown that organic matter in a form similar to humic substances, plays a major role in the accumulation of chlorinated hydrocarbons in sediment. In fact, humic substances affect the adsorption and solubilization of pesticides between interfaces (Rav-Cha and Reblun,

1992) and pesticides sorbed to sediments appear to be more persistent in the environment. This suggests care be taken when considering the disappearance of pesticides in the tropical marine environment, because mangrove trees are common along tropical coastal zones and they release a significant quantity of humic substances into sea water.

Degradation of pesticides sorbed onto sediments is dependent on the activity of anaerobic microorganisms, sediment pH, and the organic matter content of the sediment. The high capacity of pesticides to adsorb to sediments contributes to concern that the marine environment is threatened by pesticide residue accumulation, particularly in tropical regions where rainfall is usually great. As a result of heavy rains, erosion could transport large amounts of sediment from contaminated areas to the sea. Hence, estuarine waters are potentially high impact areas.

Nhan *et al.* (1998) found the concentration of γ-HCH and $\sum$ DDT insecticides in sediment with 7 to 10 percent organic carbon content from the Red River's mouth ranged between 0.1 to 5.0 ng g^{-1} dw and 4.0 to 7.0 ng g^{-1} dw, respectively, during the dry season (November to April). In the rainy season (June to September), the concentration of lindane and $\sum$ DDT was between 0.03 to 0.1 ng g^{-1} and 3.0 to 5.0 ng g^{-1} dw, respectively (see Table 8.6). Of all the DDT metabolites, *p,p'*-DDE is the most significant; its contribution to $\sum$ DDT represents almost 70 percent of the total. These levels of lindane and $\sum$ DDT are comparable to the concentration of the contaminants in sediment from other coastal regions of South East Asia (Everaarts *et al.*, 1991).

Both dry season and wet season concentrations of all cyclodienes, e.g. aldrin, dieldrin, and endrin (collectively, the 'drins'), were found to be below detection limits for GC-ECD (electron capture detector) (Nhan *et al.*, 1998), confirming that these insecticides were not used in Vietnam. Further evidence for this conclusion can be found in the study of sediments from freshwater canals in the Hanoi region (Nhan *et al.*, 2001). Only trace levels of the 'drins' were found in the 5 cm surface layer sediment samples from 12 sampling stations – 0.01 to 0.15 ng g^{-1} dw dieldrin, 0.01 to 0.10 ng g^{-1} dw endrin, and 0.01 ng g^{-1} dw aldrin – and these levels may be attributed to atmospheric transport and deposition of these persistent pesticides. However, it is possible that endosulfan and chlordane may have been used in Vietnam, but the amounts applied would have been small (Nhan *et al.*, 2001).

Generally sediment from rice paddies contains higher levels of insecticides compared to estuarine waters during the rainy season. During the 1996 rainy season, the concentration of $\sum$ DDT in paddy sediment ranged between 13 and 15 ng g^{-1} dw, a level two-fold higher than that for the dry season (Nhan *et al.*, 1998). This suggests that DDT was used during the 1996 rainy season in the Red River Valley (Nhan *et al.*, 1998).

Nhan *et al.* (1999) determined OC pesticide concentrations in sediments collected during the rainy season from the northern coast between the Balat Estuary and the Ca long River, primarily a rice growing region (Table 8.7). A higher than expected proportion of γ-HCH, i.e. 30 percent vs. 8 to 15 percent, suggested current usage of lindane in Vietnam. Also higher concentrations of $\sum$ HCHs and

Table 8.6 Seasonal (dry vs rainy) changes in concentrations of chlorinated pesticides and PCBs (as Arochlor 1254) in surface sediment samples from the Red River Delta, Vietnam (units are ng g^{-1} dw)

Sample location	Lindane		p,p'-DDE		p,p'-DDT		∑ DDT		Arochlor 1254		Reference
	DS[a]	RS	DS	RS	DS	RS	DS	RS	DS	RS	
Irrigation canal NW of Hanoi	0.14	0.07	3.90	8.52	2.43	1.20	6.95	12.95	5.64	3.83	Nhan et al., 1998
Rice paddy SE of Hanoi	0.41	0.16	4.60	10.17	1.71	1.03	7.46	14.30	10.56	2.17	
Mouth of Red River	0.50	0.05	4.60	1.52	1.30	0.51	7.11	3.04	3.27	1.23	
Intertidal mudflats, Thai Binh Province[b]	0.49 (0.13)	0.09 (0.06)	4.80 (0.89)	3.67 (0.71)	1.16 (0.42)	0.55 (0.10)	6.41 (0.77)	4.87 (0.23)	2.10 (0.46)	1.11 (0.21)	

Notes:
a DS indicates dry season sample; RS indicates rainy season sample.
b Mean (± standard deviation) concentration from three sample sites – Cua lan, Traly, and Diem dien estuaries north of the Red River Estuary.

Table 8.7 Concentrations of chlorinated pesticides and PCBs (as Arochlor 1254) in surface sediment samples collected from various locations around Vietnam (units are ng g⁻¹ dw)

Sample location	Season	HCB	Lindane (γ-HCH)	∑HCH	p,p′-DDE	p,p′-DDT	∑DDT	Arochlor 1254	Reference
Saigon River, 15 km above Ho Chi Minh City	Rainy	–ᵃ	–	–	0.98	0.42	1.76	NDᵇ	Phuong et al., 1998
Nhieu Loc-Thi Nghe canal Ho Chi Minh City	Rainy	–	–	–	73.69	98.50	253.62	381.7	
Confluence of Saigon River and Dong Nai River	Rainy	–	–	–	1.09	0.88	4.10	9.3	
Tan Hoa-Lo Gom canal in Ho Chi Minh City	Rainy	–	–	–	27.00	53.90	129.40	590.5	
Suburban Ho Chi Minh City	Rainy	–	–	–	8.82	2.18	18.18	38.8	Nhan et al.,1999
Mouth of Ca long River	Rainy	6.53	8.12	33.74	6.12	3.16	10.41	28.15	
Ha long Bay	Rainy	0.15	0.68	1.84	5.12	1.69	7.24	14.12	
Hai phong Harbor	Rainy	0.17	0.53	1.73	4.82	1.32	6.70	24.89	
Balat Estuary	Rainy	0.16	0.36	1.23	4.63	1.11	6.25	0.47	
Viettri City (5 industrial and agricultural sites)	–	–	0.68	–	–	–	5.2	2.3ᶜ	Viet et al., 2000
Hanoi (11 industrial and non-industrial sites)	–	–	0.35	–	–	–	5.0	45ᶜ	
Ha long Bay (6 sites)	–	–	6.06	–	–	–	28	37ᶜ	
Cau Dien Hanoi, Nhue River (urban)	Rainy	ND	0.04	0.58	50.0	10.20	71.29	0.79	Nhan et al., 2001
Rural area on Red River below Hanoi	Rainy	0.01	0.21	0.65	5.17	0.73	7.40	2.55	
Hanoi city center (urban)	Rainy	ND	0.06	0.44	8.15	1.05	12.54	6.31	
Paint factory Cau Buou on To Lich River, Hanoi (urban)	Rainy	0.13	0.35	3.12	37.50	8.30	80.55	39.50	
Agricultural, Gia Lam East of Hanoi	Rainy	ND	0.13	0.46	13.10	2.67	16.99	1.86	

Notes: a En dash (–) indicates no data available. b ND indicates compound not detected. c Reported as total PCBs.

HCB were found near the mouth of the Ca long River indicating higher usage of both pesticides in that valley's rice fields. The same trend, although less pronounced, was observed for $\sum$ DDT concentrations, which ranged from 6.2 to 10.4 ng g^{-1} dw. These values are comparable to $\sum$ DDT measured in sediments of the Red River Valley – 5.8 to 14.3 ng g^{-1} dw (Nhan et al., 1998). Viet et al. (2000) detected DDT and its metabolites and lindane in all sediment samples analyzed with the highest levels found in samples from Ha long Bay (highest concentrations were 34 ng g^{-1} dw $\sum$ DDT and 9.5 ng g^{-1} dw lindane) followed by Viettri City and Hanoi.

Around Hanoi, Nhan et al. (2001) detected relatively high concentrations of DDT in all sediment samples. Concentrations of $\sum$ DDTs from sampling locations located close to densely populated areas ranged from 15 to 81 ng g^{-1} compared to 7 and 8 ng g^{-1} in samples from more rural areas. The metabolite p,p'-DDE always accounted for the largest percentage (reaching 70 to 90 percent) of $\sum$ DDTs. The fraction of p,p'-DDT in sediment samples ranged from 0.1 to 0.15 compared to a ratio of 0.2 to 0.3 for samples from Ho Chi Minh City as determined by Phuong et al. (1998). Phuong et al. (1998) concluded that recent application of DDT had occurred in the Ho Chi Minh City area but they could not identify whether the use was agricultural or for public health purposes. Nhan et al. (2001) found o,p-isomers of DDT, which undergo a relatively rapid isomerization to the p,p'-isomers, present in all sediment samples except for a rice paddy located far from villages. The o,p-isomers were more abundant in samples from densely populated sites than those from rural areas further suggesting that DDT is currently being used but primarily for mosquito and fly control in public health campaigns, not as a crop protection chemical (Nhan et al., 2001). In contrast to DDT, concentrations of $\sum$ HCHs in sediment samples from the Hanoi region were much lower ranging from 0.1 to 3.1 ng g^{-1} dw (Table 8.7). While noting that there was no reason to believe that a lesser amount of technical HCH than DDT had been used, they suggested that the likely explanation for the currently lower concentrations of HCHs in sediments of Hanoi's canals is related to the relatively short environmental half-lives of HCHs in soils, its lower log K_{ow}, and its orders of magnitude higher water solubility. The proportions of the three HCH isomers are roughly similar to those of the technical HCH mixture.

In the south of Vietnam, results from analyses of sediment samples from Ho Chi Minh City canals and the Saigon River in 1996 by Phuong et al. (1998) show that $\sum$ DDT concentrations (ranging from 1.76 to 253.62 ng g^{-1} dw) were higher in urban and industrial areas and lowest in rural and suburban zones (Table 8.7). In 1999 Do Hong et al. (2000) evaluated the toxicity of both aqueous and organic solvent extracts of sediment samples collected from the same sites around Ho Chi Minh City. They found that surface sediments were more toxic to Daphnia magna Straus (Crustacea: Cladocera: Daphniidae) and Vibrio fischeri Beijerinck (Proteobacteria: Vibrionaceae) than those from deep sediments and that the toxicity of the organic sediment extracts was higher than in 1996, suggesting an increase in the pollution load of canals and the Saigon River.

In 1996, Nhan *et al.* (1998) conducted a survey of PCB pollution of the Red River Delta and the coastal area of Thai Binh Province, Vietnam. They found that the concentration of Aroclor 1254 (see Table 8.6) in sediment from rice paddies was several fold higher than the concentration in sediments from estuaries. The highest concentrations were found in the dry season and the lowest in the rainy season. PCB concentrations in sediment from estuaries ranged between 0.1 to 1.5 ng g^{-1} dw, but the concentration of individual congeners – particularly heavily chlorinated PCBs – was found to be very low (0.01 to 0.07 ng g^{-1} dw). Washout seems to play an important role in the dilution of the contaminants. Nhan *et al.* (1999) found concentrations of PCBs in sediments from the Balat Estuary to the Ca long River to be generally low, with a highest concentration of 28.15 ng g^{-1} dw. Higher concentrations were recorded near large cities, whereas the lowest levels were from rural areas. Also in 1997, Nhan *et al.* (2001) measured the concentration of 13 individual PCB congeners and two mixtures (Aroclor 1254 and 1260) in sediments from freshwater canals in the Hanoi region (Table 8.7). The concentration of summed PCB congeners and mixtures was relatively low throughout Hanoi ranging from 0.97 to 33.68 ng g^{-1} dw PCB, 0.79 to 39.50 ng g^{-1} dw Aroclor 1254, and 0.35 to 6.3 ng g^{-1} dw Aroclor 1260. However, concentrations of PCBs in sediments collected near industrial sites were 25 times higher than concentrations measured in agricultural areas, clearly pointing to industrial sites around Hanoi as sources of these contaminants. In other areas of northern Vietnam, Viet *et al.* (2000) also found PCB residues were generally higher in surface sediment samples than in soil samples and ranged from 0.64 to 120 ng g^{-1} dw in sediments with the highest concentrations recorded from Hai long Bay, probably due to the high level of mining activities in the area.

Phuong *et al.* (1998) measured levels of six PCB congeners in surface sediments from the Saigon River and major urban waterways of Ho Chi Minh City during the rainy season of 1996. PCBs were found in all samples except the upstream (~15 km above the city) sampling site, which is also above the South China Sea's tidal influence (Table 8.7). The lowest PCB concentrations – extrapolated based on the $\sum$ of six measured congeners and their theoretical contribution to Arochlor 1254 – were found in rural and suburban sites (9.3 to 38.8 ng g^{-1} dw) and the highest (146.5 to 398.6 ng g^{-1} dw) from the Nhieu Loc-Thi Nghe waterway, which drains the city center and part of the airport, and the Gom-Tan Hoa canal (590.5 ng g^{-1} dw). To a large extent, they attributed the contamination to the urban center with its high concentration of large factories (~700), untreated wastewater discharges, accidental spills, and direct solid waste dumping. They also noted that the similar distribution pattern of PCB congeners at most sites may be due to tidal influence.

Residues in biota

Natural and synthetic organic chemicals or xenobiotics including pesticides that are present in soil, sediment, water, and the atmosphere may be taken up by biota.

Xenobiotic uptake, distribution, and accumulation are dependent on the physical and chemical properties of the compounds including their molecular weight, solubility, vapor pressure, K_{ow}, and the physiology and other characteristics of the biological species, including its feeding method and environment, lipid content, etc. Bioconcentration of pesticides has been characterized by a BCF expressed in kg L^{-1} that is defined as the ratio of the concentration of a pesticide in biological tissues (μg kg^{-1}) to the concentration in water (μg L^{-1}).

OC pesticides and PCBs are known to be highly hydrophobic, extremely lipophilic, and persistent substances in the environment. These chemicals accumulate in nonpolar animal lipids. Of all biota, benthic organisms seem to be good biological indicators for monitoring the level of PCB and OC contamination in the aquatic environment because they take up contaminants from all environmental compartments including water, sediment, suspended matter, and food (McLee et al., 1980; Marinucci and Bartha, 1982).

OC and PCB residues in freshwater organisms

Kucklick et al. (1994) determined the level of OC pesticides in biota from Lake Baikal (Central Siberia, Russian Republic). They found that the concentration of $\sum$ DDT, $\sum$ HCH, $\sum$ PCBs, and toxaphene in omul or whitefish Coregonus autumnalis migratorius Georgi (Salmoniformes: Salmonidae) was 0.61, 0.026, 1.2, and 1.1 mg kg^{-1} lipid, respectively. For Lake Baikal seals or nerpa Phoca siberica (Pinnipedae: Phocidae), concentrations were 62, 0.21, 26, and 2.3 mg kg^{-1} lipid, respectively. Of all DDT metabolites, p,p'-DDE constituted about 40 percent of the total in both fish and seals. The primary PCBs accumulating in omul and Baikal seal lipids were congeners 101, 118 + 143, 105 + 132 + 153, and 138, mostly penta- and hexa-congeners. They observed a linear relationship between the log BCF and log K_{ow} with a slope of 0.47 for the omul and 0.88 for the Baikal seal.

In contrast, Nhan et al. (1998) found that clams Hyriopsis cummingii Lea (Bivalvia: Unionoidae) collected from rice paddies in northern Vietnam during the 1996 rainy season had concentrations of HCB, lindane, $\sum$ DDT, Aroclor 1254, and Aroclor 1260 of 0.11, 0.62, 78.3, 7.7, and 9.5 ng g^{-1} tissue dw, respectively (Table 8.8). The hexane extractable lipid content (HEOM) of the tissue was 12.2 and 12.0 mg g^{-1} for dry season and wet season samples, respectively. The concentrations of the OC contaminants HCB, lindane, DDT, Aroclor 1254, and Aroclor 1260 found in carp Cyprinus carpio L. (Cypriniformes: Cyprinidae) were 0.15, 1.76, 179.8, 34.0, and 16.0 ng g^{-1} dw tissue, respectively. HEOM in carp was 8.5 and 8.8 mg g^{-1} for dry season and wet season samples, respectively. The relatively high concentration of o,p'-DDT in freshwater biota collected during the rainy season strongly indicates continued usage of DDT in the Red River Valley during the rainy season. The concentration of p,p'-DDT in fresh water organisms from rice paddies of the Red River Valley are well above the maximum admissible concentrations for food (3.5 μg g^{-1} lipid) according to European standards.

Table 8.8 Seasonal (dry vs rainy) changes in concentrations of chlorinated pesticides and PCBs (as Arochlor 1254) in freshwater and marine biota from the Red River Delta, Vietnam (units are ng g⁻¹ dw)

Location	Sample	HCB		Lindane		p,p´-DDE		p,p´-DDT		Arochlor 1254	
		DS[a]	RS	DS	RS	DS	RS	DS	RS	DS	RS
Irrigation canal NW of Hanoi	Clam Hyriopsis cummingi	0.11	0.06	0.62	0.54	41.3	35.0	25.7	36.0	–[b]	7.7
Rice paddy SE of Hanoi	Carp Cyprinus carpio	0.15	0.23	1.76	1.50	109.7	120.0	27.5	21.0	–	34.0
Balat Estuary[c]	Shrimp Metapenaeus ensis	0.50	0.092	1.20	0.81	3.6	3.1	1.1	0.83	–	5.1
	Clam Meretrix meretrix	0.51	0.083	0.72	0.54	10.3	9.0	3.25	4.20	–	6.8
Diem dien Estuary[c]	Clam Mactra quadrangularis	0.13	–	0.69	–	15.5	–	5.3	–	8.1	–

Source: Adapted from Nhan et al., 1998.

Notes:
a DS indicates dry season sample; RS indicates rainy season sample.
b En dash (–) indicates no data.
c Marine estuary located in the Red River Delta.

Nhan *et al.* (2001) found a positive correlation between the accumulation of DDT and DDE in freshwater snail (*Angulyagra* sp.) soft tissues from canals around Hanoi and their level in sediment samples from the same canals, indicating the reservoir of DDTs in sediments is the main source of these contaminants to biota (Table 8.9). Conversely, the concentrations of HCHs and lindane in snails were not correlated with concentrations measured in sediments, suggesting that the accumulation of these compounds by snails is not related to the reservoir in sediments but is a function of absorption from the water. They also noted that lindane in molluscs accounts for a much higher percentage of total HCHs (63 ± 9 percent, n = 8) than measured in sediments. Lindane, under the trade name of '666' became available in Vietnam as a replacement for technical HCH mixture, being imported from the People's Republic of China. Particularly high pesticide concentrations in snails' soft tissues were measured in samples from highly urban areas of Hanoi. When HEOM content (~2 percent for snails) was taken into account, the concentrations of p,p'-DDE were about 10 times higher than maximum admissible values for food accepted by EU countries.

The concentration of 13 PCB congeners in snails' soft tissues from canals located near a transformer and electrical isolater materials factory or a painting factory were the highest values measured, 43 and 58 ng g^{-1}, respectively (Nhan *et al.*, 2001) (see Table 8.9). Despite the range of PCB concentrations found, the levels in Hanoi canals are relatively low. In a long-term study, de Boer *et al.* (1994) monitored PCB accumulation and excretion in eel *Anguilla anguilla* L. (Anguilliformes: Anguillidae) caught in the Rhine River, The Netherlands. They found eels accumulated all PCB congeners in their tissues, up to 10 mg kg^{-1}. This high level of PCBs was thought to be caused by PCB leakage from hydraulic mining equipment and the resultant heavily polluted waters of this area. After eight years, eels transferred from the river to the relatively clean waters of Lake Milligensteeg showed practically no excretion of PCBs and OCs into the environment; their body burden remained almost unchanged. A decrease in PCB or OC concentration in biota tissues was attributed to dilution of the pollutants from increased lipid content generated by normal eel growth (de Boer *et al.*, 1994).

Residues in estuarine and marine organisms

The concentration of p,p'-DDT in shrimp *Metapenaeus ensis* de Haan (Decapoda: Penaeidae) and the bivalve molluscs *Meretrix meretrix* L. (Bivalva: Veneridae) and *Mactra quadrangularis* Deshayes (Bivalva: Mactridae) taken from the mouth of the Red River in northern Vietnam was measured at 4.86, 18.75, and 24.3 ng g^{-1} dw tissue, respectively (Am *et al.*, 1995; Nhan *et al.*, 1998). The hexane extractable organic matter (HEOM) for tissues from each of the three species was 4.25, 20.3, and 23.3 mg g^{-1}, respectively. Extrapolating from these data, the concentration of p,p'-DDT residue in epibenthic and benthic organisms from the coastal zones of the northern Vietnam ranged from 92 to 115 ng g^{-1} HEOM. The concentration of p,p'-DDT and PCB contaminants in marine biota samples are comparable to the concentrations measured elsewhere in South East Asia (Everaarts *et al.*, 1991).

Table 8.9 Concentrations of chlorinated pesticides and PCBs (as Arochlor 1254) in marine and freshwater biota collected during the rainy season in various locations in northern Vietnam (units are ng g⁻¹ dw)

Location	Sample	HCB	Lindane (γ-HCH)	∑ HCH	p,p´-DDE	p,p´-DDT	∑ DDT	Arochlor 1254	Reference
Mong cai at mouth of Ca long River	Clam *Meretrix meretrix*[a] (marine)	12.21	10.15	46.67	15.20	3.21	23.33	25.30	Nhan et al., 1999
Mong dong Estuary		1.35	2.06	8.02	7.96	1.89	12.00	16.70	
Ha long Bay		0.15	0.75	1.95	8.63	2.03	13.81	16.30	
Hai phong Harbor		0.10	0.70	1.61	9.96	3.15	17.70	20.52	
Balat Estuary		0.08	0.72	1.45	10.30	3.25	18.75	5.10	
Cau Dien Hanoi, Nhue River (urban)	Snail *Angulyagra* sp. (freshwater)	0.02	0.41	0.51	363.60	3.90	393.87	8.13	Nhan et al., 2001
Cau Dien Hanoi, Nhue River (urban)		0.15	0.21	0.40	798.40	6.30	863.95	59.23	
Rural rice paddy near Hanoi (rural)		0.27	0.10	0.17	4.13	0.73	5.62	5.63	
Hanoi city center (urban)		0.38	0.16	0.29	52.10	1.13	60.43	25.17	
Thuong Dinh Hanoi on To Lich River (urban)		0.21	0.23	0.36	89.63	2.57	103.01	75.62	
Rural area on Red River below Hanoi		0.18	0.11	0.15	13.52	1.46	16.68	5.55	
Agricultural area, Gia Lam E of Hanoi		0.21	0.18	0.28	16.56	1.66	19.22	7.52	
Agricultural area, Dong Anh NE of Hanoi		0.25	0.13	0.22	10.17	1.13	12.4	6.83	

Note:
a One dozen clams obtained from local fishermen at each sampling site (of ~ 4–5 cm shell length).

A seasonal change was observed in the accumulation of the p,p'-DDT group in marine organisms. Clams collected in the 1994 dry season contained 20 to 90 ng g^{-1} p,p'-DDT dw, but this was reduced in the 1996 rainy season to 17 to 24 ng g^{-1} dw, reflecting a rainy season dilution effect of pollutants in estuarine waters (Nhan et al., 1998) (see also Table 8.8).

In a 1997 study, Nhan et al. (1999) measured chlorinated hydrocarbons in soft tissues of the clam *Mer. meretrix* (Table 8.9). Concentrations closely reflected contamination of the sediments at the sampling sites and HCB and $\sum$ HCH concentrations followed a similar pattern of decreasing from North to South. For $\sum$ DDT the pattern was less pronounced and $\sum$ DDT in the Balat Estuary approached the value measured in the Ca long Estuary. The constant ratio of p,p'-DDT (and DDE) to $\sum$ DDT in clam tissues – in sharp contrast to the variable ratios found in sediments – suggested that DDE was a result of metabolic degradation rather than a result of uptake from the medium. $\sum$ DDT concentrations varied from 12.0 to 23.3 ng g^{-1} dw and were very similar to those found previously by Nhan et al. (1998) but much lower than concentrations of DDT found in freshwater clams, 78 to 90 ng g^{-1} dw (Nhan et al., 1998), or freshwater snails from urban or suburban areas, 60 to 864 ng g^{-1} dw (Nhan et al., 2001).

Ship-scrapping activities, sewage disposal, and antifouling paints are major sources of tributyltins in Southeast Asia. Kannan et al. (1995b) determined butyltin residues in muscle tissue of fish collected from local markets in India, Bangladesh, Thailand, Indonesia, Australia, Papua New Guinea and the Solomon Islands, Vietnam and Taiwan. They reported that concentrations in fish from Asia and Oceania were lower than those reported for Japan, Canada, and the USA and tentatively suggested that intake of butyltins via fish consumption in Southeast Asia was <25 percent of the recommended maximum daily intake of 250 ng kg body weight^{-1} day^{-1}.

Pesticide and PCB residues in soils

Few studies have looked for persistent organochlorine pesticide and PCB residues in the soils of Vietnam (Thao et al., 1993a,b; Viet et al., 2000). Concentrations of $\sum$ DDT were relatively high in cultivated soil samples from both southwestern Vietnam, in Tayninh and Songbe Provinces, and northern Vietnam around Hanoi; Viettri City, Phu Tho Province; and Hai long Bay near Haiphong. Viet et al. (2000) detected DDT and its metabolites and lindane in all samples analyzed and suggested that recent input of DDT throughout Vietnam is likely. Thao et al. (1993a,b) found $\sum$ DDT concentrations as high as 290 ng g^{-1} dw (mean 110 ng g^{-1} dw) in agricultural soils of southwestern Vietnam but lower concentrations in noncultivated soils. DDT concentrations in soil samples from Vietnam were higher than in Thailand's or Taiwanese soil samples (Thao et al., 1993a). Thao et al. (1993a,b) found concentrations of HCH were low – although they were highest in samples from Vietnam – in both cultivated and noncultivated soils, ranging from 0.09 to 2.3 ng g^{-1} dw (mean 4.8 ng g^{-1} dw).

Thao *et al.* (1993a,b) found relatively high PCB concentrations in rural agri-cultural soils (high recorded was 150 with a mean of 25 ng g^{-1} dw) and in non-cultivated soils found the highest concentration on a former US airfield. In northern Vietnam, Viet *et al.* (2000) found PCB residues were generally higher in sediment samples than in soil samples. Mean levels of $\sum$ PCBs were 26 ng g^{-1} dw in soils compared to 45 ng g^{-1} dw in sediments around Hanoi and 2.6 ng g^{-1} dw in soils compared to 2.3 ng g^{-1} dw in sediments around Viettri City.

MODEL MICRO-ECOSYSTEM EXPERIMENTS

To predict the fate and distribution of agrochemicals under field conditions, it is useful to conduct experiments with the agrochemicals in the laboratory. Studies on the environmental behavior of agrochemicals should answer the following:

1   their fate and distribution when they enter the environment;
2   their bioaccumulation and biomagnification potential; and
3   their impact on the environment.

The fate and distribution of pesticide residues in the environment

Like other organic chemicals, agrochemicals in the aquatic environment are sub-jected to different phase transfer, transport, and transformation mechanisms. Phase transfer includes dissolution (solid or liquid phase to aqueous solution), sorption (aqueous solution to sorption on or in a solid phase), volatilization (aqueous solution to the atmosphere (vapor phase), and atmospheric deposition (vapor phase or sorbed from atmosphere to earth's surface).

Transport processes include bulk transport with dispersion (chemical substances move within the water column due to movement of the water itself) where the water's movement creates turbulence and mixing, which in turn leads to dispersion and dilution of the chemicals within an ever-increasing volume of water, and transport with sediment where chemicals are adsorbed onto sediment or suspended matter and transported within a water column or near the bottom of a water column. Also, transport includes diffusion; that process is important in the case of pore bottom sediment, where chemical-pollutant molecules under a concentration gradient can diffuse from the water to the bulk sediment. This process is significant only if water flow is below 2.5×10^{-4} cm sec^{-1} (Tucker and Nelken, 1982).

Chemical transformation in the aquatic environment consists of abiotic and biotic processes. Abiotic transformation includes hydrolysis, photolysis, and dissocia-tion, whereas biotic transformation is a biodegradation process. Lyman (1995) reviews the fundamental aspects of phase transfer, transport, and transformation of pesticides in the environment and provides a comprehensive list of references.

Various studies have shown that the environmental behavior of chemical pollutants, particularly pesticides, may be predicted based on experiments carried out in the laboratory using microcosms (Edwards *et al.*, 1996; Raghu *et al.*, 1996). Microcosms are set up to simulate field conditions, e.g. those of estuarine or coastal zones. Experiments to simulate pesticide behavior can be carried out with unlabeled or isotope labeled chemicals. Some of the isotopes used to label various chemicals include such radioactive isotopes as ^{14}C, tritium (^3H), ^{32}P or stable isotopes such as ^{15}N, ^{13}C, and deuterium (^2H). Because the carbon atom(s) is always present in pesticide molecules and ^{14}C isotopes can be easily synthesized in nuclear reactors by the nuclear reaction of $^{14}N(n,p)$ ^{14}C, it is commonly used as the isotope to label pesticides for fate, transport, and impact studies. Moreover, the half-life of ^{14}C isotope is great (5,630 years) making it possible to conduct long-term experiments, particularly studies to examine the bioaccumulation and environmental impact of pesticides. Consequently ^{14}C isotope is widely used for labeling different pesticides in microcosm simulation experiments. The fate and distribution of the pesticide in the environment can be followed by assaying the radioactivity of ^{14}C in different environmental compartments using liquid scintillation counting (LSC) techniques.

The use of radioisotope labeled pesticides in model ecosystem experiments has advantages over normal pesticides because specialized clean up procedures are not necessary before identification and quantification of the residues and because radioactivity assays are not interfered with by contaminants coexisting in the analyzed samples. Moreover, the bound form of residues in samples can be determined. To accomplish this, samples are decomposed using appropriate solvents and oxidizers after extracting extractable residues and the bound residues are assayed by determining the isotope content of the samples.

To monitor the fate and distribution of pesticides under prescribed conditions in experiments with commonly available pesticides, there are well-developed analytical techniques, e.g. GC, HPLC, and GC-MS, that follow well-established procedures for sample preparation and clean up before analysis (Chiron *et al.*, 1994; Holden and Marsden, 1996). These techniques always require proper clean up procedures.

Villarosa *et al.* (1984) showed microparticles originating from sludge played a significant role in lowering the partition coefficients of pesticides in sludge suspensions under laboratory conditions. There was surprisingly low extraction efficiency for recovery of chlorpyrifos from suspensions prepared by washing sludge that had been freeze-dried. They attributed the decrease in the partition coefficient to either the high concentration of surfactants or humic substances, which could solubilize the pesticide in the aqueous phase, or to the presence of particles that sorb chlorpyrifos, which is then unavailable for measurement by the analytical method. Analysis by laser microscopy shows the presence of particle clusters with a mean size of about 1,000 nm in the solution (Villarosa *et al.*, 1984). Experiments with propentamphos, a more polar OP insecticide with higher water solubility than chlorpyrifos – 110 compared to 40 mg L^{-1}, respectively – gave recoveries greater than those measured with chlorpyrifos. This indicates that for more polar substances including pesticides, the interaction between the pollutants and water (as the solvent) competes with those interactions between the pollutants and microparticles of surfactants or humic substances resulting in higher recoveries from the aqueous phase. Barcelo (1996) found similar patterns for temephos and fenitrothion when he studied their persistence and transformation under natural rice field conditions. Consequently when considering the distribution behavior of pesticides between water and sediment phases, the sorption of pesticides to microparticles suspended in the system should be taken into account.

Two studies (Karickhoff *et al.*, 1979; Carvalho *et al.*, 1992) found that partitioning between water and sediment plus suspended matter depends on the water solubility of the pesticide and the organic carbon content of the sediment. Because of their higher water solubility, OP pesticides, e.g. chlorpyrifos and parathion, can remain in the water phase to a greater extent – up to 40 to 60 percent of their initial concentrations – as compared to OC pesticides, e.g. lindane or DDT. Experiments with ^{14}C-DDT, parathion, and chlorpyrifos in models simulating the conditions of fresh, brackish, and sea water environments (Carvalho *et al.*, 1992) show that degradation of these pesticides with biota (bacteria) and suspended matter present does not follow first-order kinetics. The OP pesticides can rapidly associate with suspended particles present in the aqueous environment and this can stabilize the compounds against degradation by chemical hydrolysis, i.e. abiotic degradation. These studies also suggested that the rapid sorption rate of these pesticides to bottom sediments and suspended matter can offer protection against biotic degradation. Carvalho *et al.* (1992) showed that OP pesticides sorbed to sediment not treated with formalin degraded at a much faster rate – nearly a factor of two – compared to when they sorbed to formalin-treated sediment. They calculated the half-life of chlorpyrifos, parathion, and DDT in sea water during the first 60 days of exposure under biotic conditions at 32°C as 1.4, 9, and 130 days, respectively. But after 60 days, the half-life of chlorpyrifos and parathion became 89 and 95 days, respectively, while the half-life of DDT remained unchanged. The absence of DDT-degrading bacteria in the environment suggests that the degradation rate of DDT would be unchanged.

Numerous microcosm studies (Lalah *et al.*, 1995; Matin *et al.*, 1996; Nhan *et al.*, 1996) have been conducted with ^{14}C-DDT examining multiple abiotic factors, e.g. water and sediment, and biota, e.g. algae, oysters, and shrimp, to elucidate ^{14}C-DDT's distribution pattern among different marine environment compartments. Findings indicate that almost 90 to 95 percent of the ^{14}C-DDT introduced during the first 24 h was deposited in the bottom sediment and concentrated in the surface sediment layer (0–5 cm) (Lalah *et al.*, 1995; Matin *et al.*, 1996; Nhan *et al.*, 1996). Pesticides associated with sediment can exist in extractable and bound forms. In a seven-day exposure experiment with estuarine sediment (organic content of 8–10 percent) in an aquarium, bound DDT was 0.10–0.15 percent of the initial concentration of 30 ppb DDT (Nhan *et al.*, 1996).

Lindane was transformed into benzene in head-space gas of soil treated with lindane, using a model based on flooded soil with and without green manure as a soil amendment (Drego *et al.*, 1990). CO_2 was detected as 10.7 percent and 12.2 percent of the total ^{14}C-lindane applied to unamended and green manure amended soil, respectively. McRae *et al.* (1969) also reported that benzene was a product of lindane metabolism by *Escherichia coli* and *Clostridium sporogenese*. Siddaramappa and Sethunathan (1975) suggested that high organic matter content accelerated pesticide degradation by lowering reduction potential.

Temperature also appears to be an important factor in the degradation of pesticides. The half life of chlorpyrifos in sea water at 32°C was eight days compared to 40 days at 15°C, while DDT's half-life was 105 days at 32°C compared to 518 days at 15°C (Carvalho *et al.*, 1992).

Bioaccumulation of pesticides

Pesticides deposited in sediment, sorbed onto suspended particulates, or dissolved in water undoubtedly affect aquatic biota. Rapid accumulation of OC pesticides in biological tissues has been observed (Nhan *et al.*, 1996). Using ^{14}C -DDT, Nhan *et al.* (1996) found that the insecticide accumulated in aquaria raised shrimp (*M. ensis*) muscle to a level of 1.2–1.5 percent of the initial concentration of 30 ppb within two to three days. After one week, the BCF of DDT in shrimp rose to 2×10^3 kg L^{-1}. Lalah *et al.* (1995) found BCF values for oyster (Bivalvia) and fish *Dascyllus aruanus* L. (Perciformes: Pomacentridae) were a maximum of 6.3×10^3 and 1.5×10^3 kg L^{-1}, respectively, after one day culture in a microcosm treated with ^{14}C-DDT. Carvalho *et al.* (1992) reported the BCF for fish, molluscs, and algae was 150–650 kg L^{-1} for chlorpyrifos, 62–340 kg L^{-1} for parathion, and 4×10^3 to 2.6×10^4 kg L^{-1} for DDT, depending on species. Abdullah and Shanmugam (1995) found the BCF for lindane was greater than 1.6×10^3 kg L^{-1} in cockle (Bivalvia) after a six-day culture period.

The bioaccumulation of pesticides should depend mainly on the physico-chemical properties of the specific pesticide. Fundamental aspects of the bioaccumulation and transformation of pesticides have been considered in detail (Spacie *et al.*, 1995). Biota living in bottom sediment are expected to demonstrate higher bioconcentration than biota suspended in the water column.

Pesticides that accumulate in biotic tissues can become bound in a form that is not extractable with organic solvents, the so-called 'bound form' (Carvalho *et al.*, 1992). After one month culture in microcosms, the bound form of ^{14}C-DDT in shrimp and mussels was found to be as high as 0.6 percent and 6.0 percent, respectively, of the 30 ppb and 5.4 ppb initial concentration of the insecticide (Lalah *et al.*, 1995; Nhan *et al.*, 1996). Model microecosystem experiments could provide valuable data for determining the environmental behavior of pesticides and the rate of their excretion from biological tissues. However, data obtained from laboratory experiments cannot be extrapolated to predict field behavior because exact natural conditions are not predictable and short duration models do not account for growth over the long term (de Boer *et al.*, 1994).

ENVIRONMENTAL IMPACT AND APPROACHES FOR PESTICIDE MANAGEMENT

Environmental impact

Organisms that reside in or come in contact with contaminated environments, including those contaminated with pesticides, can respond by developing either (or both) acute and chronic toxic effects. At high exposure levels, pesticides cause acute physiological and behavioral disorders in living organisms that have ingested or absorbed them. Teratogenic, carcinogenic, and genetic effects may develop when organisms are chronically exposed to high concentrations (Stewart, 1994).

The acute toxicity response most often measured from exposure to pesticides is the 96 h or 48 h LD_{50} or LC_{50} (the lethal dose or lethal concentration that would kill 50 percent of the population of a test animal during a 96 h or 48 h period after exposure to the pesticide). For aquatic organisms, the LC_{50} for a specific pesticide can be obtained from experiments using fresh water fish. Tejada et al. (1993) have determined the 48 h LC_{50}s of many OP, carbamate, and pyrethroid insecticides, fungicides, and herbicides for the Nile tilapia Oreochromis niloticus L. (Perciformes: Cichlidae). Among insecticides, monocrotophos has the highest 48 h LC_{50} (13.80 mg L^{-1}) for O. niloticus and azinphos ethyl has the lowest (1×10^{-6} mg L^{-1}) (Tejada et al., 1993).

Acute toxic effects of pesticides also appear as physiological changes in tissue morphology, biochemistry, and behavior. Many of the major OP, OC, carbamate, and pyrethroid insecticides target the nervous system of insects; but effects can also occur on the nervous system and physiology of nontarget organisms. At sublethal exposure levels ranging from 0.5 to 1.5 mg L^{-1}, methamidophos caused inhibition of brain and liver acetylcholinesterase (AChE) of carp fingerlings Cyprinus carpio L. (Cypriniformes: Cyprinidae) (Yew and Sudderuddin, 1979). The degree of inhibition increased with increasing insecticide concentration and period of exposure to the insecticide. Moribund paddy fish, red talapia Talapia mossambica Peters × T. nilotica L. (Perciformes: Cichlidae), which had been exposed to a potentially lethal concentration of malathion for 24 h, also exhibited depressed brain AChE activity (Sulaiman et al., 1989). The degree of inhibition ranged from 93.8 percent in muscle AChE to 64.6 percent in liver AChE.

All living organisms contain intracellular enzymatic systems that are capable of transforming foreign chemicals (xenobiotics), ingested or absorbed by the organism, such as pesticides. Usually the transformed xenobiotics are less toxic to the organism or are more readily excreted. The major detoxifying mechanisms in organisms involve oxidation, hydroxylation, or conjugation (by transferases), and are concentrated in the liver and kidney. In some cases, biochemical transformation may result in products that are more toxic than the original substances (Stewart, 1994: Di Jiulio et al., 1995). At low concentration and over long periods of time, the effects of pesticides are less certain and by no means uniform, though some have been implicated in genetic and carcinogenic outcomes.

The cellular mechanism of genotoxic chemicals, once these agents become bioavailable, can be a relatively complex phenomenon and, in some cases, incomplete detoxification leads to the formation of highly reactive electrophilic metabolites (Shugart, 1995). These intermediates can attack nucleophilic centers in macromolecules, such as lipids, proteins, RNA, and DNA, which often results in cellular toxicity. Their interaction with DNA is manifested primarily by structural alterations in the DNA molecule and can take the form of adducts (where the chemical or its metabolite becomes covalently attached to the DNA), strand breakage, or of chemically altered bases. All these processes might lead to genotoxicity by the xenobiotic (Shugart, 1995).

It is generally recognized that contamination of aquatic environments has led to increased occurrence of neoplasia in aquatic species. Results from field studies with English sole *Parophris vetula* Girard (Pleuronectiformes: Pleuronectidae) taken from Puget Sound, Washington, USA suggest a strong correlation between the levels of polycyclic aromatic hydrocarbons (PAHs) and polychlorinated biphenyls (PCBs) in sediments and the prevalence of liver neoplasms in the species studied (Malins *et al.*, 1984).

Approaches for pesticide use and management

There are several ways by which farmers might increase their crop yield, including:

1   the introduction into farming practice of new high yielding crop varieties;
2   the improvement of the fertilization and irrigation regime;
3   the use of more effective pesticides; and
4   the use of integrated pest management practices, which may in some cases result in increased yield, but will always result in lower production costs through a combination of savings in pesticide costs, savings in labor costs for pest scouting and application, and savings from spray equipment maintenance and replacement.

Recent experience with new crop cultivars demonstrates that while they can increase production yields, they are usually only weakly resistant to pest insects or diseases unless resistance to an insect or disease pest has been specifically selected by the plant breeder in addition to selecting traits for increased yield. Consequently new cultivars are frequently adversely affected by poor control of insect or disease pests. Because plants, like other living organisms, can produce defensive substances against predators or diseases, plant breeders and crop scientists throughout the world are moving in the direction of creating new crop hybrids with both high yield and good defense function.

Cuong *et al.* (1997) conducted a study designed to examine the relationship between crop yield and populations of brown planthoppers *Nilaparvata lugens* Stal, whitebacked planthoppers *Sogatella furcifer* Horvath (Homoptera: Delphacidae), green leafhoppers *Nephotettis* spp. (Homoptera: Cicadellidae), and hemipteran and arachnid predators on rice varieties that were either susceptible, moderately resistant, or highly resistant to *N. lugens* under insecticide-treated and insecticide-free conditions. In insecticide-free plots, the susceptible rice strain had lower yield than resistant varieties in only one of four seasons while for moderately resistant and highly resistant varieties both yield and insect populations – predator and prey species – were similar in both insecticide-treated and insecticide-free plots. They concluded that for farmers who do not overuse insecticides, susceptible varieties of rice will not necessarily be damaged by *N. lugens* even if outbreaks occur in adjacent paddies and moderate to high levels of resistance in rice strains do not appear to be incompatible with biological control of *N. lugens* and other homopteran pests.

Application of pesticides has proven to be a highly effective means of controlling pests, and thus it contributes significantly to increasing crop production. However, improper use of pesticides can violate the concept of ecological balance. Inappropriate high rates of pesticide application can kill useful natural insect and disease enemies (Lam and Son, 1994). Endosulfan, methamidophos, methyl parathion, and chlorpyrifos, to name a few, have been shown to exert their harmful effects on important insect predators, e.g. lady beetles *Micraspis* spp. (Coleoptera: Coccinellidae), mirid bugs *Cyrtorhinus lividipennis* Reuter (Hemiptera: Miridae), and spiders (Arachnida: Araneae) in rice fields (Lam and Son, 1994). The populations of these predators were noticeably reduced three to five days after insecticidal application. The recovery of lady beetle and mirid populations occurred 10 days after spraying the insecticides, but the recovery of the spider population was very slow, and only very low numbers of the predators could be recovered (Lam and Son, 1994). Proper pesticide application requires rational management to reduce unintended adverse effects. In this respect, the use of pesticides with high efficacy, a narrow spectrum of effects, and rapid environmental decay is now widely recommended as proper farming practice in Vietnam.

Heong *et al.* (1994) compared the use of pesticides by rice farmers in the Philippines and Vietnam. They found that the majority of pesticide applications made in the Mekong Delta, Vietnam and Leyte, Philippines were insecticides and that farmers in Vietnam applied more insecticides per season (~6.2 applications) than Filipino farmers (~2.6 applications). About half of the insecticides were OPs, primarily methyl parathion, monocrotophos, and methamidophos, with 17 percent and 22 percent of the chemicals in Vietnam and the Philippines, respectively, classified as extremely hazardous, toxicity Class Ia, by the WHO. Another 20 percent and 17 percent in Vietnam and the Philippines, respectively, were classified as highly hazardous or Class Ib. In Vietnam, 42 percent of the applications were targeted at leaf-feeding insects, a practice that has been shown not to increase yields. Thus, a large proportion of the insecticides that are currently being used may be unnecessary (Heong *et al.*, 1994).

The detrimental effects of synthetic pesticides on human health and wildlife and ecosystem health have necessitated the search for alternative control strategies in Vietnam and elsewhere. Since 1970, Vietnam has synthesized the insect attractant methyleugenol and its analogues for mass trapping adults of the fruitfly *Bactrocera* (formerly *Dacus*) *dorsalis* Hendel (Diptera: Tephritidae) to reduce damage on oranges grown on many plantations in Vietnam (Hao *et al.*, 1996). During the 1978 to 1980 period, insect pheromones were tested and provided evidence for the unsuspected presence of six species of moth pests in southeast Vietnam. In the following decade, 15 pheromones of moths and weevils were prepared in the Institute of Agrochemistry and the Institute of Tropical Biology for mass trapping and monitoring of numerous pest species (Hao *et al.*, 1996). This work has become the foundation of alternative strategies for insect control in Vietnam.

Biocontrol of pests of sapodilla *Manilkara zapota* L. (Sapotaceae) – a fruit tree, native to Yucatan, nearby parts of southern Mexico, northern Belize and northeastern Guatemala, whose fruit has a sweet and pleasant taste, ranging from a pear flavor to crunchy brown sugar, perhaps best known for chicle, the latex obtained from its bark and used as a chewing gum base – is also under study in Vietnam (van Mele and Cuc, 2001). In orchards where the black ant *Dolichoderus thoracicus* Smith (Hymenoptera: Formicidae) was present, 25 percent fewer farmers sprayed insecticides to control damage by the fruit borer *Alophia* sp. The authors recommended increased farmer-to-farmer training and mass media campaigns to promote wider use of this ant as a biological control agent and to reduce the use of pesticides in sapodilla orchards.

In citrus orchards of the Mekong Delta, farmers have a long tradition of managing the weaver ant *Oecophylla smaragdina* Fabricius (Hymenoptera: Formicidae) to control citrus pests. Despite a significant rise (from 66 percent to 84 percent) in insecticide use between 1994 and 1998 in orchards where *O. smaragdina* occurred, ca. 75 percent of sweet orange, *Citrus sinensis*, and 25 percent of Tieu mandarin, *C. reticulata*, orchards had large weaver ant populations (van Mele and Cuc, 2000). They found that in orchards with weaver ants, farmers sprayed less often and used fewer highly hazardous class insecticides – major insecticides for sweet orange were monocrotophos and α-cypermethrin and for Tieu mandarin oranges methidathion, imidacloprid, and fenpropathrin – and expended half as much for pesticides with no reduction in either yield or farm income. Unfortunately, they noted that farmers practicing ant husbandry were significantly older than those farmers not practicing ant husbandry, adding that this traditional practice should be incorporated into IPM training programs for citrus farmers.

Many studies examining the use of preparations extracted from plants as natural pesticides have been conducted by researchers at various institutions in Vietnam and abroad. An extract of *Artemisia annua* L. (Compositae: Asteraceae), with 5 percent a.i., was shown to kill up to 66.7 percent of a population of *Leptocorisa acuta* Thunb. (Hemiptera: Alydidae) and 37.2 percent of a *Cletus* sp. (Hemiptera: Coreidae) 24 h after application in field trials. These insect species feed on young rice grains and cause decreased grain quality or empty hulls, particularly if damage occurs near harvest time (Con *et al.*, 1993). More recently, 15 new natural insecticidal compounds have been isolated from various parts – twigs of *Aglaia oligophylla*, roots and flowers of *Aglaia duperreana*, and bark of *Aglaia spectabilis* – of three species of

Meliaceane Juss. found in Vietnam (Chaidir *et al.*, 1999; Hiort *et al.*, 1999; Schneider *et al.*, 2000; Dreyer *et al.*, 2001). The Family Meliaceane contains 50–51 genera and 550 species, many of which are rainforest understory trees or shrubs and is pantropical to subtropical in its distribution. Members of the family serve as a source of oils used in soapmaking, as insecticides, and as ornamental and food species. It is perhaps most well-known as the source of the highly prized wood, mahogany, and the insecticide azadirachtin from the neem tree. Of the 15 new natural products, all but one were cyclopentatetrahydrobenzofurans of the rocaglamide type exhibiting strong to moderate insecticidal activity toward neonate larvae of the polyphageous pest insect *Spodoptera littoralis* Boisduval (Lepidoptera: Noctuidae). The remaining new product was of the aglain and aglaforbesin types. The most active compounds were similar in insecticidal activity to azadirachtin.

Another aspect of pesticide use and management beyond yield increases lies in production costs of food crops. An excellent example of this is the peri-urban production of fresh vegetables. Urban populations in Vietnam are growing up to four times as fast as the rural population, fueling a greater demand for timely supplies of fresh vegetables (Jansen *et al.*, 1996). This demand has been met by a significant increase in small peri-urban market economy farms over centrally-planned collective farms. Typically, those around Ho Chi Minh City are small, averaging only 0.8 ha, de facto owned (~two-thirds of them), and 60 percent are dedicated to vegetable production. Added value per day per hectare of vegetables is double or greater than that for rice and employment is five or more times as high as rice, despite rice's very high labor use. Labor is thus the limiting factor for vegetable production enterprises, primarily for supplying irrigation to the crop. Jansen *et al.* (1996) found that vegetable production's major input costs are organic materials, chemical fertilizers, and pesticides and that there is an apparent overuse of both chemical fertilizers and pesticides. However, the introduction of some IPM practices into vegetable farming is an encouraging sign for the future success of market-oriented vegetable farms in Vietnam.

In 1991, Vietnam joined the FAO Intercountry Programme for Integrated Pest Control in South and Southeast Asia, initiated by the FAO Development Program in 1975. This was an education program to help farmers understand the importance of biological control, crop ecosystems, and other principles of IPM through a non-formal education process of learning by experimentation and discovery (Ooi, 1996). This was a complete break from the practice of the previous 40 years, when technical information was packaged in simplistic messages for delivery to Asian farmers and prophylactic chemical control was the common approach to pest control. In the new program crop fields are used as classrooms with farmers participating in Farmer Field Schools (FFS) (for further description of FFS, see Chapter 10, The Philippines, Section E, Pesticide Management). Here they were taught crop physiology, agronomy, health risks from pesticides, group dynamics, and the principles of IPM. Emphasis was placed on growing a healthy crop; visiting fields regularly, preferably once a week for monitoring; understanding and conserving natural enemies; and becoming an expert in pest management (Ooi, 1996).

The adoption of this approach for rice farming has led to a 60 percent drop in the use of insecticides and a 13 percent rise in yield in Vietnam, Indonesia, and the Philippines (Ooi, 1996). Much has been accomplished by the IPM Programs in Vietnam to train local farmers in combining biological, physical, and chemical methods to control pests and diseases, and this training has resulted in some successes. As early as 1994, Nhat (1994) reported at a meeting organized by the National IPM Programs in Hanoi, that due to the application of IPM strategies, farmers from the Red River and Mekong River deltas were saving 10–12 percent of their expenditures per unit of rice growing area while maintaining crop yield. At the same meeting, however, it was recognized that there was still a lack of understanding of the integrated pest management concept among many rice growing farmers. Consequently, facilitating farmers' understanding of biological control through field investigations should be considered the key to successful implementation of IPM.

Effective training of local farmers to reduce pesticide use may be accomplished through other approaches in addition to FFS. Huan *et al.* (1999) and Heong *et al.* (1998) evaluated insecticide reduction intervention for early season rice leaf folder *Cnaphalocrocis medinalis* Guenee (Lepidoptera: Pyraustidae) control in the Mekong Delta between 1992 and 1997. A media campaign, reaching about 92 percent of the 2.3 million farmer households in the Mekong, was launched to motivate farmers to experiment with whether early season spraying for leaf folders was necessary or not. Leaflets, radio drama, and posters had the most effective reach in the media campaign (Heong *et al.*, 1998). Previously, farmers usually decided to spray insecticides based on their perception of potential damage and losses caused by the pest. They also generally overestimated the seriousness of rice leaf-folder from the visible damage and, therefore, applied insecticides early. Huan *et al.* (1999) compared media with FFS training, which reached about 108,000 farmers or 4.3 percent. Farmers' insecticide use reduced markedly over the five-year period, down from 3.4 to 1.0 applications per season, a 70 percent reduction (Heong *et al.*, 1998; Huan *et al.*, 1999). The number of farmers who did not use any insecticides increased from 1 percent to 32 percent (Heong *et al.*, 1998). Farmers spraying in the seedling, tillering, and booting stages changed from 18, 65, and 45 percent, respectively, to 1, 12, and 22 percent, respectively. Changes in farmers' beliefs were also significant over the period. Huan *et al.* (1999) found significant differences between farmers reached by the media campaign and trained by FFS, farmers reached only by the media campaign, and those farmers not reached by either intervention. Pesticide application frequencies were 0.5, 1.2, and 2.1, respectively, and similar differences were found for early season spraying and changes in farmers' beliefs. In a survey of Long An Province, Heong *et al.* (1998) found that the savings in production costs (insecticide cost and labor costs) was the most important incentive for farmers to stop early season spraying, a fact cited by 89 percent of farmers surveyed. It was evident that the media campaign and FFS played complementary roles in changing farmers' beliefs and in their insecticide usage patterns in the Mekong Delta.

For crops not covered by FFS or media campaigns, farmers' knowledge, perceptions, and practices in pest management remains mired in the old pest control paradigm of prophylactic chemical control – calendar spraying, exaggerated yield loss figures to scare farmers and justify use of synthetic chemicals, and maintaining a monopoly of information to persuade farmers that modern agriculture is equal to chemical control (Ooi, 1996). Van Mele *et al.* (2001) surveyed mango farmers during the dry season of 1998 in the Mekong Delta. Pest identification and control were often based on damage symptoms rather than on sightings and recordings of causal agents, e.g. damage caused by the seed borer *Noorda (Deanolis) albizonalis* Hampson (Lepidoptera: Pyralidae) was often attributed to the Oriental fruit fly Bactrocera dorsalis. Nearly all mango farmers applied calendar sprays of insecticides (97 percent) and fungicides (79 percent) from the preflower stage until harvest, applying an average of 13.4 and 11.6 applications per year, respectively (van Mele *et al.*, 2001). About 20 percent of insecticide applications used highly toxic WHO Class I compounds while the remainder were nearly all moderately toxic Class II compounds. Pyrethroids were the most popular (57 percent) – half of all target sprays used only three pyrethroid products – followed by OPs (25 percent) and carbamates (15 percent). Recommendations by pesticide sellers increased farmers' applications from 26 to 37 per year and the number of products used per farmer was increased from 2.6 to 3.9 with advice from extension staff and media. As expected, estimated yield loss correlated with estimated pest severity and pesticide expenditures were correlated with fertilizer expenditures; however, there was no relationship between the quantity of pesticides used and yield. Furthermore, only about 10 percent of the 93 participating farmers were knowledgeable about natural pest enemies (van Mele *et al.*, 2001), suggesting that significant reduction in pesticide use could result from an intervention campaign.

Heong and Escalada (1999) analyzed farmers' decision in rice stem borer management. They found that farmers spent an average of US$39 ha⁻¹ for insecticides, while believing they were preventing a US$402 ha⁻¹ loss. Perceived benefits were directly related with use and perceived severity. While farmers recognized that insecticides could destroy natural enemies, they placed only moderate importance in conserving them. Significantly, they had mixed beliefs about whether spraying insecticides could result in poor health. High peer pressure affected farmers' spray decisions and influenced perceived benefits, spending, and application frequency. All of these studies highlight the importance of spreading the word about IPM through FFS, media campaigns, extension personnel continuing education, and controls on pesticide sellers. The result can be a significant reduction in pesticide applications – leading to reduced environmental contamination by pesticides – with no loss in crop yields, lowered production costs, better health for farm workers, and higher returns for farms and farmers.

REFERENCES

Abdullah, R.A. and Shanmugam, S.S. 1995. Distribution of lindane in a model mudflat ecosystem. *Fresenius Environ Bull.* 4:497–502.

Am, N.M., Nhan, D.D., Thuan, V.V., Cu, N.D., Dieu, L.V. and Hoi, N.C. 1995. *Evaluation of the Level of Organochlorinated Pesticide Contamination in the Environment of the Red River and its Balat Estuary.* Preprint: VINATOM, VAEC-C-027.

Barcelo, D. 1996. Persistence of temephos and fenitrothion and their transformation products in rice field waters. In: *Environmental Behavior of Crop Protection Chemicals.* Proc. Int. Symp. on Use of Nuclear and Related Techniques for Studying the Environmental Behavior of Crop Protection Chemicals,1–5 July 1996. Vienna: IAEA/FAO pp. 331–42.

Budavari, S., O'Neil, M.J., Smith, A. and Heckelman, P.E. (eds) 1989. *The Merck Index,* 11th edn. Rahway, NJ: Merck.

Carvalho, F.P., Fowler, S.W., Readman, J.W. and Mee, L.D. 1992. Pesticide residues in tropical coastal lagoons: use of ^{14}C labeled compounds to study the cycling and fate of agrochemicals. In: Proc. Int. Symp. on Applications of Isotopes and Radiation in Conservation of the Environment held 9–13 March 1992 in Karlsruhe, Baden-Württemberg, Germany. Vienna: IAEA, pp. 613–23.

Chaidir, Hiort, J., Nugroho, B.W., Bohnenstengel, F.I., Wray, V., Witte, L., Hung, P.D., Kiet, L.C. and Proksch, P. 1999. New insecticidal rocaglamide derivatives from flowers of *Aglaia duperreana* (Meliaceae). *Phytochemistry.* 52:837–42.

Chien, T.D. 1994. The rice damage caused by brownplant hopper insect in 1993. *Plant Protect Bull (Vietnam).* 6:23–7.

Chiron, S., Dupas, S., Scribe, P. and Barcelo, D. 1994. Application of on-line solid-phase extraction followed by liquid chromatography-thermospray mass spectrometry to the determination of pesticides in environmental waters. *J Chromatogr.* 665:295–305.

Chuong, N.I. 1995. The agroproduct processing industry. In: *Developing Vietnam.* Hanoi, Vietnam: UNDP Office of Vietnam in association with the Chamber of Commerce and Industry of Vietnam, pp. 19–24.

Con, V.Q., Muu, L.P., Thinh, T.H., An, D.T. and Lam, T.X. 1993. Results of the application of plant origin insecticide extracted from *Artemisia annua* L. for controlling pests at nearly harvest period. *Plant Protect Bull (Vietnam).* 5:11–4.

Cung, H.A. 1995. Research into measures to control pest insects and plant diseases in different ecoregions. National Research Program (NRP) Report Nr KN 01.08. Hanoi, Vietnam: NRP. 217 p.

Cuong, N.L., Ben, P.T., Phuong, L.T., Chau, L.M. and Cohen, M.B. 1997. Effect of host plant resistance and insecticide on brown planthopper *Nilaparvata lugens* (Stal) and predator population development in the Mekong Delta, Vietnam. *Crop Prot.* 16(8):707–15.

de Boer, J., van der Valk, F., Kerkhoff, M.A.T. and Hagel, P. 1994. 8-year study on the elimination of PCBs and other organochlorine compounds from eel (*Anguilla anguilla*) under natural conditions. *Environ Sci Technol.* 13:2242–8.

Di Jiulio, R.T., Benson, W.H., Sander, B.M. and van Veld, P.A. 1995. Biochemical mechanism: metabolism, adaptation, and toxicity, Chapter 17. In: Rand, G.M. (ed.) *Fundamentals of Aquatic Toxicology: Effects, Environmental Fate, and Risk Assessment,* 2nd edn. Washington, DC: Taylor & Francis, pp. 601–53.

Do Hong, L.C., Becker-van Slooten, K., Sauvain, J.J., Minh, T.L. and Tarradellas, J. 2000. Toxicity of sediments from the Ho Chi Minh City canals and Saigon River, Viet Nam. *Environ Toxicol.* 15(5):269–75.

Drego, J., Murthy, N.B.K. and Raghu, K. 1990. [^{14}C]-γ-Hexacyclohexane in a flooded soil with green manuring. *J Agric Food Chem.* 38:266–8.

Dreyer, M., Nugroho, B.W., Bohnenstengel, F.I., Ebel, R., Wray, V., Witte, L., Bringmann, G., Muhlbacher, J., Herold, M., Hung, P.D., Kiet, L.C. and Proksch, P. 2001. New insecticidal rocaglamide derivatives and related compounds from *Aglaia oligophylla.* *J Nat Prod.* 64(4):415–20.

Duy Nghi. 1991. Agrochemicals at domestic markets and related issues to be solved. *Plant Protect Bull (Vietnam).* 3:9–22.

Edwards, C.A., Knacker, T.T., Pokarzhevski, A.A., Subler, S. and Parmelee, R. 1996. The use of microcosms in assessing the effects of pesticides on soil ecosystems. In: *Environmental Behavior of Crop Protection Chemicals.* Proc. Int. Symp. on Use of Nuclear and Related Techniques for Studying the Environmental Behavior of Crop Protection Chemicals, 1–5 July 1996. Vienna: IAEA/FAO pp. 435–51.

Everaarts, J.M., Nasreen, B., Swennen, C. and Hillebrand, M.T.J. 1991. Cyclic chlorinated hydrocarbons in benthic invertebrates from three coastal areas in Thailand and Malaysia. *J Sci Soc Thailand.* 17:31–49.

Gonzalez-Farias, F. and Mee, L.D. 1988. Effect of humic-like substances on biodegradation rate of detritus. *J Exper Marine Biol and Ecol.* 119:1–13.

Ha, M.C., Cordier, S., Bard, D., Thuy, L.T.B., Hao, H.A., Quinh, H.T., Dai, L.C., Abenhaim, L. and Phuong, N.T.N. 1996. Agent orange and the risk of gestational trophoblastic disease in Vietnam. *Arch Environ Health.* 51(5):368–74.

Heong, K.L., Escalada, M.M. and Mai, V. 1994. An analysis of insecticide use in rice: case-studies in the Philippines and Vietnam. *Int J Pest Manage.* 40(2):173–8.

Heong, K.L., Escalada, M.M., Huan, N.H. and Mai, V. 1998. Use of communication media in changing rice farmers' pest management in the Mekong delta, Vietnam. *Crop Prot.* 17(5):413–25.

Heong, K.L. and Escalada, M.M. 1999. Quantifying rice farmers' pest management decisions: beliefs and subjective norms in stem borer control. *Crop Prot.* 18(5):315–22.

Hiort, J., Chaidir, Bohnenstengel, F.I., Nugroho, B.W., Schneider, C., Wray, V., Witte, L., Hung, P.D., Kiet, L.C. and Proksch, P. 1999. New insecticidal rocaglamide derivatives from the roots of *Aglaia duperreana.* *J Nat Prod.* 62(12):1632–5.

Hoa, N.T.Q. 1995. GC-ECD for the determination of OC residues in water of the Nhue River [PhD dissertation]. Hanoi University. Hanoi, Vietnam.

Hao, N.C., Giang, N.C.T.H., Khoa, N.C. and Son, N.T. 1996. Synthesis and application of insect attractants in Vietnam. *Resour Conserv Recy.* 18(1–4):59–68.

Hoi, P.C. 1994. The safe use of pesticides. In: Abstracts of the Workshop on the Impact of Pesticides on Human Health in Vietnam held 27–28 April 1994 in Hanoi, Vietnam. Hanoi: National Institute for Protection of Children's Health and the National Institute of Occupational and Environmental Health.

Holden, A.V. and Marsden, K. 1996. Single-stage clean-up of animal tissue extracts for organochlorine residue analysis. *J Chromatogr.* 44:481–92.

Huan, N.H., Mai, V., Escalada, M.M. and Heong, K.L. 1999. Changes in rice farmers' pest management in the Mekong Delta, Vietnam. *Crop Prot.* 18(9):557–63.

Huber, K., Le Loan, L., Hoang, T.H., Tien, T.K., Rodhain, F. and Failloux, A.B. 2000. *Aedes aegypti* in Vietnam: ecology, genetic structure, vectorial competence and resistance to insecticides. *Ann Soc Entomol Fr.* 36(2):109–20.

Jansen, H.G.P., Midmore, D.J., Binh, P.T., Valasayya, S. and Tru, L.C. 1996. Profitability and sustainability of peri-urban vegetable production systems in Vietnam. *Neth J Agri Sci.* 44(2):125–43.

Kannan, K., Tanabe, S., Quynh, H.T., Hue, N.D. and Tatsukawa, R. 1992. Residue pattern and dietary intake of persistent organochlorine compounds in foodstuffs from Vietnam. *Arch Environ Contam Toxicol.* 22(4):367–74.

Kannan, K., Tanabe, S. and Tatsukawa, R. 1995a. Geographical distribution and accumulation features of organochlorine residues in fish in tropical Asia and Oceania. *Environ Sci Technol.* 29(10):2673–83.

Kannan, K., Tanabe, S., Iwata, H. and Tatsukawa, R. 1995b. Butyltins in muscle and liver of fish collected from certain Asian and Oceanian countries. *Environ Poll.* 90(3):279–90.

Karickhoff, S.W., Brown, D.S. and Scott, T.A. 1979. Sorption of hydrophobic pollutants on natural sediment. *Water Res.* 13:241–8.

Kucklick, R.J., Bidlemam, F.T., McConnell, L.L., Walla, D.M. and Ivanov, G.P. 1994. Organochlorines in the water and biota of Lake Baikal, Siberia. *Environ Sci Technol.* 28:31–7.

Lalah, J.O., Wandiga, S.O., Mbuvi, L. and Yobe, A.C. 1995. Experiments on the accumulation of ^{14}C-DDT residues in fish (*Dascyllus aruanus*), oysters and sediment in a model ecosystem glass tank with forced aeration. Report presented at 2nd Research Coordination Meeting (RCM) of IAEA Cooperative Research Program (CRP) held 12–16 June 1995 in Kuala Lumpur, Malaysia.

Lam, P.V. and Son, B.H. 1994. Influence of broad spectrum insecticides on main population of pain predators in the rice fields. *Plant Protect Bull (Vietnam).* 6:7–12.

Le Trung. 1994. Occupational poisoning due to pesticides in Vietnam. In: Abstracts of the Workshop on the Impact of Pesticides on Human Health in Vietnam held 27–28 April 1994 in Hanoi, Vietnam. Hanoi: National Institute for Protection of Children's Health and the National Institute of Occupational and Environmental Health, pp. 45–6.

Lyman, W.J. 1995. Transport and transformation processes, Chapter 15. In: Rand, G.M. (ed.) *Fundamentals of Aquatic Toxicology: Effects, Environmental Fate, and Risk Assessment*, 2nd ed. Washington, DC: Taylor and Francis, pp. 449–92.

McLee, D.W., Metcalfe, C.D. and Pezak, D.S. 1980. Bioaccumulation of chlorbiphenyls and endrin from food by lobsters (*Homarus americanus*). *Bull Environ Contam Toxicol.* 25:161–8.

McRae, I.C., Raghu, K. and Bautista, E.M. 1969. Anaerobic degradation of lindane by *Clostridium* sp. *Nature* 221:859–60.

Malins, D.C., McCain, B., Brown, D.W., Chan, S.L., Meyers, M.S., Landahl, J.T., Prohaska, P.G., Friedman, A.J., Rhodes, L.D., Burrow, D.G., Gronlund, W.D. and Lodgins, H.O. 1984. Chemical pollutants in sediments and diseases in bottom dwelling fish in Puget Sound, Washington. *Environ Sci Technol.* 18:705–13.

Marinucci, A.C. and Bartha, R. 1982. Accumulation of the polychlorinated biphenyl Arochlor 1242 from contaminated sediment and water by the saltmarsh detritivore (*Uca pugnax*). *Bull Environ Contam Toxicol.* 29:326.

Matin, M.A., Hoque, E., Khatoon, J., Khan, Y.S.A., Ahmed, M. and Mian, A.J. 1996. Distribution and fate of ^{14}C-DDT in microcosm experiments simulating tropical marine environment of Bay Bengal. In: *Environmental Behavior of Crop Protection Chemicals*. Proc. Int. Symp. on Use of Nuclear and Related Techniques for Studying the Environmental Behavior of Crop Protection Chemicals, 1–5 July 1996. Vienna: IAEA/FAO, pp. 279–87.

Ministry of Agriculture and Rural Development (MARD). 1996. List of pesticides permitted, restricted, and banned for use in Vietnam (update dated 22 May 1996). Hanoi, Vietnam: MARD.

Nhan, D.D., Thuan, V.V. and Ain, N.M. 1996. Distribution and fate of ^{14}C-DDT in the estuarine environment of the north of Vietnam. In: *Environmental behavior of crop protection chemicals*. Proc. Int. Symp. on Use of Nuclear and Related Techniques for Studying the Environmental Behavior of Crop Protection Chemicals, 1–5 July 1996. Vienna: IAEA/FAO, pp. 313–9.

Nhan, D.D., Am, N.M., Hoi, N.C., Dieu, L.V., Carvalho, F.P., Villeneuve, J.-P. and Cattini, C. 1998. Organochlorine pesticides and PCBs in the Red River Delta, north of Vietnam. *Mar Poll Bull*. 36:742–9.

Nhan, D.D., Am, N.M., Hoi, N.C., Carvalho, F.P., Villeneuve, J.-P. and Cattini, C. 1999. Organochlorine pesticides and PCBs along the coast of north Vietnam. *Sci Total Environ.* 237/238:363–71.

Nhan, D.D., Carvalho, F.P., Am, N.M., Tuan, N.Q., Hai Yen, N.T., Villeneuve, J.-P. and Cattini, C. 2001. Chlorinated pesticides and PCBs in sediments and molluscs from freshwater canals in the Hanoi region. *Environ Poll.* 112:311–20.

Nhat, P.T. 1994. The status of the implementation of IPM programme into Vietnam during 1992–1994 period. In: Proc. National IPM Programme-Policy Meeting held 29–30 March 1994 in Hanoi, Vietnam. Hanoi, Vietnam: MARD, pp. 1–15.

Ooi, P.A.C. 1996. Experiences in educating rice farmers to understand biological control. *Entomophaga.* 41(3–4):375–85.

Pesticide Management Section, Plant Protection Department of Vietnam. 1997. Pesticide Production Capability of the Vietnam-South Korea Joint Venture in Song Be Province. Hanoi, Vietnam: MARD.

Phuong, P.K., Son, C.P.N., Sauvain, J.J. and Tarradellas, J. 1998. Contamination by PCB's, DDT's, and heavy metals in sediments of Ho Chi Minh City's canals, Vietnam. *Bull Environ Contam Toxicol.* 60(3):347–54.

Quyen, P.B. and Van San, N. 1994. In: Abstracts of the Workshop on the Impact of Pesticides on Human Health in Vietnam held 27–28 April 1994 in Hanoi, Vietnam. Hanoi: National Institute for Protection of Children's Health and the National Institute of Occupational and Environmental Health, pp. 38–41.

Quyen, P.B., Van San, N. and Lan, T.N. 1994. *Vietnam Plant Prot.* 3:31–4.

Quyen, P.B., Nhan, D.D. and Van San, N. 1995. Environmental pollution in Vietnam: analytical estimation and environmental priorities. *Trends Anal Chem.* 14(8):383–8.

Raghu, K., Murthy, N.B.K., Kale, S.P. and Kulkarni, M.G. 1996. Model ecosystems for predicting the behavior of pesticides in environment. In: *Environmental Behavior of Crop Protection Chemicals.* Proc. Int. Symp. on Use of Nuclear and Related Techniques for Studying the Environmental Behavior of Crop Protection Chemicals, 1–5 July 1996. Vienna: IAEA/FAO, pp. 205–13.

Rav-Cha, Ch. and Reblun, M. 1992. Building of organic solutes to dissolved humic substances and its effects on adsorption and transport in the aquatic environment. *Water Res.* 26:645–654.

Sethunathan, N. and Yoshida, T. 1969. Fate of diazinon in submerged soil: accumulation of hydrolysis product. *J Agri Food Chem.* 17:1192–5.

Schecter, A., Dai, L.C., Papke, O., Prange, J., Constable, J.D., Matsuda, M., Thao, V.D. and Piskac, A.L. 2001. Recent dioxin contamination from Agent Orange in residents of a southern Vietnam city. *J Occup Environ Med.* 43(5):435–43.

Schecter, A., Toniolo, P., Dai, L.C., Thuy, L.T.B. and Wolff, M.S. 1997. Blood levels of DDT and breast cancer risk among women living in the North of Vietnam. *Arch Environ Contam Toxicol.* 33(4):453–6.

Schneider, C., Bohnenstengel, F.I., Nugroho, B.W., Wray, V., Witte, L., Hung, P.D., Kiet, L.C. and Proksch, P. 2000. Insecticidal rocaglamide derivatives from *Aglaia spectabilis* (Meliaceae). *Phytochemistry.* 54(8):731–6.

Shugart, L.R. 1995. Environmental genotoxicology, Chapter 13. In: Rand, G.M. (ed.) *Fundamentals of Aquatic Toxicology: Effects, Environmental Fate, and Risk Assessment,* 2nd edn. Washington, DC: Taylor and Francis, pp. 351–95.

Siddaramappa, R. and Sethunathan, N. 1975. Persistence of gamma-BHC and beta-BHC in Indian rice soil under flooded conditions. *Pest Sci.* 6:395–403.

Spacie, A., McCarty, L.S. and Rand, G.M. 1995. Bioaccumulation and bioavailability in multiphase systems, Chapter 16. In: Rand, G.M. (ed.) *Fundamentals of Aquatic Toxicology: Effects, Environmental Fate, and Risk Assessment,* 2nd edn. Washington, DC: Taylor and Francis, pp. 537–93.

Stewart, P.B. 1994. The biochemistry of pesticide action and toxicity. In: Abstracts of the Workshop on the Impact of Pesticides on Human Health in Vietnam held 27–28 April 1994 in Hanoi, Vietnam. Hanoi: National Institute for Protection of Children's Health and the National Institute of Occupational and Environmental Health.

Sulaiman, A.H., Abdullah, A.R. and Ahmad, S.K. 1989. Toxicity of malathion to red talapia (hybrid *Talapia mossambica* × *Talapia nilotica*): behavioural, hispathological and anticholinesterase studies. *Malays Appl Biol.* 18:163–70.

Technical Section, 1st Vietnam Pesticide Supply Company (VIPESCO I). 1997. Pesticide Usage in Vietnam. Hanoi, Vietnam: VIPESCO I.

Tejada, A.W., Bajet, C.M., Magallona, E.D., Magbanua, M.G., Gambalan, N.B. and Araez, L.C. 1993. Toxicity and toxicity indices of pesticides to some fauna of the lowland rice fish ecosystem. *Philippine J Agric.* 76:373–82.

Thanh Tu, P.T. 1995. Assessment on the agrochemical residues in the central part of Vietnam [PhD dissertation]. Hanoi National University. Hanoi, Vietnam.

Thao, V.D., Kawano, M. and Tatkusawa, R. 1993a. Persistent organochlorine residues in soils from tropical and subtropical Asian countries. *Environ Poll.* 81(1):61–71.

Thao, V.D., Kawano, M., Matsuda, M., Wakimono, T., Tatkusawa, R., Cau, H.D. and Quynh, H.T. 1993b. Chlorinated-hydrocarbon insecticide and polychlorinated biphenyl residues in soils from southern provinces of Vietnam. *Int J Environ Anal Chem.* 50(3):147–59.

Thiet, N.D. 1994. Some aspects about the storage and use of pesticides in households. In: Abstracts of the Workshop on the Impact of Pesticides on Human Health in Vietnam held 27–28 April 1994 in Hanoi, Vietnam. Hanoi: National Institute for Protection of Children's Health and the National Institute of Occupational and Environmental Health.

Tuan, N.Q., Doanh, B.S. and Dung, NX. 1992. Determination of trace of crop protection chemicals using solid phase extraction (SPE) technique. *J Chem (Vietnam).* 4:51–4.

Tucker, W.A. and Nelken, L.H. 1982. Diffusion coefficient in air and water, Chapter 17. In: Lyman, W.J., Reehl, W.F. and Rosenblatt, D.H. (eds) *Handbook of Chemical Property Estimation Methods.* New York: McGraw-Hill, pp. 1–25.

Van Mele, P. and Cuc, N.T.T. 2000. Evolution and status of Oecophylla smaragdina (Fabricius) as a pest control agent in citrus in the Mekong Delta, Vietnam. *Int J Pest Manage.* 46(4):295–301.

Van Mele, P. and Cuc, N.T.T. 2001. Farmers' perceptions and practices in use of *Dolichoderus thoracicus* (Smith) (Hymenoptera: Formicidae) for biological control of pests of sapodilla. *Biological Control.* 20(1):23–9.

Van Mele, P., Cuc, N.T.T. and Van Huis, A. 2001. Farmers' knowledge, perceptions and practices in mango pest management in the Mekong Delta, Vietnam. *Int J Pest Manage.* 47(1):7–16.

Verle, P., Lieu, T.T.T., Kongs, A., Van der Stuyft, P. and Coosemans, M. 1999. Control of malaria vectors: cost analysis in a province of northern Vietnam. *Trop Med Int Health.* 4(2):139–45.

Viet, P.H., Hoai, P.M., Minh, N.H., Ngoc, N.T. and Hung, P.T. 2000. Persistent organochlorine pesticides and polychlorinated biphenyls in some agricultural and industrial areas in northern Vietnam. *Water Sci Techol.* 42(7–8):223–9.

Villarosa, L., McCormic, J.M., Carpenter, D.P. and Marriott, J.P. 1984. Effect of activated sludge microparticles on pesticide partitioning behavior. *Environ Sci Technol.* 28:1916–20.

Yew, N.C. and Sudderuddin, K.I. 1979. Effect of methamidofos on the growth rate and esterase activity of the common carp, *Cyprius carpiro* L. *Environ Poll.* 18:213–21.

Chapter 9

Pesticides in the People's Republic of China

Zhong Chuangguang, Chen Shunhua, Cai Fulong, Liao Yuanqi, Pen Yefang, and Zhao Xiaokui

INTRODUCTION

There are currently more than 1.26 billion people in the People's Republic of China and providing an adequate food supply for such a large population is one of the nation's biggest problems. To meet the challenges of a rapidly increasing population and a noticeable shortage of major natural resources for agriculture, China has had to develop more sustainable and productive agricultural systems (Wen *et al.*, 1992). The food supply problem may be solved by controlling population growth, increasing agricultural production through enhanced use of hybrid seeds and fossil-energy-derived inputs such as synthetic fertilizers and pesticides, or through some combination of the two. Given the current agricultural cultivation practices in China, the most effective method for increasing the grain crop yield is to use pesticides for crop protection. China has 100 M ha of cultivated land and 140 M ha of sown land. There are more than 1,350 kinds of pests – these include > 770 insect species, > 550 diseases, > 80 weed species, and > 20 rodent species – that may harm crops. With this in mind, it is not surprising that using pesticide for pest control is the most popular method for limiting pest damage to crops.

In 1950, China began to produce DDT and BHC, the beginning of organic pesticide synthesis in China. In the subsequent half century more than 700 pesticide factories have been established with an annual production capacity in 1994 of 555,000 T of a.i.(s). In recent years, the actual output of pesticides was approximately 210,000 T, making China second in the world in pesticide production. There are more than 170 pesticide a.i.(s) with more than 600 formulations based on them currently in production in China. Annual export of pesticides is about 30,000 T comprising 30 a.i.(s), while annual imports total about 10,000 T. In recent years, insecticides accounted for 73 percent of the total domestic production, fungicides 12 percent, herbicides 13 percent, and plant growth regulators 1.3 percent. Of these, 37 percent was used to control pests in rice (*Oryza sativa* L.) paddy fields, 14 percent was applied on other grain crops, 32 percent was used in fruit and vegetable production, and the remaining 17 percent was used in other crops.

Pesticide use is closely related to the level of agricultural education and training. Agricultural production methods in China have not been standardized so that, consequently, farmer quality remains low. Crop cultivation and management systems, such as for rotation cycles, fertility management, pest management, etc., have not been perfected or standardized for individual crops. Consequently, pesticides are used indiscriminately and the primary pesticide application technique in the countryside remains hand-application. This is one of the primary reasons extensive environmental pesticide pollution exists in China. It is estimated that pesticide waste during application is 50 to 70 percent. In addition, pollutants from pesticide factories increase environmental pollution. Fortunately environmental protection awareness has been raised in the past two decades. Laws and regulations issued by the central and local governments guide the production and application of pesticides so as to minimize their impact on the environment.

PESTICIDE MANAGEMENT IN CHINA

The regulatory framework for the control of pesticide use in the People's Republic of China has not been developed. However, some pesticide management guidelines have been issued by the central and local governments. In the 1950s and 1960s, the key goal of pesticide management was to guard against acute human poisoning and to control production quality. In the 1970s, pesticide residue problems emerged, and all uses of mercury were prohibited. DDT, BHC, mercury and arsenic formulations, and chlordimeform were prohibited from use on tea, tobacco, fruit, and vegetable crops. Use standards began to be adopted in the late 1970s by the Institute for the Control of Agrochemicals in the Ministry of Agriculture (ICAMA). In the early 1980s, pesticide registration was established and comprehensive pesticide evaluation was required. This process examined and evaluated control efficacy, product quality, pesticide toxicity, residue levels, and environmental impacts of pesticides to be registered. In 1982, concurrent with the establishment of the pesticide registration system, the Pesticide Regulation and Evaluation committee was formed. The committee was administered by ICAMA and had five branches: toxicity, environmental protection, production, circulation and effect, and residues. Next, in 1984, the 'Standards for Safe Application of Pesticides' was promulgated, followed by the 'Guidelines for Safe Application of Pesticides I, II, III, ...' commencing in 1987. An inspection system for post registration pesticide evaluation and monitoring was also developed, showing that pesticide management in China had begun to become regulated and standardized.

Chronological summary of pesticide regulations in China

Operative Rules for the Safe Use of '1605' and '1059' Pesticides (draft) issued by the Ministries of Agriculture and Public Health, and China's National Supply and Marketing General Cooperative, 26 March 1957.

This revision of the 'Ways' for the safe use of the pesticides '1605' (parathion) and '1059' stipulated that these two OP pesticides should not be used for the control of pests on vegetables. The revision was prompted by the recognition of instances of human poisonings by the insecticide '1605'. Research had demonstrated the chemical structure, poisoning mechanism, uptake routes, symptoms, clinical and experimental diagnosis, first aid, treatment, and poisoning prevention for this pesticide.

'Ways for Safe Use of "1605" and "1059" (draft)' issued by the Ministries of Public Health and Agriculture, and China's National Supply and Marketing General Cooperative, 11 July 1959.

The 'Ways' limited the scope of use of these two OP pesticides, stipulating that they must not be used on fruit trees whose fruit were nearly mature and on vegetables just prior to harvest (no set number of days before harvest was specified). Their use for controlling medical and veterinary pests, e.g. mosquitoes, flies, and bedbugs, was also strictly prohibited. The 'Ways' stipulated details of pesticide transport and storage, preparation, application, and other matters requiring attention. Attachments to this regulation included: a) temporary first-aid methods for pesticide poisoning; advanced emergency methods in the case of pesticide poisoning by arsenic preparations, BHC, '1605', mercury preparations, sodium fluoride, fluorine sodium silicate, etc.; and a listing of general antidotes; b) symptoms of '1605' poisoning and treatment methods (for reference); and c) poisoning symptoms of OP formulations, prevention, and emergency treatment methods (for reference).

Regulations to Strengthen Safe Management Practices for Pesticides (draft) jointly issued by eight ministries including the Agriculture Ministry, 4 September 1959.

Detailed regulations were published about pesticide production, supply, transport, management, and use. It declared, for the first time, that pesticide factories must be placed some distance from sources of drinking water and civilian houses. Siting of a pesticide manufacturer's facilities must be approved by the Chemical Industry Ministry and with the consent of local government. Equipment for processing poisonous gases, wastewater, and hazardous chemicals must be installed inside the factory. It was also stipulated that the product be securely packed; exhibit an eye-catching special mark; and be accompanied by a detailed booklet of directions about properties, uses, safe storage, and application of the pesticide.

Several regulations about strengthening management of quality and price for chemical pesticides issued by the Ministries of Chemical Industry, Commerce, and Agriculture, 17 November 1959.

As a result of the establishment of large numbers of chemical pesticide factories in various parts of China, production increased rapidly, but quality control fell short of expectations. Several regulations were approved to improve quality control and regularize the price of chemical pesticides. It was proposed that state, regional, and factory pesticide standards be shown on the label and the true composition also be listed on the package. New products or existing chemical pesticides produced by a new factory must be approved at the provincial level prior to production and sale.

Rules for Safe Use of Highly Toxic OP Pesticides (Revised draft) issued by the Ministries of Agriculture and Public Health, and China's National Supply and Marketing General Cooperative, 10 March 1964.

Regulations for Trial Implementation of Engagement, Management, and Safety of Highly Toxic Pesticides issued by the Bureau of Agricultural Means of Production of the National Supply and Marketing General Cooperative, 4 March 1964.

Matters Applicable to the Safe Use of Highly Toxic Pesticides issued by the Ministries of Agriculture and Forestry, 19 April 1971.

A report establishing a national leadership group for pesticides jointly advanced by the Ministries of Commerce, Foreign Trade, Public Health, Chemical Industry, Agriculture, and Forestry, and the China Academy of Sciences promulgated 17 June 1971.

The State Council approved establishment of this six-member group with responsibility to: a) strengthen collaboration among production, use, and research departments and suggest to the State Council pesticide programs, production planning, and the future overall arrangement and development direction of pesticide production; b) stop the importation and production of mercury-containing pesticides and organize related units to cooperate on the development of pesticides with high performance and low toxicity to replace highly toxic pesticides such as mercury preparations; c) strengthen the work promoting the safe use of pesticides; d) energetically develop the production of new pesticides with high performance and low toxicity; e) strengthen research on biological pesticides; f) develop pesticides from plant and microbial sources; g) advance standards limiting pesticide residues post application; h) strengthen management of the details of pesticide transport, supply, and storage.

Methods for Trial Implementation of a New Pesticide's Use and Management issued by the Ministry of Commerce, 1 January 1973.

Suggestions about Safe and Reasonable Use of Pesticides promulgated by the Ministries of Agriculture, Forestry, Fuels, Chemical Industry, and Commerce, 12 December 1973.

This proposed that pesticides with high performance and low toxicity should be aggressively used in food, tea, tobacco, vegetable, melon, and fruit crops instead of pesticides with high residues and toxicity. Further the scope of use of each pesticide should be stipulated.

Report on Preventing Pesticides from Contaminating Food by the National Planning Committee on 20 August 1974, approved by the National Council, and promulgated to each province and autonomous region.

The problem of pesticide pollution and accumulation was addressed. Related departments were requested to organize trial production of new pesticides with high performance and low toxicity. The application of pesticides with high residual toxicity, e.g. DDT, BHC, Hg preparations, and As preparations, to crops of tea, tobacco, fruits, and vegetables should be forbidden or severely limited.

Announcement on the Prohibition of the Use of Pesticides with High Residues on Crops promulgated by the Ministries of Agriculture and Forestry, and the National Supply and Marketing General Cooperative and issued 1 January 1978.

During 1977, DDT and BHC residues were measured in 334 lots of tea from

10 provinces. The results showed that residues from BHC and DDT in tea were a very serious problem. For 301 lots (90 percent), BHC residues exceeded the then current standard of 0.2 ppm – the highest concentration measured was 1.772 ppm. DDT exceeded the then current standard of 0.2 ppm in 154 lots or 40 percent. The highest concentration measured was 10.966 ppm. The announcement requested all provinces to implement the National Committee's report on preventing food contamination, which banned the use of DDT, BHC, and other pesticides with high residues on crops. It further requested provinces to strengthen management of the safe use of pesticides and to encourage development of pesticides with high performance and low toxicity to quickly solve the problem of high pesticide residues on food crops.

Prevention Methods for Insects, Molds, Rodents, and Sparrows in Stored Grains issued by the Ministry of Commerce and implemented 1 August 1978.

Strict limits were set for the use of chemical preparations to control insect, mold, rodent, and sparrow pests of stored grains. Directions were issued for using chemical preparations and application safety standards were set for individuals and the public.

Regulations on the Management of Pesticide Quality (a draft for trial implementation) issued by the Ministries of Chemical Industry, Agriculture, and Forestry, and the National Supply and Marketing General Cooperative on 25 November 1978.

Methods for Trial Implementation of the Scientific Use of Pesticides promulgated by the Ministries of Agriculture and Chemical Industry, and the National Supply and Marketing General Cooperative on 27 October 1980.

Suggestions were issued about the scientific use, development, production, supply and marketing, and labeling of pesticides. The document encouraged users to achieve the maximum insect, disease, and weed control for the greatest economic benefit using the minimum amount of pesticide while ensuring normal crop growth without harm to humans and livestock. It also stated that environmental pollution should be limited as much as possible.

Regulations for the Safe Use of Chlordimeform issued by the Ministry of Agriculture on 9 December 1980.

Because of the potential teratogenicity of chlordimeform, its use was strictly limited to one application per rice crop cycle. For applications of 25 g ha^{-1} a.i., the time of application must be no less than 40 days prior to harvest and for applications of 50 g ha^{-1} a.i., the time of application must be not less than 70 days from harvest. Chlordimeform was banned from use in other food crops, oil crops, fruits, vegetables, medicinal materials, tea, tobacco, sugarcane, and beet crops.

Management Methods of Foreign Company's Tests of Pesticide Performance Carried out in Chinese Fields (for trial implementation) implemented by the Ministry of Agriculture, 1 June 1981.

Detailed regulations were issued about requirements for foreign companies to carry out field pesticide performance experiments in China. ICAMA was designated to examine and verify data submitted from field tests.

Standards for Safe Use of Pesticides issued by the Ministry of Agriculture in April 1981.

The standards were developed on the basis of many years of research and field trials organized by the Ministry of Agriculture and conducted by 43 universities and institutes. The goal of this effort was to minimize pesticide residues on farm produce and prevent soil and water pollution while at the same time effectively control disease, insect, and weed pests. The standard listed the recommended application rate, the maximum application rate, the maximum number of applications, and the safe interval for multiple applications for various pesticides.

Regulations for Pesticide Registration by the Ministries of Agriculture, Forestry, Chemical Industry, Public Health, and Commerce, and the Lead Group for Environmental Protection of the State Council issued 10 April 1982 and implemented 1 October 1982.

The regulation was formulated in accordance with the 'Law of Environmental Protection of the People's Republic of China (for trial implementation)' to protect the environment; safeguard people's health; promote the development of agriculture, forestry and animal husbandry; and strengthen pesticide management. Three classes of pesticide registration exist: a) regular or variety registration for pesticides with a.i.(s) that have not previously been registered; b) supplemental or additional registration for pesticides whose a.i.(s) have been registered but their scope of use, content, or formulation has changed; c) temporary registration for pesticides used in field trials to gather performance data or pesticides used under special conditions.

When applying for pesticide registration, certain informational materials must be submitted along with the application. These include the pesticide name, structure and formula, and the primary physical and chemical properties of the pesticide. Also, the pesticide's production and manufacturing process must be described with a brief synopsis of the raw materials used and waste management procedures for any wastes and byproducts, termed 'three-waste' management for the three compartments affected – soil, water, and air. Product information must be submitted including technical product description; efficacy test conditions, test methods, and test results; packaging; package labeling; product storage conditions and expiration date; and transportation and safety requirements. The application techniques for the pesticide should be described along with its effectiveness, potential harmful effects, use method and scope, target organisms, and its effect on non-target organisms, if any. A sample booklet of use directions should be submitted. Toxicity test results from acute, subacute, and chronic tests should be submitted with information about the pesticide's potential to cause carcinogenic, teratogenic, or mutagenic effects in organisms. Pesticide residue data, metabolism studies, and degradation pathways and degradation products in crops and soils must be described with the analytical methods used in the studies. Also, suggestions on standards for food hygiene, labor hygiene, and safe use must be included. The pesticide's effect on environmental quality, its potential for soil and water pollution, and its fate and transport in air, water, soil, plants, and ecosystems must be described.

Regulations for Safe Use of Pesticides issued by the Ministries of Agriculture, Animal Husbandry and Fishery, and Public Health, 5 June 1982.

Pesticides were classified according to a comprehensive toxicity evaluation: highly-toxic, medium-toxicity and low-toxicity. Many pesticides were included in the general 'Standard for the Safe Use of Pesticides' while others had specific regulations. Highly toxic pesticides cannot be used on vegetables, tea, fruit trees, medicinal materials, and other food crops. They must not be used for medical and veterinary purposes and must not be used to kill rodents (except rats). High-residue pesticides, e.g. BHC, DDT, and chlordane, must not be used on such crops as fruit trees, vegetables, tea, medicinal materials, tobacco, coffee, taro, and others. Chlordimeform may be used for pest control only once per rice crop cycle and only under a stipulated safe pre-harvest interval.

In addition, this regulation also stipulated rules for pesticide purchase, use (with precautionary measures, if any), transport, storage, selection of qualified application staff, and personal protection procedures.

Temporary Regulation for Management of the Pesticide Industry issued by the Ministry of Chemical Industry, 17 July 1982 and implemented 1 January 1983.

All pesticide products and units producing these products were brought under the management of this regulation. Every pesticide enterprise must operate in a safe, responsible manner, enthusiastically carry forward the 'three-waste' concept of waste management, prevent environmental pollution, and build and maintain clean, safe factories. They must also strive to improve their products, e.g. reformulation. The output of pesticides with high residue levels, e.g. BHC, must be limited yearly. It established the system of licenses for pesticide production.

Detailed Rules for Implementation of Regulations for Pesticide Registration promulgated by the Ministries of Agriculture and Animal Husbandry and Fishery, September 1982.

Published as 'Examination and Approval Methods for Pesticide Registration', this regulation specified forming an examination and approval committee for pesticide registration. Several annexes were included: a) Request for Data from Residue Tests in Pesticide Registration; b) Request for Data from Field Tests of Performance in Pesticide Registration; c) Temporary Regulation for Test Methods of Pesticide Toxicity (for trial); and d) Procedures for Toxicological Evaluation of Food Safety (for trial).

Law for Environmental Protection of the People's Republic of China (for trial) passed 13 September 1979.

Chapter 3 section 21 of the 'Environmental Protection' law directs the pesticide industry to actively develop high-performance, low-toxicity, and low-residue pesticides. It further directs expansion of integrated pest management practices, biological pest management, and the reasonable use of wastewater for irrigation and directs the industry to prevent pollution of soil and crops.

Chapter 3 section 24 of the law specifies that toxic chemicals must be strictly registered and managed. Highly toxic materials must be strictly sealed to prevent leakage during storage and transport.

Law of Food Hygiene of the People's Republic of China (for trial) issued 19 November 1982 and implemented 1 July 1983.

Chapter 5 section 16 of this law directs that the safety of chemicals such as pesticides and fertilizers must be examined by the hygiene administrative departments of the State Council.

Standards for pesticide regulation in China

Standard for the safe use of pesticides, BG4285-84, issued by the Ministry of Environmental Protection and Urban and Rural Construction, 18 May 1984.

Measurement of OP pesticides in water by gas chromatography, GB13192-91, issued by the National Bureau of Environmental Protection, approved 31 August 1991 and implemented 1 June 1992.

Hygiene standard for drinking water, GB5749-85, issued by the Public Health Ministry 16 August 1985 and implemented 1 October 1986. The peak concentration standard for DDT and BHC was stipulated as DDT at 1 ppb and BHC at 5 ppb.

Water quality standard for fisheries (trial), TJ 35-79, issued by the Lead Group for Environmental Protection of the State Council, the National Construction Committee, the National Economics Committee, and the General Bureau of Aquatic Products in March 1979. The standard set was DDT <1 ppb, BHC <2 ppb, and malathion <5 ppb.

Water quality standard for sea water, GBH2.2-82 and GB3097-82, issued by the Lead Group for Environmental Protection of the State Council 6 April 1982. The highest permitted concentration of OCs was set as 1st kind at 1 ppb, 2nd kind at 2 ppb, and 3rd kind at 4 ppb.

Water quality standard for wastewater in city sewers, GJ18-86, issued by the Ministry of Environmental Protection and Urban and Rural Construction 11 December 1986 and implemented 1 July 1989. The highest concentration of OPs allowed is 0.5 ppm.

The residue content of BHC and DDT in food, GBn53-77, issued by the National Bureau of Standard Measurement and trial implemented since 1 May 1978. Maximum residue levels presented in Table 9.1.

Sanitary standard for design of industrial enterprises, TJ 36-79, issued by the Public Health Ministry. The highest permitted concentration of harmful substances in the air of residential areas is 0.1 mg m^{-3} trichlorphon. The highest permitted concentration (mg L^{-1}) in surface waters is 0.25 for malathion, 0.02 for BHC, 0.05 for γ-BHC, 0.003 for parathion, and 0.08 for dimethoate. Air quality standards (mg m^{-3}) for harmful substances in the workplace are presented in Table 9.2.

Table 9.1 Maximum permitted residue levels (MRLs) of BHC and DDT in food from GBn53-77

Food	Maximum BHC[a] residue level (mg kg⁻¹)	Maximum DDT[b] residue level (mg kg⁻¹)
Grains	0.3	0.2
Fruits and vegetables	0.2	0.1
Fats and meats (based on fw)	0.5	0.5
Pure fat from meat	4.0	2.0
Fish	2.0	1.0
Egg (without shell)	1.0	1.0
Egg products	Converted as egg	
Milk	0.1	0.1
Milk products	Converted as milk	

Notes:
a BHC residues calculated as $\sum$ of $\alpha, \beta, \gamma, \delta$ isomers.
b DDT residue levels calculated as the $\sum$ of p,p'-DDT, o,p'-DDT, p,p'-DDD and p,p'-DDE.

Table 9.2 China's air quality standards (mg m⁻³) for harmful substances in the workplace

Pesticide	Standard (mg m⁻³)	Pesticide	Standard (mg m⁻³)
BHC	0.10	γ-BHC	0.05
Phorate	0.01	Malathion	2.00
Dimethoate	1.00	Trichlorfon	1.00
DDT	0.30	Parathion	0.05
Dichlorvos	0.30	Methyl-parathion	0.10

EXECUTION OF PESTICIDE MANAGEMENT IN CHINA

Pre-registration management

According to the 'Pesticide Management Regulation', if a pesticide has not been registered in China, it cannot be produced and used. Also imported pesticides will not be allowed to be produced and used in China without registration, even if registered in other countries. Pesticide registration takes place in three to four stages, moving from field trial registration (1), to temporary registration (2), then regular registration (3), and additional registration (4), if required. Stages (1), (2), and (3) are successive steps for pesticide product registration. Field trial registration is designed for field-plot testing before temporary registration and this license is valid for three years. Temporary registration occurs when the field-test plots reach 1 ha or total >3 ha or when the pesticide is to enter trial sales or be used for special circumstances (emergency use). Temporary registration licenses are valid for one to two years. Regular or variety registration is required before a pesticide enters commercial use and this licence lasts for five years. Supplemental or additional registration may be required if the formulation changes or the application range (target pest species, use rate, or other significant change) alters and this occurs after regular registration. The new license is also valid for five years. As for regular product registration, information about product toxicity, environmental ecology, residue levels, and product efficacy (residue and efficacy data must be based on tests conducted in China) must be submitted for the judgment of the Pesticide Registration and Evaluation Committee.

Post-registration management

The management of pesticide labeling is accomplished by requiring producers to provide a sample copy of the label for a pesticide when applying for registration. The sample label must be ratified by ICAMA and no changes are allowed after approval. Label content must include the name of the pesticide, its chemical specifications, registration number, product license, net weight, the name of the manufacturer, pesticide classification, application directions, the toxicity mark, points for attention by applicators, data on production, and batch number.

Pesticide advertisement management is based on the 'Advertisement Management Rules'. Advertisement content must be checked in by the Agricultural Administrative Department. If the pesticide product has not received registration approval, it is not allowed to be advertised. The content of advertisements must not contain information that is inconsistent or contrary with what is in the announcement of pesticide registration and the registration certificate. Deception in advertising or hiding the truth from consumers is not allowed.

China currently has two national centers for pesticide quality supervision and monitoring. One is in ICAMA and the other is in the Shenyang Chemical Research Institute. Each year the National Technique Supervision Bureau audits the pesticide monitoring program and publishes its results. In most provinces, cities, and autonomous regions, branch units for pesticide monitoring have been set up. Local regulations for pesticide management have been implemented in some cities, e.g. Guangzhou and Shanghai. Through 1995, 1,489 domestic pesticide products and more than 170 foreign products were officially registered (including additional registrations), and 1,631 domestic pesticide products and 130 foreign products had received temporary registration.

China's Agriculture Vice Minister recently announced plans to progressively ban the production of five OP pesticides beginning in 2001 (Anonymous, 2000). The first step will be to disallow production by new companies followed by reduction in the production levels by current manufacturers.

Organization and functions of pesticide quality management in China

Organization	Function(s)
National Technique Supervision Bureau	Sets national standards
Pesticide Standardization Technique Committee	Set standards
National Standard Bureau	Examine and issue national standards
National Chemical Department	Examine and issue special standards
Local Chemical Department	Examine and issue enterprise standards
ICAMA	Examine and issue registration certificate
Chemistry Department	Sign and issue the product license
Quality Supervision Department	Sign and issue certificate of product quality
Techniques Supervision Department	Monitor markets for pesticide quality
Standard Measure Bureau	
Industrial and Commercial Administration Bureau	
Consumer	

Pesticide management in China is still imperfect. The key problem is that the 'Pesticide Registration Regulation' has no legal force; it is only a recommended process. As a result, illegal or poor quality pesticides can and often do enter the market and there is no legal way to punish transgressors. Therefore, it is urgent for China to pass enforceable pesticide legislation as soon as possible.

Government efforts to promote green products

To improve the people's quality of life and strengthen producers' consciousness of the need for environmental protection, China's government has advocated production of 'green food' since the late 1980s. Various rules have been formulated over time to standardize the production of 'green food'. In May 1991, the Agriculture Ministry promulgated three sets of rules including 'Temporary Provisions for the Management of Green Products', 'Provisional Means for Management of the Green-Product Mark', and 'Coverage of Commodities Using the Green-Product Mark'. These rules stipulate that the Green-Product Mark is a mark of quality for safe, non-harmful products raised using environmentally-friendly (sound) management practices. In addition to meeting the nourishment and hygiene standards for ordinary food, food products that obtain the Green-Product Mark must conform to four basic conditions. The production site for the product's major raw materials must come from an ecologically good environment that has been examined by the supervisory department for environmental protection as designated by the Agriculture Ministry. It must also meet the production and operation norms for raw crops in accordance with the standard for production of green food. Further, the enterprises must submit documentation when applying for the Green Food Mark including the 'Monitoring Report of Agricultural Ecological Environment' and 'Situation Tables to Control Public Harm During Production', among others. A policy of 'high quality that commands high prices' is carried out for qualified green food producers.

In the south China city of Guangzhou (formerly Canton), a project called 'Technical Norms for Green Vegetable Production', cosponsored by the Agriculture Ministry and the governments of Guangdong Province and Guangzhou, has achieved marked success under the direct leadership of the city government. This research project was initiated in 1991 with twin goals of decreasing pesticide residues on vegetables and progressively expanding the production base for green vegetables. This is accomplished through research into integrated pest management techniques for green vegetables using both standard research methods and demonstration projects. As research progresses, the lessons learned are applied and production expanded through the issuance of technical bulletins to the farming community. More than five years of successful work has seen the production base for green vegetables expand from an initial 200 ha to the current 8,400 ha. Production standards have now been implemented in all vegetable production areas of Guangzhou. These standards ban the use of highly toxic and highly persistent pesticides; specify that residues of other pesticides may not exceed national or international standards; and decrease the application quantity of chemical pesticides by 30 percent while increasing use of biological pesticides by >30 percent, resulting in a net savings of 25 percent of the cost of pesticides.

Guangzhou's government has made this a priority project since its beginning. In 1993 Guangzhou issued directives including the 'Announcement about Strengthening Management of Pesticides and Preventing Pesticides from Polluting Vegetables' and the 'Announcement about Further Enforcing Management of Pesticides'. It has efficiently organized implementation of this project, establishing four levels of leadership groups consisting of city, district, township, and village members. Leadership groups are headed by local people but include expert members from many government units, e.g. the Agriculture Committee, the Bureau of Industry and Commerce, the Supply and Marketing Cooperative, the Bureau of Agriculture, the Public Health Bureau, and the National Bureau of Environmental Protection. Leadership groups from all four levels regularly supervise the implementation of green vegetable production guidelines and check all pesticide marketing outlets. They also confer with industry officials and commercial enterprises to uncover and deal with illegal pesticide use and illegal sales. Scientific and technical networks have been established. Farmer education and training are also conducted. During this time, investment in vegetable production has been increased; from 1991 to 1994, funds provided from city government increased 73.6 million yuan. As new techniques and practices have been put into effect, obvious improvements have occurred. Banned pesticide residues were not detected in vegetables from demonstration villages in 1991 or 1992. Furthermore, residues of other pesticides did not exceed set standards. Each year commodity vegetables are randomly sampled and tested, typically this amounts to 340 samples per year. The frequency of detection of highly toxic pesticides has decreased year after year, from 73 percent of samples in 1991 to 18 percent in 1992 and 15 percent in 1993.

The establishment of a substantial production base for green vegetables has resulted in favorable impacts upon society, which have been reported in newspapers and on television. Many organizations and groups from other parts of the country have come to Guangzhou to visit and learn. Officials from the fisheries and agriculture departments of Hong Kong have also visited to conduct on-the-spot investigations. After basic changes in China's agricultural system and output-united family contracting – similar to farmer cooperatives – were established, questions remained about how to efficiently organize and expand agricultural education techniques to enable farmers to grasp methods for the reasonable use of pesticides. Also officials were concerned with how farmers could reduce their environmental pollution and how they could increase the quality of their farm produce. Answers to these questions have undoubtedly been found in the production of green vegetables in Guangzhou.

DIFFICULTIES AND CHALLENGES AHEAD FOR CHINA'S GOVERNMENT

Production of pesticides: waste and scale

Generally, pesticide production in China is conducted by medium and small-sized enterprises, most of which continue to use production techniques from the 1950s and 1960s. These techniques require high levels of investment and high consumption of raw chemicals but output quantities are low. As a result many raw materials become 'three-waste', potentially entering the environment and causing serious pollution problems. Current production facilities in China's pesticide industry are commensurate with the level of the 1950s and 1960s of developed countries with consumption of raw materials being high and the synthesis rate being low. The conversion rate of raw materials for the entire industry is only 30 to 40 percent, much below that of pesticide production facilities in developed countries, which surpasses 70 percent. Production data indicates that 4 T of raw materials are consumed to produce 1 T of pesticide. The remainder is drained off as unreacted material and by-products, leading to a large amount of 'three-waste' from pesticide production and a serious pollution problem in the environment. In addition, there are many water-cleaning processes during pesticide synthesis and many factories use new water for each process. The result is a large volume of wastewater discharged. This leads not only to a requirement for large-scale wastewater treatment facilities but also, because the unreacted intermediates and by-products left over during synthesis are all difficult to biodegrade, many difficulties in the waste treatment processing. Therefore it is very important to promote the adoption of modern manufacturing processes and strengthen research into improving currently used production techniques, decreasing the consumption of raw materials, and developing techniques yielding little or no waste. Meanwhile, the current use of methylbenzene and dimethylbenzene as pesticide solvents should be changed to decrease solvent pollution during pesticide application by adopting the use of aqueous solvent.

By June 1995, there were nearly 1,000 factories that had applied for pesticide registration. However, only 15 of these businesses could be termed key national mainstay enterprises, with an output of >1,000 T. The others are all small factories distributed around the country with annual outputs of tens to hundreds of tonnes. Small-scale production is synonymous with antiquated production techniques, very low material recovery rates, serious problems with loss of materials, and the production of low quality pesticides that perform poorly. This leads to environmental damage from both the production and application of pesticides that are produced in small-scale factories. The best solution to this problem is to implement large-scale production.

The vastness of China makes the transportation of pesticides to distant regions relatively difficult, especially during the application season. Thus, for reasons of reasonable distribution distances and appropriate scale, pesticide production remains regional. Developing intensive, large-scale production of pesticides requires a substantial planning process. Therefore a nationally supported pesticide project must be designed that considers both economic conditions and market potential. This would encourage the pesticide industry to develop or adopt greater production capacity, with concurrent synthesis of many pesticide varieties; a relatively large unified scale; advanced management techniques; and modern equipment. Such a project would encourage the development of large pesticide manufacturing plants, which in turn would form the foundation of a modern pesticide production industry.

Development of new varieties

Since the production of BHC and DDT was stopped in 1983, most pesticides produced in China have been OPs and these constitute the basic type of pesticide produced today. OPs currently constitute more than 50 percent of the total output and include those varieties produced in greatest quantities. Among the nine pesticides with production >500 T, viz. trichlorphon, dichlorvos, dimethoate, omethoate, methyl parathion, methamidophos, chlordimeform, Shachonsuan, and nitrofen, six are OPs. Of the four with annual output of more than 1,000 T, viz. dichlorvos, dimethoate, methamidophos, and Shachonsuan, three are OPs. They are widely used and their performance is relatively ideal but they do have shortcomings, including their higher toxicity compared to other pesticides, difficulty in handling their wastewater and the byproducts from their manufacture, and serious pollution potential. In recent years, new pyrethroid pesticides have been developed but they also have problems, e.g. many steps in their synthesis, low recovery rates, high price, and rapid development of pest resistance. When resistance develops too rapidly in pest populations, it becomes difficult to replace OPs with pyrethroids. Therefore, China must energetically develop new pesticide classes and pesticide varieties with high performance, novel modes of action, low toxicity, and low residues to replace the older pesticide varieties that can cause serious pollution of agricultural ecosystems and leave high residue levels on farm produce. Concurrently China must also pursue research, development, and production of biological pesticides and pursue subsidized use of biological and 'safe' chemical pesticides. This policy will force product structures to tend toward becoming more ecologically friendly.

Reasonable use of pesticides

For many years, China's pesticide industry has attached a higher priority to the production of crude pesticides while ignoring preparatory processing to increase technical purity and advanced formulation technology. This attitude ignores the relationship between how the final product is used in the field and the life-span or life-cycle of the technical product. Also, this has lead to fewer pesticide varieties and single formulations of them, which has shortened the life-span of some good pesticides. Pesticide manufacturers frequently place their emphasis on increasing the output of crude pesticide to increase profits while ignoring the actions of end users and the effect on raw pesticide production. Naturally, farmers always hope that pests will die as soon as the pesticide is applied and, therefore, tend to continuously use the pesticide that gives them the best performance, i.e. dead pests and better crops. Continuous use of single pesticides leads to rapid resistance development in pests and, ultimately, to failure of the pesticide from pest resistance. In China, manufacturers and farmers seldom investigate the causes of such failures – whether from how the pesticide was used or from how it was prepared – but blindly increase the concentration or frequency of use, further inducing resistance by pests and polluting the environment.

However, in recent years this situation has progressively improved. Many new pesticide mixtures and new formulations have been introduced into agricultural production and have demonstrated beneficial effects. Nevertheless, the government still has a great deal of work to do in the standardization and technical appraisal of pesticide mixtures to ensure that they meet health safety and efficacy standards based on scientific studies. China is a large agricultural country, but her farmers' concept of safely and reasonable use of pesticides is relatively tenuous. Because of the implementation of output-united family contracting (similar to family member-oriented farmer cooperatives), each farm family has become an independent production unit and, so, there exist many difficulties in regulating and supervising the reasonable and safe use of pesticides. As a national standard 'Norms for the Reasonable Use of Pesticides' has been promulgated for years but the phenomenon of indiscreetly using and abusing pesticides is still common in the countryside, resulting in many serious consequences.

Between March and May 1994, testing of vegetable samples taken from markets in Beijing indicated that, in the 81 samples of 11 kinds of vegetables, 41 samples had pesticide residue levels exceeding the national standard, a failure rate of 50.6 percent. The most serious problem was with celery; 100 percent of samples exceeded the standard. Several highly-toxic pesticides exceeded the standard for vegetables, e.g. phorate, omethoate, and dichlorvos. The first two are actually forbidden for use on vegetable crops in the national standard. In China's southern regions, farmers use methamidophos to control vegetable pests but the interval between final pesticide application and harvesting is too short. Consequently, this sometimes results in serious accidental poisonings. Residues (157 ppb) of fenvalerate have been found in tea exported from China to Japan (Miyata *et al.*, 1993) suggesting that the problems of pesticide residues could have effects on the export market for some crops. In a survey to monitor OC residues in milk available in Hong Kong markets during 1993 through 1995, Wong and Lee (1997) found 16.7 percent contained residues exceeding the MRLs. DDE and HCH isomer levels were substantially higher than those found in a 1984 to 1987 survey – dairy production had shifted to mainland sources during the interim. The situation with regard to pesticide residues in and on food products does not appear to have improved. The increased use and misuse of pesticides for crop protection, notably in vegetable production, have led to worrisome levels of pesticide residues on agricultural produce according to a recent study by Wang, J. *et al.* (1999). They examined agricultural produce from two villages of Zianjiang municipality, Hubei Provinec, sampling six food groups from the fields prior to harvesting. OC residues were detected in almost all food with mean residue levels for BHC at 31.7 μg kg^{-1} and 102.5 μg kg^{-1} for DDT. OP residues were detected at levels exceeding the MRL of phoxim and methamidophos in vegetables. Mean residue levels were 89.9 and 36.5 μg kg^{-1} for phoxim and methamidophos, respectively. Wang J *et al.* (1999) estimated daily intakes of pesticide residues per person of 4.88 mg for $\sum$ DDT, 2.04 μg for $\sum$ BHC, and 19.33 μg for methamidophos.

Thus, while the State promulgates relevant pesticide regulations, it must also assume the task of educating people on safe pesticide use and establishing an efficient means of supervising pesticide use to safeguard people's health. In addition, farmers' use of inappropriate tools for pesticide application results in serious waste of pesticides and pollutes the environment.

Technical training and spreading the concept of environment protection

For historical reasons, the quality of China's farmers is relatively low; their understanding of the scientific basis for the use of pesticides is incomplete and their concept of environmental protection is minimal. Because farmers directly use pesticides, it is very important to increase their knowledge of the reasons behind protecting the environment and minimizing pesticide use. Moreover, it is also essential to conduct technical and environmental awareness training for policy-makers at different political levels in addition to training the technicians and workers involved in pesticide production and application.

Policymakers, pesticide production workers, and end users should understand the following problems and concepts. Side effects of pesticides caused by poor production and poor application techniques may include serious pollution and other environmental problems in addition to their toxic effects on wildlife and human beings. The strategic importance of implementing environmentally clean production and reasonable use of pesticides in China must be considered. Clean production techniques will decrease or limit pollution while simultaneously increasing output and improving product quality. Reasonable use practices will lower residue pollution of crops and decrease food production costs. The basic methodology of clean production includes the appropriate selection of raw materials; proper design and production of products; and careful, responsible operation, maintenance, and management of the production system. There are many lessons from both home and abroad to be learned from others' experiences with clean production and responsible use of pesticides.

HISTORY OF THE PESTICIDE INDUSTRY IN CHINA

The history of pesticide application in China is rather long. As early as 1,800 years ago, ancient Chinese used mercury formulations, arsenic formulations, and plant pesticides for pest control. In 1944, China began to synthesize DDT and the commercial product was made widely available in 1946. The pesticide industry developed rapidly after the 1950s as the government of the People's Republic of China became aware of the importance of pesticides in the development of agriculture. Therefore, China established a wide array of pesticide research institutes and manufactures at different political levels, e.g. provinces, cities, and counties. In the 1950s and 1960s, OCs were the primary pesticide produced followed by OPs in the late 1960s to the present. After 1980, some low-toxicity and low-persistency pesticides such as fenvalerate were produced. The expansion of biological pesticide production also developed rapidly after this time.

After more than 40 years of development, China has constructed a rather integrated pesticide industry, including the manufacture of technical products, formulations, intermediates, adjuvants, and a pesticide research system. After BHC and DDT were banned in China, OPs became the predominant pesticide class in the mid-1980s. Later, following development of pyrethroids, the pesticide industry in China entered its most active period. Many new products were produced and new pesticide enterprises were established. In the period from 1983 to 1995, tens of new varieties were commercially produced. Government statistics indicate that, as of June 1995, there were about 1,000 registered pesticide factories throughout China, among which there were more than 200 national manufacturers. There are 15 key national factories with production capacity of 1,000 to 5,000 T y^{-1}. The technical staffs, management systems, and facilities of these factories are much larger than most smaller factories. This guarantees sufficient pesticide production for the nation's agriculture system.

In 1979, pesticide production in China was 201,900 T comprising 110 pesticides. About one quarter were highly persistent pesticides such as BHC and DDT. By

June 1996, the number of registered pesticides had risen to 218 technical products and 839 formulations – of these, 434 were mixed formulations (two or more pesticides). This mix of products, formulations, and production levels almost satisfied the requests of agriculture in China. Table 9.3 lists the quantities and types of pesticides produced from 1993 through mid-1995.

China's most outstanding achievement has come in the production of pyrethroids. Research into pyrethroids began in 1972, with commercial production beginning in the 1980s. In less than 10 years, China completed the process of research, synthesis, commercial production, and production expansion. Currently, China produces more than 10 technical products and >150 formulations of pyrethroids. Fenvalerate is produced in the greatest quantity (Table 9.4).

Research and development of biological pesticides has also been fruitful. Jingan meisu – Jianganmycin, a biological fungicide developed in the 1970s – has become the first choice for protecting rice from bacterial blight caused by *Xanthomonas oryzae* Ishiyama. In recent years *Bacillus thuringensis* Berliner, Yutenqin (rotenone), Yinbieqin (diapropetryn), Kuliansu (tooosederin), Kusen (materine), Yanjian (nicotine), and Chuchongjuzhu (pyrethrin) have been developed. Biological pesticides have certain desirable properties that make their use preferable to chemical pesticides. These include high biological activity, better crop protection from pest damage, low or no toxicity to humans and animals, little or no environmental pollution, few if any harmful effects to pests' natural enemies, and little potential for resistence development by target pests. Although biological pesticides are difficult to place in large-scale production, they are likely to become the primary type of pesticide in the future. Table 9.5 lists many of the pesticides currently under commercial production in China and the year they were placed in production.

Pesticide mixtures have undergone rapid development during the same period. By June 1995, 839 domestic pesticide formulations had been registered of which 434 were mixtures. Most mixed formulations sold in China are ECs but some are WPs, dusts, dispersible granules, MCs (miscible concentrates), etc. Before mixed formulations are registered, the following steps should be completed: laboratory toxicity testing; optimization of component proportions; field efficacy trials and field toxicity tests; analytical methodology research; establishment of environmental monitoring methods; determination of optimal application techniques, equipment recommendations, and spray intervals; and establishment of integrated information files for the mixtures. The production of mixed pesticide formulations has yielded significant economic benefit to China's agriculture industry and, thereby, is a significant benefit to society.

PESTICIDE RESEARCH IN CHINA

Pesticide research began in China in the 1940s. Some pioneers studied the synthesis of DDT in 1944 and other studies were conducted not only to investigate the toxicological aspect of pesticides to insects and other animals but also how to

Table 9.3 The quantities and types of pesticides produced in China for the period 1993 to mid-1995 ($\times 10^7$ T)

Year	Type	Single ingredient formulations				Mixture formulations						
						Binary formulations			Trinary formulations			
		I^a	F^b	H^c	Plant regulators	I^a	F^b	H^c	I^a	F^b	H^c	$I^a +$ F^b
1993	Technical	94	58	39	12	91	35	13	12	2	3	8
	Formulated	184	93	54	18							
1994	Technical	103	60	41	13	164	50	18	93	8	9	8
	Formulated	208	101	60	21							
Through June 1995	Technical	103	60	42	13	200	56	25	102	11	18	22
	Formulated	216	104	63	22							

Notes:
a I = insecticides.
b F = fungicides.
c H = herbicides.

Table 9.4 Some pyrethroid pesticides produced in China during 1991–94 (100 percent a.i. T)

Year	Fenvalerate	Delta-methrin	Fenpro-pathrin	Cyper-methrin	Permethrin	Jiamijuzhi (methrothrin)
1991	420	–	8	32	19	13
1992	484	–	49	30	4	6
1993	358	5	116	25	29	25
1994	533	35	146	40	31	47

effectively apply pesticides. Research on the behavior of pesticides in the environment began in the early 1980s.

Insecticides

OC pesticides – Many studies have focused on the behavior of OC pesticides especially BHC and DDT in agricultural ecosystems. Chen and Xu (1982) used ^{14}C-lindane (γ-BHC) to study its adsorption in different soil types and to correlate residues in wheat and respective soils. They showed that adsorption of γ-BHC in soils is closely related to organic matter content, extractable aluminum, and soil pH. Soil temperature also has an effect on adsorption. Their data indicates that there is a significant relationship between adsorption of γ-BHC in soil and residue levels in the wheat itself. If the physical and chemical properties of a specific soil are known, the adsorption of γ-BHC can be predicted and thus the residue levels of γ-BHC in the soil and the wheat growing therein can be estimated.

Zhang, S. *et al.* (1983) collected 350 rice paddy soil samples from representative districts throughout China and measured BHC content in the plowed layer (1–15 cm). They found that in 83.1 percent of the samples BHC residue in soil was <0.5 ppm and ranged from 0.021 to 1.96 ppm with an average of 0.307 ppm. Isomers were found in the order β- >α- >δ- >γ-BHC. There was no correlation between the BHC content of brown rice and soil (r = 0.379, n = 25); however, BHC content in brown rice did increase with an increase in the amount of applied BHC over the growing season. In upland soils, BHC was more persistent than in paddy soil (Zhang, S. *et al.*, 1988) and the ratio of the β-isomer to the γ-isomer increased. Also, the absorption by various crops was different from that found for rice. Peanut BHC content during harvest was significantly correlated with soil residues. The rate of BHC degradation increased with increasing moisture, organic content, and temperature.

Migration of BHC, DDT, and their isomers in soil and crops has also been studied (Xia *et al.*, 1981). The residual concentration of $\sum$ BHC in different parts of plants decreases rapidly with the distance of migration. The mobility of BHC's α- and γ-isomers in crops increases with the distance of migration, but is opposite for its β-isomer. Thus, α- and γ-BHC easily migrate and accumulate in grains, but β-BHC does not. In soil, DDT and BHC distribute primarily in the top 20 cm of

Table 9.5 Listing of pesticides produced in China by year of commercial production[a]

Pesticide	Commercial production (Y)	Pesticide	Commercial production (Y)	Pesticide	Commercial production (Y)
DDT	1945	Carbaryl	1966	Tsumacide	1973
Lead arsenate	1950?	Coumaphos	1966	Alar, daminozide	1974
Arab mothproof	1950s	CCC	1966	Phenazine	1974
BHC	1951	Gesatop, simazine	1966	Phoxim	1974
Methyl bromide	1955	Menazon	1966	TCE-S	1974
Gibberellin	1955	Phaltan, folpet	1966	Macbal	1975
Chloropicrin	1956	Phosmet	1966	Monocrotophos	1975
Parathion	1957	Phosphamidon	1966	Chlormequat	1975
2,4-DB	1958	Phostoxin	1966	BPMC	1976
Demeton	1958	Sulphenone	1966	Ofunack	1976
Dibaichong	1958	Carbetamide	1966	Carboxin	1976
PCP	1958	Fenchlorphos	1967	Cartap	1976
Phorate	1958	Fussol	1968	Ethyl-chlordimeform	1976
Captan	1959	Prometryn	1968	Dicofol	1976
Chlorfenson	1959	Barbane	1969	HCB	1976
DNOC, Sinox	1959	Chlorodencone	1969	Omethoate	1976
Ethion	1959	Diazoben, fenaminosulf	1969	CPMC	1977
Dimethoate	1960	Phenthoate	1969	Hydroprene	1977
Dichlorvos	1961	Phenthoate-ethyl	1969	Mebenil	1977
Carbophenothion	1963	PSP	1969	Shachonsuan	1977
Malathion	1963	Kitazin EBP	1969	Parathion	1977
MCPA, Agroxone	1963	Atrazine	1970	Dimelon	1977
NAA	1963	Dibromo-chloropropane	1970	Fosetyl-Al, Aliette	1979
Camphechlor	1964	Diphacinone	1970	Isoprocarb	1979
Fenitrothion	1964	Gliftor	1970	MO-338	1979
Methyl parathion	1964	Kasugamycin	1970	Povamyein M	1979
Propanil	1964	Blasticidin-S	1970	Jianganmycin	1979

Pesticide	Commercial production (Y)	Pesticide	Commercial production (Y)	Pesticide	Commercial production (Y)
Thiram	1964	Ethachlor	1970	Azinphos ethyl	after 1979
Ziram	1964	Di-allate	1970	N-23	1980
Chlordane	1965	Hinosan (edifenphos)	1971	Tetramethrin	1980
Dalapon	1965	Propachlor	1971	Fenvalerate	after 1980
Heptachlor	1965	E-701	1972	Permethrin	after 1980
Maneb	1965	Fenthion	1972	Trifluralin	after 1980
Parate zineb	1965	Chlordimeform	1972	Difenzoquat	not known
Sulfotep	1965	D-204	1972	Ziram+Thiram+ Urbazid	not known
Swep	1965	Acephate	1973	Diallate	not available commercially
Tetradifon	1965	Ethephon	1973		
Amobam	1966	IBP, Kitazin-P	1973		
Baomianfen	1966	Methamidophos	1973		

Note:
a Pesticides designated by letter and number or other abbreviation may not be listed in the Appendix.

soil. The permeability of BHC's α- and γ-isomers is also higher than that of its β-isomer so the permeability property of BHC in soil depends mainly upon the α- and γ-isomers. Residues and degradation of BHC in soil are affected by many factors, including soil type, physical and chemical conditions in the soil, biological factors, climate, and application technique. Studies have shown that the soil compartment is the main fate of both BHC and DDT. Under normal conditions, DDT is very stable in soil (Xia et al., 1981). Yao et al. (1987) suggested that six years after cessation of DDT application, agricultural ecosystems can be restored to preapplication conditions. The decomposition of BHC in soil is relatively rapid at first and then decreases; thus, BHC remains in soils for a long time.

The movement and fate of BHC in aquatic environments have also been investigated (Chang, Y. et al., 1981b). They concluded that some of the major reasons for biodegradation of BHC in oxidation ponds included a pH increase due to CO_2 consumption by algae during photosynthesis, accumulation in plankton and transfer to sediment, and anaerobic degradation in sediments. BHC in water can enter fish via their gills and so the BHC residue level in fish is determined by the distribution equilibrium between water and body fat. When BHC concentration is low in water, higher BHC residue levels in body tissues can result in release back into the water. The release rate is closely related to the ambient temperature. Studies have also looked at the transfer and accumulation of BHC in food chains (Li, Z. et al., 1985; Huang, S. et al., 1985). Huang S et al. (1985) in a laboratory cage study determined the transfer and bioaccumulation of residual BHC in soil through the food chain of earthworms to quail. When earthworms Eisenia foetida Savigny (Opisthopora: Lumbricidae), were raised for 45 d in soil with a BHC concentration of 0.507 ppm, the BHC content in the earthworms was 1.63 ppm. Subsequently when these earthworms were used to feed Japanese quail Coturnix coturnix japonica Temminck and Schlegel (Galliformes: Phasianidae) for 10 d, the BHC content in quail fat reached 2.36 ppm. They also determined the distribution of BHC in various tissues of the quail. BHC accumulated in the order: fat (2.36 ppm) > brain (0.227 ppm) > liver (0.079 ppm) > muscle (0.071 ppm) > blood plasma (0.062 ppm), illustrating that fat tissue is the major site for BHC bioaccumulation in quail. Li, Z. et al. (1985) used both field studies and laboratory studies to examine the absorption and accumulation of BHC by earthworms from soil and to discuss bioaccumulation of BHC in a terrestrial food chain. They found that BHC is bio-concentrated in living organisms and can be transferred via soil to earthworms and on to quail as well as from soil to maize and on to quail.

OP pesticides – Chang, Y. et al. (1981b) conducted research to reveal the mechanism of biodegradation of OP pesticides in aquatic ecosystems. They also investigated the possibility of treating wastewater from OP pesticide factories in oxidation ponds. Their results showed that malathion, parathion, dimethoate, dimethyldithiophosphate (DMDTP), and diethylthiophosphate (DETP) can be degraded in an algae-bacteria system. The half-life for these compounds was 2, 5, 2, 42, and 62 d, respectively. Results of simulation experiments for oxidation ponds in series showed that removal efficiency of TOC and COD in wastewater was 65.9 percent and

67.8 percent respectively. The effluent toxicity to fish decreased successively down the series of ponds; fish could grow and reproduce normally by the third pond.

Further research by Chang, Y. *et al.* (1981a) isolated two strains of bacteria, identified as *Pseudomonas* sp. CTP-01 and CTP-02, respectively, from wastewater of the oxidation ponds. These bacteria were able to grow using parathion and *p*-nitrophenol as sole carbon sources. Parathion was rapidly degraded by *P.* CTP-01 to produce diethylthiophosphate and *p*-nitrophenol with the latter product being further metabolized. Enzymatic hydrolysis of parathion was investigated using a cell-flee enzyme preparation of *P.*CTP-01. This was found to hydrolyze parathion at a maximum rate of 1×10^4 nmoles mg^{-1} $protein^{-1}$ min^{-1} at an optimum temperature of 45 to 50°C. The optimal pH was 7.0 to 7.5 and, in the presence of 10^{-3} molar Cu^{2+} ion, enzyme activity was increased about 20-fold. *Pseudomonas* CTP-02 utilized *p*-nitrophenol as sole carbon source with an optimum temperature of 35°C and optimum pH of 7.5. When the cultures of *P.* CTP-02 were supplied with *p*-nitrophenol, stoichiometric quantities of nitrite were released and an aromatic nitro group was detached before ring fission.

The effect of parathion and its degradation products on photosynthesis by *Scenedesmus obliquus* Turpin (Chlorophyceae: Scenedesmaceae) was also investigated by Chang, Y. *et al.* (1981a). The toxicity of *p*-nitrophenol was much greater than that of the sodium salts of nitrophenol, diethylthiophosphate, and parathion. An artificial algae-bacteria system – consisting of *P.* sp. CTP-02 and *S. obliquus* and using *p*-nitrophenol as the substrate – indicated that the oxygen required by aerobic bacteria can be provided from algae photosynthesis.

Zhang, Z. *et al.* (1991) studied the residue level and distribution pattern of ^{14}C-fenitrothion in an artificial rice-fish ecosystem. The pesticide was applied to rice plants at low (116 mg a.i. per 1.08 m^2) and high (2X) application rates. Results showed that fenitrothion residues in rice flood water and in rice leaves and stems initially increased in soil, fish, and rice roots but decreased thereafter. At harvest, fenitrothion residues remained in different parts of the ecosystem and were distributed as follows: flood water (0.0027 ppm); upper level of soil (0.2653 to 0.4994 ppm); lower level of the soil (0.0380 to 0.0993 ppm); unpolished rice (0.9633 to 2.1024 ppm); rice leaves and stems (1.7818 to 4.2429 ppm); fish (2.1469 to 4.3400 ppm). About 60 to 90 percent of the pesticide remained in the soil and plants as bound residues, which tended to increase over time.

Carbamates – Guo *et al.* (1996) studied the behavior of pirimicarb in an artificial aquatic ecosystem. The accumulation of pirimicarb in sediment, grass carp, duckweed, and water lettuce increased with time while its concentration in the water column decreased continuously over time. They found nine degradation products for pirimicarb in the aquatic ecosystem.

Pyrethroids – The application of pyrethroids in China began in the early 1980s concurrent with research on these insecticides. Sun *et al.* (1986a) studied the degradation of fenvalerate in lowland rice fields. They found that ^{14}C-fenvalerate was degraded with the peak release of $^{14}CO_2$ occurring 63 to 70 d post application. However, some soil-bound residues were also detectable. They also studied the adsorption of fenvalerate to soil and found that the rate was correlated with organic content of the soil (Sun et al., 1986b). Gan and Chen (1986) used an artificial rice–water–fish system and described the dynamics of fenvalerate with a two-compartment model. The maximum residual levels of fenvalerate in goldfish *Carassius auratus* L. (Pisces: Cyprinidae) tissues were estimated to occur 1.3 to 1.9 d post application and the accumulation and persistence of fenvalerate in edible parts of the fish were rather low. The residue level of fenvalerate in both the water column and fish tissues was low because of the high adsorption capacity of this pesticide to sediment.

Fungicides

Fungicides are less used in China's agricultural production and, thus, are little studied. Peng et al. (1995) examined the mobility and adsorption of metalaxyl in soil using ^{14}C radio-labeled tracer technique. They found the distribution coefficient of metalaxyl between n-octyl alcohol and water was 12.01 and it was therefore easy for metalaxyl to accumulate in living organisms. TLC of soil showed that metalaxyl was barely mobile in black soil, but showed moderate mobility in sandy soil and brown soils. They also showed that adsorption in soil increased proportionally with the concentration of metalaxyl and that adsorption curves were similar for the same soil and different for different soils.

Xiao et al. (1990) examined residue levels and the movement of tricyclazole in the rice-soil-water ecosystem of southern China's rice-growing areas using field tests in conjunction with laboratory tests. They found that tricyclazole could transfer into nearby pond water through evaporative concentration even faster than it could reach ground water by vertical migration. Under laboratory conditions, rice seedlings could absorb tricyclazole and absorption was positively correlated with the pesticide concentration in water ($P < 0.01$).

Herbicides

Usually the effects of herbicides on field ecosystems are less than those of insecticides, primarily because herbicides have higher selectivity. Even non-selective herbicides can attain some selectivity by the choice of application method (Zhang, S. et al., 1988). The relationship between field environmental conditions and the degradation rate of some herbicides have been extensively studied. The half-life of butachlor, acifluorfen-Na, quizalofop-ethyl, and fluazifop-butyl applied to several crops and soils is shown in Table 9.6. In all samples involving rice stalks, unpolished rice, husks, paddy field water, and soil, the final residue content is below the detection limit (Yu et al., 1988). In both plant and seed samples of late stage soybean *Glycine max* (L.) Merril, acifluorfen-Na cannot be detected (Mo, T. et al., 1990).

Min (1993) studied the effects of trifluralin on soil microorganisms and earthworms. He showed that low levels of trifluralin can stimulate both the growth and growth rate of soil bacteria (actinomycetes and molds) but it has no obvious

Table 9.6 The half-life of some herbicides in crops and soils

Herbicide	Preparation type	Sample	Half-life (d)	Reference
Butachlor	60% Milk oil emulsion	Rice stem and leaf	1.07–1.28	Yu et al., 1988; 1993
		Soil	2.67–5.33	
		Field water	1.65–2.48	
	5% Pellet	Rice stem and leaf	1.19–1.41	
		Soil	4.95–6.30	
		Field water	5.79–6.30	
	NA[a]	Paddy soil	3.17–3.61	Chen and
	NA	Paddy water	1.11–1.12	Fan, 1988
Tackle	NA	Soybean plants	1–3	Mo T et al.,
(Acifluorfen-Na)	NA	Soil of Heibei Province	3–5	1990
	NA	Soil of Jilin Province	6–10	
Quizalofop-ethyl	NA	Beet growing soil *Beta vulgaris* L. (Chenopodiaceae)	15	Bao et al., 1991
Fluazifop-butyl	35% Milk oil	Peanut leaf and stem *Arachis hypogaea* L. (Leguminosae: Fabaceae)	1.26–1.65	Tang et al., 1987
		Soil	1.07–1.58	

Note:
a NA indicates preparation type not applicable.

selectivity for the species of microorganism. Trifluralin can promote the growth of nodule bacteria and nitrogen-fixing bacteria of soybean. If trifluralin is added to the culture medium at the same time as nitrogen-fixing bacteria, it can inhibit the acetylene-reducing activity of the bacteria. However, if it is added to culture medium with a healthy, established population of nitrogen-fixing bacteria, it can markedly stimulate the acetylene-reducing activity of the bacteria. Soil microorganisms can use trifluralin as their sole carbon and nitrogen source to rapidly degrade trifluralin. However, it is toxic to earthworms more than 2.0 ppm.

An and Chen (1993) investigated the effects of soil-bound trifluralin on the growth of wheat *Triticum aestivum* L. (Poaceae) and rye *Secale cereale* L.(Poaceae) using ^{14}C-trifluralin in various soils (phaeozem and paddy soil). They found that phaeozem (black soil), which has a higher organic matter and clay content compared to paddy soil, can bind more ^{14}C-residues – binding >20 percent of added trifluralin – and that it is primarily bound with soil as its metabolites. Pot tests also showed that bound residues of trifluralin can be taken up by wheat and rye plants and potentially inhibit the growth of agricultural crops.

Both Chen and Fan (1988) and Yu et al. (1993) found soil microorganisms are responsible for the degradation of butachlor. Chen and Fan (1988) determined the residual half-life of butachlor in both sterilized and unsterilized soils. Its half-life in sterilized and unsterilized soil was 433 d and 18.5 to 29.4 d, respectively. Wang, F. et al. (1995) found that the use in rice culture of controlled release formulations of benthiocarb and butachlor has the expected advantages over commercial formulations of the same herbicides, namely higher weed control efficiency, better targeting of pest species, lower toxicity to non-target species, increased safety for applicators, and less contamination to the environment. The controlled release herbicides provided increased weed control later in the crop cycle and this was highly correlated with an increased yield.

Other pesticides

In recent years, several studies have examined the behavior of various other pesticides in the environment. Wu (1993) studied the degradation dynamics and residue levels of chlorbenside in tea, tea plants, and their soils. The half-life in fresh leaves and finished tea was 3 and 3.3 d, respectively. The preharvest safe application interval is about 7 d. Huang, Y. et al. (1994) studied the effects of monoformamidine insecticide on a model grassy pond ecosystem. After a one-week exposure to 12.5, 25.0, and 50 mg L^{-1} monoformamidine, they found large aquatic plants, plankton, and benthos (animals) were all subject to the pesticide's harmful effects but to different degrees. Species variety, density, and species diversity index all decreased. In the high dose group, all plankton died and the number of heterotrophs in water markedly increased. This pesticide also increased the water column content of organic nitrogen and phosphate salts. They proposed that the artificial grassy pond ecosystem is an efficient tool for evaluating the holistic ecological effects of insecticides and other related chemicals. Mo, H. et al. (1995) studied residue levels and persistence of DMAH (a formamidine insecticide) in the cotton field ecosystem and suggested that its reasonable use would not harm people's health. The half-life values of DMAH were 4.4 to 8.1 d in cotton and 11.4 to 12.3 d in soil and the pesticide was non-persistent and showed little residual accumulation – less than 0.17 ppm in cotton seeds.

RESIDUES AND BEHAVIOR OF PESTICIDES IN MARINE BIOTA

BHC and DDT were the dominant pesticides used in China from 1950 through the 1970s. In eastern Guangdong, 64,793 T of pesticides were used during the period 1978 to 1982; of these 33,654 T were OCs. In 1982, a total of 16,757 T of pesticides was used in this area. Estimates suggest that about 16,092 T of OCs, both BHC and DDT, drained into the sea from runoff into the Hanjiang and Rongjiang rivers.

Zhang and Zen (1987) summarized their survey of BHC residues in coastal regions of Guangdong Province from 1976 to 1985. They concluded that Guang-dong's coastal regions were contaminated by BHC to some extent. The highest level of BHC was found in sediment followed by bio-organisms. However, seawater contained only low levels of BHC. They found a trend showing nearshore sediment contamination was higher than seaward areas, and that sediment contamination in the western part of Guangdong was much higher than in the eastern part (Table 9.7). Liao (1986) reported on a 1983 survey of the pesticides BHC and DDT in seawater, sediments and organisms along the eastern coast of Guangdong. In six bays and harbors, the mean levels of $\sum$ DDT and $\sum$ BHC in surface water were in the range of 0.4 to 1.0 ng L^{-1} and 4.5 to 6.9 ng L^{-1}, respectively (Table 9.8). The concentrations of $\sum$ DDT and $\sum$ BHC in sediment were in the range of 0.3 to 94.3 μg kg^{-1} and 4.2 to 90 μg kg^{-1}, respectively (Table 9.7). Some exceptional samples were found such as in seawater samples from Haimen Bay (Table 9.8) that had concentrations of $\sum$ BHC in surface water of 6.9 ng L^{-1} and in bottom water of 13.37 μg L^{-1}. Bivalves were seriously contaminated with DDT having an average concentration of 253.5 μg kg^{-1} ww.

From a study of OC insecticides in water and sediments of Daya Bay in the eastern part of Guangdong Province, Zhou *et al.* (2001) detected BHCs, DDTs (Tables 9.7 and 9.8), heptachlor, aldrin, endosulfan, dieldrin, endrin, and methoxychlor in water and sediment samples. In water samples, $\sum$ insecticides ranged from 143.4 to 5,104.8 ng L^{-1} with a mean of 576.6 $\pm$ 407.14 ng L^{-1} while insecticide concentrations in sediment were much lower and less variable, ranging from 2.43 to 86.25 ng g^{-1} with a mean of 15.24 $\pm$ 26.80 ng g^{-1} dw. Methoxychlor also contributed substantially to $\sum$ insecticide concentrations (Table 9.9). In water samples, the mean methoxychlor level was 330.6 $\pm$ 575.49 ng L^{-1} with a range of 18.2 to 2,179.0 ng L^{-1}. For sediment samples, the mean methoxychlor level was 7.8 $\pm$ 12.70 ng g^{-1} dw with a range of 0.81 to 40.8 ng g^{-1} dw.

During 1980 and 1981, Chen, X. *et al.* (1983) studied the transfer of BHC and DDT in the Bohai Bay ecosystem located at the western end of the Yellow Sea. They measured residues of BHC and DDT in 11 species of fish, clams, shrimp, and sea birds collected from Bohai Bay (Table 9.10). Concentrations of BHC in fish ranged from 2.0 ng g^{-1} to 2.14 μg g^{-1} ww and DDT ranged from 2.0 to 542.7 ng g^{-1} ww. Concentration factors showed that DDT had a much higher accumulation capacity in organisms. The order of concentration factors for BHC and DDT respectively were as follows: sea birds, 8,600 and 33,000; fish, 1,080 and 7,020; shrimp 240 and 2,200; clams 3,700 and 16,000. Similar research further south in Hangzhou Bay, Zhejiang Province was conducted by Le and Liu (1985). Their survey conducted from 1981 to 1982 found that the concentration factors of BHC in marine organisms were in the range of 50 to 680. Although concentrations of BHC in both sediment and sea water were higher than that of DDT, the DDT content of all fish samples was higher than that of BHC.

Though China had already banned the production and use of OCs like DDT and BHC by the early 1980s, Cai *et al.* (1997) found residues in samples from the Zhujiang (Pearl) River mouth in Guangdong Province over a decade later. They measured residues of $\sum$ DDT and $\sum$ BHC in surface sediment samples from the 1994 to 1995 dry season and the 1995 wet season. Dry season $\sum$ DDT and $\sum$ BHC mean concentrations were 33.46 (range: 17.79 to 51.71) μg kg^{-1} dw and 11.15 (range: 4.98 to 20.27) μg kg^{-1} dw, respectively. Mean $\sum$ DDT levels for the wet

Table 9.7 Residue levels of DDT and BHC in sediment of some coastal regions of China (in ng g^{-1} dw)

Province or municipality	Location	Time of survey	ΣDDT		ΣBHC		Reference
			Range	Mean	Range	Mean	
Guangdong	Zhujiang (River) Mouth	–[a]	6.5–14.5	11.1	41.9–101.4	72.5	Liao, 1983
Guangdong	Zhujiang (River) Mouth	1983	0.3–72.3	6.2	4.2–90.0	37.0	Liao, 1986
Guangdong	Eastern Guangdong	1983	16.9–94.3	6.4	9.5–62.4	34.9	
Guangdong	Eastern Guangdong	1980–81	–	–	0.56–68.0	24.6	Zhang and Zen, 1987
Guangdong	Zhujiang (River) Mouth	1976–78	–	–	4.0–48.0	17.6	
Guangdong	Western Guangdong	1978–80	–	–	0.8–208.0	65.5	
Zhejiang	Hangzhou Bay	1981–82	2.0–27.0	10.0	4.0–43.0	22.0	Le and Liu, 1985
Fujian	Shachen Harbor	1983–85	10.3–37.2	27.2	8.0–17.2	11.6	Reference not available
Fujian	Funin Bay	–	4.4–38.6	20.9	5.2–7.9	6.0	
Fujian	Sansha Bay	–	9.0–75.8	20.2	10.3–23.6	10.3	
Fujian	Luoyuan Bay	–	25.4–27.2	26.3	17.8–18.4	18.1	
Fujian	Minjiang (River) Mouth	–	ND[b]–45.4	25.5	0.3–31.5	17.6	
Fujian	Haitan Strait	–	5.3–16.8	9.0	5.2–20.9	10.8	
Fujian	Xinbhua Bay	–	20.7–32.5	26.0	18.5–28.9	22.4	
Fujian	Meizhou Bay	–	5.2–78.9	25.6	2.4–26.5	11.8	
Fujian	Quanzhou Bay	–	17.1–44.0	28.7	20.5–182.0	64.0	
Fujian	Shenghu Bay	–	15.0–48.5	26.0	4.0–20.0	12.9	
Fujian	Xiamen eastern harbor	–	1.0–33.8	17.9	6.0–15.0	13.4	
Fujian	Xiamen western harbor	–	22.0–75.8	49.9	13.9–23.3	18.4	
Fujian	Jiulongjiang (River) Mouth	–	3.5–54.0	25.8	9.0–75.0	33.3	
Fujian	Futou Bay	–	11.0–20.0	15.0	4.9–16.0	9.0	
Fujian	Dongshan Bay	–	0.2–128.0	68.5	4.1–23.0	14.2	
Fujian	Zhaoan Bay	–	11.0–130.0	52.3	17.0–310.0	116.3	

Province or municipality	Location	Time of survey	ΣDDT		ΣBHC		Reference
			Range	Mean	Range	Mean	
Fujian	Jiulongjiang (River)	–	4.1–6.1	5.2	0.29–0.69	0.49	Zhang et al., 1996
Fujian	Juilongjiang Estuary	–	8.7–69.0	26	3.7–13.0	9.1	Chen et al., 1986
Fujian	Minjiang (River)	–	6.9–13.1	10	4.2–9.4	6.8	Hu et al., 1996
Zhejiang	Qiantangjiang (River)	–	–	0.1	–	0.7	Wu et al., 1999
Shanghai	Huangpujiang (River)	–	–	1.3	–	2.9	
Jiangsu	Changjiang (Yangtze River)	–	0.1–0.2	0.2	0.4–0.7	0.6	
Shandong	Huanghe (Yellow River)	–	0.7–2.4	1.3	1–5	3	
Tianjin (Tientsin)	Haihe	–	9.5–11.5	10.5	7.8–10.8	9.3	
Jiangsu	Changjiang (Yangtze River)	1998	0.21–4.5	–	0.25–1.41	–	Xu et al., 2000
Hong Kong	Hong Kong	1997–98	0.27–14.8	5.06	0.1–16.7	5.02	Richardson and Zheng, 1999
Fujian	Xiamen Western Bay	1993	4.45–311.0	42.8	0.14–1.12	0.45	Hong et al., 1995
Hong Kong	Victoria Harbor	1992	1.38–30.3	12.46	< 0.1–2.3	1.27	
Guangdong	Zhujiang (Pearl River) Estuary	1996–97	1.36–8.99	2.84	0.28–1.23	0.68	Hong et al., 1999
Fujian	Xiamen Harbor	1998	ND[b]–0.06	0.04	ND–0.14	0.09	Zhou et al., 2000
Hong Kong	Mai Po Marshes Nature Reserve	–	3.4–14.2[c]	8.29[c]	9.7–28.5	16.21	Zheng et al., 2000
Guangdong	Daya Bay	1999	0.14–20.27	2.7	0.32–4.16	1.45	Zhou et al., 2001
Guangdong	Zhujiang (Pearl River) Mouth	1994–95 Dry season 1995	17.79–51.71	33.46	4.98–20.27	11.15	Cai et al., 1997
		Rainy season	4.13–83.84	17.88	2.13–24.65	11.07	

Notes:
a En dash (–) indicates no data.
b ND indicates not detected.
c Sum of p,p'-DDE and p,p'-DDT only.

Table 9.8 Residue levels (ng L^{-1}) of ∑ DDT and ∑ BHC in sea water of coastal areas of the South and East China seas

Province or municipality	Location	Time of survey	∑ DDT		∑ BHC		Reference
			Range	Mean	Range	Mean	
Guangdong	Haimen Bay	1983–84		1.0		6.9	Liao, 1986
Guangdong	Shanwei Harbor			0.7		6.2	
Guangdong	Shantou Harbor			0.9		5.6	
Guangdong	Zhilin Bay			0.4		5.4	
Guangdong	Jiazi Harbor			0.5		5.2	
Guangdong	Jieshi Bay			0.4		4.5	
Guangdong	Daya Bay	1999	26.8–975.9	188.4	35.5–1,228.6	285.0	Zhou et al. 2001
Guangdong	Eastern Guangdong	1976–85			10–1,040	100	Zhang and Zen, 1987
Guangdong	Zhujiang (River) Mouth	1994–95			50–4,740	1,010	Cai et al., 1997
Guangdong	West Guangdong				50–4,800	1,250	
Guangdong	Zhujiang (River) Mouth	Dry season	ND^a–263 (S)^b	80	57– 156 (S)	87	
			ND–1,221 (B)^c	506	36–305 (B)	117	
		1995					
		Rainy season	ND–86 (S)	41	21–84 (S)	45	
			10–72 (B)	35	28–85 (B)	48	
Fujian	Xiamen Harbor	1998	0.9–2.3	1.5	3.5–27.8	8.6	Zhou et al., 2001
Zhejiang	Hangzhou Bay	1981–82 Dry season			110–360 (S)	220	Le and Liu, 1985
					80–370 (B)	200	
		1981–82 Rainy season			50–430 (S)	190	
					50–330 (B)	170	

Notes: a ND represents not detected. b S indicates surface water. c B indicates bottom water.

Table 9.9 Residue levels of OCs (excluding the DDTs and BHCs) in water and sediment of some coastal regions of China

Location	Time of survey	Sample matrix	Σ Endosulfan	HCB	Σ Heptachlor	Dieldrin	Aldrin	Chlordane	Methoxychlor	Reference
Hong Kong	1997–98	Sediment	–[a]	1.75 (ND–9.78)[b]	–	3.02 (ND–19.4)	0.85 (ND–4.27)	2.46 (ND–11.3)	–	Richardson and Zheng 1999
Mai Po Marshes	–	Sediment	–	–	7.47 (2.9–16.2)	5.75 (2.4–11.0)	3.9 (1.3–9.2)	3.85 (0.9–6.8)	–	Zheng et al., 2000
Xiamen Harbor	1998	Sediment	0.03 (<0.01–0.09)	–	0.03 (<0.01–0.15)	0.01 (<0.01–0.03)	0.02 (<0.01–0.1)	–	0.01 (<0.01–0.05)	Zhou et al., 2000
	1998	Water	0.35 (0.1–1.0)	–	0.52 (<0.1–3.1)	0.28 (<0.1–0.5)	0.42 (0.2–1.0)	–	1.13 (0.6–1.9)	
Daya Bay	1999	Sediment	0.67 (0.08–3.79)	–	1.12 (0.06–8.80)	0.04 (<0.01–0.13)	1.08 (0.07–7.00)	–	7.81 (0.81–40.80)	Zhou et al., 2001
	1999	Water	51.79 (1.0–283.1)	–	26.0 (1.0–121.3)	0.75 (0–2.3)	27.76 (1.0–227.0)	–	330.6 (18.2–2,179)	

Notes:
a En dash (–) indicates no data and ND indicates OC not detected.
b A (B–C) indicates the mean concentration followed by the range of concentrations with sediment residues in ng g^{-1} dw and water samples in ng L^{-1}.

Table 9.10 Residue levels of DDT and BHC in marine fish and bivalve samples from Bohai Bay and Hangzhou Bay in the early 1980s (expressed in mg kg⁻¹, fresh weight)

Sample species name	Year	DDT					BHC					Reference/Location
		ΣDDT	pp'-DDT	pp'-DDE	pp'-DDD	op'-DDT	ΣBHC	α-BHC	β-BHC	γ-BHC	δ-BHC	
Pseudosciaena crocea Richardson (Perciformes: Sciaenidae)	1981–82	77.0	30.0	23.9	12.3	10.8	22.0	8.6	7.9	2.6	2.9	Le and Liu, 1985
Trichiurus haumela Forsskål (Perciformes: Trichiuridae)		77.2	43.9	26.5	3.3	3.5	58.0	29.6	19.7	6.4	2.3	Hangzhou Bay
Pampus (Stromateoides) argenteus Euphrasen (Perciformes: Stromateidae)		64.0	23.0	17.9	14.1	9.0	52.0	27.1	15.1	6.2	2.6	
Muraenesox cinereus Forsskål (Anguilliformes: Muraenesocidae)		230.0	66.7	94.3	46.0	23.0	32.0	18.6	7.6	4.2	1.6	
Scomberomorus sp. Cuvier (Perciformes: Scombridae)		162.0	84.2	40.5	16.2	21.1	28.0	14.8	6.7	3.9	2.5	
Ilisha elongata Bennett (Cluperiformes: Clupeidae)		73.0	32.1	21.9	11.7	7.3	31.0	16.1	8.1	4.0	2.8	
Miichthys miiuy Basilewsky (Perciformes: Sciaenidae)		38.0	14.6	10.6	6.8	6.5	11.0	4.8	3.4	1.7	1.1	
Coilia mystus L. (Clupeiformes: Engraulidae)		236.0	110.9	47.2	50.0	28.3	114.0	16.0	52.4	16.0	29.6	
Setipinna taty Valenciennes (Clupeiformes: Engraulidae)		277.0	128.5	36.0	58.2	44.3	122.0	17.1	56.1	17.1	31.7	
Cynoglossus joyneri Günther (Pleuronectiformes: Cynoglossidae)		52.0	11.4	25.5	12.5	2.6	26.0	8.8	9.6	3.6	3.9	
Trypauchen vagina Bloch & Schneider (Perciformes: Gobiidae)		48.0	4.8	15.4	26.4	1.4	22.0	5.5	10.3	2.6	3.5	

Sample species name	Year	DDT					∑BHC	BHC				Reference/Location
		∑DDT	pp'-DDT	pp'-DDE	pp'-DDD	op'-DDT		α-BHC	β-BHC	γ-BHC	δ-BHC	
Harpadon nehereus Hamilton (Aulopiformes: Synodontidae)		36.0	11.5	12.2	10.1	2.2	28.0	7.3	12.6	3.6	4.5	
Collichthys lucidus Richardson (Perciformes: Sciaenidae)		110.0	31.9	31.9	37.4	8.8	55.0	11.6	27.0	7.7	8.8	
Sinonovacula constricta[a] Lamarck (bivalvia: Psammobiidae)		79.0	0.8	28.4	45.0	4.7	9.0	1.7	4.5	1.0	1.8	
Mactra sp.[a] (Bivalvia: Mactridae)		32.0	6.1	9.3	12.8	3.8	94.0	26.3	22.6	11.3	33.8	
Arca subcrenata[a] Lischke (Bivalvia: Arcidae)		38.0	4.9	17.1	13.7	2.3	38.0	14.4	10.3	7.6	5.7	
Mugil cephalus L. (Perciformes: Mugilidae) Bohai Bay	1980–81	72.7	16.1	27.5	27.2	1.9	70.3	12.4	9.5	26.5	21.9	Chen, X. et al., 1983
Mugil soiuy Basilewsky (Perciformes: Mugilidae)		83.7	21.3	29.1	28.5	4.8	403.2	102.5	56.4	101.8	142.5	
Lateolabrax japonicus Cuvier (Perciformes: Percichthyidae)		90.9	29.6	23.4	31.7	6.2	90.1	21.2	27.8	14.7	26.4	
Miichthys miiuy Basilewsky (Perciformes: Sciaenidae)		9.9	2.2	5.0	1.8	0.9	32.7	11.2	14.8	4.5	2.2	
Nibea albiflora) Richardson (Perciformes: Sciaenidae		46.1	17.4	14.2	10.5	4.0	114.7	25.8	22.7	19.8	46.4	
Sawara niphonia Valenciennes (Syn: *Scomberomorus niphonius* Cuvier) (Perciformes: Scombridae)		46.3	16.2	11.4	12.1	6.6	51.7	12.0	18.1	17.4	4.2	

Table 9.10 continued

continued…

Table 9.10 continued

Sample species name	Year	DDT						BHC				Reference/Location
		∑DDT	pp´-DDT	pp´-DDE	pp´-DDD	op´-DDT	∑BHC	α-BHC	β-BHC	γ-BHC	δ-BHC	
Pampus (Stromateoides) argenteus		30.2	12.9	5.1	8.5	3.7	83.7	26.1	30.6	14.5	12.5	
Acanthogobius hasta Temminck & Schlegel (Perciformes: Gobiidae)		56.2	11.2	24.2	15.9	4.9	100.7	30.5	23.1	18.0	29.1	
Platycephalus indicus L. (Scorpaeniformes: Platycephalidae)		22.0	6.5	7.4	4.5	3.6	97.9	24.6	18.9	21.2	33.2	
Paralichthys olivaceus Temminck & Schlegel (Pleuronectiformes: Paralichthyidae)		51.0	16.4	13.9	13.2	7.5	40.5	24.0	7.1	4.2	5.2	
Cynoglossus semilaevis Günther (Syn: Arelia rhomaleus Jordan & Starks) (Pleuronectiformes: Cynoglossidae)		63.5	8.7	29.8	22.5	2.5	41.7	23.9	8.2	3.9	5.7	

Note:
a Bivalves.

season were much lower at 17.88 (range 4.13 to 83.83) $\mu g \, kg^{-1}$ dw while $\sum$ BHC was nearly unchanged. Comparing the range of values with data obtained in 1983 by Liao (1986), one can conclude that DDT is very stable in the marine environment while BHC in the environment has declined significantly.

Laboratory experiments are useful for predicting the environmental fate of pesticides. Using these, Zhong *et al.* (1995) studied the accumulation and elimination of ^{14}C-DDT and ^{14}C-fenvalerate in marine biota. Concentration factors for various tissues are shown in Table 9.11. In green mussel *Perna viridis* L. (Bivalvia: Mytilidae), DDT and its metabolites primarily accumulated in the digestive gland or visceral mass with concentration factors reaching 17,346. Fenvalerate did not bioaccumulate to the same degree as DDT; its concentration factors were much lower. The highest concentration factor for fenvalerate in green mussel was in the gill at 211.4, which was only about 1.7 percent of DDT's factor. Higher persistence of DDT in marine organisms compared to fenvalerate was also found in clam *Paphia undulata* Born (Bivalvia: Veneridae) (Table 9.11). DDT was retained in the digestive gland for the entire period of the experiment; thus it is hard to estimate the half-life of DDT in this organ (Table 9.12). The authors suggested *P. undulata* be used as the bio-indicator for field surveys of DDT contamination. This suggestion received further support when Zhong *et al.* (1996) studied the bioavailability of DDT to marine organisms and found *P. undulata* could continuously uptake ^{14}C-DDT released from sediment into the ambient environment. Further analysis demonstrated that the extractable labeled compound was retained as DDT. However, other organisms could rapidly degrade DDT into its metabolites. *Tilapia* spp. rapidly accumulate ^{14}C-DDT from sea water but quickly metabolize this pesticide (Zhong *et al.*, 1997). The metabolites could be detected in viscera within 2 h. By 6 h, extractable residues in the viscera were DDT (31.05 percent), DDD (43 .52 percent), DDE (5.6 percent), DDMU (3.3 percent), and DDA (2.32 percent), with 14.21 percent unknown compounds. These results demonstrated that tilapia could degrade DDT rapidly, and the degradation might be completed by the liver. The possible degradation pathway was DDT to DDD and then to more polar compounds and excretion.

In the case of the tiger shrimp *Penaeus monodon* Fabricius (Decapoda: Penaeoidae), DDE was the major metabolite of DDT (Zhong *et al.*, 1998c). When shrimp were exposed to ^{14}C-DDT for seven days, DDE was 61.02 percent of the total extractable radioactivity from the digestive gland. An 'S' accumulation pattern was observed in some organs of the shrimp. In a long-term experiment with the clam *Meretrix meretrix* L. (Bivalva: Veneridae), Zhong *et al.* (1998d) demonstrated that DDT was very stable in this clam. In an excretion experiment, the extractable labeled compounds after 46 d were mostly DDTs, among which 78.54 percent was DDT, 4.86 percent DDE, 8.85 percent DDD, 0.96 percent DDMU, and 2.93 percent DDA. Even at 179 d, DDT constituted 21.51 percent of extractable labeled compound, while 13.14 percent was DDE, 34.66 percent DDD, 3.18 percent DDMU, and 6.77 percent DDA.

Zhong *et al.* (1998b) developed a mini-model ecosystem to study the fate of radiolabeled pesticides (Figure 9.1). In this system, 1 ml of sea water contaminated

Table 9.11 Concentration factors of DDT and fenvalerate in organs of some marine organisms after 24 h exposure to DDT and 12 h exposure to fenvalerate[a]

Organism	Pesticide	Concentration factor of the organ							
		Shell	Siphon	Foot	Mantle	Gill	Digestive gland	Operculum	Visceral mass
Perna viridis L. (Bivalvia: Mytilidae)	DDT	49		8,054	8,086	12,656	17,346		
	Fenvalerate	2.4		28.6	17.6	211.4	149.7	6.3	
Paphia undulata Born (Bivalvia: Veneridae)	DDT	31	954	907	1,367	3,419	9,197		
	Fenvalerate	2.0	12.5	15.1	18.0	270.8	81.9		
Gafrarium tumidum Roeding (Bivalvia: Veneridae)	DDT	38	1,120	860	922	1,763	1,581		
Anomalocardia flexuosa (Bivalvia: Veneridae)	DDT	28	1,853	4,360	1,705				1,680
Chione isabellina (Bivalvia: Veneridae)	DDT	9	362	383	268				1,098
Batillaria zonalis Bruguiere (Caenogastropoda: Potamididae)	DDT	35		1,543				356	2,450
	Fenvalerate	2.3		21.3				16.6	38.4
Cerithidea (Cerithideopsilla) cingulata Gmelin (Caenogastropoda: Potamididae)	DDT	57		3,884				945	9,098

Note:
a Taken from Zhong et al., 1995.

Table 9.12 Half-life of DDT and fenvalerate in the organs of some marine organisms[a]

Organism	Pesticide	Shell	Siphon	Foot	Mantle	Gill	Digestive gland	Operculum	Visceral mass
Perna viridis	DDT	8.4		12.8	8.9	7.4	7.9		
	Fenvalerate	3.5		4.2	11.4	3.1	4.1	5.1	
Paphia undulata	DDT	8.2	30.2	34.4	49.8	14.0	NA[b]		
	Fenvalerate		14.0	11.2	7.7	7.3	6.3		
Gafarium tumidum	DDT	6.9	7.6	7.8	6.2	6.2	21.7		
Batillaria zonalis	DDT	5.1		20.6				5.6	10.4
	Fenvalerate	11.2		18.3					9.1
Cerithidea cingulata	DDT	6.0		9.9				6.3	10.3

Notes:
a Taken from Zhong et al., 1995.
b NA indicates data not available.

with radiolabeled pesticide and 0.5 g of sediment were placed in a 3 ml test tube inside an LSC vial. Volatile compounds were trapped by polyurethane cotton placed above the test tube and NaOH in the bottom of the vial absorbed $^{14}CO_2$ released from the ecosystem. This system can be used to conduct comparative studies of the fate of pesticides in both aerobic and anaerobic environments and under both sterile and non-sterile conditions. Table 9.13 shows results from a comparative study of the fate of nine labeled pesticides, ^{14}C-DDT, ^{14}C-BHC, ^{14}C-chlorpyrifos, ^{14}C-trichlorphon, ^{14}C-fenvalerate, ^{14}C-chlordimeform, ^{14}C-monocrotophos, ^{14}C-bensulfuron methyl, and ^{14}C-chlorsulfuron, in this marine mini-model ecosystem (Zhong et al., 1998b).

Chen, S. et al. (1998) studied the absorption, distribution, and dynamics of ^{14}C-chlorpyrifos in five freshwater organisms using a 50 × 50 × 30 cm model ecosystem. All organisms tested absorbed ^{14}C-chlorpyrifos rapidly for 4 h following exposure. Concentration factors were in the order *Gambusia affinis* > *Bellarnya purificata* > *Planorbis* sp. > *Lemna polyrrhiza* > *Nasturtium officinale* after 48 h. Concentration factors and time of peak concentration of the pesticide were 375 (at 48 h), 249.7 (at 48 h), and 30 (at 24 h) for *G. affinis*, *B. purificata* and *Planorbis* sp., respectively. The absorption peaks for *L. polyrrhiza* and *N. officinale* were 28.5 and 7.8 at 4 and 24 h, respectively. After peak absorption, radioactivity decreased over time in all species with a slight increase in ^{14}C-radioactivity in the water due to excretion by the organisms.

During the course of the IAEA CRP in the People's Republic of China, the Isotope Laboratory (School of Life Sciences, Zhongshan University in Guangzhou) studied the fate, uptake dynamics, loss, metabolism, bound-extractable ratio, and bioavailability of 17 ^{14}C-labeled pesticides in more than 30 marine and freshwater species (Zhong, 1999; Table 9.14). Their researchers have developed new techniques based on these studies including a way to collect labeled pesticides and their metabolites in water and a new method to determine both extractable and bound residues of ^{14}C-labeled pesticides in sediments. They have also drawn several conclusions from their work, e.g. that direct conversion is not a reasonable way to convert radioactivity into pesticide concentrations because of the breakdown of the pesticides; that concentration factors of pesticides are difficult to calculate in the traditional way with simulation experiments without analyzing for breakdown products in the water medium; and that, based on the bioavailability of sediment sorbed ^{14}C-DDT to some marine organisms, ratios and concentrations of the pesticide and its metabolites in biota do not always reflect the ratio and concentrations in the sediment. They found that the ratios and concentrations of the pesticide and its metabolites in biota are dependent on the uptake capability of the organism for both the pesticide and its metabolites, the bioavailability of the pesticide and its metabolites, and the rates of metabolism of these compounds in the organism.

Chan et al. (1999) determined OC residue levels in muscle samples from 15 species of fish purchased from Hong Kong markets in 1997 and 10 liver samples of tilapia collected from the Shing Mun River, Hong Kong. All fish had detectable

Table 9.13 Fate of C^{14} labeled pesticides in miniature model ecosystems (see Figure 9.1)[a]

Pesticide	Total initial activity DPM	Compartment	After 1 week		After 2 weeks		After 4 weeks	
			DPM	Percentage	DPM	Percentage	DPM	Percentage
Trichlorphon	74,828	Volatile	519	0.69	475	0.64	211	0.28
		CO$_2$	31,534	28.78	33,042	44.16	42,559	55.54
		Bound	0	0	965	1.29	19,198	25.66
BHC	118,294	Volatile	22,238	18.80	35,826	30.29	37,590	31.78
		CO$_2$	242	0.20	1,024	0.87	1,049	0.89
		Bound	12,624	10.67	33,407	28.24	38,757	32.76
Chlorpyrifos	189,962	Volatile	14,900	7.40	15,517	8.69	14,748	7.76
		CO$_2$	407	0.21	274	0.14	592	0.31
		Bound	35,024	18.44	98,938	52.08	102,890	54.16
DDT	128,360	Volatile	350	0.27	505	0.39	968	0.77
		CO$_2$	69	0.05	508	0.40	1,179	0.92
		Bound	49,159	38.30	56,460	43.99	54,598	42.54
Chlorsulfuron	149,378	Volatile	0	0	0	0	0	0
		CO$_2$	125	0.10	267	0.18	309	0.21
		Bound	13,863	9.28	17,265	11.56	23,266	15.58
Bensulfuron methyl	93,502	Volatile	420	0.45	829	0.89	1,755	1.88
		CO$_2$	74	0.08	147	0.16	364	0.39
		Bound	8,054	8.61	15,218	16.28	17,536	18.75
Monocrotophos	180,525	Volatile	37	0.02	130	0.07	12	0.01
		CO$_2$	19,552	10.83	21,404	11.86	34,032	18.85
		Bound	333	0.02	28,760	15.91	39,604	21.94
Fenvalerate	35,376	Volatile	0	0	0	0	0	0
		CO$_2$	5,604	15.84	8,702	24.60	9,601	27.14
		Bound	6,398	18.09	9,427	26.65	9,712	27.45
Chlordimeform	121,664	Volatile	599	0.49	469	0.39	2,596	2.13
		CO$_2$	6,035	4.96	16,968	13.95	38,516	31.66
		Bound	26,033	21.40	19,374	15.92	26,587	21.85

Note: a Taken from Zhong et al., 1998b.

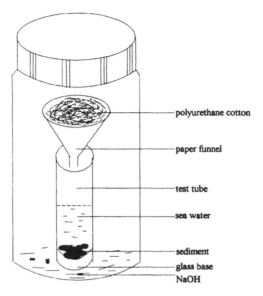

polyurethane cotton

paper funnel

test tube

sea water

sediment
glass base
NaOH

Figure 9.1 Schematic drawing of a miniature model ecosystem (after Zhong et al., 1998)

$\sum$ DDTs, ranging from 3.3 to 75.6 ng g^{-1} ww for market fish and 7.1 to 88.8 ng g^{-1} for tilapia. The DDT/DDE ratios in the market fish (12 to 75 percent) showed higher variability than did tilapia (30 to 46 percent). Lindane residues in market fish were below 0.1 ng g^{-1} ww for all but two carp species (15 and 0.47) and mean dieldrin levels were 0.53 ± 0.18 ng g^{-1} ww. However, mean chlordane levels in market fish were slightly higher at 1.2 ± 0.28 (excluding one sample at 7.6) ng g^{-1} ww. The levels of all seven groups of OCs measured were very low (<0.1 μg g^{-1} ww) and should pose no human health hazard even with daily fish consumption.

Parsons and Chan (1998) determined OC residues in blubber samples from eight Indo-Pacific hump-backed dolphins *Sotalia (Sousa) chinensis* Osbeck (Cetacea: Stenidae) and eleven finless porpoises *Neophocaena phocaenoides* Cuvier (Cetacea: Phocoenidae) stranded between 1993 and 1996 in Hong Kong waters. Levels of HCB, HCHs, dieldrin, chlordane, and mirex were low, generally below 1 mg g^{-1} lipid. However, DDT levels were quite high with median levels of 26.7 mg g^{-1} lipid for *S. chinensis* and 50.3 mg g^{-1} lipid for *N. phocaenoides*. Minh *et al.* (1999) also measured OC concentrations in blubber samples of small cetaceans from coastal Hong Kong waters collected from stranded animals found between 1993 and 1997. They found mean (range) levels of $\sum$ DDTs of 46 (5.1 to 80) and 39 (2.6 to 160) μg g^{-1} ww for eleven hump-backed dolphins and nine finless porpoises, respectively. Mean concentrations for HCHs, chlordanes, and HCB (in ng g^{-1} ww) were 760, 380, and 71 for hump-backed dolphins and 240, 210, and 71 for finless porpoises, respectively. The relatively high concentrations of DDTs can most likely be attributed to continued environmental inputs from the region. Another study of persistent chlorinated endocrine disrupters in cetaceans by Minh *et al.* (2000) found

Table 9.14 Studies conducted using 17 ¹⁴C-labeled pesticides in marine and freshwater species of the People's Republic of China[a]

Pesticide	Organisms studied[b]	Fate	Uptake dynamics	Loss	Metabolism	Bound
DDT	Tilapia sp. (Smith) (F)	X	X		X	
	Perna viridis L. (B)		X	X		
	Paphia undulata Born (B)		X	X		
	Gafrarium tumidum Röding (B)		X	X		
	Anomalocardia flexuosa L. (B)		X	X		
	Clausinella (Chione) isabellina Philippi (B)		X	X		
	Batillaria zonalis Bruguière (G)		X	X		
	Cerithidea cingulata Gmelin (G)		X	X		
	Penaeus monodon Fabricius (S)		X		X	
	Macrobrachium nipponense De Haan (S*)		X		X	
	Platymonas subcordiformis Wille (A)		X	X		
	Cirrhina molitorella Valenciennes (F*)		X		X	X
	Anodonta calypygos (B*)		X	X	X	X
	Frog tadpole		X		X	X
	Rana japonica Gunther (Fr)		X		X	X
	Other marine gastropods (G)		Bioavailability from sediment		X	X
DDE	Tilapia sp.(F)		X			
	Frog tadpole		X		X	X
	Cirr. molitorella (F*)		X	X	X	X
	Bellarnya purificata (G*)		X	X	X	X
DDD	Tilapia sp. (F)		X			
	Frog tadpole		X		X	X
DDA	Tilapia sp. (F)		X			
	Frog tadpole		X		X	X
BHC	Periophthalmus cantonensis Osbeck (F)	X	X			
	Sesarma plicata (C)					
	Alectryonella plicatula Gmelin (B)		X			X
	Per. viridis (B)		X	X		X
	Pap. undulata (B)		X	X		X
	Meretrix meretrix L. (B)		X			X
	Libitina japonica (B)		X			
	Brachydontes marginatus (B)		X	X		
Lindane	Pap. undulata (B)		X			
	Sinonovacula constricta Lamarck (B)		X			
	Ampullaria gigas Spix (G*)		X		X	X

continued...

Table 9.14 continued

Pesticide	Organisms studied[b]	Fate	Uptake dynamics	Loss	Metabolism	Bound
	Clam (B*)		X		X	X
	Misgurnus anguillicaudatus Cantor (F*)		X		X	X
Aldrin	Sino. constricta (B)		X		X	X
	Pen. monodon (S)		X		X	X
Parathion	Tilapia sp. (F)		X		X	X
	Pen. monodon (S)		X		X	X
	Quail (Galliformes: Phasianoidea)		X		X	X
	Snake		X		X	X
	Sino. constricta (B)		X		X	X
	Cirr. molitorella (F*)		X	X	X	X
	Bella. purificata (G*)		X	X	X	X
Chlorpyrifos	Perio. cantonensis (F)	X	X		X	
	Tilapia sp. (F)		X		X	X
	Ses. plicata (C)					
	Scapharca subcrenata Lischke (B)		X		X	
	Alec. plicatula (B)		X		X	X
	Per. viridis (B)		X		X	X
	Pap. undulata (B)		X		X	X
	Mer. meretrix (B)		X		X	X
	Lib. japonica (B)					
	Sino. constricta (B)		X		X	X
	Bra. marginatus (B)		X	X		
	Gambusia affinis Baird & Girard (F*)		X			
	Bella. purificata (G*)		X			
	Planorbis sp. L. (G*)		X			
	Nasturtium officinale R. Brown (P*)		X			
	Lemna (Spirodela) polyrrhiza L. (P*)		X			
	Cirr. molitorella (F*)		X		X	X
	Clam (B*)		X		X	X
	Amp. gigas (G*)		X		X	X
	R. japonica (Fr)		X		X	X
Trichlorphon	Perio. cantonensis (F)	X	X			
	Ses. plicata (C)		X			
	Scap. subcrenata (B)		X			
	Alec. plicatula (B)		X			
	Per. viridis (B)		X			
	Pap. undulata (B)		X			
	Mer. meretrix (B)		X			
	Lib. japonica (B)		X			

continued...

Table 9.14 continued

Pesticide	Organisms studied[b]	Fate	Uptake dynamics	Loss	Meta-bolism	Bound
Mono-crotophos	Ses. plicata (C)	X	X			
	Scap. subcrenata (B)		X			
	Alec. plicatula (B)		X			
	Per. viridis (B)		X			
	Mer. meretrix (B)		X			
	Lib. japonica (B)		X			
Chlordime-form	Perio. cantonensis (F)	X	X			
	Ses. plicata (C)					
	Scap. subcrenata (B)		X			
	Alec. plicatula (B)		X			
	Per. viridis (B)		X	X		
	Pap. undulata (B)		X	X		
	Mer. meretrix (B)		X			
	Lib. japonica (B)					
	Bra. marginatus (B)		X	X		
Fenvalerate	Platy. subcordiformis (A)	X	X	X		
	Per. viridis (B)		X	X		
	Pap. undulata (B)		X	X		
	Gaf. tumidum (B)		X	X		
	Anom. flexuosa (B)		X	X		
	Chione. isabellina (B)		X	X		
	Bat. zonalis (G)		X	X		
	Cerith. cingulata (G)		X	X		
	Certophyllum demersum (P*)		X			
	Clam (B*)		X			
	Macro. nipponense (S*)		X			
	Cirr. molitorella (F*)		X			
	Bella. purificata (G*)		X			
Chlorsulfuron	Mer. meretrix (B)	X	X		X	X
	Sino. constricta (B)		X		X	X
	Cirr. molitorella (F*)		X		X	X
	Bella. purificata (G*)		X		X	X
Bensulfuron methyl	Cirr. molitorella (F*)	X	X		X	X
	Bella. purificata (G*)		X		X	X
Butachlor	Pap. undulata (B)		X		X	X
	Tilapia sp. (F)		X		X	X
	Cirr. molitorella (F*)		X		X	X
	Clam (B*)		X		X	X
	Misg. anguillicaudatus (F*)		X		X	X
	Snake		X		X	X
	Quail		X		X	X
	R. japonica (Fr)		X		X	X
Atrazine	Cirr. molitorella (F*)		X		X	X
	Bella. purificata (G*)		X		X	X

continued...

Table 9.14 continued

Notes:
a Studies conducted by the Isotope Laboratory, School of Life Sciences, Zhongshan University, Guangzhou, PRC (Chen S *et al.*, 1998; Zhang Y *et al.*, 1998; Zhong *et al.*, 1997; 1998a,b,c,d,e). The following papers from this work have been submitted for publication: 1) Zhong Chuangguang, A new method for the study of the proportion of bound and extractable labeled pesticides by sediment; 2) Zhong Chuangguang, Carvalho FP, Zhang Yiqiang, Zhao Xiaokui, Chen Shunhua and Jiang Shigui, Accumulation and metabolism of ^{14}C-parathion, ^{14}C-chlorpyrifos and ^{14}C-butchlor by marine *Tilapia* sp.; 3) Zhong Chuangguang, Chen Shunhua, Zhao Xiaokui, He Jianguo, Jiang Shigui and Liao Yuanqi, Accumulation and metabolism of ^{14}C-DDT by *Penaeus monodon*; 4) Zhong Chuangguang and Shi Jun, Carbon column absorption to collect radioactivity labeled compounds in water; 5) Zhong Chuangguang, Shi Jun, Carvalho F P, Chen Shunhua, Zhao Xiaokui and Yan Xudong, Accumulation of ^{14}C-chlorpyrifos, ^{14}C-lindane and ^{14}C-butchlor by *Paphia undulata*; 6) Zhong Chuangguang, Chen Shunhua, Carvalho FP and Zhao Xiaokui, Bioavailability of sediment bound ^{14}C-DDT to tilapia and two bivalves; 7) Zhong Chuangguang, Liu Jingxuan, Chen Shunhua, Zhao Xiaokui, Jiang Shigui and Yan Yuanyi, Accumulation and elimination of ^{14}C-HCH, ^{14}C-chlorpyrifos and ^{14}C-chlordimeform in some marine mollusks; 8) Zhou Shinin, Lu Yongjun, Huan Lei, Cai Chuanghua, Zhou Yiping, Zeng Xiaohui, Fang Yanchang, Gu Jiang and Zhong Chuangguang, Preliminary study on the resistance to pesticides, antibiotics and heavy metals of marine bacteria isolated from Daya Bay; 9) Zhao Xiaokui, Zhong Chuangguang, Chen Shunhua and Carvalho FP, Uptake and metabolism of ^{14}C- chlorotoluron by two marine bivalves; 10) Zhong Chuangguang, Zhao Xiaokui, Chen Shunhua and Jiang Shigui, Accumulation of ^{14}C-DDT and ^{14}C-fenvalerate by a marine crab, *Sesarma plicata*; 11) Chen Shunhua, Peng Yefang and Zhong Chuangguang, Preliminary study on the absorption and desorption of ^{14}C-DDT and ^{14}C-fenvalerate by a single cell alga, *Platymonas subcordiformis*; 12) Zhang Yiqiang, Zhang Haiqing, Zhong Chuangguang, Zhao Xiaokui and Chen Shunhua, Behaviors of ^{14}C-butchlor, ^{14}C-chlorpyrifos and ^{14}C-DDT in *Rana japonica* Gunther; 13) Zhang Yiqiang, Zhong Chuangguang, Carvalho FP, Chen Shunhua, Zhao Xiaokui and Zhang Haiqing, Behaviors of ^{14}C-butchlor and ^{14}C-chlorpyrifos in *Cirrhina molitorella*; and 14) Zhong Chuangguang, Zhang Yiqiang, Carvalho FP, Chen Shunhua, Zhao Xiaokui and Zhang Haiqing. Uptake and metabolism of ^{14}C-DDT in *Cirrhina molitorella*.
b Letters in parenthesis indicate species grouping as follows: 1) A–marine algae, A*–freshwater algae; 2) B–marine bivalve, B*–freshwater bivalve; 3) C–crab; 4) F–marine fish, F*–freshwater fish; 5) Fr–frog; 6) G–marine gastropod, G*–freshwater gastropod; 7) S–marine shrimp, S*–freshwater shrimp; P*–freshwater plant.

the highest concentrations of DDTs in northern North Pacific dolphins and hump-backed dolphins and finless porpoises from Hong Kong. They also noted that greater residue levels of HCHs found in cetaceans from Hong Kong compared to cetaceans from high latitudes may suggest recent usage of these compounds in southern China.

CASE STUDY

IPM for rice in Dasha Township

Led by the internationally renowned entomologist Professor Pu Zhelong of the Institute of Entomology at Zhongshan University, a research group has carried out IPM experiments in Dasha Township of Sihui County, Guangdong Province since 1973. Principal techniques used in IPM included: 1) cultural practices such as leveling rice fields, improving irrigation methods, ploughing fields in early spring followed by immersion to kill overwintering larvae of white rice stem borers in the rice stubble; 2) rearing ducks for controlling pest insects; 3) applying a preparation of *Bacillus thuringiensis* Berliner for biocontrol of insect pests; 4) releasing insect egg endoparasitoids, e.g. *Trichogramma* spp. Westwood (Hymenoptera: Trichogrammatidae) to control several species of lepidopterous rice insects, chiefly rice leaf rollers; and 5) applying chemical insecticides rationally. To achieve rational use of pesticides, this IPM program has had to overcome the traditional heavy reliance by farmers on chemical pesticides. By monitoring population levels of both rice pests and their natural enemies, farmers now apply chemical pesticides only in specific areas seriously infested with the pest, instead of spraying the entire rice field as before. They also use pesticides on rice seedlings to avoid pest infection after re-planting, and, hence, reduce the quantity of pesticides used in the rice field and pesticide residues in rice.

Adoption of IPM solved the problem presented by insect pests and diseases in rice production, leading to significantly reduced pesticide use (Table 9.15). It also changed the traditional orientation of pest and disease control in rice-farming operations. Both the number of species and population size of natural enemies increased, improving field ecology (Table 9.16). Combining those IPM techniques described above with another alternative, viz. increased host-plant resistance, is an effective substitute for pesticide use (Widawsky *et al.*, 1998). Under the intensive rice production systems in eastern China, pesticide productivity is low – indeed, returns to pesticide use are negative at the margin – compared to the productivity of host-plant resistance. This suggests that pesticides are being overused in the

Table 9.15 Quantity of pesticides used in two rice fields of Dasha township during the period 1970–83 before and after introduction of IPM (in kg of formulated product)

	Year	Quantity used in field 1	% of pre-IPM[a]	Quantity used in field 2	% of pre-IPM
Pre-IPM	1970	138,698	100	118,259	100
	1971	139,401		112,995	
	1972	149,197		114,354	
IPM implemented in some areas	1973	132,847	93.27	92,228	80.06
	1974	101,728	71.42	69,475	60.31
IPM implemented in all rice fields	1975	46,307	32.51	25,537	22.17
	1976	59,116	51.54	37,970	32.96
	1977	70,570	59.55	58,249	50.57
	1978	50,176	35.23	33,793	29.33
	1979	46,940	30.98	41,843	36.01
	1980	35,481	24.91	26,227	22.77
	1981	35,147	24.68	21,100	18.32
	1982	32,942	23.13	20,284	17.61
	1983	50,796	35.64	25,658	22.27

Note:
a Mean of usage for 1970–72 equals 100 percent.

Table 9.16 Comparison of pesticide residues in IPM area of Dasha township and a nearby chemical pest management (CPM) area based on a 1978 study

		Residues (mg kg⁻¹)		
Sample	Pesticide	Dasha IPM area	Nearby CPM area	IPM ÷ CPM
Soil	BHC	0.0091	0.0410	4.51
	Parathion	0.0017	0.0019	1.117
	DDT or trichlorphon	0.0123	0.0103	0.837
Rice	BHC	0.0088	0.0400	4.545
	Parathion	0.0015	0.0025	1.67
	DDT or trichlorphon	0.0050	0.0080	1.60
Straw	BHC	0.0525	0.9320	17.15
	Parathion	0.0020	0.0015	0.75
	DDT or trichlorphon	0.0075	0.0150	2

region and host-plant resistance is being underutilized. Government policy to promote development of increased host-plant resistance could prove the means to further reduce pesticide use.

Based on 1983 data, BHC residues in rice decreased to 5.4 μg kg⁻¹ from 16 μg kg⁻¹ in 1977 and 60 μg kg⁻¹ in 1974. BHC residues in soil samples from 1983 were less than half the 1977 level. Dasha Township was the earliest large-scale demonstration project of IPM in China. Through more than 20 years of practice, IPM has been shown to be very effective and these techniques have taken root in local farm practices. Analysis of 1994 samples did not detect other insecticide, fungicide, and herbicide residues except 0.04 to 0.05 ppm of Sachongshuan. In 1995, The Green Rice Company of Dasha was established, and the township has begun batch processing of 'environmentally green' rice.

REFERENCES

(*) Indicates incomplete citation. For further information please contact Chapter authors through the Isotope Laboratory, School of Life Sciences, Zhongshan University, Guangzhou, People's Republic.

An Qiong and Chen Zuyi. 1993. Bound residue of trifluralin in soil and its bioavailability to plants. *Huanjing Kexue Xuebao* or (*Acta Scientiae Circumstantiae*). 13(3):295–303. (in Chinese; English abstract)

Anonymous. 2000. China to curb certain pesticides. *Chemical and Engineering News*. 78(21):13 May 22 issue.

*Bao, Y.P. *et al.* 1991.

Cai Fulong, Lin Zhifeng, Chen Ying, Chen Shumei, Yiang Jiadong, Cai Feng and Qian Lumin. 1997. Behavior of BHC and DDT at the mouth of the Zhujiang River, China. In: *Environmental Behavior of Crop Protection Chemicals*. Proc. Int. Symp. on Use of Nuclear and Related Techniques for Studying Environmental Behavior of Crop Protection Chemicals, 1–5 July 1996. Vienna: IAEA, pp. 349–58.

Chan Hing Man, Chan King Ming and Dickman, Michael. 1999. Organochlorines in Hong Kong fish. *Mar Poll Bull.* 39(1–12):346–51.

Chang Yung-yuan, Tan Yu-yun and Sun Mei-juan. 1981a. Mechanism of biodegradation of organophosphate pesticides in aquatic ecosystem. *Huanjing Kexue Xuebao* or (*Acta Scientiae Circumstantiae*). 1(2):115–25 (in Chinese; English abstract).

Chang Yung-yuan, Sun Mei-juan and Tan Yu-yun. 1981b. Transport and fate of BHC in aquatic environment. *Huanjing Kexue Xuebao* or (*Acta Scientiae Circumstantiae*) 1(3):242–9.

Chen, S., Lin, Z. and Lin, M. 1986. Distributions of BHC and DDT concentration in the surface sediments of Xiamen Harbour. *Taiwan Strait* 5:32–7 (in Chinese).

Chen Shunhua, Zhong Chuangguang and Zhao Xiaokui. 1998. Absorption, distribution, dynamics of ¹⁴C-chlorpyrifos in several kinds of animals and plants in fresh water ecosystem. *Acta Agriculturae Nucleatae Sinica* 10(12):5–13.

*Chen Xulong *et al.* 1983. *Haiyang Huanjing Kexue* or (*Marine Environmental Science*) 2(3):14–9.

Chen Zhongxiao and Fan Defang. 1988. Leaching, persistence and degradation of the herbicide butachlor in soils. *Huanjing Huaxue* or (*Environmental Chemistry*) 7(2):30–6 (in Chinese; English abstract).

Chen Ziyuan and Xu Bujin 1982. Adsorbability of lindane on soil and its correlation with residual lindane in wheat and soil. *Huanjing Kexue Xuebao* or (*Acta Scientiae Circumstantiae*) 2(4):299–306 (in Chinese; English abstract).

Gan Jianying and Chen Ziyuan 1986. Fate of fenvalerate in an agricultural ecosystem. *Huanjing Kexue Xuebao* or (*Acta Scientiae Circumstantiae*) 6(3):263–72 (in Chinese; English abstract).

Guo Jiangfeng, Sun Jinhe and Zhai Linlin. 1996. Behavior of pirimicarb in the aquatic ecosystem. *Henong Xuebao* or (*Acta Agriculturae Nucleatae Sinica*) 10(3):177–80 (in Chinese; English abstract).

Hong, H.S., Xu, L., Zhang, L.P., Chen, J.C., Wong, Y.S. and Wan, T.S.M. 1995. Environmental fate and chemistry of organic pollutants in the sediment of Xiamen and Victoria harbours. *Mar Poll Bull.* 31(4–12):229–36.

Hong, H.S., Chen, W.Q., Xu, L., Wang, X.H. and Zhang, L.P. 1999. Distribution and fate of organochlorine pollutants in the Pearl River Estuary. *Mar Poll Bull.* 39(1–12):376–82.

Hu, M.H., Yang, Y.P. and Xu, S.H. 1996. Estuarine geochemistry of the Minjiang River. In: Zang, J. (ed.) *The Biogeochemistry of Major Chinese Estuaries.* Beijing, PRC: Ocean Press, p. 241 (in Chinese).

Huang Shizhong, Li Zhixiang, Wang Yiru and Tian Zuofeng. 1985. Transmission and accumulation of BHC in food chains. *Huanjing Kexue* or (*Chinese Journal of Environmental Science*). 6(4):15–8 (in Chinese; English abstract).

Huang Yuyao, Gao Yurong, Cao Hong, Ren Shuzhi and Zhu Jiang. 1994. Preliminary study on the effects of monoformamidine insecticide on a model grassy pond ecosystem. *Huanjing Kexue Xuebao* or (*Acta Scientiae Circumstantiae*) 14(4):466–74 (in Chinese; English abstract).

*Le Zhongkui and Liu Zhigang 1985. *Donghai Marine Science* 3(2):69–75.

*Li Zhixiang *et al.* 1985. *Journal of Zhejiang Agriculture University* (2):151–7.

Liao Qiang. 1983. Contamination of organochlorines in waters, sediments and biotoms from western Guangdong coastal area. *Haiyang Huanjing Kexue* or (*Marine Environmental Science*) 2(4):37–40.

Liao Qiang. 1986. Investigation of the residue of pesticides (BHC, DDT) in sea water, sediments and organisms along the East Coast of Guangdong. *Haiyang Huanjing Kexue* or (*Marine Environmental Science*) 5(2):8–14.

*Min Huang 1993. *Agricultural Ecological Environment* (3):40–3.

Minh, T.B., Watanabe, M., Nakata, H., Tanabe, S. and Jefferson, T.A. 1999. Contamination by persistent organochlorines in small cetaceans from Hong Kong coastal waters. *Mar Poll Bull.* 39(1–12):383–92.

Minh, T.B., Prudente, M.S., Watanabe, M., Tanabe, S., Nakata, H., Miyazaki, N., Jefferson, T.A. and Subramanian, A. 2000. Recent contamination of persistent chlorinated endocrine disrupters in cetaceans from the North Pacific and Asian coastal waters. *Water Science and Technology* 42(7–8):231–40.

Miyata, M., Kamakura, K., Hirahara, Y., Narita, M., Okamoto, K., Hasegawa, M., Koiguchi, S., Yamana, T., Tonogai, Y. and Ito, Y. 1993. Studies on simultaneous determination of 12 pyrethroid and 29 organophosphorus pesticides in agricultural products.

Mo Hanhong, An Fengchun, Yang Kewu and Liu Ye. 1995. Residue and persistence of the pesticide N´-(2,4-dimethylphenyl)-N-methylformamidine hydrochloride in cotton fields. *Huanjing Kexue* or (*Chinese Journal of Environmental Science*) 16(1):5–7, 22 (in Chinese; English abstract).

Mo Tao, Zhang Shaojun, Chen Shuxin and Zheng Shanqiang. 1990. Residues of the herbicide Tackle in soybean and soil. *Huanjing Wuran Yu Fangzhi* or (*Environment Pollution and Prevention*) 12(3):2–5 (in Chinese; English abstract).

Parsons, E.C.M. and Chan, H.N. 1998. Organochlorines in Indo-Pacific hump-backed dolphins (*Sousa chinensis*) and finless porpoises (*Neophocaena phocaenoides*) from Hong Kong. In: Morton, B. (ed.) *The Marine Biology of the South China Sea III*. Hong Kong, PRC: Hong Kong University Press, pp.423–37.

*Peng Genyuan et al. 1995. *Henong Xuebao* or (*Acta Agriculturae Nucleatae Sinica*) 9 (supplement):107–11.

Richardson, B.J. and Zheng, G.J. 1999. Chlorinated hydrocarbon contaminants in Hong Kong surficial sediments. *Chemosphere* 39(6):913–23.

Sun Jing-he, Zhang Yong-xi, Chen Zi-yuan and Li Xin-min. 1986a. Dynamics of degradation and residue of ^{14}C-fenvalerate in flooded paddy soils. *Yuan Zi Neng Nong Ye Ying Yong* or (*Application of Atomic Energy in Agriculture*) (3):8–14 (in Chinese; English abstract).

*Sun Jing-he, Zhang Yong-xi, Chen Zi-yuan and Li Xin-min. 1986b. *Yuan Zi Neng Nong Ye Ying Yong* or (*Applications of Atomic Energy in Agriculture*) (5).

*Tang Ke et al. 1987. *Pesticides* (2):33–5.

Wang Fujun, Qi Mengwen, Yang Genhai, Wang Huaguo and Gou Qianhua. 1995. Studies on the efficacy of controlled release formulations of herbicides against weeds in transplanted rice field. *Henong Xuebao* or (*Acta Agriculturae Nucleatae Sinica*) 9(2):107–12 (in English).

Wang, J., Yang, J.B., Han, C.R., Cai, L.W., Li, X.B., Bukkens, S.G.F. and Paoletti, M.G. 1999. Pesticide residues in agricultural produce in Hubei Province, PR China. *Critical Reviews in Plant Sciences* 18(3):403–16.

Wen, D.Z., Tang, Y.X., Zheng, X.H. and He, Y.Z. 1992. Sustainable and productive agricultural development in China. *Agricul Ecosys and Environ.* 39(1–2):55–70.

Widawsky, D., Rozelle, S., Jin, S.Q. and Huang, J.K. 1998. Pesticide productivity, host-plant resistance and productivity in China. *Agricultural Economics* 19(1–2):203–17.

Wong, S.K. and Lee, W.O. 1997. Survey of organochlorine pesticide residues in milk in Hong Kong (1993–1995). *Journal of AOAC International* 80(6):1332–5.

*Wu, T.Q. 1993.

Wu, Y., Zhang, J. and Zhou, Q. 1999. Persistent organochlorine residues in sediments from Chinese river/estuary systems. *Environ Poll.* 105(1):143–50.

*Xia Zheng Lu *et al.* 1981. *Collection of Environmental Science.* (4):24–9.

*Xiao Yunxiang *et al.* 1990. *Huanjing Kexue* or (*Chinese Journal of Environmental Science*). 11(5):6–12.

Xu, S., Jiang, X., Wang, X., Tan, Y., Sun, C., Feng, J., Wang, L., Martens, D. and Gawlik, B.M. 2000. Persistent pollutants in sediments of the Yangtse River. *Bull Environ Contam Toxicol.* 64(2):176–83.

*Yao Jianren *et al.* 1987. *Chinese Journal of Ecology* (1):33–6.

*Yu Kangning *et al.* 1988. *Pesticides* (6):28–9.

Yu Kangning, Qi Chengjiu, Tang Ke and Chen Zhongxiao. 1993. Relation between degradation rate of butachlor and conditions of paddy fields. *Huanjing Kexue Xuebao* or (*Acta Scientiae Circumstantiae*) 13(2):169–73 (in Chinese; English abstract).

*Zhang Hejing and Zen Yupin 1987. *Haiyang Huanjing Kexue* or (*Marine Environmental Science*) 6(2):19–23.

Zhang, L.P., Chen, W., Lin, L.M. and Hong, H.S. 1996. Concentrations and distribution of HCHs, DDTs and PCBs in surface sediments of Xiamen Western Bay. *Tropic Oceanol.* 15(1):91–5 (in Chinese; English abstract).

Zhang Shuiming, Ma Xingfa, An Qiong and Li Xunguang. 1983. BHC residue in paddy soil and in rice contamination. *Turang Xuebao* or (*Acta Pedologia Sinica*) 20(1):79–84 (in Chinese; English abstract).

Zhang Shuiming, Ma Xingfa and An Qiong. 1988. Persistence and degradation of BHC in soil. *Turang Xuebao* or (*Acta Pedologia Sinica*) 25(1):81–8 (in Chinese; English abstract).

Zhang Yiqiang, Zhong Chuangguang, Carvalho FP, Chen Shunhua, Zhao Xiaokui and Zhang Haiqing. 1998. Behaviors of ^{14}C-butachlor, ^{14}C-DDT and ^{14}C-chlorpyrifos in *Cirrhina molitorella*, I Accumulation dynamics in the fish. Joint Conference on Environmental Science held 26–28 November 1998, Guangzhou, PRC.

Zhang Zhongliang, Wang Huaxin, Guo Dazhi, Chen Zhiyu and Wu Suqiong. 1991. Residues of ^{14}C-fenitrothion in a simulated rice/fish ecosystem. *Henong Xuebao* or (*Acta Agriculturae Nucleatae Sinica*) 5(3):163–8.

Zheng, G.J., Lam, M.H.W., Lam, P.K.S., Richardson, B.J., Man, B.K.W. and Li, A.M.Y. 2000. Concentrations of persistent organic pollutants in surface sediments of the mudflat and mangroves at Mai Po Marshes Nature Reserve, Hong Kong. *Mar Poll Bull.* 40(12):1210–14.

Zhong Chuangguang *et al.* 1995. Report presented at 2nd Research Coordination Meeting (RCM) of IAEA Cooperative Research Program (CRP) held 12–16 June 1995 in Kuala Lumpur, Malaysia.

Zhong Chuangguang *et al.* 1996. Report presented at 3rd Research Coordination Meeting (RCM) of IAEA Cooperative Research Program (CRP) held 9–13 September 1996 at Universidad Nacional, San Jose/Heredia, Costa Rica on Pesticides in the Marine Environment.

Zhong Chuangguang, Carvalho, F.P., Chen Shunhua and Zhao Xiaokui. 1997. Studies on the accumulation, distribution and degradation of ^{14}C-DDT in marine *Tilapia* sp. *Acta Scientiarum Naturalium Universitatis Sunyatseni* 36(Suppl 2):52–7 (in English).

Zhong Chuangguang, Zhao Xiaokui, Chen Shunhua and Fu Yun. 1998a. Accumulation and metabolism of four ^{14}C-labeled pesticides by some marine animals. *Tropic Oceanol.* 17(4):51–6 (in Chinese; English abstract).

Zhong Chuangguang, Chen Shunhua, Carvalho, F.P. and Zhao Xiaokui. 1998b. A mini model ecosystem about the fate of labeled pesticides. *Acta Scientiarum Naturalium Universitatis Sunyatseni* 37(1):93–7 (in English).

Zhong Chuangguang, Chen Shunhua, Zhao Xiaokui, Jiang Shigui and Yan Xudong. 1998c. Accumulation and elimination of ^{14}C-DDT and ^{14}C-fenvalerate by some marine organisms. *Tropic Oceanol.* 17(3):80–8 (in Chinese; English abstract).

Zhong Chuangguang, Carvalho, F.P., Chen Shunhua, Zhao Xiaokui and Shi Jun. 1998d. Uptake and metabolism of ^{14}C-chlorpyrifos by marine bivalves. International Symposium on Marine Pollution held 5–9 October 1998, Monaco.

Zhong Chuangguang, Zhang Yiqiang, Carvalho, F.P., Chen Shunhua, Zhao Xiaokui and Zhang Haiqing. 1998e. Behaviors of ^{14}C-butachlor, ^{14}C-DDT and ^{14}C-chlorpyrifos in *Cirrhina molitorella*, II Metabolism patterns. Joint Conference on Environmental Science held 26–28 November 1998, Guangzhou, PRC.

Zhong Chuangguang. 1999. The distribution, fate and effects of pesticides on biota in Daya Bay, South China Sea. IAEA CRP Report 7933/MC.

Zhou, J.L., Hong, H., Zhang, Z., Maskaoui, K. and Chen, W. 2000. Multi-phase distribution of organic micropollutants in Xiamen Harbour, China. *Water Research* 34(7):2132–50.

Zhou, J.L., Maskaoui, K., Qiu, Y.W., Hong, H.S. and Wang, Z.D. 2001. Polychlorinated biphenyl congeners and organochlorine insecticides in the water column and sediments of Daya Bay, China. *Environ Poll.* 113(3):373–84.

Chapter 10

Ecotoxicology of pesticides in Philippine aquatic ecosystems

Cristina M. Bajet

INTRODUCTION

The Philippines is an archipelago, found off the southeastern coast of the Asian mainland, composed of 7,107 islands divided into three groups: Luzon, Visayas, and Mindanao with a total land area of approximately 300,000 km^2 (*The Philippine Atlas*, 1975). The Philippines is further subdivided into 16 regions; each composed of several provinces growing rice, corn, bananas, pineapples, coconuts, sugarcane, mangos, and vegetables as major crops (Figure 10.1). The total water area of the Philippines is more than seven times the land area. This extensive coastline of about 17,460 km offers livelihoods to more than one million people. Consequently the Philippines is the 12th largest fish-producing country in the world (Aypa, 1991, Munoz, 1997; BFAR, 2001). The 2.2 M km^2 of marine resources are composed of 12 percent coastal and 88 percent oceanic waters (BFAR, 1996; 2001). The aquatic environment is not limited to the sea, as there is also vast hectarage of inland waters consisting of swamplands, lakes, rivers, reservoirs, and small impoundments. These impoundments include fish ponds and irrigated rice fields with an additional aggregate area of 1.87 M ha where fish are produced (Aypa, 1991). In 1995, fish production from commercial (33.8 percent), municipal (36.1 percent), and aquaculture (30.1 percent) fisheries was 2.74 M T with a value of US$3.225 billion (BFAR, 1996) (see Figure 10.2). Production increased to 33.0, 32.9, and 34.1 percent respectively in 2000 equivalent to 2.87 M T and a 1.64 percent positive growth over the previous year. Municipal fisheries production includes fishing in coastal and inland waters with or without boats (of <3 tons displacement) whereas commercial fisheries production is from offshore waters with the use of fishing vessels of >3 tons. Among the major regional contributors to total fish production, Region IV was the largest producer (21.24 percent) followed by ARMM (15.03 percent) where seaweed is the major product and region IX (14.15 percent) ranked third (BFAR, 2001). Filipinos are fish eaters with a per capita consumption of 36 kg year^{-1} (99 g day^{-1}), equivalent to 12.3 percent of the total food intake per person per day (BFAR, 1996).

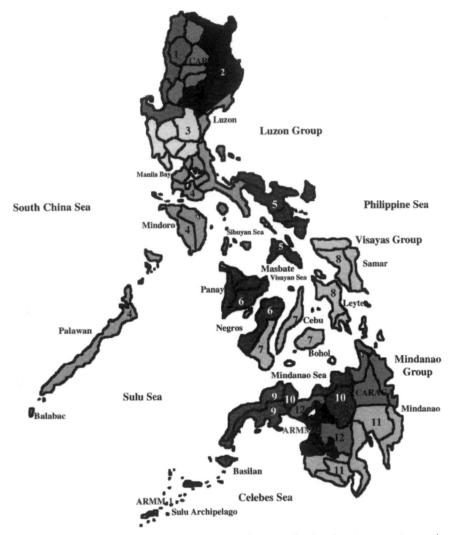

Figure 10.1 Geopolitical map of the Philippines showing island and region groupings and surrounding east Asian seas

PESTICIDE REGULATION

In the Philippines, pesticide use continues to be a significant component of agricultural practice. The pesticide industry is regulated by the Fertilizer and Pesticide Authority (FPA), which was created in May 1977 pursuant to Presidential Decree 1144. FPA is a regulatory agency under the Department of Agriculture (DA) and is in charge of pesticide registration; issuance of experimental use permits; regulation of labeling, packaging, and disposal of pesticides; establishment of MRLs in

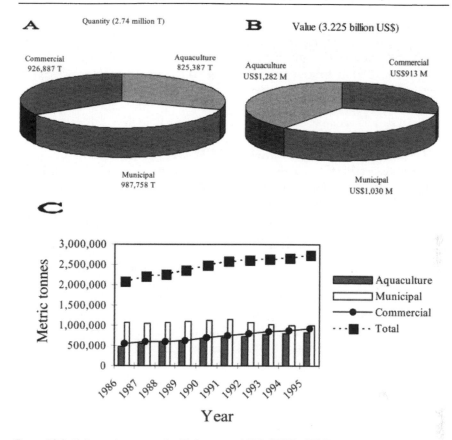

Figure 10.2 Fish production in the Philippines, 1995 (BFAR, 1996)

feed and foodstuffs; post registration activities including monitoring pesticide importation, sales outlets, and formulation plants; pesticide information systems; licensing of commercial pesticide applicators; and agromedical program activities. For the implementation of certain facets of registration, FPA has formed an inter-agency Pesticide Technical Advisory Committee (PTAC) composed of experts in various fields of pesticide management. For residue and formulation analysis, FPA draws expertise mainly from the National Crop Protection Center (NCPC) and the Bureau of Plant Industry (BPI) while for bioefficacy evaluations, it calls on experts within NCPC, BPI, the University of the Philippines Los Baños (UPLB), and other government institutions. For assistance in monitoring and extension, FPA utilizes personnel from the National Food and Agriculture Council (NFAC) and the Bureau of Agricultural Extension (BAEx) and, for statistical data about pesticide use, the Bureau of Agricultural Economics (BAEcon). The bulk of FPA's work is registration; import permits; training of pesticide applicators, dealers, and designated responsible care officers of the pesticide companies; inspection of formulation plants, dealers, and distributors of pesticides; and training of medical doctors regarding pesticide poisoning.

The FPA also enjoins members of the Crop Protection Association of the Philippines (CPAP), which is composed of representatives from the major pesticide companies in the Philippines, to take charge of product stewardship, to minimize pesticide misuse, and to train farmers about safety procedures for pesticide application. Each company adopts a province and within their adopted province takes charge of training farmers about the safe use of pesticides.

The institutional and legal framework of FPA is adequate but its implementation of policies especially for regulating pesticide residues in foods and feed is deficient. There is currently no implementation of FPA guidelines for MRLs in food and feed products in the country. The Philippines adopted the MRLs set by FAO/WHO, which are primarily generated in temperate countries using agricultural practices accepted in those countries.

The official laboratory of the FPA is BPI-National Pesticide Analytical Laboratory (BPI-NPAL), which is also under the DA, and its mandate is to monitor pesticide residues in market produce. However, it has little or no power to confiscate and destroy food found to exceed the established limits. BPI is independent of FPA, although both institutions are under the DA. This arrangement lengthens the time from the analysis of pesticide residues in food to regulatory and control action by FPA. Likewise, produce and other food has been sold by the time the results of analyses are available. However, pesticide residue monitoring data show pesticide residue profiles and the extent of food contamination, which is a basis for regulatory decisions. These include the banning or restriction of products found to be causing health risks, environmental contamination, or unacceptable residues and requirements for more training for pesticide dealers, retailers, and farmers. BPI also analyzes formulations submitted by the pesticide companies to determine if the products are within specifications. FPA uses BPI data for official legal action if any pesticide formulations are found in the marketplace that are counterfeit or not within specifications.

The Pesticide Chemistry and Toxicology Laboratory (PTCL) of the National Crop Protection Center, University of the Philippines, Los Baños does extensive research on the impact of pesticides in the environment. NCPC is an independent research center under the College of Agriculture, University of the Philippines, Los Baños. Although NCPC personnel are on the University payroll, the Center and the Laboratory subsist on research grants from local and international sources. Because the DA head sits as one of the board members of NCPC, data generated by the Center is utilized and channeled to the government and the private sector, partly through the DA and the Regional Crop Protection Centers nationwide.

Current research of the PTCL, NCPC on the environmental aspect of pesticides involves determining pesticide residues in food and the environment as a consequence of crop protection practices; generating residue data for setting Philippine MRLs; monitoring pesticide residues in sediments collected in rivers draining into Manila Bay; determining the bioavailability of sediment-bound pesticide residues;

evaluating farmers' exposure to pesticides used in mango cultivation and for other crops; evaluating a rapid test kit for determining residues in foodstuffs; measuring bioaccumulation of pesticides in selected food chains; determining the effects of continued use of pesticides on biodiversity, toxicity, and remediation of pesticides in waterways using aquatic plants and tilapia as bioindicators; and identifying and testing botanical pesticides for pest control.

The Environment Management Bureau (EMB) of the Department of Environment and Natural Resources (DENR) is mandated to monitor environmental pollutants; however they focus their work on industrial pollutants, e.g. heavy metals, polyaromatic hydrocarbons, oils, etc. Monitoring of pesticide residues by the EMB is performed sparingly. EMB has monitored pesticide residues in biota collected from Manila Bay only once.

PESTICIDE DISTRIBUTION, USAGE PATTERN, USE, AND MISUSE

Pesticide sales in 1993 were equivalent to approximately US$120 million, 46.17 percent of which was used in rice, followed by bananas (15.25 percent), vegetables (15.14 percent), mangos (9.30 percent), and pineapples (7.01 percent) (CPAP, 1993). There are 16 major formulation plants employing a total of 75 a.i.(s) in their products. Most pesticide formulation is done locally except for fungicides. FPA importation data for technical materials ranged from 3,165 to 4,453 T while pesticide finished products ranged from 6,705 to 11,514 T in the years 1985 to 1992 (Ocampo and Bajet, 1997) (see Figure 10.3). In 1996, 5,641 T of 100 percent a.i.(s) were sold, an average annual growth of 6 percent from 1993 to 1996 (Dave, 1997) (see Figure 10.4).

The bulk of pesticides used in 1993 and in 1996 were insecticides and herbicides. The most commonly used insecticides in 1993, in decreasing order, were endosulfan followed by monocrotophos, methyl parathion, cypermethrin, and azinphos ethyl. Herbicides in decreasing order of use were butachlor, pretilachlor, a mixture of piperophos + 2,4-D, and 2,4-D. Although applied in limited quantities, the most commonly used molluscicides were niclosamide, metaldehyde, and isazofos (CPAP, 1993).

However, with the banning by the Fertilizer and Pesticide Authority of methyl parathion, azinphos ethyl, organotin compounds, and endosulfan 35 EC and the restriction of monocrotophos use to beanfly control in legumes, there has been a considerable change in pesticide use patterns. In Nueva Ecija, one of the provinces in the rice producing area, there has been a shift to the use of pyrethroids, e.g. cypermethrin and deltamethrin. However, endosulfan, carbofuran and isoprocarb are still widely used (Rola, 1995).

Nationwide in 1996, pyrethroid use increased to 98 T a.i. from 44 T in 1993 with 87 T a.i. of new pesticides (Dave, 1997). For rice, pretilachlor, butachlor, and its mixture with propanil are the most popular herbicides while niclosamide is the

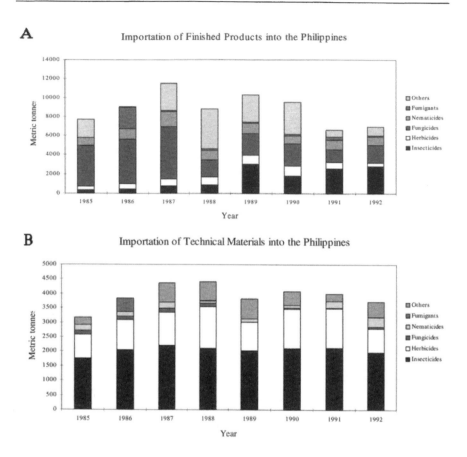

Figure 10.3 Pesticide importation statistics in the Philippines for the period 1985–92 (Ocampo and Bajet, 1997). The category 'others' includes acaricides, molluscicides, and rodenticides

most commonly used molluscicide (Rola, 1995). However, triazophos, diazinon, and chlorpyrifos continue to be the most popular OP insecticides in terms of quantity sold for use in the Philippines and deltamethrin and cypermethrin are the most commonly used pyrethroids (CPAP, 1996).

In spite of the banning of a number of pesticides by the FPA in 1993, a 15 percent growth in sales volume was reported in 1995 with 978 T of banned a.i.(s) sold. The growth in sales was attributed to growth in low toxicity and low dose products. Changes from 1993 to 1996 showed a decrease of insecticide usage equivalent to 512 T a.i.; whereas there was a drastic increase in the use of fungicides, equivalent to 1090 T a.i. (Dave, 1997). However, from 1997 to 1999, the 27,845 M T of formulated products sold at the distributors' level first decreased to 20,649 M T in 1998 and then increased slightly to 23,165 M T in 1999 (CPAP, 2000). This was an overall decrease of 16.78 percent over the period. Forty-seven percent

A 1986 Pesticide sales (× 1000 L or kg of formulated product)

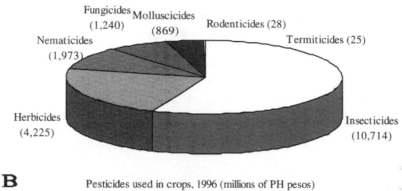

B

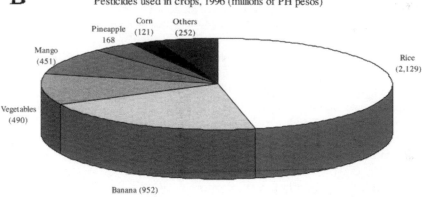

C

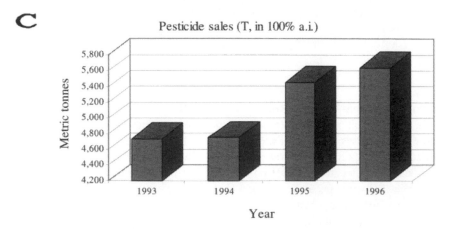

Figure 10.4 Pesticide sales, usage, and trends in the Philippines for the period 1993–96 (Dave, 1997). Average growth of pesticide sales was 6 percent

of the pesticides sold in the Philippines in 1996 were used in the rice crop followed by banana, vegetable, mango, pineapple, and corn crops (Figure 10.4). However, the average cost of pesticides used per hectare was highest in bananas (US$1220 ha^{-1}) followed by mangos (US$310 ha^{-1}), pineapples (US$217 ha^{-1}), vegetables (US$118 ha^{-1}), and rice (US$25 ha^{-1}) (Dave, 1997). Likewise, in 1999, usage in rice decreased to 44.8 percent followed by plantation crops (banana and pineapple) (23.5 percent), fruits (14.20 percent) and vegetables (11.80 percent) (CPAP, 2000). This indicates that although the usage in rice is more widespread, the usage in banana, mango, and pineapple plantations is more intensive and that, in these ecosystems, pesticide residues could be a problem. However, these crops, although concentrated in isolated areas mostly in the Mindanao Islands, are planted in upland areas with a limited number of watercourses nearby, which could serve to transport the residues into significant portions of the aquatic environment.

A nationwide survey of rice farmers in 1996 by the NCPC likewise shows an indication of a shift to the use of pyrethroids like cypermethrin, deltamethrin, and λ-cyhalothrin. Based on the survey, butachlor, niclosamide, and coumatetralyl were the most popular herbicide, molluscicide, and rodenticide, respectively (NCPC, 1996). Initial data also show that farmers tend to spray at greater than the recommended rate as well as spraying crops on which the pesticide is not recommended. For example, the registered uses of carbofuran are for bananas, ornamentals, corn, eggplant, and rice but 3.64 percent of the farmers interviewed in Region III used it for little-leaf disease of bitter gourd. Diazinon, also not recommended for bitter gourd, was used by 4.55 percent of the farmers in Region II. Monocrotophos, which is registered only for legumes, was also applied in corn for the control of a wide range of corn pests from corn-borer to armyworm and cutworm; in rice for Tungro (a disease caused by the green leafhopper *Nephotettix* spp. (Homoptera: Cicadellidae) and the zigzag leafhopper *Recilia dorsalis* Motschulsky (Homoptera: Cicadellidae)), mole cricket, rice bug *Leptocorisa oratorius* Fabricius (Hemiptera: Alydidae), and semi-looper *Naranga aenescens* Moore (Lepidoptera: Noctuidae); and in mangos for leafhoppers and aphids. In these cases, the pesticides are not recommended for these crops and there is doubt the pesticides that were used were effective on the pests mentioned.

There is probably a high occurrence of pesticide misuse in the country. A case study done during the 1991 wet season in Leyte (eastern Visayas, Region 8) found that of the 841 sprays applied to rice by 300 farmers with an average of 2.8 applications per farmer, only 190 or 23 percent may be considered to have been applied at the appropriate time and for the intended targets (Heong *et al.*, 1995). This was due to farmers' perception of pests that are highly visible and are often perceived to be damaging when they have yet to reach the EIL threshold. Farmers tend to overreact and apply pesticides as soon as the insect is noticed, which may be too early in the season or population cycle or not during the most susceptible stage of the insect. Pesticide application patterns reflect no alleviation of the farmer's current pest situation because only 160 of the 841 applications (19 percent) used the correct pesticide for the intended pest, which could have prevented yield loss (Heong *et al.*,

1995). Sometimes, pesticide(s) purchased by the farmer was sprayed on whatever standing crop or pest was present regardless of whether the label recommended it for that crop or pest and this practice was continued until the farmer was ready to buy a new supply of pesticide(s). When production data from a farm-level survey were integrated with health data collected from the same population of farmers, pesticide use was shown to have a negative effect on farmers' health and that farmers' health had a positive effect on productivity (Antle and Pingali, 1994). Thus, there are likely to be social gains from a reduction in insecticide use in Philippine rice production.

There is also difficulty controlling illegal entry of pesticides into the Philippines, the volume and extent of which is undocumented. These pesticides are used by farmers without proper labeling and sometimes the instructions are not in readable or understandable language except for the percentage a.i. A number of pesticides are known to be used in fishing activities and aquaculture but the extent of their usage remains undocumented because no one will acknowledge their use. This inability to control the misuse of pesticides is expected to result in residues of misused pesticides reaching foodstuffs and environmental compartments. It is fortunate that under tropical conditions, pesticide degradation and dissipation is faster due to intense sunlight, relatively high temperatures, high rainfall, active microorganisms, etc. However, most of these same factors allow pest species to multiply rapidly and continually, resulting in pests in all different stages of development, e.g. egg, larvae or nymph, pupae, and adult. This situation necessitates frequent pesticide applications.

PESTICIDE RESIDUES IN FOOD AND AGROECOSYSTEM WITH PARTICULAR REFERENCE TO THE RICE ECOSYSTEM

Pesticide residues in food and feeds

Pesticide residues were monitored in food from 1990 to 1994 with a total of 10,251 samples being taken. These samples consisted of various vegetables, fruits, and other food commodities from markets nationwide. Results of analyses showed 1,831 samples were positive for pesticide residues and the yearly percentage of positive samples ranged from 10.17 to 22.52 percent (Pacaba, 1995) (see Table 10.1). However, the data does not indicate if the residues were above the recommended MRLs. In 1995, 181 samples of the 2,277 samples analyzed were positive (7.95 percent) for residues. In decreasing order of frequency, methamidophos, α-endosulfan, chlorpyrifos, and diazinon were most frequently detected (BPI, 1995). This suggests that endosulfan is still marketed and being used despite the ban on the 35 EC formulation by the FPA in 1993. Sabularse (2001) reported that from 1997 to 2000, 1,393 to 2,472 food samples were analyzed for pesticide residues and most of the positive samples analyzed were collected at CAR (the Cordillera Autonomous Region) except in 1999 where Region V had the greatest number of

Table 10.1 Nationwide monitoring for pesticide residues in markets of the Philippines[a]

Year	No. of samples analyzed				No. samples positive to residues				% Samples positive to residues			
	F	V	O[b]	Total	F	V	O	Total	F	V	O	Grand total
1990	126	1,852	135	2,113	28	439	9	476	22.2	23.7	6.7	22.52
1991	85	1,993	227	2,305	19	325	35	379	22.3	16.3	15.4	16.44
1992	81	1,760	199	2,040	3	182	132	317	3.7	10.3	66.3	15.54
1993	49	1,654	135	1,838	3	143	21	187	46.9	8.6	15.6	10.17
1994	94	1,735	126	1,955	12	423	37	472	12.9	24.3	29.4	24.14
Total	435	8,994	822	10,251	85	1,521	234	1,831	19.54	16.81	28.46	17.86

Notes:
a Adapted from Pacaba, 1995.
b Abbreviations represent: F = fruits, V = vegetables, O = others.

positive samples. For the period, cypermethrin was the most frequently detected (every year) followed by chlorpyrifos (1998, 2000). Endosulfan and metamidophos (1997), chlorothalonil (1998), deltamethrin and λ-cyhalothrin (1999) and profenophos (2000) residues were also detected.

Tipa *et al.* (1995) monitored pesticides in foodstuffs from markets in Laguna, a province in the southern Tagalog region of Luzon (Region IV). Data showed that a number of vegetables exceeded the maximum permissible residue limits. Those exceeding limits included detections for methomyl in eggplant *Solanum melongena* L. (Solanaceae) and tomato *Lycopersicum esculentum* L. (Solanaceae), for diazinon in pechay or heading Chinese cabbage *Brassica rapa* L. subsp. *pekinensis* (de Loureiro) Hanelt (Brassicaceae), for triazophos in eggplant, and for methyl parathion in eggplant, cabbage *Brassica oleracea* var. *capitata* L. (Cruciferae), and Chinese pechay (non-heading Chinese cabbage or bok choy) *Brassica rapa* L. subsp. *chinensis* (L.) Hanelt (Table 10.2). Tejada (1995) likewise monitored residues from string beans *Phaseolus vulgaris* L. (Fabaceae) in different markets located in Laguna and Manila because string bean production requires intensive use of pesticides due to heavy pest pressure compared to other vegetables. Residues of fenvalerate and methyl parathion were found to exceed the MRLs set by the FAO/WHO in three of the markets monitored.

Table 10.2 Monitoring of pesticide residues in vegetables (based on fresh weight) of Laguna Province, Philippines[a]

Pesticide	Vegetable	Concentration (mg kg⁻¹ ww)	MRL (mg kg⁻¹)
malathion	Baguio beans	0.11	8.0 (beans,dry)
chlorpyrifos	Baguio beans	0.012	0.2 (common beans)
methomyl	Baguio beans	0.142	2.0 (common beans)
	String beans	1.04	_[b]
	Eggplant	0.533, 0.200	0.2 (eggplant)
	Tomato	0.211–1.30	0.5 (tomato)
diazinon	Baguio beans	0.248	0.5 (vegetables)
triazophos	Pechay	0.671	–
methyl parathion	Eggplant	0.285	0.2 (common beans)
	Eggplant	0.285	0.2 (Bromica, vegetables, tomato)
	Cabbage	0.458–0.482	
	Chinese pechay	0.420	–

Notes:
a Adapted from Tipa *et al.*, 1995.
b En dash (–) indicates no MRL has been established.

Pesticide residues in the rice ecosystem

The rice paddy ecosystem is probably the contributor, or at least a major one, to pesticide pollution in aquatic ecosystems through its irrigation canals, which carry off water from the paddy fields. Almost half of the pesticides sold in the Philippines are used in rice production (CPAP, 1993; 1996). However, with the government's effort to promote IPM schemes, pesticide usage should be considerably reduced as IPM implementation increases. Rice farmers currently apply an average of 2.1 applications/crop season at an average of 1.57 kg a.i. ha^{-1} (Rola and Pingali, 1993).

Pesticide residues in the rice grain as a consequence of crop protection may not be significant as compared to post-harvest treatment. No residues were detected for chlorpyrifos, BPMC, methyl parathion, diazinon, monocrotophos, and endosulfan in dehusked and rough rice after the 30 to 40 d pre harvest interval (PHI) in farmer-cooperator's fields (Tejada et al., 1993b). In another study, no residues of azinphos ethyl, benomyl, and butachlor were detected but carbofuran and chlorpyrifos + BPMC were found at 0.01 mg kg^{-1} in the whole grain (Tejada, 1995). Residues are usually not detected in the rice grain because of the long PHI from last spray to harvest and also due to the protection afforded by the rice hull.

The danger from pesticide residues in rice culture lies in the practice of draining the paddy water into irrigation canals at the milk dough stage of rice. These canals eventually empty into rivers that may flow into lakes but ultimately enter marine waters. The bulk of spray deposits does not remain on the plant but goes into the paddy water where it partitions between the soil; plants, e.g. kangkong or water spinach *Ipomoea aquatica* Forsskål (Convolvulaceae); and animals, e.g. tilapia, tadpoles, etc., present in the paddy. Field collected tadpoles were found to be more resistant to rice pesticides, e.g. monocrotophos, BPMC + chlorpyrifos, methyl parathion, malathion, and mancozeb, than tilapia fingerlings (Tejada et al., 1994).

Kangkong is the poor man's vegetable in the Philippines and probably in much of Asia. It grows well in rice paddies, waterways, and irrigation canals and, while often considered a weed, it is eaten by farmers and livestock. Thirty-four percent of Laguna farmers (Region IV) and 63 percent of Nueva Ecija farmers (Region III) consume vegetables, root crops, fish, frogs, snails, birds, and crabs living within the paddy and 90.6 percent and 88.3 percent, respectively, harvest feed commodities including straw, grass, snails, and kangkong (Warburton et al., 1995). Using radio-labeled tracers, Bajet and Magallona (1982) found isoprocarb is taken up by components of the rice ecosystem, including tilapia, kangkong, snails, and the rice plant itself. Most of the ^{14}C isoprocarb was taken up by kangkong within one day following application and was concentrated in the veins of the old leaves rather than in the young shoots. Residues in the rice plant were discharged through gustation – the excretion of excess water as droplets from glands on the leaves of many plants, commonly occurring in periods of high humidity (Bajet and Magallona, 1982).

Snails *Pila luzonica* Reeve (Gastropoda: Pilidae) found at non-pestiferous levels in the rice paddy were formerly part of the protein in a farmer's diet. However, following the introduction of the golden snail *Pomacea caniculata* Lamarck

(Gastropoda: Ampullariidae) as a good human protein source and feed supplement to ducks, golden snails became pests of newly transplanted rice and were found to be more resistant to herbicides than tilapia (Tejada *et al.*, 1994). In decreasing order of toxicity, Tejada *et al.* (1994) found endosulfan, chlorpyrifos, butachlor, thiobencarb, thiodicarb, azinphos ethyl, 2,4-D, malathion, mancozeb, oxadiazon, and monocrotophos were toxic to the golden snail.

Some pesticides also affect non-target organisms in the paddy water as indicated by the colonization and population profile of zooplankton in the rice paddy ecosystem. Zooplankton colonization, e.g. copepods, cladocerans, and ostracods, is dependent on the season and agricultural practice. Their fluctuating populations respond to flooding, plowing, field drainage, and pest control (Tejada *et al.*, 1995). Ostracods seem to be least affected by pesticide usage followed by copepods, with cladocerans the most sensitive. During the stage of rapid production of tillers (lateral branches that form at below ground nodes) in rice – the maximum tillering stage – tillers filtered the pesticides because of the dense plant canopy and, hence, did not affect the microflora of the flood water (Tejada *et al.*, 1995). In comparison, the impact of pesticides on algal density was instantaneous. Further, the greater the frequency of pesticide application, the more depressive pesticides were on algal population, regardless of the type of algae (Tejada *et al.*, 1995).

PESTICIDE RESIDUES IN FRESHWATER, BRACKISH WATER, SEA WATER, THE RICE FISH ECOSYSTEM, AND AQUACULTURE[1]

Pesticide residues in the rice fish ecosystem

The culture of Nile tilapia *Oreochromis niloticus* L. (Perciformes: Cichlidae) and (or) common carp *Cyprinus carpio* L. (Cypriniformes: Cyprinidae) with rice is practiced by farmers in an effort to diversify their production and augment their income. It also makes for efficient utilization of their resources. This is accomplished by constructing a fish trench in the middle of, at the sides of, or peripheral to the rice paddy. Fish are driven out of the paddy and into the canals by draining the paddy water before pesticide application (Figure 10.5). To avoid the pesticide's toxic effects, fish are released back into the paddy one week after spraying or fish are stocked in the paddy one week after spraying. In some instances the fish trench is located adjacent to the rice field without direct contact with paddy water. However, contamination can still occur via seepage, leaching, and drift. Residues can also be transported via other organisms to whom the soil residues are bioavailable, especially those animals living in close contact with the soil.

Nile tilapia was found to be adaptable to rice management practices and may be considered as a good indicator species or test organism for pesticide toxicity in the tropics (Tejada *et al.*, 1994). Soil application of carbofuran, due to low levels of its residues in fish fillets, and 2,4-D herbicide were the initial pesticides recommended for rice fish culture when the technique was first promoted in 1985

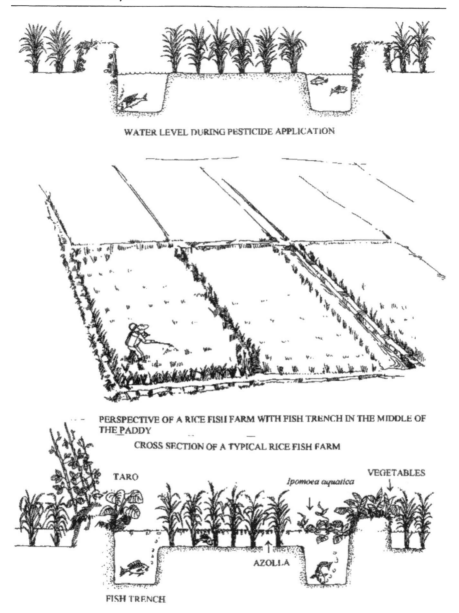

WATER LEVEL DURING PESTICIDE APPLICATION

PERSPECTIVE OF A RICE FISH FARM WITH FISH TRENCH IN THE MIDDLE OF THE PADDY

CROSS SECTION OF A TYPICAL RICE FISH FARM

TARO

Ipomoea aquatica

VEGETABLES

AZOLLA

FISH TRENCH

Figure 10.5 Cross-section of a typical rice fish field (after Sevilleja, 1982)

(Sevilleja, 1992). Pesticides currently reported as used in rice fish culture are carbofuran, carbaryl, methomyl, MIPC, methyl parathion, monocrotophos, cypermethrin, and permethrin (Cagauan and Arce, 1992). Many pesticides used for integrated rice fish culture experiments in Nueva Ecija Province (northern Luzon, Region III) were observed to cause some fish mortality (Cagauan, 1990).

These included carbofuran, carbaryl, MIPC or isoprocarb, BPMC, MTMC (tsumacide, metolcarb, or m-tolyl methylcarbamate), monocrotophos, chlorpyrifos, azinphos ethyl, methyl parathion, triazophos, permethrin, and cypermethrin with mortality depending on the time and rate of application. No mortality was observed with λ-cyhalothrin and cyfluthrin at the application rates and times tested (Cagauan, 1990).

Tejada *et al.* (1993a) sprayed the recommended rate of monocrotophos, azinphos ethyl, and endosulfan separately at 40 d after transplanting rice. Paddy water was collected 3 h after spraying and was bioassayed with 50 mm long tilapia fingerlings. No mortality was observed with monocroptophos because the concentration in paddy water (0.58 mg L^{-1}) was below the LC_{50} (13.8 mg L^{-1}) for Nile tilapia (Tejada *et al.*, 1993a). However, endosulfan and azinphos ethyl, with concentrations in paddy water exceeding their LC_{50}s, produced 100 percent mortality. The maximum concentration of endosulfan in paddy water in rice fish culture was 0.12 mg L^{-1} for both the α and β isomers. Endosulfan in paddy water was toxic to tilapia up to 18 d after spraying (DAS). Bioaccumulation factors (BAF) were 9 (head) and 82 (flesh) for the isomer and 171 (head) and 279 (flesh) for the β isomer (Medina *et al.*, 1993a). On the other hand, when azinphos ethyl was sprayed at the recommended rate 40 d after transplanting rice, the concentration in paddy water was 0.068 μg L^{-1}, which exceeded the LC_{50} in tilapia (0.001 μg L^{-1}) (Tejada *et al.*, 1993a). Likewise, Cagauan (1990) reported a higher concentration (0.132 to 0.271 μg L^{-1}) when azinphos ethyl was sprayed 14 d after transplanting rice. Data indicate that both endosulfan and azinphos ethyl are not suitable for rice fish culture.

In rice fish culture, the maximum concentration of monocrotophos measured in paddy water was 0.58 mg ml^{-1} and in paddy soil, 0.034 mg g^{-1} with a half-life of 0.88 and 9.69 d, respectively. Residues were detected on rice and in water up to 5 DAS and in soil up to 10 DAS. There was no toxic effect of monocrotophos on tilapia survival and no residues were detected in rice grain and fish at harvest (Tejada and Bajet, 1990). Residues of methyl parathion were detected in farmers' rice fish fields both in paddy water (0.026 mg L^{-1}) and in trench water (0.021 mg L^{-1}) up to 3 DAS and in fish (0.003 mg kg^{-1}) up to 2 DAS (Tejada and Bajet, 1990). Half-life values in paddy and trench water were 0.58 and 0.62 d, respectively (Tejada and Bajet, 1990).

Isoprocarb was found to accumulate in the head of tilapia (Bajet and Magallona, 1982) while BPMC residues (Zulkifli *et al.*, 1983; Varca and Magallona, 1987) and carbofuran (Tejada and Magallona, 1985) were concentrated in the entrails. Waterborne carbofuran was biomagnified 100-fold in tilapia and was concentrated to the highest degree in the entrails, followed by the fillet, and least concentrated in the head (Tejada and Magallona, 1985). Residues of BPMC in the fillet peaked at 2 DAS at a biomagnification factor of 5.8 (Zulkifli *et al.*, 1983).

No fish mortality was observed when cypermethrin and cyfluthrin were applied separately in paddy fields because water quality parameters were within the ideal range for fish culture (Monje, 1989). Residues were observed in paddy water two days after the second application but declined after four days with no accumulation remaining at 7 DAS (Monje, 1989). There was a die-off in plankton and benthic organisms at 1 DAS but an increase in density of the organisms was observed one week after application.

Farmers' acceptance of rice fish culture has been limited due to the problems associated with pesticide use, the added labor and expense required to build bigger trenches for fish, the small fish size at harvest – due to the short culture period – (Sevilleja, 1992), and the limited availability of fingerlings. Although some refinement of recommendations has been achieved in terms of pesticide selection, application techniques, and application rates, rice fish culture remains underutilized because some farmers still use pesticides indiscriminately, which causes damage to both fish and human health.

Ecotoxicological impact of pesticide residues in freshwater ecosystems

Inland water resources, consisting of swamp lands (40.18 percent); existing fish ponds (30.14 percent); and lakes, rivers, and reservoirs (29.68 percent) with an aggregate area of 84,225 km^2, could potentially be contaminated by pesticides used on nearby farms (BFAR, 1996). This area decreased to 74,992 km^2 in 2000 with areas of the three classifications mentioned above equivalent to 32.81 percent, 33.85 percent and 33.33 percent, respectively. Agricultural wastes and chemicals used for farming and fish culture enter the irrigation canals and then river systems through leaching and runoff. Floods with their associated runoff carry large quantities of silt that contain adsorbed and (or) absorbed pesticides. Most pesticides are absorbed into organic particles (Bajet, 1996a) or taken up by organisms such as filter-feeding animals (Bajet, 1997). Soil-bound pesticides that are extractable are also bioavailable especially to biota living near or within the sediment (Bajet, 1997). Varca (1996) found that 52 percent of bound residues of DDT in soil were released by 28 d after the addition of fresh soil. The release of soil-bound DDT residues increased to 81 percent after 84 d. Biological release of soil-bound residues of DDT was attributed to microbial action as a result of the addition of fresh soil containing microorganisms.

In natural tropical freshwater, Varca (1996) found that DDT dissipated by 36 percent of the applied DDT on the first day. The decline in residual ^{14}C-DDT became gradual after day 2, declining from 45.23 percent to 22 percent after 14 d, with 1.69 percent ^{14}C-DDT remaining after six months. The overall half-life for DDT dissipation in water was 31 d (Varca 1996). In Philippine soils, Varca and Magallona (1994) found the half-lives of ^{14}C-DDT and ^{14}C-DDE were 105 and 151 d, respectively and concluded that p,p'-DDT and p,p'-DDE dissipate from Philippine soils at rates that may preclude accumulation of residues.

An environment impact assessment of pesticides was conducted by Gregorio (1995) in one of the national irrigation systems in Camarines Sur Province (southern Luzon, Region V). The assessment covered an area of 2,800 ha and included 4,927 farmers. Residues of endosulfan in sediments – 0.52 mg kg^{-1} total endosulfan and 0.28 mg kg^{-1} α-endosulfan, respectively – were detected in the inflow and outflow of the Barit River (Gregorio, 1995). In the Waras River, no residues were detected in sediment in the inflow but 0.03 mg kg^{-1} α-endosulfan was detected in sediment in the outflow. However, no residues were detected in waters of both rivers from either the outflow or inflow (Gregorio, 1995). These data indicate that endosulfan can be carried by sediments through the irrigation system and that this pesticide is still in use in spite of its banning in 1993.

Laguna Lake, an inland freshwater lake southeast of Manila (Region IV) with an area of 89,076 ha, was monitored in the early 1970s by Barril (1973). Pesticide residues were reported for a number of species of fish in the lake but at levels that were considered safe for human consumption. Organic contaminants were generally lowest in milkfish *Chanos chanos* Forsskål (Gonorynchiformes: Chanidae) and Nile tilapia compared to native species such as kanduli *Arius manillensis* Valenciennes (Siluriformes: Ariidae). In 1993, residues of aldrin (0.1 to 5 mg L^{-1}), dieldrin (0.2 to 1.5 mg L^{-1}), BHC (0.3 to 9 mg L^{-1}), and DDT (0.1 to 180 mg L^{-1}) were detected in the east and west bay of Laguna Lake (Bajet and Tejada, 1995). In 1994, one of the tributaries of Laguna Lake contained residues of chlorpyrifos and methyl parathion (Tejada, 1995) indicating that these river systems can carry agricultural pollutants from inland farming activities. Recently, Sta Ana (2001) confirmed the presence of a number of OC pesticides, including DDT and its metabolites, in sediments collected in Laguna Lake and selected tributaries. However, Sly (1993) reported that persistent contaminants in Laguna Lake are not a major factor affecting fish health or a concern for human consumption of fish but the data is limited and long-term monitoring of the lake is recommended as agricultural activities become more intensive and as diets become more heavily dependent on fish. Pen culture of milkfish, which occupies one third of the surface area of Laguna Lake, also uses chemicals for pests, diseases, anti-fouling, etc. and has disrupted natural fish spawning and nursery grounds (Beveridge and Phillips, 1993).

Bhuiyan and Castaneda (1995) studied drainage outflows from the Santa Cruz River Irrigation System before discharge into Laguna de Bay. During the wet season, the highest pesticide concentrations found were 0.54 ng ml^{-1} for carbofuran in irrigation water and 3.46 ng ml^{-1} for methyl parathion in drainage water. Endosulfan, butachlor, methyl parathion, and carbofuran were all detected in drainage outflows from both irrigated and rain-fed areas. During both wet and dry seasons, average pesticide concentration was higher in drainage water (except for butachlor during the wet season) than in irrigation water indicating that applied pesticides were transported into drainage water.

Carbofuran 3 G was applied to paddy rice and the water was drained into a fish pond three hours after application to determine movement, distribution, and loss in an aquatic ecosystem (Tejada *et al.*, 1990). Residues in fish were concentrated in the entrails but some residue was detected in the fillet – however, levels were below the MRL of 0.02 mg kg^{-1} for meat – and the lowest concentration was measured in the head. Residues in pond water were absorbed in the kangkong plant up to 15 DAS but declined thereafter. However, the highest residues in kangkong were within the Acceptable Daily Intake (ADI) for humans (0.003 mg kg^{-1} for cabbage) (Tejada *et al.*, 1990). In 1991, no pesticide residues were detected in sediment, water, and fish samples taken from Lake Buhi and Lake Bato, which are located in southern Luzon (Tejada, 1995).

Effect of pesticides in marine ecosystems

When pesticides reach the marine ecosystem, they are usually in the form of sediment bound residues providing the source of contamination is inland. This is not the case, however, when the pesticide is added directly and deliberately to marine waters to enhance fish catch. In the former case, the sediment burrowers and filter feeders such as oysters and mussels are exposed to these sediment bound residues, whereas in the later, water residues directly affect fish (who could get away if the concentration is not lethal). The significance of bound pesticide residues should be addressed based on their bioavailability, both in their quantities and forms of uptake (Khan, 1982). Little is known about the significance of bound pesticide residues, their bioavailability, toxicity, and accumulative nature.

In the east Asian seas, which surround the Philippines, Jacinto (1997) reported that marine contaminants have been analyzed only sporadically in water, sediments, and biota due to limited analytical and manpower capacity for determining contaminant concentrations. Raw sewage, discharged into coastal waters through rivers and waterways, and eutrophication, due to high nutrient loads, have been considered the major source of pollution in these areas. Eutrophication is pronoun-ced in Manila Bay and other bay areas where a very large population leads to discharges of untreated sewage in addition to significant inputs from nearby agriculture and aquaculture activities (Jacinto, 1997). Although pesticide residues were detected in sediments collected in Manila Bay, the overall priority problem is sewage; nutrient loading that causes eutrophication and harmful algal blooms; accumulation of heavy metals; and, to a lesser degree, oil spills and plastics pollution (Jacinto, 1997). A recent study (Minh *et al.*, 2000) has analyzed for residues of DDTs, PCBs, and HCHs in blubber samples from two species of dolphins, the spinner dolphin *Stenella longirostris* Gray (Cetacea: Stenidae) and Fraser's dolphin *Lagenodelphis hosei* Fraser (Cetacea: Delphinidae), taken in the Mindanao Sea. Levels of each of these persistent chlorinated endocrine disrupting chemicals were low compared to other locations.

A risk assessment by Calamari and Delos Reyes (1997) found carbaryl to be a potential risk to aquatic fauna in Batangas Bay (southern Luzon, Region IV). However, carbaryl has a low probability of affecting marine life due to its short half-life, especially in the marine environment. Cypermethrin has strong affinity for soil and sediments with low probability of reaching marine waters. Butachlor and chlorpyrifos may be present in marine waters based on the risk assessment procedures used in this study but, because of low application rates, these pesticides represent a low risk to the marine environment, in particular to Batangas Bay.

Manila Bay, with an area of $1,510 \text{ km}^2$, is one of the historic fishing grounds for Luzon, the largest island in the Philippines. It is a semi-closed body of water with large rivers emptying into it and is bounded by the provinces of Cavite, Rizal, Bulacan, Pampanga, and Bataan. In 1995, 25,046 T of fish from commercial fishing and 11,649 T from municipal fishing were harvested in Manila Bay (BFAR, 1996). Water samples collected from various tributaries of Manila Bay contained 2.0 to 2.5 mg L^{-1} DDT, 0.2 to 0.8 mg L^{-1} DDE, and 0.2 to 1.2 mg L^{-1} DDD (Tejada, 1995). That DDT concentration was higher than its metabolites implies that the waters were recently contaminated with DDT. At the time of sampling, DDT was still used for the control of malaria vector mosquitoes. In 1996, sediment samples were collected from the mouths of 17 tributaries emptying into Manila Bay and DDE, DDT, dieldrin, and lindane residues were detected (Bajet, 1997, Table 10.3). Most tributaries in coastal areas are adjacent to aquaculture farms but, further inland, these tributaries traverse Regions II and III, which are the major rice-producing areas of the Philippines. Similarly in 1983, the EMB (Environment Management Bureau) monitored pesticides from fish liver removed from catch taken from Manila Bay and detected residues of BHC, aldrin, and

Table 10.3 Monitoring of pesticide residues in sediment and water from tributaries of Manila Bay, Philippines (residue levels reported in ng g^{-1} dw sediment or ng ml^{-1} water)

Tributary	DDT	DDE	DDD	Others	Sample	Reference
Batang	2.5	0.4–0.6	1.2	–[a]	water	Tejada, 1995
Orani	2.0	0.2–0.8	1.2	–	water	
Pangilisan	–	0.3	–	–	water	
Mamata	–	–	0.2	–	water	
Orani	ND[b]	3.13	–	1.32 lindane	sediment	Bajet, 1997
Batang	ND	0.94	–	ND[b]	sediment	
Madamo	0.83	0.34	–	ND	sediment	
Makangkong	ND	ND	–	ND	sediment	
Hagonoy	0.60	0.96	–	1.10 dieldrin	sediment	
				1.23 aldrin	sediment	
Gumitna	ND	ND	–	ND	sediment	
Nabao (big)	ND	8.86	–	ND	sediment	
Nabao (small)	ND	2.84	–	ND	sediment	
Manuhol	ND	1.30	–	1.23 aldrin	sediment	
Pugad	ND	ND	–	0.36 dieldrin	sediment	
Tibagin	ND	ND	–	ND	sediment	
Masukol	ND	4.61	–	ND	sediment	
Sta. Cruz	ND	5.12	–	1.40 dieldrin	sediment	
Pamarawan	ND	0.29	–	ND	sediment	
Matilakin	1.17	ND	–	ND	sediment	
Salambaw	0.53	4.66	–	ND	sediment	
Taliptip	ND	6.76	–	ND	sediment	

Notes:
a En dash (–) indicates no analysis performed.
b ND indicates no detectable residues.

dieldrin (Table 10.4). However, the levels were within the safe limits set by FAO/ WHO for acceptable daily intake of pesticides in foodstuffs (Hingco, 1990).

In 1990, of the 59 marine products analyzed coming from four different regions in the Philippines, three fish samples were found positive for endosulfan with concentrations of 0.01 to 4.16 mg kg⁻¹ (BPI, 1990). In 1996, 11 dried, salted fish samples taken from different barangays (the smallest geopolitical unit in the Philippines) in Guimaras province, an island in the Visayas (Region VI), were found to contain chlorpyrifos residues with concentrations ranging from 0.001 to 0.012 mg g⁻¹ (Bajet, 1996b). Bajet suspected the pesticide was added as a post harvest treatment to minimize flies during the sun-drying process. Suva (1995) likewise reported that the Bureau of Food and Drugs investigated a case of pesticide contamination of dried fish that had allegedly had insecticide used to control insect infestation during storage. Golob *et al.* (1987) likewise reported the use of pirimiphos methyl, fenitrothion, diflubenzuron, and deltamethrin in Kenya for the control of beetle infestations in dried tilapia.

When ^{14}C-DDT was added to an open marine microecosystem, residues in sediment were extractable with an increasing trend for bound residues amounting to 22.1 percent of the applied DDT at 15 d after application (DAA) (Bajet, 1996a). In oysters and mussels sampled 30 DAA, 2–3 ng ml⁻¹ had entered biota fluids, 3–4 ng g⁻¹ had been incorporated into shell, and 8–9 ng g⁻¹ had entered the animal's muscle tissue. Slightly higher concentrations were detected in the oyster than the mussel (Bajet, 1996a). Extractable DDT residues in the sediment were found to be released into brackish water and were taken up by oysters who degraded 45 percent of the residues to DDE. Nonextractable DDT residues in the sediment (after exhaustive solvent extraction) were also transformed partly to extractable residues, released into brackish water, and taken up by oysters (Bajet, 1997). Bajet proposed

Table 10.4 Some pesticide levels in fish liver samples from Manila Bay, Philippines[a]

Species	Sample size	Concentration (ng g⁻¹)		
		BHC	Dieldrin	Aldrin
1. *Epinephelus tauvina* Forsskål (Perciformes: Serranidae)				
Greasy grouper	3	2.201	0.820	–[b]
Greasy grouper	2	0.268	–	trace
2. *Trichirius lepturus* L.				
Hairtail or ribbonfish	19	trace	trace	0.048
3. *Sillago sihama* Forsskål				
Whiting	72	0.041	0.016	–
4. *Rastrilleger faughani*				
Faughans mackerel	15	0.024	0.011	–

Notes:
a Adapted from Environment Management Bureau, formerly National Pollution Control Council, 1983.
b En dash (–) indicates no analysis performed.

the mechanism of release of sediment residues was partly due to microorganisms in the brackish water.

In a similar study with ^{14}C-chlorpyrifos added to brackish water, residues were partitioned between water (44–57 percent of the applied radioactivity) and sediment (42–55 percent) during the first 3 DAA (Bajet, 1996a). Chlorpyrifos residues gradually moved into the sediment, though greater than 90 percent of the sediment residues were extractable and only 2 percent were bound at 15 DAA. The concentration of ^{14}C-chlorpyrifos was 0.5 ng g^{-1} and 2.2 ng g^{-1} in oysters and mussels, respectively, after one month of exposure (Bajet, 1996a).

Pesticide use in aquaculture and fishing activities

The major aquaculture crops cultivated in the Philippines are seaweeds *Euchema* spp.; milkfish *Ch. chanos*; prawns and shrimp, e.g. giant tiger prawn *Penaeus monodon* Fabricius (Decapoda: Penaeoidae); tilapia *O. niloticus*; other fish including bighead carp *Aristichthys nobilis* Richardson (Cypriniformes: Cypinidae), silver carp *Hypophthalmichthys molitrix* Valenciennes (Cypriniformes: Cypinidae), seabass *Lates calcarifer* Bloch (Perciformes: Latidae); and shellfish (oysters and mussels) as minor products (Figure 10.6). In 1995, inland and coastal aquaculture, with a total area of 162,234 ha, had production of 825,387 T equivalent to 30.1 percent of the total fish production (BFAR 1996). Production increased in 1998 to 954,396 T, equivalent to 34.2 percent of total fish production (BFAR, 1999). Most aquaculture production comes from mariculture and brackish water fish ponds, although the area devoted to the latter is more than 20 times the former (Table 10.5). Most brackish water fish ponds are located in the coastal areas of Regions III and VI (producing primarily milkfish and shrimp) where the use of chemicals in aquaculture becomes a source of pollution for the marine environment. The bulk of aquaculture production is dominated by seaweed production from the Autonomous Region for Muslim Mindanao (ARMM) and Region IX (Figure 10.6). However, in 1998 and 2000 Region IV placed second next to ARMM in production with Region IX in third place (BFAR, 1999; 2001). Oysters and mussels are mainly produced in Regions IV and VI, whereas freshwater pond culture of tilapia is found in Region III (BFAR, 1996; 1999; 2001). Fishpen culture of milkfish and fishcage culture of tilapia is found mainly in Region IV where the largest inland lake Laguna de Bay is located (BFAR, 1996; 1999; 2001). Regions III, IV, VI, IX, and ARMM are the most active in terms of aquaculture production (BFAR, 2001). However, in terms of value, the high growth in brackish water fish ponds in Regions III and VI (BFAR, 2001) is due to high prices for the milkfish and shrimp grown in these areas (Camacho and Lagua, 1988).

Chemicals used in aquaculture include fertilizers and lime to improve productivity; pesticides to control pests, diseases, and parasitoids; disinfectants; anaesthetics; and antifoulants. Ponds are prepared by draining, drying, and leveling. Pesticides are applied to small pools of water in the pond bottom where snails and unwanted

A   Major species produced in Philippine aquaculture for 1995

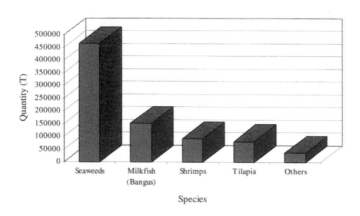

B   Philippine aquaculture production by region for 1995

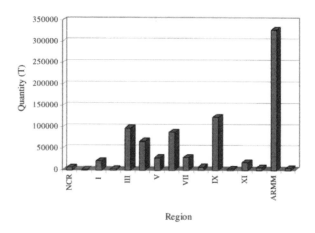

Figure 10.6 Aquaculture in the Philippines (Bureau of fish and aquatic resources, 1996)

fish tend to congregate. Liming, followed by fertilization using chicken or animal manure with or without chemical fertilizers (Anonymous, 1993), constitutes the required preparation for the culture of tilapia or growing lablab (a microbenthic plant and animal complex used as the food base for the culture of milkfish).

A survey of intensively managed prawn ponds in western Visayas and northern Mindanao indicated that some forty chemical and biological products were utilized, including therapeutics and disinfectants, soil conditioners, bacteria-enzyme

Table 10.5 Aquaculture production and area under cultivation by type of culture, 1995[a]

Type	Area (ha)	Production (T)
Brackish water fish pond	143,197	237,056
Freshwater fish pond	6,523	42,944
Fishpen[b]	4,206	20,658
Fishcage[b]	1,369	33,520
Mariculture	6,940	491,209
Total	162,234	825,387

Notes:
a Adapted from BFAR, 1996.
b In fresh and marine waters.

preparations, algicides and piscicides, plankton growth promoters, and feed additives (Primavera *et al.*, 1993). They also found that 43 percent of the farms discharge effluents directly to bays or the sea. Tongguthai (1996) reported that in Thailand the pesticides trichlorfon, trifluralin, rotenone, and fentin acetate are used in aquaculture. Bagarinao and Lantin-Olaguer (2000) developed a responsible IPM strategy to eliminate the use of the triphenyltin compounds Aquatin and Brestan for control of the potamidid snail *Cerithidea cingulata* Gmelin (Caenogastropoda: Potamididae), a pest in brackishwater milkfish ponds derived from mangrove areas. See the research findings of the Aquaculture Department, Southeast Asian Fisheries Development Center (SEAFDEC, 2000) for additional information.

Pesticides have been extensively utilized in fish ponds and net pens (Chua, 1993) to control unwanted species, pests, and predators, e.g. fish including tilapia *Oreochromis mossambicus* Peters (Perciformes: Cichlidae), Pacific tarpon *Megalops cyprinoides* Broussonet (Elopiformes: Megalopidae), tenpounder *Elops machnata* Forsskål (Elopiformes: Elopidae), and gobies (Perciformes: Gobiidae); polychete worms (Annelida: Polychaeta); and snails *Telescopuim* spp. and *Cirithidia* spp. (Andalecio and Pahilanga, 1995). Endosulfan, lindane, paraquat, trichlorfon, malathion, azinphos ethyl, and the organotins fentin acetate and fentin chloride are some of the pesticides used for pest and predator control (Andalecio and Pahilanga, 1995).

Predator control is considered essential in nursery and grow-out operations in the aquaculture industry. Drying the bottom after each harvest and using fine mesh nets in gates and pipes are common precautionary measures, although the application of pesticides like azinphos ethyl, fentin acetate, and fentin chloride is widespread (Camacho and Lagua, 1988). Tea seed cake (a.i. is saponin), tobacco (a.i. is nicotine), tuba root *Derris elliptica* Benth. (Fabaceae) (a.i. is rotenone), chemical molluscicides, and other synthetic chemicals are also used to control unwanted snails including *Thiara* sp. (Caenogastropoda: Thiaridae), *Clypeomarus* sp., and *Telescopium* spp. (Caenogastropoda: Potamididae); unwanted fish; and annelids. Some chemicals like trichlorfon (Chua, 1993) and dichlorvos (Chua, 1993; Gowen and Rosenthal, 1993) are toxic to crustaceans living in the vicinity of the farm site and can cause resistance to parasitic copepods and increased sensitivity to the compound in crop fish, thus reducing its efficacy (Gowen and Rosenthal, 1993).

OP insecticides are used in aquaculture to control argulosis and dactylogyrosis – the former is caused by a fish louse, a parasitic copepod; the latter is caused by ectoparasites – and to efficiently eliminate copepod and cladoceran populations without affecting rotifers. Azinphos ethyl was used as both a molluscicide and a piscicide in fish ponds before it was banned in the Philippines. At 0.0125 mg L^{-1}, azinphos ethyl would eradicate tilapia within 6.5 h and guppy fish *Poecilia reticulata* Peters (Cyprinodontiformes: Poeciliidae) within 19 h with 100 percent survival of milkfish during the 96 h observation period (Andalecio and Pahilanga, 1995). Pyrethroids like deltamethrin, and sometimes cyanide, are used to stun fish or shrimp in lakes and rivers to facilitate catching (harvesting) these species. Fortunately the organisms were observed to recover after a period of time. Rampant illegal fishing using poisons, and in some cases, pesticides, occurs (Aypa, 1991) but this remains largely undocumented.

Organotin compounds are used in fish and shrimp farms (Chua, 1993) and also on aquaculture facilities, e.g. cage nets, as anti-fouling compounds (Tanabe *et al.*, 2000). They have been shown to be toxic to non-target organisms, to accumulate in fish flesh, and to cause mortality of farmed fish (Gowen and Rosenthal, 1993). Rotifers and crustaceans appeared to be most affected by fentin acetate and fentin chloride while flagellates were least affected (Andalecio and Pahilanga, 1995). Cruz *et al.* (1988) found that *T. mossambicus* can tolerate the levels of fentin chloride and fentin acetate being used in fish ponds. However, histological analysis of gills, intestine, liver, and kidney showed pathological changes even at sub-lethal levels. Estimated safe concentrations of fentin chloride and fentin acetate were 0.06 and 0.006 mg a.i. L^{-1}, respectively.

After the banning of organotin compounds, endosulfan, and azinphos ethyl, the only commercially available pesticides for aquaculture were niclosamide, metaldehyde, and a botanically-derived pesticide composed of tannins, glycosides, etc. Plant-based biodegradable pesticides such as tobacco dust (nicotine), tea seed cake (saponin), and *Derris* root extract (rotenone) are popular in Asian countries in both fish and shrimp farms (Chua, 1993). On intensive prawn farms, excessive algal growth is controlled by the use of copper compounds and unwanted fish are eradicated using teaseed cake. Teaseed, copper, and lime are used to induce molting when shell related diseases like tail rot or black spot are observed (Andalecio and Pahilanga, 1995; Primavera *et al.*, 1993). A 25 percent formulation of niclosamide applied in fish ponds at the rate of 4 L ha^{-1} reached a maximum concentration in water of 1.14 mg L^{-1} two hours after application. The resultant residues in dead tilapia *O. niloticus* and tenpounder *E. hawaiiensis* Regan were 2.13 and 0.32 mg kg^{-1}, respectively, 6 h after treatment (Calumpang, 1994). However milkfish, reared in the pond and collected 127 d after treatment, contained <0.03 mg kg^{-1} niclosamide. Similarly niclosamide applied at a different site (rate: 3 L ha^{-1}) resulted in a maximum concentration of 2.54 mg L^{-1} in water and 4.0 mg kg^{-1} in mud (Calumpang, 1994). Niclosamide residues in biota collected 2 h after application were 5.45, 0.88, and 0.4 mg kg^{-1} for the banded goby *Amblygobius phalaena* Valenciennes (Perciformes: Gobiidae), tilapia, and snail *Telescopium telescopium* (Caenogastropoda: Potamididae), respectively (Calumpang, 1994). The effluent water from the pond treated with 4 L ha^{-1} niclosamide was drained through a channel 5 d post treatment. Caged tilapia and milkfish were stationed at 0, 5, 10, 25, 50, and 100 m from the discharge point but no mortalities were observed over a 24 h period after draining (Calumpang, 1994).

Pesticides are known to be used to facilitate fishing but this practice remains undocumented. Other destructive practices are cyanide fishing (Rubec, 1988; Munoz, 1991), blast fishing with explosives (Galvez and Sadorra, 1989; Munoz, 1991), and electro-fishing, which uses electric current to kill the fish (Munoz, 1991). In the Philippines, where control of the use of pesticides and other poisons in fish culture and fishing activities are less stringent or sometimes absent, abuses may be frequent.

PESTICIDE AND BIOTA INTERACTIONS – UPTAKE AND TOXICITY IN FRESHWATER, AQUACULTURE, AND MARINE ECOSYSTEMS

LC$_{50}$ values for tilapia fingerlings to most pesticides used in rice indicate that the pesticides are highly toxic (Table 10.6), when tested in the laboratory. However, when these pesticides are employed in integrated rice fish culture, their toxicity is reduced due to the presence of environmental substrates that tend to decrease the concentration of the pesticide in the paddy or trench water and by the various pesticide application techniques (Tejada and Bajet, 1990). Cruz *et al.* (1988) determined the LC$_{50}$s of fentin acetate and fentin chloride in tilapia and found them to be 0.036 to 0.210 and 0.516 to 0.802 mg a.i L^{-1}, respectively, and classified the two compounds as extremely and moderately toxic (Table 10.6). The calculated 48 h LC$_{50}$ for malathion was 3.10 mg a.i. L^{-1} and the calculated safe level was 0.302 mg L^{-1} (Capinpin, 1992). After a 96 h exposure, tilapia recovery time for gills and liver was 6 d and for kidney, 72 h (Capinpin, 1992).

Guppy fish *P. reticulata* are usually found in waterways and can be affected by pesticides applied in the rice field. Thus, they may indicate a potential for pesticide pollution. Generally rice insecticides were less toxic to guppy fish than tilapia fingerlings. For example, Nile tilapia was more sensitive to azinphos ethyl than guppy fish with LC$_{50}$ values of 1.0×10^{-6} mg L^{-1} and 0.02 mg L^{-1}, respectively (Tejada *et al.*, 1994). Azinphos ethyl was found to be most toxic to guppy fish followed by methomyl, chlorpyrifos, carbofuran, cypermethrin, thiobencarb, BPMC + cypermethrin, isoprocarb, carbaryl, thiodicarb, diazinon, permethrin, BPMC, MTMC, and oxadiazon in order of decreasing toxicity (Tejada *et al.*, 1994).

In toxicity studies conducted in a rice paddy ecosystem, niclosamide applied at field rates of 0.250 and 0.375 kg a.i. ha^{-1} resulted in 100 percent mortality of caged carp located at 0, 1, 5, 20, and 100 m from the discharge point when the

Table 10.6 Toxicity of some insecticides to Nile tilapia (Oreochromis niloticus L.) fingerlings under laboratory conditions

Toxicity category[a]	48 hr LC_{50}(mg L^{-1})	Pesticide	References
Extremely toxic	<0.5 ppm	azinphos ethyl, cypermethrin	Tejada et al., 1993b
			Cagauan and Arce, 1992[b]
		endosulfan, cyfluthrin, chlorpyrifos, fenvalerate, triazophos, etofenprox, thiodicarb, carbosulfan, BPMC, fenitrothion, α-cypermethrin + BPMC, monocrotophos + fenvalerate	
		BPMC + chlorpyrifos	Tejada et al., 1993b
		fentin acetate	Cruz, 1988[b]
		permethrin	Cagauan and Arce, 1992[b]
		fenitrothion	Bajet, 1995
Moderately toxic	0.5–10 ppm	malathion, methamidophos, methyl parathion, carbaryl	Tejada et al., 1993b
		fentin chloride	Cruz, 1988
		malathion	Capinpin, 1992[b]
		MTMC (tsumacide)	Cagauan and Arce, 1992[b]
		diazinon, thiobencarb	Bajet, 1995
Low toxicity	>10 ppm	monocrotophos	Tejada et al., 1993b

Notes:
a Ranking based on Nishiuchi, 1974.
b Calculated values, these are expressed as mg a.i. L^{-1} vice mg of formulated product L^{-1} as reported.

paddy water was drained at the time of application (Calumpang et al., 1995). It was also slightly toxic to tilapia found in treated paddies with 20–33 percent mortality occurring at the discharge point on the day of application (Calumpang et al., 1995).

Studies of the uptake and detoxification of diazinon and fenitrothion showed that diazinon was accumulated in tilapia over one week with a BCF of 12 (Table 10.7) but could be reduced to trace levels 7 d after transferring fish to clean water (Bajet, 1995). Fenitrothion, on the other hand, was taken up by tilapia for a maximum of 5 d with a BCF of 3. There was no accumulation of fenitrothion in tilapia at 7 d after treatment and no residues detected in water during this period as result of rapid dissipation and degradation of the pesticide (Bajet, 1995). In rice fish culture, no mortality was observed in tilapia when diazinon and fenitrothion were applied at their recommended rate. The maximum concentrations in paddy water reached 0.025 and 0.012 mg L^{-1}, respectively, for diazinon and fenitrothion and were below their LC$_{50}$ values for tilapia.

BCFs were calculated for tilapia for a number of pesticides used in rice (Tejada, 1995). Methyl parathion and α-endosulfan had the lowest BCF and carbosulfan and β-endosulfan the highest (Tejada, 1995) (see Table 10.7). Carbofuran and chlorpyrifos were concentrated more in the entrails than in the fillet suggesting that a portion of these pesticides were taken up through oral ingestion. Celino et al. (1988) applied ^{14}C-carbofuran at the rate of 0.75 kg a.i. ha^{-1} to the soil in a model rice fish microecosystem and found about 7 percent of the total amount of carbofuran was released into the paddy water. The disappearance of radioactivity and therefore residues from the soil gave a half-life of 89 d for carbofuran. Tilapia had taken up 0.002 percent of the applied carbofuran after 28 d but the pesticide was not detectable 56 d after application. No fish mortality was observed during the experiment (Celino et al., 1988).

Table 10.7 Bioconcentration of pesticides in fish from the Philippines

Insecticide	Tissue	Fish (μg kg^{-1})	Water (μg kg^{-1})	BCF	Reference
Methyl parathion		8.30	1.0	8.3	Tejada, 1995
Carbofuran	fillet	10.50	0.09	117	
	entrails	17.00	0.09	189	
Carbosulfan		1.04	0.004	239	
α-Endosulfan		0.90	0.011	82	
β-Endosulfan		1.50	0.007	214	
Chlorpyrifos	fillet	1.76	0.027	65	
	entrails	4.1	0.027	152	
Fenitrothion		0.03	0.01	3	Bajet, 1995
Diazinon		0.23	0.02	12	
Thiobencarb	fillet	0.36	0.10	36	
	head	0.48	0.10	48	

The persistence and distribution of chlorpyrifos applied at the rate of 0.5 kg a.i. ha^{-1} was also studied in a model rice fish ecosystem (Medina *et al.*, 1993b). They obtained a half-life of 3.7 d in paddy water, with a maximum residue level of 0.009 mg L^{-1} at 2 DAS. In paddy soil, a maximum residue level of 1.2 mg kg^{-1} was found at 1 DAS. Chlorpyrifos residues by ordinary and exhaustive extraction had a half-life of 1.7 and 2.8 d, respectively. Soil-bound residues started to accumulate at 19 DAS and chlorpyrifos residues were detected below 20 cm in paddy soil at 6 DAS. Half-life values in leaves, stems, and roots were 0.9, 1.5, and 1.2 d, respectively. No detectable residues were present in the grain at harvest. No toxic effects to tilapia were observed at 6 DAS and no bioaccumulation was noted (Medina *et al.*, 1993b).

A study of farmer-cooperator's fields examined organisms collected within the vicinity of the treated fields for pesticide residues (Tejada *et al.*, 1995). No residues were detected in snails and tilapia for monocrotophos, chlorpyrifos + BPMC, monocrotophos + fenvalerate, endosulfan, and chlorpyrifos. Likewise, no residues were detected in catfish *Clarias gariepinus* Burchell (Siluriformes: Clariidae), shrimp *Macrobrachium* sp. Cowles (Decapoda: Palaemonidae), mudfish *Ophicephalus striatus* Bloch (Perciformes: Channidae), and a local fish Gurami *Trichogaster pectoralis* Regan (Perciformes: Belontiidae (Gouramies)). However, residues of chlorpyrifos were detected in duck heart and muscle at 0.020 and 0.027 mg kg^{-1}, respectively (Tejada *et al.*, 1995).

In a marine model microecosystem, DDT and chlorpyrifos were taken up by oyster and mussel (Bajet, 1997). Extractable and non-extractable sediment-bound residue of DDT was released and taken up by the oysters. Release of bound residues was enhanced by the presence of oysters and microorganisms in brackish water (Bajet, 1997).

Marine pollution monitoring of butyltins and OCs in the green mussel *Perna viridis* L. (Bivalvia, Mytiloida: Mytiloidae) collected from the coastal waters of the Philippines, Thailand, and India was conducted by Tanabe *et al.* (2000) during the period 1994 through 1997. They found widespread contamination by butyltin residues along coastal waters, especially in areas of high boating activity and coastal aquaculture areas; however, contamination levels in Asian developing countries were lower than those in developed nations although they noted that butyltin contamination may become serious in the future due to increased usage of tributyltin for aquaculture activities. Comparing butyltin composition in migratory (long-billed Mongolian plover *Charadrius mongolus* Pallas sub sp. *atrifrons* (Charadriiformes: Charadriidae)) and resident (Chinese little bittern Gmelin *Ixobrychus sinensis* (Ciconiiformes: Ardeidae)) birds collected from the Philippines with birds collected from several countries, Senthilkumar *et al.* (1998) found that birds from the Philippines had greater concentrations than those from India, but less than those from developed nations. Concentrations of OCs detected in green mussels (Tanabe *et al.*, 2000) were almost always lower than butyltins (Table 10.8). The PCBs contamination pattern suggested that PCBs may be coming from discharge and runoff from the more populated and industrialized cities of the country. Higher

Table 10.8 Concentration of butyltins and OCs (ng g^{-1} ww) in green mussels from coastal waters of the Philippines

Location	Fat content %	Tributyltin	∑Butyltin	∑PCB	∑DDT	∑Chlordane	∑HCH	HCB
Freedom Island, Paranaque (south of Manila)	3.1	76	104	36	3.3	9.5	0.19	0.04
Rizal Park, Ermita (north of Manila)	2.0	640	787	32	4.2	7.2	0.11	0.02
CCP Complex, Malate (near Manila)	1.6	200	294	33	3.2	7.4	0.09	0.02
Malabon, Metro Manila	2.6	44	62	31	2.0	3.2	0.13	0.01
Samal, Bataan (Manila Bay)	1.7	<1	ND[a]	2.6	1.6	0.34	0.16	0.01
Mean of four other sites around Manila Bay[b]	1.6	31 ± 12.7	44.5 ± 15.9	16.7 ± 6.5	1.4 ± 0.28	2.9 ± 0.54	0.13 ± 0.03	0.01
Jiabong, Samar	1.5	1	4	0.69	1.3	0.19	<0.01	0.01
Villareal, Samar	1.8	28	30	2.3	0.77	0.36	0.06	0.01
Diit, Leyte	0.50	<1	ND	1.1	0.19	0.15	<0.01	0.01
Sapian Bay, Capiz, Panay	1.2	<1	1	0.91	0.27	0.22	0.08	0.02

Source: Tanabe et al., 2000.

Notes:
a ND indicates below detection limit of 1 ng g^{-1} ww.
b Sites include 2 from Bocaue, Bulacan in Manila Bay, Philippines.

p,p'-DDT levels from population centers reflect the use of DDT for sanitary control of mosquitoes. Unlike mussels from India and Thailand, $\sum$-chlordane concentrations in mussels from the Philippines were highest among the OCs analyzed.

PESTICIDE MANAGEMENT

The proper approach in pesticide management is to know what pests attack each different stage of a crop, the most susceptible and damaging stage of the pests, the natural enemies of the pests, and then to choose and apply the correct pesticide at the proper dose to the stage where the pest is most vulnerable or sensitive to the pesticide. Pesticide applications should be timed to control the most destructive stage of the insect, and only when the pest population is above the EIL. Choice of pesticide and formulation should consider what pesticide or formulation will cause the least damage to natural enemies of the pest. The basis of sound pesticide management is the integration of chemical, biological, mechanical, behavioral, and cultural control practices. However, in real field situations, chemical control is the frontline defense in terms of pest management (NCPC, 1996) and judicious use of pesticides is not practiced.

The DA, in trying to wean farmers from relying solely on the use of pesticides, introduced the concept of IPM. This is locally known as Kasakalikasan, an acronym for Kasaganaan ng Sakahan at Kalikasan, which means 'Nature is Agriculture's Bounty'. The Kasakalikasan Programme is supported by the FAO's Intercountry Intergrated Pest Control in Rice in South and Southeast Asia in collaboration with Local Government Units (LGUs) and Non-Government Organizations (NGOs) of participating provinces and municipalities. Farmers are trained throughout the crop season regarding pests and their natural enemies in farmer field schools (FFS) situated at strategic locations nationwide. Kasakalikasan uses participatory, experimental, and discovery based learning techniques to effectively enhance farmers' ecological knowledge and skills in growing healthy crops. These field schools meet one day each week for the duration of a cropping season, usually from 14 to 16 weeks. Each school has a 'learning field' containing a farmer-run comparative study of IPM. Each week, farmers practice agroecosystem analysis to make field management decisions that render their farming systems more productive, profitable, and sustainable. Training empowers farmers to be experts through direct experience of the ecological processes involved in farming. They also learn to organize themselves and create a strong network with other farmers, extension workers, and researchers. The pest management practices of untrained farmers are usually based strictly on the use of pesticides and are documented prior to FFS. About 86 percent of rice, 92 percent of corn, and 86 percent of cabbage farmers were previously heavy insecticide users but after FFS, the number of non-users increased to 46–52 percent in rice, 72–83 percent in corn and 16– 21 percent in cabbage, with a shift among users from highly hazardous to less hazardous pesticides (Navarro et al., 1998).

Kasakalikasan also supports the training of field workers as IPM Trainers (Training of Trainers or TOT) with 1,345 field workers from LGUs, NGOs, and farmer organizations having completed the IPM Trainers Course as of December 1997 (Kasakalikasan Program, 1997). About 300 IPM Specialist Training personnel, deployed as IPM Field Officers, were responsible for TOT. IPM training teams, composed of two IPM field trainers and assisted by two extension workers, conduct two FFS per season with an average of 25 farmers in each school. After training, farmer volunteers act as IPM instructors and can get involved in a farmer-run FFS (Navarro et al., 1998).

The FFS site is selected based on the extent of intensively cropped fields within an extension area, the proximity to the offices of trainers and extension workers, locations with the most active farmer groups, and areas with supportive and competent extension workers. This assures that IPM principles get into major rice, corn, and vegetable-producing areas. Each extension worker works with 5 to ten farmer groups and selects two farmer groups for FFS from their jurisdiction. The farmer groups are selected on the basis of the extension worker's knowledge of the group, how active the group is, and on the advice of the local government. A total 1,811 rice FFS, 150 corn FFS, and 175 cabbage FFS were established nationwide with 2,666 trainers, 534 specialists, and 183,829 farmers trained from 1993 to 1997 (Navarro et al., 1998).

The knowledge and skills gained by FFS graduates were significant with average test scores of 78, 68, and 74 percent for rice, corn, and cabbage farmers, respectively. Questions on insect (90 percent correctly answered), rodent (32 percent correct), and snail management (39 percent correct) as answered by rice FFS graduates indicated that their emphasis was mainly on insect pest management. After FFS, there was an increase in insecticide non-users, reduced frequency of insecticide applications, and a shift to less toxic insecticides (Kasakalikasan Program, 1997).

Another approach to pesticide management is to identify ways and means to reduce pesticide residues by both the crop producer and consumer. Pesticide residues can be reduced during crop production by using only the recommended pesticides for pests and crops as indicated on the label. The pesticide label is strictly regulated by the FPA and farmers should always follow label recommendations in terms of rate, frequency, and timing of applications. Following the recommended pre-harvest interval will ensure that the residues in the crop ready for harvest will be below the MRL. Spray techniques and the type of formulation used affect the load of pesticide reaching the crops and the environment. Granular formulations and wettable powders generally result in lower pesticide residues in the environment than emulsifiable concentrates and water soluble pesticides. Producer decisions to apply pesticides as opposed to utilizing non-chemical means of pest control, e.g. biocontrol using pest predators/parasites, IPM, and EIL trigger stratagems, are directly influenced by pesticide prices, which are in turn influenced by governmental pesticide regulatory policies and exchange rate policies (Tjornhom et al., 1998). For nine pesticides commonly applied to vegetable crops, Tjornhom et al. (1998) calculated the effective rate of protection for pesticide prices from Government policies and found a net effect of a 6 to 8 percent pesticide subsidy, primarily from an overvalued exchange rate. Both vegetable producers and consumers receive economic surplus gains from this subsidy so long as the negative effects associated with pesticide use are not included in the evaluation and analysis. Because changes in Government's policies can directly influence farmers' pesticide use decisions, this can be used to create incentives or disincentives to adopt more IPM management methods throughout the country.

Another alternative for reducing pesticide residues in the environment is through the use of controlled-release formulations, which release an effective concentration at the most critical time. An alginate formulation of thiobencarb produced at the Pesticide Toxicology and Chemistry Laboratory, NCPC was evaluated under field conditions and was found to be compatible with rice fish culture because the thiobencarb was released slowly insuring that pesticide concentrations were below toxic levels (Bajet and Araez, 1994). The maximum concentration measured in water (0.47 mg L^{-1}) was reached at 7 DAA and in soil (2.01 mg kg^{-1}) at 20 DAA (Bajet and Araez, 1994) and was below the toxic level for *Tilapia*.

Draining the paddy field, usually performed at the milk dough stage of rice, should be done at least seven to ten days after last application of pesticide or the drainage water should be impounded before it is discharged into irrigation canals. This procedure would allow degradation of the pesticide to occur and reduce the contamination of waterways. In cases of rice fish culture, a separate pond for fish is ideal to minimize toxicity to fish. Pesticides that are not toxic to fish should be used in the rice paddies to prevent accidental contamination of the fish ponds and consequent loss of fish productivity.

Pesticides harmful to fish and other non-target organisms should be restricted from use in areas near aquatic environments and should not be used as an alternative to fishing or maintenance of aquaculture ponds. Suitable management practices should be developed to minimize environmental risk. For example, Bagarinao and Lantin-Olaguer (2000) have formulated an IPM strategy that can replace the use of triphenyltin compounds to manage populations of *C. cingulata* in milkfish ponds. Their strategy includes removal of spawning snails for shell-craft and other enterprises, dry-down of ponds to kill adults and egg strings, use of nitrogen fertilizers and lime during pond preparation to kill snails in puddles, and timing water input to periods of low veliger (free-swimming larval stage) counts in the supply water. Draining of pond water from aquaculture farms treated with pesticides should follow label recommendations to minimize risk. Likewise, the use of pesticides for fishing should be monitored and fishermen should be punished if found guilty of violating regulations. The use of water soluble pesticides should not be allowed near aquatic ecosystems nor spraying pesticides near water sources. After crops are harvested, residues in food can be further reduced by the consumer in the kitchen through removal of the outer leaves of cabbage; thorough washing of leaf and fruit vegetables with water and/or soap or detergent; cooking, boiling, drying, pulverizing, and other food processing methods, especially those involving heat treatment and refining procedures.

Legislative control of pesticide residues in food is accomplished by FPA through regulating the contents of the pesticide label, formulating policies that result in a reduction of residues such as the training of farmers, banning or restricting the use of pesticides found to leave persistent residues in food or to pose unacceptable health risks, and implementing MRLs in the country. Importation of pesticide finished products and technical grade materials used in formulation is also highly regulated. Inspection of pesticides sales outlets is also done by FPA coordinators nationwide to minimize the sale of illegal pesticides.

SUMMARY

The situation in the Philippines is far from ideal and awareness among the general public about pesticide residues and their potential for contaminating the environment is lacking. The capability of Philippine Government agencies to generate and analyze data on pesticide contamination in food, feedstuffs, and the environment is limited by a lack of sophisticated instrumentation, the requisite equipment maintenance funding and staff, and adequate manpower with the expertise needed to run such nationwide programs. With limited analytical capability, implementation of controls by regulatory agencies would be nil or non-existent.

Government funds for environmental impact studies of pesticides and other toxicants are limited, with the government's highest priority geared toward increasing production in the agriculture and fisheries sectors. Support from international sources is also limited or non-continuing in nature and, therefore, pesticide residue data are few and far between. Government agencies with specific and compartmentalized mandates (FPA: pesticide registration and regulatory responsibilities; DA: crop and livestock production; BPI-NPAL: pesticide residue monitoring of food; BFAR/DA: fish production; EMB/DENR: environmental pollutants; NCPC: crop protection and pesticide residue research) should work hand-in-hand on issues of common concern but this is not always true. Lack of coordination and integration of government effort and the funding required for implementation is lacking especially in the area of environmental management.

Currently the Japanese government, through the Japan International Cooperation Agency (JICA) Project, is building up the analytical capabilities of the BPI-NPAL with four satellite laboratories at strategic locations nationwide. However, most of the analyses will be performed on foodstuffs collected in markets nationwide and will begin the generation of residue data for Philippine MRLs, especially for crops important to the South East Asia Region. The current work on pesticides in the environment at the EMB/DENR is limited to non-existent. Most of the current residue data is generated by the PTCL, NCPC. This is insufficient for a country truly concerned about pesticide residues in food and the environment.

An assessment of the Philippine fisheries sector identifies resource depletion, environmental damage, poverty among municipal fisherfolk, low productivity in aquaculture, and limited utilization of offshore waters by commercial fishermen as the major problems in this sector of the economy (Munoz, 1997). Over fishing and habitat degeneration has resulted in no substantial increase in fish capture in near-shore areas. An increment in fish production is expected to come from aquaculture but aquaculture's use of chemicals should be adequately researched, evaluated, and managed.

Research is needed within high risk areas where pesticides are intensively used. Documentation and evaluation of pesticide use in aquaculture and fishing activities is necessary to provide baseline information. Research on the environmental consequences of chemical use in inland and coastal aquaculture should examine pesticide impact on non-target organisms, chemical fate, effects, accumulation, degradation, movement in the environment, and resistance development. The capability of monitoring for contaminants should be strengthened and should include quantifying inputs and outputs of contaminants and assessing the environmental and social costs of inland and coastal aquaculture, rice fish culture, etc. Pesticides and other contaminants should be evaluated not only for localized effects but in the context of a nationwide risk assessment. This will provide data for new legislative initiatives to protect the environment from unnecessary risks from pesticide use.

The Philippine Government has many limitations to developing a truly ideal situation where pesticide residues in food and the environment are constantly monitored for the implementation of regulatory controls. There are a limited number of laboratories with the capability of analyzing pesticide residues and limited manpower and monetary support for continued and sustained monitoring of foods in markets, much less the necessary monetary support to do quality residue work. Also, local produce goes to market through a middleman or directly through the farmer. Market inspectors only inspect meat and meat products and do not inspect the quality of agricultural produce in terms of non-conformities. Non-conformities in food products to pesticide regulations are the responsibility of the importer (for incoming shipments) and the exporter (for outgoing products) with government intervention occurring only if a problem arises or is perceived. Supply and price take precedence over high quality expensive imports. Most of the time the importer relies on residue analyses conducted by certified laboratories at the port of departure. Currently the Quarantine Section of BPI requires residue analysis only for incoming shipments of garlic and onions because these crops are perceived to have received heavy applications of pesticides. Finally there is neither the political will nor sufficient money in the national budget to police all local markets and imported products.

Considering the current limitations of the Philippine Government as discussed above, the most likely immediate action would be a nationwide information dissemination and training program especially for farmers and aquaculturists on the effects of pesticides in the environment. This should involve all government agencies concerned, with FPA as the lead agency. A nationwide campaign about environmental contaminants is imperative and must include school children to provide continuity for the program. Pingali *et al.* (1995) suggested that local government officials should be invested with police power to confiscate food found to be contaminated with pesticide residues and FPA should constantly monitor, train, and inform pesticide dealers of banned and illegal pesticide products. They also suggested that pesticides be reregistered with a review of current data on the pesticide and a requirement that the registering company submit residue data collected under Philippine conditions. Currently once registration is given, a pesticide could be marketed forever until restricted or banned by FPA because of its hazards (Pingali *et al.*, 1995).

Although human and industrial sources of pollution are highly visible and can be easily attributed to a source, pollution contributed from the agriculture and fisheries sector cannot be discounted as intensive food production naturally follows a growing population. Every time pesticides are used, a certain risk is involved especially if the usage is not judicious. Perceived personal benefits should not be the overriding factor in the decision to apply pesticides. Filipinos should also begin to reorient their concept of the environment so that it is not limited to the house, the yard, the place of work, and the immediate community but focuses on the national and global scale. The increased productivity of Philippine resources should proceed hand-in-hand with preservation of those same resources for future generations. The idea of sustainable development along with the responsibility of the current generation to preserve its resources for the future should be ingrained in every Filipino and, indeed, in every citizen of the world.

ACKNOWLEDGMENTS

The author wishes to acknowledge the contribution of Ms Marife P. Navarro for helping out in the search for scientific names of local species and Philippine IPM program strategies. Credits also go to Drs Milton Taylor and Stephen Klaine and Fernando Carvalho for patiently editing and giving their suggestions for the improvement of this chapter with special mention of Dr Taylor's invaluable help in acquiring updated literature.

NOTE

1 Generally if the analyte is soil, sediment, oyster, or mussel, the residues are expressed on a dry weight basis. However, for vegetables, fruits, fish/salted fish, and rice (containing an average of 12 percent moisture), the residues are expressed on a fresh weight or wet weight basis.

REFERENCES

Andalecio, M.N. and Pahilanga, J.H. 1995. Effect of some pesticides on aquatic organisms: a review. Paper presented at the Workshop on Pesticide Residue Control Program, 22–23 August 1995. Cavite, Philippines: Development Academy of the Philippines.

[Anonymous]. 1993. Pond culture of tilapia. *Aqua Farm News.* 11(3):8.

Antle, J.M. and Pingali, P.L. 1994. Pesticides, productivity, and farmer health: a Philippine case-study. *Am J Agricul Econ.* 76(3):418–30.

Aypa, S.M. 1991. Fisheries and environmental issues. Paper presented to Workshop on River Rehabilitation Program for the Manila Bay Region. 24–25 July 1991. Diliman, Quezon City, Philippines: UNEP-EMB/DENR.

Bagarinao, T. and Lantin-Olaguer, I. 2000. From triphenyltins to integrated management of the 'pest' snail *Cerithidea cingulata* in mangrove-derived milkfish ponds in the Philippines. *Hydrobiologia* 437(1–3):1–16.

Bajet, C.M. 1995. Toxicity and bioaccumulation of pesticides used in rice. Final Report. College, Laguna, Philippines: National Crop Protection Center/National Agriculture Food Council.

Bajet, C.M. 1996a. Fate and effects of pesticides in Manila Bay and in microecosystem using radiolabeled tracers. Report presented at 3rd Research Coordination Meeting (RCM) of IAEA Cooperative Research Program (CRP) held 9–13 September 1996 at Universidad Nacional, San Jose/Heredia, Costa Rica on Pesticides in the Marine Environment.

Bajet, C.M. 1996b. Residues of pesticides in dried fish. Report submitted to the Small Islands Agriculture Support Service Program. Guimaras, Philippines.

Bajet, C.M. 1997. Behaviour and effects of pesticides in Manila Bay and in model microecosystem using radiolabeled techniques. Report presented at 4th Research Coordination Meeting (RCM) of IAEA Cooperative Research Program (CRP) held 16–20 June 1997 at University of Nairobi. Nairobi, Kenya.

Bajet, C.M. and Araez, L.C. 1994. Environmental study of controlled release thiobencarb using radiochemical techniques. *Phil Agric.* 77(2):169–79.

Bajet, C.M. and Magallona, E.D. 1982. Chemodynamics of isoprocarb in the rice paddy environment. *Phil Entomol.* 5(4):355–71.

Bajet, C.M. and Tejada, A.W. 1995. Pesticide residues in the Philippines: an analytical perspective. *Trends Anal Chem.* 14(9):430–4.

Barril, C.R. 1973. Analysis of chlorinated insecticide residues in aquatic fauna of Laguna de Bay. In: Proc. 1st Regional Seminar on Ecology. Manila, Philippines: National Research Council of the Philippines, pp. 71–89.

Beveridge, M.C.M. and Phillips, M.J. 1993. Environmental impact of tropical inland aquaculture. In: Pullin, R.S.V., Rosenthal, H. and Maclean, J.L. (eds) *Environment and Aquaculture in Developing Countries. International Center for Living Aquatic Resources Management (ICLARM) Conference Proceedings* 31:213–36.

BFAR. 1996. Philippine fisheries profile. Quezon City, Philippines: Department of Agriculture.

BFAR. 1999. The 1998 Philippine fisheries profile. Quezon City, Philippines: Department of Agriculture.

BFAR. 2001. The 2000 Philippine fisheries profile. Quezon City, Philippines: Department of Agriculture.

Bhuiyan, S.I. and Castaneda, A.R. 1995. The impact of ricefield pesticides on the quality of freshwater resources. In: Pingali, P.L. and Roger, P.A. (eds) *Impact of Pesticides on Farmer Health and the Rice Environment.* Dordrecht, Netherlands: Kluwer Academic Publishers, pp. 181–202.

BPI. 1990. A final report on contamination of foods found in the Philippines. Manila, Philippines: FAO. Unpublished report.

BPI. 1995. A report on food contamination. Unpublished data. Manila, Philippines.

Cagauan, A.G. 1990. Fish toxicity, degradation period and residues of selected pesticides in rice fish culture. Report presented at the Workshop on the Environmental and Health Impacts of Pesticide Use in Rice Culture held 28–30 March 1990 at the International Rice Research Institute, Los Baños, Philippines.

Cagauan, A.G. and Arce, R.G. 1992. Overview of pesticide use in rice fish farming in Southeast Asia. In: de la Cruz, C.R., Lightfoot, C., Costa Pierce, B.A., Carangal, V.R. and Bimbao, M.P. (eds) *Rice Fish Research and Development in Asia. ICLARM Conf Proc.* 24:217–33.

Calamari, D. and Delos Reyes, M. 1997. Appraisal of environmental risk from pesticide pollution in Batangas Bay and Xiamen waters. *Trop Coasts.* 4(2):13–5.

Calumpang, S.M.F.C. 1994. Fate of niclosamide in various components of a fish and prawn pond ecosystem. *Phil Agric.* 77(3):393–401.

Calumpang, S.M.F.C., Medina, M.J.B., Tejada, A.W. and Medina, J.R. 1995. Environmental impact of two mollucicides: niclosamide and metaldehyde in rice paddy ecosystem. *Bull Environ Contam Toxicol.* 55:494–501.

Camacho, A.S. and Lagua, N.M. 1988. The Philippine aquaculture industry. In: Juario, J.V. and Benitez, L.V. (eds) *Perspectives in Aquaculture Development in Southeast Asia.* Iloilo, Philippines: Southeast Asian Fisheries Development and Education Center (SEAFDEC), pp, 91–116.

Capinpin, E.F.C. 1992. Acute toxicity and histopathological effects of malathion on Nile tilapia (*Oreochromis niloticus*) fingerlings [MSc thesis]. College of Fisheries, University of the Philippines. Visayas, Iloilo, Philippines.

Celino, L.P, Gambalan, N.B. and Magallona, E.D. 1988. Fate of carbofuran in a simulated rice fish ecosystem: Manila, Philippines. *Proc. 11th Int. Congress Plant Protection.* 1:393–5.

Chua, T.E. 1993. Environmental management of coastal aquaculture development. In: Pullin, R.S.V., Rosenthal, H. and Maclean, J.L. (eds) *Environment and Aquaculture in Developing Countries. ICLARM Conf Proc.* 31:199–212.

CPAP. 1993. Sales statistics. Unpublished data. Manila, Philippines.

CPAP. 1996. Sales statistics. Unpublished data. Manila, Philippines.

CPAP. 2000. Sales statistics. Unpublished data. Manila, Philippines.

Cruz, E.R., de la Cruz, M.C. and Sunaz, N.A. 1988. Hematological and histopathological changes in *Oreochromis mossambicus* after exposure to the mollusicides Aquatin and Brestan. In: Pullin, R.S.V., Bhukaswa, T., Tonguthai, K. and Maclean, J.L. (eds) 2nd Int. Symp. on Tilapia in Aquaculture. *ICLARM Conf Proc.* 15:99–110.

Dave, O. 1997. Crop protection campaign: industry sector. Report presented at Nat Symp on Crop Protection Strategies: A Unified Movement under the Gintong Ani Programme held 17–19 April 1997 in Asia Pacific Economic Community (APEC) Building, Los Baños, Laguna, Philippines.

Galvez, R. and Sadorra, M.S.M. 1989. Blast fishing: a Philippine case study. *Tropical Coastal Area Management (ICLARM).* 3(1):9–10.

Golob, P., Cox, J.R. and Kilminster, K. 1987. Evaluation of insecticide dips as protectant of stored dried fish from dermestid beetle infestation. *J Stored Prod Res.* 23(1):47–56.

Gowen, R.J. and Rosenthal, H. 1993. The environmental consequence of intensive coastal aquaculture in developed countries: What lessons to be learnt. In: Pullin, R.S.V., Rosenthal, H. and Maclean, J.L. (eds) *Environment and Aquaculture in Developing Countries. ICLARM Conf Proc.* 31:102–15.

Gregorio, R.M. 1995. Agrochemical impact in soil and water at Barit River irrigation system. Report presented at Workshop on Pesticide Residue Control Programme held 22–23 August 1995 at Development Academy of the Philippines, Cavite, Philippines.

Heong, K.L., Escalada, M.M. and Lazaro, A.A. 1995. Misuse of pesticides among rice farmers in Leyte, Philippines. In: Pingali, P.L. and Roger, P.A. (eds) *Impact of Pesticides on Farmer Health and the Rice Environment*. Dordrecht, Netherlands: Kluwer Academic Publishers, pp. 97–108.

Hingco, T.G. 1990. *Pollution in Manila Bay*. Diliman, Quezon City, Philippines: Philippine Social Science Center, University of the Philippines.

Jacinto, G.S. 1997. Preliminary assessment of marine pollution issues in the East Asian Seas region at the end of the milennium. *Trop Coasts*. 4(2):3–7.

Kasakalikasan Program-Department of Agriculture. 1997. Pre-project completion impact evaluation project. SEAMEO Regional Center for Graduate Study and Research in Agriculture. College, Laguna, Philippines, p. 115.

Khan, K.U. 1982. Bound pesticide residues in soils and plants. *Res Reviews*. 84:1–25.

Medina, M.J.B., Calumpang, S.M.F., Tejada, A.W., Medina, J.R. and Magallona, E.D. 1993a. *Fate of Endosulfan in Rice Fish Ecosystem*. Vienna: IAEA. IAEA TECDOC 695, pp. 33–45.

Medina, M.J.B., Calumpang, S.M.F., Tejada, A.W., Medina, J.R. and Magallona, E.D. 1993b. *Fate of Chlorpyrifos in a Model Rice Fish Ecosystem*. Vienna: IAEA. IAEA TECDOC 695, pp. 83–93.

Minh, T.B., Prudente, M.S., Watanabe, M., Tanabe, S., Nakata, H., Miyazaki, N., Jefferson, T.A. and Subramanian, A. 2000. Recent contamination of persistent chlorinated endocrine disrupters in cetaceans from the North Pacific and Asian coastal waters. *Water Science and Technology* 42(7–8):231–40.

Monje, P.M. 1989. Fish toxicity evaluation of selected synthetic pyrethroid under paddy field conditions using Nile tilapia (*Oreochromis niloticus*) and common carp (*Cyprinus carpio*) [MSc thesis]. Central Luzon State University. Munoz, Nueva Ecija, Philippines.

Munoz, J.C. 1991. Manila Bay: status of its fisheries and management. In: Proc. Int. Conf. on Environmental Management of Enclosed Coastal Systems held 3–6 August 1990 in Kobe, Japan, pp. 311–14.

Munoz, J.C. 1997. Coastal resource management in the Philippines. In: Proc. Regional Workshop on Coastal Fisheries Management Based on Southeast Asian Experiences held 19–22 November 1996 in Changmai, Thailand, pp. 80–5.

NCPC. 1996. National Pest Survey Monograph. University of the Philippines at Los Baños. College, Laguna, Philippines.

Navarro, R.L., Medina, J.R. and Callo, D.P. (eds) 1998. Updates and perspectives in sustainable agriculture in empowering farmers: the Philippine national Integrated Pest Management program. SEAMEO Regional Center for Graduate Study and Research in Agriculture. Los Baños. Laguna, Philippines.

Nishiuchi, Y. 1974. Testing methods for the toxicity of agricultural chemicals. *Japan Pestic Info*. 19:16–9.

Ocampo, V.R. and Bajet, C.M. 1997. Pesticides in the Philippine environment. In: Proc. 27th Annual Convention of the Pest Management Council of the Philippines held 7–10 May 1996 in Davao City, Philippines.

Pacaba, V.T.D. 1995. Pesticide residues in fruit and vegetables. Report presented at the Workshop on Pesticide Residue Control Programme held 22–23 August 1995 at Development Academy of the Philippines, Cavite, Philippines.

Pingali, P.L., Gerpacio, R.V. and Rola, A.C. 1995. Persistent pesticide pollution: synthesis of evidences from the Philippine rice ecosystem. In: Proc. Workshop on Pesticide Control Program held 22–23 August 1995 in Tagaytay, Philippines.

Primavera, J.H., Pitogo, C.R.L., Ladja, J.M. and de la Pena, M.R. 1993. A survey of chemical and biological products used in intensive prawn farms in the Philippines. *Mar Poll Bull.* 26(1):35–40.

Rola, A.C. 1995. Estimating net benefits of pesticides at the farm and society levels. In: Final report on the evaluation of current Philippine pesticide data as an input to policy decisions. University of the Philippines at Los Baños and Fertilizer and Pesticide Authority of the Philippines.

Rola, A.C. and Pingali, P.L. 1993. Pesticides, rice productivity and farmer's health: an economic assessment. College, Laguna, Philippines: World Resources Institute/International Rice Research Institute.

Rubec, P.J. 1988. Cyanide fishing and the Marine-life Alliance National Training Programme. *Tropical Coastal Area Management (ICLARM).* 3(1):11–3.

Sabularse, D.R. 2001. Pesticide usage in vegetables and rice. Paper presented at the Seminar/Workshop on Minimizing the Off Site Impact of Pesticides from Agricultural Systems: A Risk Based Approach. 29 August 2001. NCPC, College, Laguna Philippines.

SEAFDEC. 2000. Research findings: milkfish. Aquaculture Department of the Southeast Asian Fisheries Development Center. Available at http://www.seafdec.org.ph/home.html.

Senthilkumar, K., Kannan, K., Tanabe, S. and Prudente, M. 1998. Butyltin compounds in resident and migrant birds collected from Philippines. *Fresenius Environ Bull.* 7(9–10):561–71.

Sevilleja, R.C. 1992. Rice fish farming development in the Philippines: past, present, and future. In: de la Cruz, C.R., Lightfoot, C., Costa Pierce, B.A., Carangal, V.R. and Bimbao, M.P. (eds) *Rice Fish Research and Development in Asia. ICLARM Conf Proc.* 24:77–89.

Sly, P.G. 1993. Major environmental problems in Laguna Lake, Philippines: a summary and synthesis. In: *Laguna Lake Basin, Philippines: Problems and Opportunities.* Manila, Philippines: Environ. Res. Management Project Report 7, pp. 304–30.

Sta Ana, J. 2001. LLDA's role in Laguna Bay. Paper presented at the Seminar/Workshop on Minimizing the Off Site Impact of Pesticides from Agricultural Systems: a Risk Based Approach. 29 August 2001. NCPC, College, Laguna Philippines.

Suva, L. 1995. Reaction paper: pesticide residue analysis in the Philippines with particular reference to the work of the Bureau of Food and Drugs. In: Proc. Workshop on Pesticide Control Program held 22–23 August 1995 in Tagaytay, Philippines.

Tanabe, S., Prudente, M.S., Kan-atireklap, S. and Subramanian, A. 2000. Mussel watch: marine pollution monitoring of butyltins and organochlorines in coastal waters of Thailand, Philippines and India. *Ocean Coastal Manage.* 43(8–9):819–39.

Tejada, A.W. 1995. Pesticide residues in foods and in the environment as a consequence of crop protection. *Phil Agric.* 78(1):63–79.

Tejada, A.W. and Bajet, C.M. 1990. Fate of pesticides in rice fish culture. *Phil Agric.* 73(2):153–63.

Tejada, A.W. and Magallona, E.D. 1985. Fate of carbosulfan in rice paddy environment. *Phil Entomol.* 6(3):255–73.

Tejada, A.W., Calumpang, S.M.F. and Magallona, E.D. 1990. Fate of carbofuran in rice fish and livestock farming. *Tropical Pest Manage.* 36(3):237–43.

Tejada, A.W., Varca, L.M., Ocampo, P.P., Bajet, C.M. and Magallona, E.D. 1993a. Fate and residues of pesticides in rice production. *Int J Pest Manage.* 39(3):281–7.

Tejada, A.W., Bajet, C.M., Magallona, E.D., Magbanua, M.G., Gambalan, N.B. and Araez, L.C. 1993b. Toxicity and toxicity indices of pesticides to some fauna of the lowland rice fish ecosystem. *Phil Agric.* 76(4):373–82.

Tejada, A.W., Bajet, C.M., Magbanua, M.G., Gambalan, N.B., Araez, L.C. and Magallona, E.D. 1994. Toxicity of pesticides to target and non target fauna of the lowland rice ecosystem. In: Widianarko, B., Vink, K. and Straalen, M.N. (eds) *Environmental Toxicology in Southeast Asia.* Amsterdam, Netherlands: Vrije University Press, pp. 89–104.

Tejada, A.W., Varca, L.M., Calumpang, S.M.F., Ocampo, P.P., Medina, M.J.B., Bajet, C.M., Paningbatan, E.B., Medina, J.R., Justo, V.P., de Leon Habito, C., Martinez, M.R. and Magallona, E.D. 1995. Assessment of the environmental impact of pesticides in paddy rice production. In: Pingali, P.L. and Roger, P.A. (eds) *Impact of Pesticides on Farmer Health and the Rice Environment.* Dordrecht, Netherlands: Kluwer Academic Publishers, pp. 149–80.

The Philippine Atlas. 1975. A historical, economic and educational profile of the Philippines, Volume 1. Manila, Philippines: Funds for Assistance for Private Education (FAPE).

Tipa, E.V., Tejada, A.W., Barril, C.R., Merca, E. and Quintana, B.B. 1995. Rapid bioassay of pesticide residues in vegetable samples from Laguna, Philippines. *Phil Agric.* 80(1–2): 1–12.

Tjornhom, J.D., Norton, G.W. and Gapud, V. 1998. Impacts of price and exchange rate policies on pesticide use in the Philippines. *Agricultural Economics* 18(2):167–75.

Tonguthai, K. 1996. Use of chemicals in aquaculture. *Aquaculture Asia.* Jul/Sept:42–5.

Varca, L.M. 1996. The fate, degradation and dissipation of DDT in a tropical environment [PhD dissertation]. University of the Philippines at Los Baños. College, Laguna, Philippines.

Varca, L.M. and Magallona, E.D. 1987. Fate of BPMC in paddy rice components. *Phil Entomol.* 7(2):177–89.

Varca, L.M. and Magallona, E.D. 1994. Dissipation and degradation of DDT and DDE in Philippine soil under field conditions. *J Environ Sci Health Part B Pestic Food Contam Agric Wastes* 29(1):25–35.

Warburton, H., Palis, F.G. and Pingali, P.L. 1995. Farmer perceptions: knowledge, and pesticide use patterns. In: Pingali, P.L. and Roger, P.A. (eds) *Impact of Pesticides on Farmer Health and the Rice Environment.* Dordrecht, Netherlands: Kluwer Academic Publishers, pp. 59–95.

Zulkifli, M., Tejada, A.W. and Magallona, E.D. 1983. The fate of BPMC and chlorpyrifos in some components of a rice paddy ecosystem. *Phil Entomol.* 6(5–6):555–65.

Chapter 11

Pesticides in the coastal zone of Mexico

F. González-Farias

PESTICIDE REGULATION IN MEXICO

History of pesticide usage

The use of pesticides in Mexico began at the end of the 19th century. In 1898, approximately 38 different compounds were in use to protect crops and grains from pests. The most commonly used pesticides were lead arsenate, copper aceto-arsenate (Paris green), hydrogen cyanide, carboxylic acid, and a mixture of copper sulfate and calcium hydroxide (Bordeaux mixture) (Restrepo, 1988).

Intensive application of pesticides began in the middle of the 1940 to 1950 decade when these compounds were used to augment crop production and to satisfy the quality requirements of the export market. The main crops cultivated in Mexico since those years are wheat, maize, beans, soybean, rice, carthamus (safflower), chickpeas, sorghum, cotton, sugarcane, coffee, fruits (oranges, bananas, mangoes, etc.), and horticultural products (tomato, cucumber, eggplant, chili, squash, etc.), all of which demand the use of great quantities of agrochemicals (CAADES, 1987). Economic conditions in Mexico at the start of intensive pesticide application and the very positive economic impact of crop exportation generated a huge demand for pesticides, which gave rise to the development of several industries dedicated to their importation, production, and formulation (Restrepo, 1988).

In 1947, the first inorganic insecticide (copper arsenate) was produced in Mexico, and in 1959 the production of OCs (DDT, HCB) and thiocarbamates was initiated. In addition to the OCs being produced in Mexico, a great quantity was imported. By the end of the 1960s, around 165 pesticides were being imported from several different countries (Restrepo, 1988). The quantity of imported OCs reached a peak of 12,000 T in 1958, and, thereafter, it declined sharply due to increased domestic production (Albert, 1996).

In 1968, Mexico's government was strongly promoting the production of the OC pesticides DDT, HCB, and toxaphene by a government-owned company (FERTIMEX) despite the fact that those compounds were at that time no longer being applied in other parts of the world. Albert (1996) reported that, during the

1969 to 1979 period, around 9,000 T per year of OCs were used in Mexico and, of these, domestic production was 7,800 T (3,900 T of DDT, 2,000 T of toxaphene, 1,600 T of HCB, and 300 T of endrin). In the 1970s, the production of ethyl-parathion, methyl-parathion, and malathion began in Mexico (Narro, 1979). The Instituto Nacional de Estadística, Geografía e Informática or INEGI (1994a) reported that the domestic production of pesticides had increased to 11,200 T in 1992. Overall, the use of pesticides in Mexico has increased considerably since the 1950 to 1960 decade. An estimate, based on Albert, 1996; Berrelleza and Sosa Pérez, 1995; Carvalho *et al.*, 1996; Diaz Simental, 1995; López Pérez, 1995; Núñez Cebreros, 1995; Restrepo, 1988; and Rosales, 1979, of the total quantity of pesticides used through the period 1950 to 1995 is presented in Figure 11.1.

This level of pesticide usage resulted in 35 pesticide-related companies engaged in business in 1988, with 11 of them being importers of a.i.(s) and 24 producers of a.i.(s). Of these companies, 16 were Mexican companies and 19 were international consortiums based in the United States, Switzerland, Germany, Great Britain, Italy, France, and Japan (Restrepo, 1988).

According to the Mexican Ecology Minister, at present there is no synthesis of new a.i.(s) for pesticides in Mexico. The a.i.(s) for new compounds are imported from several different countries (INEGI, 1995). The principal exporters of pesticides to Mexico and the percentage of the different compounds, e.g. insecticides, herbicides, fungicides, etc., imported are presented in Figure 11.2. The tropical conditions prevalent in Mexico are especially favorable for insect and fungal

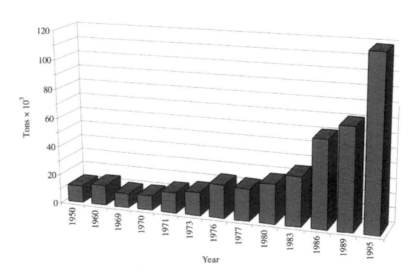

Figure 11.1 Estimate of the quantity of pesticides used in Mexico from 1950–95 (adapted from Albert, 1996; Berrelleza and Sosa Pérez, 1995; Carvalho *et al.*, 1996; Díaz Simental, 1995; López Perez, 1995; Nuñez Cebreros, 1995; Restrepo, 1988; Rosales, 1979)

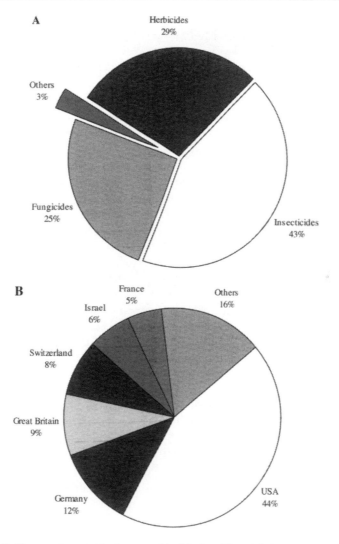

Figure 11.2 Classes of pesticides imported by Mexico (A) and their countries of origin (B)

development. Thus, it is not surprising that insecticides and fungicides make up almost 70 percent of all pesticide imports.

Current pesticide regulations

There is international concern for governments to develop appropriate legislation with management regulations for agrochemicals. Unfortunately, in developing countries like Mexico, there is a critical lack of information about the environmental presence and fate of these pollutants in local ecosystems. It is for precisely these reasons that, in most developing countries, the use of agrochemicals has not been controlled, despite the growing scientific evidence of the health and environmental hazards associated with them (Albert, 1996).

Although pesticides have been applied in Mexico since the end of the last century, the first pesticide legislation, the Law of Plagues, was not promulgated until 1924, after the Mexican Revolution. The main focus of this document was to protect crops from the economic point of view, with neither health nor environmental aspects being addressed in this law (Restrepo, 1988). In 1940, the Law of Sanity for Crops and Cattle (Ley de Sanidad Fitopecuaria) was published. However, this law did not consider the environmental hazards associated with the use of organic pesticides as those hazards were unknown at the time. Obviously this law was obsolete in the 1940 to 1950 decade when there was heavy application of OCs in Mexico (Albert, 1996). It was not until 1974 that pesticide regulatory legislation – taking into account the potential for environmental impacts – was enacted.

In 1971, the Federal Law for Preventing and Controlling Environmental Pollution and its guiding regulations were published. Even though the environmental aspects of pesticide usage were considered, the published regulations were broad and non-specific. For example, the criterion for toxic substances in certain bodies of water permitted the presence of toxic compounds based on the declared (by the government) use of that water (Diario Oficial de la Federación, 1971). However, in 1988, the General Law of Ecology and Environmental Protection stated that environmental impact studies are required for every human action that can potentially harm the environment (Diario Oficial de la Federación, 1988).

In addition to the Law of Sanity for Crops and Cattle and the General Law of Ecology and Environmental Protection, pesticide usage is also regulated by the Federal Law for Preventing and Controlling Environmental Pollution; the General Law of Health; Federal Law of Labor; Federal Law of Roads; Bridges, and Federal Transportation; the Law of Customs; and the General Law of Weights and Measurements. All of these laws are supported by a set of official standards with guidelines and regulations. Control of the production, importation, formulation, transportation, commercialization, and application of pesticides is accomplished through the Organic Law of Federal Administration, within the jurisdiction of the Agriculture (Secretaría de Agricultura, Ganadería y Desarrollo Rural), Health (Secretaría de Salud), Commerce (Secretaría de Comercio y Fomento Industrial), and Environment (Secretaría de Medio Ambiente, Recursos Naturales y Pesca) ministers.

In 1988, with the participation of all the ministers mentioned above, the Interministerial Commission for the Control of the Process and Use of Pesticides, Fertilizers, and Toxic Chemicals (Comisión Intersecretarial para el Control del Proceso y Uso de Plaguicidas, Fertilizantes y Sustancias Tóxicas, known as CICOPLAFEST) was created. This commission is responsible for periodic publication of the Official Catalog of Pesticides, which presents those agrochemicals whose importation, production, formulation, commercialization, transportation, and use are permitted in Mexico. This catalog also includes a list of those pesticides whose use is banned or restricted (Table 11.1). Furthermore, there are notes included on the labeling, commercialization, application, crop residue levels, toxicity, and other environmental aspects of the use of these products. The commission has published a yearly update of the Official Catalog of Pesticides since 1988. Figure 11.3 presents an example of the changes in the number of insecticides for various classes of pesticides permitted in the first and last editions of the catalog (published by CICOPLAFEST in 1988 and 1996). The establishment of CICOPLAFEST has been the most important action instituted by the government to control, in a broad sense, the use of agrochemicals.

Table 11.1 Pesticides banned from importation, fabrication, formulation, commercialization, and use or restricted in Mexico[a]

Banned pesticides	Restricted use pesticides[b]
Aldrin	Alachlor
Cyanophos	Aldicarb
Chloranil	Chlordane
DBCP	Chloropicrin
Dialiafor (Dialifos)	Chlorothalonil
Dieldrin	1,3-dichloropropene
Dinoseb	DDT[c]
Endrin	Dicofol
Erbon	Phorate
Formothion	Fosethyl-aluminium
Fumisel (Phosphine)	Methyl isotiocyanate
Kepone/Chlordecone	Lindane
Mercury phenyl acetate	Methamidophos
Mercury phenyl propionate	Metham-sodium
Mirex	Methoxychlor
Monuron	Methyl bromide
Nitrofen	Mevinphos
Schradan	Paraquat
Sodium fluoracetate	Pentachlorophenol
Triamifos	Quintozene
2,4,5-T	
HCB[d]	
EPN[d]	
Parathion-ethyl[d]	
Thalium sulfate	
Toxaphene[d] (Camphechlor)	

Notes:
a Adapted from CICOPLAFEST, 1996.
b These pesticides can only be purchased with a written recommendation of an agriculture technician who has been authorized by the federal government.
c Used only by government sanitary (Public Health) authorities in campaigns against malaria vectors.
d Banned for commercialization and use, no reference is made about their importation, fabrication, and formulation in the Official Catalog of Pesticides.

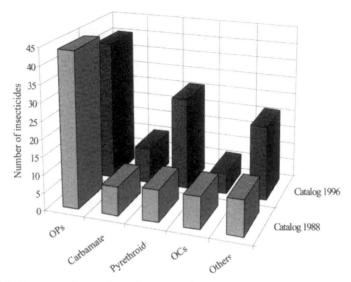

Figure 11.3 Change in the number of authorized insecticides for the different chemical groups for 1988 and 1996 (adapted from CICOPLAFEST 1988; 1996)

Limits of pesticide regulation and implementation

According to Albert and Aranda (1986), legislation governing the control of pesticides is the most complete of those laws regulating hazardous and toxic substances in Mexico, but it has deficiencies that prevent its enforcement. In Mexico, the law must be complemented with defined regulations and respective official standards (Norma Oficial Mexicana, know as NOM). These should specify the technical details for every action in which pesticides are involved, e.g. importation, exportation, characteristics of the labels, packing, transportation, analysis of active ingredients, production, formulation, application, disposition of containers, residues in food, residues in water, etc. NOMs are published as updated, but compliance certification by the authorities requires an adequate administrative and technical infrastructure of trained personnel that is lacking. Proper enforcement of pesticide regulations is mainly limited by the lack of knowledge and awareness of the hazards these compounds pose by both end-users and authorities. This problem has developed because governmental authorities have determined pesticide usage to be a low priority problem. This results in an insufficient budget to generate the appropriate infrastructure and maintain the required trained personnel. This, in turn, contributes to loose pesticide enforcement and regulatory supervision by the government.

The only way to improve pesticide regulatory control is to clearly inform end-users about the health risks and environmental damage that misuse of pesticides can cause, and to generate data documenting the presence and dynamics of pesticide residues in the environment, food, and organisms. This could generate sufficient public and governmental awareness to lead the government to rigorously enforce the current legislation and their attendant regulations. Albert (1996) considers that, due to a deficit of data on the use and, in particular, the presence of pesticide residues in the environment, it is not currently possible either to establish valid environmental impacts or to tighten the law.

With respect to the presence of pesticide residues in the coastal zone, it should be said that Mexico has written legislation for preventing and controlling pollution in the aquatic environment, but, nevertheless, the timely issuance of licenses and the enforcement of the regulations is inadequate (Saavedra-Vázquez, 1996). This results in the user – who pollutes the environment with pesticide residues – having de facto oversight of the law. Even though great advances in legislation and control of pesticide usage have been made in the last few years, there are still many aspects of legislation, infrastructure, training, and research that need to be improved.

PESTICIDE USE AND DISTRIBUTION IN MEXICO

Past and current usage patterns

Since the first time pesticides were applied in Mexico, their use has increased significantly (Figure 11.1). At the end of last century, inorganic pesticides predominated. In the 1940s through the 1960s, the OCs (mainly DDT) were most heavily used, but, then, in the 1970s, due to the growing global concern about the environmental and health problems caused by the use of OC pesticides, a shift to OP and carbamate pesticides took place in most of the agricultural regions of Mexico. In general, OC pesticides are no longer used for agricultural purposes and most of them are banned. Furthermore, the ones that are still labeled are restricted and their use limited to public health campaigns against mosquitoes, seed and soil treatment, and some restricted agricultural uses. The banning of several OC pesticides took place recently as can be seen in Table 11.2. It should be noted that endosulfan is the only OC currently listed for use in agriculture with no restrictions (except for coffee plantations).

The Official Catalog of Pesticides (CICOPLAFEST, 1996) permits the use of 279 compounds classified in 22 different chemical families, e.g. OCs, OPs, carbamates, pyrethroids, thiocarbamates, phthalimides, organic tin compounds, organic arsenic compounds, urea derived compounds, triazines, compounds derived from biological origin, etc. Presently the most commonly used pesticides are the OPs. These insecticides listed in Table 11.3 are used in two zones with heavy agricultural activity, the coastal zone of the Gulf of Mexico and the Culiacán River Valley in Sinaloa in northwest Mexico.

It is difficult to present an accurate usage pattern for pesticides in different regions of Mexico due to the scarcity of information about this aspect of pesticide use and the difficulty in obtaining it. Moreover, the impact of these compounds on the coastal zone is also difficult to evaluate due to insufficient information. There are only around 20 research papers that include some information on the

Table 11.2 Change in legal status of OC insecticides between 1988 and 1996 as listed in Mexico's *Official Catalog of Pesticides*[a]

Pesticide	1988	1996
BHC	restricted (sanitary campaigns)	banned
Chlordane	restricted (urban and industrial use)	restricted (urban and industrial use)
Chlorobenzilate	restricted (only ornamental flowers)	non authorized
DDT	restricted (sanitary campaigns)	restricted (sanitary campaigns)
Dicofol	agriculture	restricted (agriculture use)
Dienochlor	agriculture	agriculture (only ornamental flowers)
Endosulfan	agriculture	agriculture[b]
Lindane	agriculture, forestry, cattle	restricted (seed and soil treatments)
Methoxychlor	agriculture	restricted (seed treatment)
Toxaphene	restricted	banned

Notes:
a Taken from CICOPLAFEST, 1988; 1996.
b Restricted use for coffee plantations.

distribution of pesticides in the coastal environments of Mexico (see References). Based on the information that is available, it seems that OPs, which are more toxic but less persistent than OCs, are the most used pesticides (around 50 percent of the total), then carbamates (approximately 15 percent), and next the pyrethroids (10–30 percent). Unfortunately it appears that certain banned OCs are still in use, based on the presence of OC residues in samples obtained from different coastal areas of the country (Table 11.3).

Location of pesticide applications with respect to the marine environment

Pesticides are used in Mexico for agriculture, public health campaigns against arthropods, home and industry sanitation, home lawn care, and gardening. As seen in Figure 11.4a, agriculture consumes the greatest quantity of pesticides. Of these agrochemicals, insecticides comprise the highest quantity at 37 percent of the total, followed by fungicides and herbicides at 27 percent and 26 percent, respectively (Figure 11.4b). Estimates place 1995 usage at around 118,000 T for all of Mexico (Figure 11.1), with 90,000 T applied for agricultural activities.

Even though there are no official data by state and use category for a particular pesticide or even for pesticides in general, an estimation of the impact can be construed from the usage of pesticides in agriculture. Areas of significant pesticide application for each coastal state can be derived by taking into account the area of irrigated agricultural land, the state's gross internal product derived from agriculture (GIPA), the type of crops grown in the state, and the hectares planted in each crop.

Table 11.3 The most used insecticides in the Culiacan River Valley, Sinaloa (northwest Mexico), for the summer of 1997, and in the coastal zone of the Gulf of Mexico

Culican River Valley, Sinola[a]			
OPs 40.0%	*Pyrethoids* 26.7%	*Carbamates* 13.3%	*Others* 20.0%
Acephate	α-cypermethrin	Aldicarb	Abamectin
Azinphos-methyl	Cyfluthrin	Carbaryl	Amitraz
Chlorpyrifos	Cypermethrin	Carbofuran	*Bacillus thurigiensis*
Dimethoate	Deltamethrin	Pirimicarb	Bifenthrina
Ethion	Fluvalinate		Diflubenzuron
Malathion	λ-cyhalothrin		Endosulfan
Methamidophos	Permethrin		
Omethoate	Z-cypermethrin		
Oxydemeton-methyl			
Parathion-methyl			
Temephos			
Terbufos			
Coastal zone of the Gulf of Mexico[b]			
OPs 62.5%	*Pyrethoids* 12.5%	*Carbamates* 12.5%	*Others* 12.5%
Azinphos-methyl	Cypermethrin	Carbaryl	DDT
Chlorpyrifos	Deltamethrin	Carbofuran	Endosulfan
Diazinon	Permethrin	Metomilo	Lindane
Dimethoate			
Disulfoton			
Edifenphos			
Ethion			
Foxim			
Malathion			
Methamidophos			
Momocrotophos			
Omethoate			
Oxydemeton-methyl			
Parathion-ethyl			
Parathion-methyl			

Notes:
a F. Escobosa, personal communication.
b Adapted from Benítez and Bárcenas, 1997.

This estimation permits an evaluation of the potential pollution from pesticides in the coastal areas for the different states of Mexico.

Of Mexico's 31 geopolitical states and one federal district, 17 of them have coasts with a summed littoral zone of 11,115 km. That includes the coastline within the Gulf of California, along the Pacific Ocean, the Gulf of Mexico, and the Caribbean Sea. Diverse aquatic ecosystems (estuaries, mangrove swamps, coral reefs, coastal lagoons, etc.) with high levels of biodiversity are abundantly

A

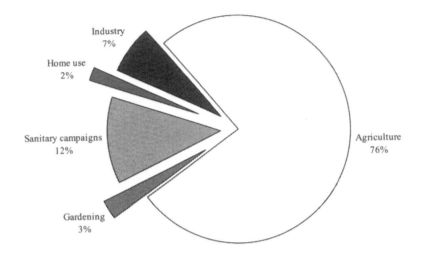

B

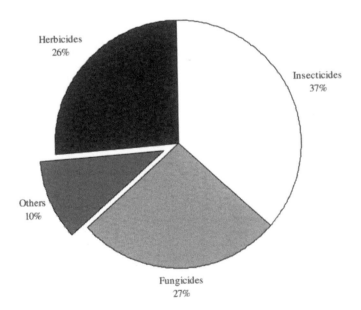

Figure 11.4 Percentage of total pesticides used in Mexico by activity (A) and by class (B)

represented in Mexico's coastal zone. The location of each state and some geographic characteristics of the state's coastal zones are presented in Figure 11.5 and Table 11.4.

Mexico has a total 30,537,701 ha of agricultural land with 15 percent under irrigation and around 85 percent dependent on natural rainfall (INEGI, 1994b). It should be emphasized that 65 percent of Mexico's total agricultural land area and 71 percent of its total irrigated land are located in its coastal states (Figure 11.6). Of the 17 coastal states, seven comprising Sonora, Sinaloa, Tamaulipas, Chiapas, Michoacan, Baja California, and Jalisco posses 64 percent (i.e. 2,931,620 ha) of Mexico's irrigated land (Figure 11.6) with large quantities of pesticides being applied every year. In general, irrigated areas have been developed mainly in Mexico's coastal plains due to climatic factors, land morphology, and fresh water availability. In such areas, pesticides are heavily applied year round and they can have important environmental impacts in coastal waters. This also affects other important economic activities, including offshore fisheries, aquaculture businesses, and tourism, among others.

Mexico's GIPA for 1993 was around US$7,073 M (INEGI, 1994b). The 17 coastal states contributed 62.7 percent of the total GIPA, with 48.2 percent from states located on the Pacific coast and 14.5 percent from those located in the Gulf of Mexico and the Caribbean. Coastal states with the highest contributions to the GIPA were: Sinaloa 8.7 percent, Jalisco 7.9 percent, Veracruz 7.4 percent, Michoacan 6.6 percent, Chiapas 5.5 percent, Sonora 5.4 percent, Oaxaca 5.4 percent, and Tamaulipas 3.9 percent. Sinaloa had the highest GIPA primarily from horticultural products (around 700,000 T per year), which command a good price in the export market. It is evident that, with the exception of Veracruz and

Figure 11.5 Location of the coastal states of Mexico (see Table 11.4 for abbreviations of state names). Triangles indicate places were samples (water, sediment, organisms) for pesticide analysis have been collected. Open triangles indicate OC analysis and filled triangle indicates OP analysis

Table 11.4 Geographic characteristics of the coastal states of Mexico[a]

State name	State abbreviation	Surface area km²	Littoral length km	Coastal lagoons[b]	
				Number	Surface area ha
Baja California	BC	69,921	1,555	11	15,890
Baja California Sur	BCS	73,475	2,230	13	133,760
Campeche	CAM	50,812	523	4	250,000
Chiapas	CHIS	74,211	256	9	100,800
Colima	COL	5,191	139	1	7,200
Guerrero	GRO	64,281	485	8	18,000
Jalisco	JAL	80,836	342	6	4,000
Michoacan	MICH	59,928	247	0	0
Nayarit	NAY	26,979	300	7	40,700
Oaxaca	OAX	93,952	597	6	43,200
Quintana Roo	QROO	50,212	865	6	11,000
Sinaloa	SIN	58,328	656	16	138,300
Sonora	SON	182,052	1,208	28	46,410
Tabasco	TAB	25,267	184	5	21,260
Tamaulipas	TAM	79,384	457	5	214,852
Veracruz	VER	71,699	745	17	113,914
Yucatan	YUC	38,402	342	8	3,100

Notes:
a Adapted from Alvarez and Gaitán, 1994; Contreras 1988; INEGI, 1995.
b From Contreras (1988).

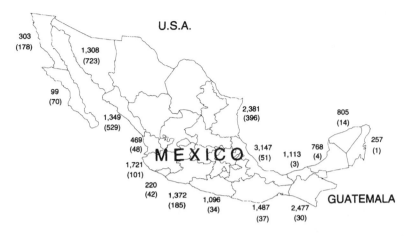

Figure 11.6 Total agricultural area (ha × 10³) and irrigated land [in brackets] (ha × 10³) of each coastal state in Mexico

Oaxaca, the states with higher GIPAs are those that have large areas of irrigated land.

With respect to total national production of important crops, it should be noted that coastal states contribute either the large majority or high percentages of all major crops (Table 11.5). For example, 99 percent of bananas, 99 percent of sesame seed, 92 percent of rice, 87 percent of sugarcane, 87 percent of chickpeas, 55 percent of coffee, 80 percent of cotton, 80 percent of oranges, 69 percent of wheat, 66 percent of maize, 63 percent of sorghum, and almost 100 percent of exported horticultural products, e.g. tomato, cucumber, eggplant, squash, etc., are produced in the coastal states. Additionally 45 percent of Mexico's pasture is within the coastal states (INEGI, 1994b). CICOPLAFEST (1996) considers that 65 percent of the total quantity of pesticides used in Mexico are applied to the crops, maize, sorghum, sugarcane, rice, horticultural products, and pasture lands. This confirms that huge quantities of pesticides are used in the coastal zones.

Based on the data presented, the main regions in which pesticides are applied correspond to the northern part of Baja California (on cotton, horticultural products, and wheat), Sonora-Sinaloa (on cotton, horticultural products for export, chickpeas, beans, maize, rice, wheat, and sugarcane), Jalisco-Michoacan (on wheat, maize, sugarcane, and chickpeas), Oaxaca-Chiapas (on bananas, maize, sugarcane, and coffee) and Tamaulipas-Veracruz (on sugarcane, sorghum, maize, bananas, oranges, rice, cotton, and pasture land). This is in good agreement with Restrepo (1988) who reports that pesticides are heavily used in the coastal states of Sonora, Sinaloa, Tamaulipas, Michoacan, Chiapas, and also in the inland states of Durango and Cohahuila (located in the northern part of Mexico). Benítez and Bárcenas (1997) also reported that 93 percent of the pesticides used in the Gulf of Mexico coastal states are applied in Tamaulipas and Veracruz with only 7 percent being applied in Tabasco and Campeche.

Table 11.5 Coastal state's crop production exceeding 10 percent of total country production (TCP[a])[b]

	Bananas	Coffee	Orange	Pasture	Sorghum	Sugar cane	Rice	Cotton	Sesame seed	Chick-pea	Wheat	Beans	Maize
TCP by crop	1,051	1,947	1,503	457	3,690	35,541	294	534	53	170	3,475	1,279	10,228
State abbrev.[c]													
SON								31.4	27.2	17.3	35.7		
SIN							19.1		26.0	25.3	17.3	13.5	
JAL						11.8				29.8			
MICH										10.9			10.6
GRO									14.5				
OAX		12.8							13.0				
CHIS	31.8	33.2											
TAM					38.3			27.0					
VER	13.8	32.1	51.9	20.2		35.5	26.4						
TAB	29.2												
CAM							15.0						

Notes:
a TCP in T × 10³.
b Calculated from INEGI (1994b).
c See Table 11.4 for abbreviations of state names.

Distribution of pesticide residues in the coastal zone

Despite almost 100 years of pesticide application in Mexico, studies of pesticide residues in coastal ecosystems are scarce and dispersed in time. Albert (1996), in an excellent review of pesticides in Mexico, recorded only 15 references (published between 1977 and 1995) studying aquatic systems, which included sampling in agricultural drainage areas, rivers, coastal lagoons, and shorelines. In those studies, analyses of OC pesticides were performed on water, sediment, and organism samples. In addition to those references cited by Albert (1996), there are also the studies of Readman et al. (1992), Botello et al. (1994b), Carvalho et al. (1996), González-Farias et al. (1997), the International Mussel Watch Report (IMW, 1995), Noreña-Barroso et al. (1998), Reyes et al. (1999), and de Llasera and Bernal-Gonzàlez (2001).

Although there has been a significant shift from OCs to OPs, carbamates, and pyrethroids (Table 11.6), almost all the published works present only OC residue levels. Readman et al. (1992) and Carvalho et al. (1996) present the first data for OP residues in the coastal environments of Mexico. No references were found that included residue data for carbamates – de Llasera and Bernal-González (2001) completed a very preliminary study on the presence of carbamate pesticides in wells, irrigation channels, drains, and a dam in the Yaqui Valley, Sonora, Mexico finding methiocarb in a groundwater sample ($5.4 \mu g\, L^{-1}$) and 3-hydroxycarbofuran ($18 \mu g\, L^{-1}$) in a surface water sample – and pyrethroids in coastal ecosystems, even though these groups together constitute from 20 percent to 40 percent of the total pesticide usage in Mexico. Because there is insufficient information on pesticide residue levels in the different coastal areas, it is not possible to establish an accurate assessment of the potential environmental exposure patterns for Mexico's coastal zone. However, with the data available, the presence or absence of pesticide residues in coastal state aquatic environments (Figure 11.4) can be evaluated. A determination of the frequency of occurrence of the most common compounds per sample was derived from published works between 1977 and 1997. As can be seen in Figure 11.7, the compounds occurring with higher frequency are DDE (64 percent of samples), DDT (50 percent), lindane (38 percent), heptachlor epoxide (25 percent), HCB (23 percent), and the 'drins' (aldrin 34 percent, dieldrin 27 percent, and endrin 23 percent). These findings confirm that DDT was the most commonly used insecticide in the country for agricultural purposes and the presence of fresh DDT reported in recent works (Readman et al., 1992; IMW, 1995; Carvalho et al., 1996) is probably a result of its use in public health campaigns against mosquitoes. With reference to the OPs, it is very interesting to find a 21 percent occurrence rate for chlorpyrifos in analyzed samples, which indicates a greater persistence (Readman et al., 1992) compared to other heavily used OPs, i.e. parathion with an 8 percent occurrence rate.

The number of different pesticide residues as reported for each state is presented in Figure 11.8. Sinaloa has the greatest number (19) as well as the only site from

Table 11.6 Concentration of pesticide residues (highest values) reported from 1977 to 1998 for the coastal states of Mexico[a]

Pesticide	Baja California				Baja California Sur			
	Water $\mu g\ L^{-1}$	Sediment ng g^{-1} dw	Organism[b] ng g^{-1} dw	Ref[c]	Water $\mu g\ L^{-1}$	Sediment ng g^{-1} dw	Organism ng g^{-1} dw	Ref
Aldrin			1.66	15				
Dieldrin			2.17	6				
			3.33	9				
Endrin			0.83	6				
			6.61	9				
DDT[d]			7.53	6			16.72	9
			Σ (152.06)	9				
DDE			40.41	6			1.40	6
			105.30	9			17.78	9
DDD			12.36	6			14.43	9
			144.82	9				
Lindane			1.83	6				
Chlordane			4.91	6			0.67	6
			15.73	9				
Nonachlor			5.16	6			0.59	6
Methoxychlor			19.60	6				
α-Endosulfan			7.60	9				
Heptachlor			6.00	15				
Heptachlor			13.24	15				
epoxide			15.60	9				

Pesticide	Sonora				Sinaloa			
	Water $\mu g\ L^{-1}$	Sediment ng g^{-1} dw	Organism[b] ng g^{-1} dw	Ref[c]	Water $\mu g\ L^{-1}$	Sediment ng g^{-1} dw	Organism ng g^{-1} dw	Ref
Aldrin		1.85		7		6.95		7
						<0.01		6
						2.88		2
Dieldrin		5.85		7	0.09			1
						0.80		4
						0.04	15.49	2
						0.70		3
							2.54	6
Endrin					0.73			1
						0.16	12.00	22
DDT		Σ (7.62)		7	1.44			1
			Σ (0.01)	8		3.10		4
						Σ (16.40)		7
							2.52	6
							12.17	
DDE					1.26			1
						2.16	218.57	2
							22.00	5
							68.62	6
DDD						0.33	63.00	2
							9.12	6
DDMU						0.32	1.40	2
HCB						0.03	0.31	2
Lindane		10.45		7	1.88			1
						0.24	3.30	2
							0.88	6

continued...

Table 11.6 continued

Pesticide	Sonora				Sinaloa			
	Water $\mu g\ L^{-1}$	Sediment ng g^{-1} dw	Organism[b] ng g^{-1} dw	Ref[c]	Water $\mu g\ L^{-1}$	Sediment ng g^{-1} dw	Organism ng g^{-1} dw	Ref
Chlordane							1.35	6
Nonachlor							0.77	6
Methoxychlor							0.37	6
α-Endosulfan						1.20		2
β-Endosulfan						0.08		2
Endosulfan sulphate						1.50	140.00	2
Heptachlor		5.40		7		18.20		7
Heptachlor epoxide						8.8		7
							5.19	6
Parathion-methyl						<0.1		4
Chlorpyrifos					0.002			2
						9.50		4

Pesticide	Oaxaca				Chiapas			
	Water $\mu g\ L^{-1}$	Sediment ng g^{-1} dw	Organism[b] ng g^{-1} dw	Ref[c]	Water $\mu g\ L^{-1}$	Sediment ng g^{-1} dw	Organism ng g^{-1} dw	Ref
Aldrin			<0.01	6		5.00		10
Endrin			<0.02	6		12.00		10
DDT			30.90	6		26.00		10
DDE			135.92	6		25.00		10
DDD			21.49	6				
HCB						17.00		10
Chlordane			1.92	6				
Nonachlor			0.31	6				
Methoxychlor			2.35	6				
α-Endosulfan						14.00		10
β-Endosulfan						250.00		10
Endosulfan sulphate						12.00		10
Heptachlor						10.00		10
Heptachlor epoxide						61.00		10

Pesticide	Tamaulipas				Veracruz			
	Water $\mu g\ L^{-1}$	Sediment ng g^{-1} dw	Organism[b] ng g^{-1} dw	Ref[c]	Water $\mu g\ L^{-1}$	Sediment ng g^{-1} dw	Organism ng g^{-1} dw	Ref
Aldrin			0.82	6		2.11	6.61	12
Dieldrin						0.45		11
						2.05	2.73	12
Endrin			1.54	6		7.82	7.95	12
DDT			9.70	6		0.87		11
						2.24	1.64	12
							11.00	6
DDE			71.34	6		1.78		12
							117.85	6
DDD			29.93	6		0.89		12
							40.49	6
HCB						1.86		12
Lindane			1.08	6			0.56	6

continued...

Table 11.6 continued

Pesticide	Tamaulipas				Veracruz			
	Water μg L⁻¹	Sediment ng g⁻¹ dw	Organism[b] ng g⁻¹ dw	Ref[c]	Water μg L⁻¹	Sediment ng g⁻¹ dw	Organism ng g⁻¹ dw	Ref
Chlordane			9.53	6			2.40	6
Nonachlor			2.72	6			0.87	6
Methoxychlor							0.74	6
α-Endosulfan						1.22	1.22	12
β-Endosulfan						0.67	17.65	12
Endosulfan sulphate						1.40		12
Heptachlor			0.75	6		3.91		12
						2.91		12
Heptachlor epoxide			3.41	6		0.86	2.17	12

Pesticide	Tabasco				Campeche			
	Water μg L⁻¹	Sediment ng g⁻¹ dw	Organism[b] ng g⁻¹ dw	Ref[c]	Water μg L⁻¹	Sediment ng g⁻¹ dw	Organism ng g⁻¹ dw	Ref
Aldrin		1.15	2.56	12		0.25	0.98	18
						9.02	6.25	13
Dieldrin		0.32		11		2.29	1.19	18
		6.84		12		0.34		11
Endrin		4.91	10.61	12	$\sum (2.34)$	$\sum (13.35)$		18
						56.70		13
DDT		2.28		11	$\sum (1.45)$	$\sum (1.49)$		18
		1.47	5.60	12		0.96		11
							$\sum (6.00)$	14
							2.71	6
DDE		0.26	4.17	12		17.67		13
							13.08	6
DDD		0.26	2.11	12		0.55	16.80	13
							8.53	6
HCB		0.62	1.94	12	$\sum (0.35)$	$\sum (1.97)$		18
						0.32		13
Lindane						0.67	0.08	18
						0.57	1.25	13
							0.29	6
Chlordane							1.98	6
							23.08	18
Nonachlor							0.50	6
α-Endosulfan		0.87	0.83	12			$\sum (11.1)$	18
β-Endosulfan		0.06	14.93	12				
Endosulfan sulphate		1.62		12				
Heptachlor		5.19	2.10	12	$\sum (0.26)$	$\sum (0.63)$		18
Methoxychlor						0.95	1.6	18
Heptachlor epoxide		0.27	3.24	12				
Dieldrin		0.6		11				
Drins[e]		0.25		19				

continued...

Table 11.6 continued

Pesticide	Quintana Roo				Nayarit			
	Water $\mu g\ L^{-1}$	Sediment ng g^{-1} dw	Organism[b] ng g^{-1} dw	Ref[c]	Water $\mu g\ L^{-1}$	Sediment ng g^{-1} dw	Organism ng g^{-1} dw	Ref
Pentachloro-anisole		0.16		19				
Pentachloro-benzene		0.44		19				
HCH(s)		1.46		19				
Chlordane(s)		0.87		19				
Mirex		0.17		19				
Endosulfan(s)		0.27		19				
DDT		0.87		11	4.81			17
		$\sum$ 1.43[d]		19				
		$\sum$ OCs (60.00)		16				

Notes:
a See Figure 11.4 for the location of the states.
b Organisms are primarily bivalves (oysters and mussels) and shrimp.
c Reference numbers correspond to the following: 1) Albert and Armienta, 1977; 2) Carvalho
   et al., 1996; 3) González-Farias *et al.*, 1997; 4) Readman *et al.*, 1992; 5) Michel and Gutiérrez-
   Galindo, 1989; 6) IMW, 1995; 7) Rosales *et al.*, 1985; 8) Rosales, 1979; 9) Gutiérrez-Galindo
   et al., 1992; 10) Rueda *et al.*, 1997; 11) Rosales and Alvarez-León, 1979; 12) Botello *et al.*,
   1994a; 13) Gold-Bouchot *et al.*, 1993; 14) Rosales *et al.*, 1979; 15) Gutiérrez-Galindo *et al.*,
   1983; 16) Botello *et al.*, 1994b; 17) de la Lanza, 1986; 18) Gold-Bouchot *et al.*, 1995; and
   1993) Noreña-Barroso *et al.*, 1998 (median values).
d Indicates total DDT isomers and metabolites.
e Drins equal the sum of aldrin, endrin, and dieldrin.

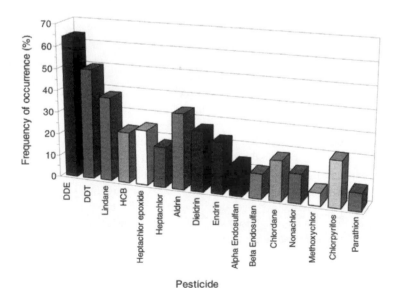

Figure 11.7 Frequency of occurrence of the most common pesticides per sample as
reported in studies published from 1977–97

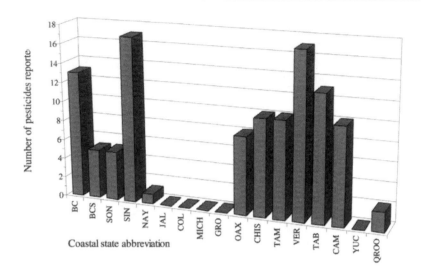

Figure 11.8 Number of different pesticide residues reported for each coastal state in Mexico

which two OPs (chlorpyrifos and parathion) are reported. The number of pesticides in sediment and organism samples from others states varies from two to seventeen different OC residues. It should be noted that for the western coast of Mexico, including Jalisco, Colima, Michoacan, and Guerrero, there are no reports of pesticide residues, even for the most important agricultural states in the region (Jalisco and Michoacan).

Significant concentrations of the various pesticides from water, sediment, or organism samples are summarized for different coastal locations (mainly coastal lagoons) in Table 11.6. Note, that in Sinaloa, analyses were performed on each type of sample (water, sediment, and organisms), but, in Sonora, Veracruz, Tabasco, and Campeche, no analyses were performed on water samples. Reports for these states cover only sediment and organism samples. In Baja California, Baja California Sur, Oaxaca, and Tamaulipas, only samples of organisms were studied and in Chiapas and Quintana Roo only sediments were analyzed. In Nayarit, one water sampling site has been analyzed and no reports exist for samples from Jalisco, Colima, Michoacan, Guerrero, and Yucatan. The highest concentrations of "drins" (aldrin, dieldrin, and endrin) were reported in sediments and organisms from Veracruz, Campeche, and Tabasco and in sediments from Sonora, Sinaloa, and Chiapas. The highest concentrations of DDT were found in water samples from Nayarit and from organism and sediment samples from Oaxaca and Chiapas. Lower, but significant concentrations were also found in organisms from Baja California, Baja California Sur, Tamaulipas, Veracruz, Tabasco, and Campeche. Total OC residues in sediments from Quintana Roo were high (Laguna Bojórquez and Cancun), probably from fumigation activities. In general, the concentration of DDT metabolites was higher than the parent compound in both sediment and organism samples (Table 11.6).

The highest concentration of lindane reported corresponded to sediments from Sonora. Chlordane and nonachlor were present at high concentrations in organism samples from Baja California and Tamaulipas. Methoxychlor was present in high concentrations only in organisms from Baja California. Endosulfan (α, β, and sulfate) was detected in organism and sediment samples from Veracruz and Tabasco, in organisms from Baja California, and in sediments from Chiapas and Sinaloa. Heptachlor and heptachlor epoxide were present mainly in organism samples from Baja California and Tabasco, and in sediments from Sinaloa, Chiapas, and Veracruz (Table 11.6). The values presented in Table 11.6 demonstrate that the main coastal areas impacted by pesticides are those of Baja California, Sonora-Sinaloa, Oaxaca-Chiapas, and Tamaulipas-Veracruz. Based on the data presented and in agreement with Botello et al. (1994b), the coastal area of Tabasco-Campeche should also be considered as pesticide impacted. Because of the large quantity of pesticides applied to recreational (golf courses) and landscaped areas and for mosquito control around tourist resort developments, future pesticide studies should examine samples from developments in coastal areas. At Laguna Bojórquez in Cancun (the most important tourist area of Quintana Roo), total OC residues in sediment samples of around 60 ng g^{-1} have been reported (Botello et al., 1994b).

PESTICIDE IMPACT IN COASTAL ZONES OF MEXICO

Agriculture is one of the most important of human activities through the production of food, fiber, and livestock. Undoubtedly the use of agrochemicals (fertilizers and pesticides) has dramatically increased crop yield while minimizing loss from weed, insect, and vertebrate pests; however, the misuse of agrochemicals has generated pollution problems on a global scale, namely OC pesticide effects (Simonich and Hites, 1995). Carvalho et al. (1997) suggested that the increased food demand by a constantly growing world population renders it unlikely that agrochemicals will be replaced by other control methods in the near future. Consequently the appropriate use of pesticides is necessary to avoid the impact of pollutants in coastal ecosystems, which also happen to be an important source of food.

Coastal areas, primarily coastal lagoons, are considered to be the most productive areas of the world's oceans. In fact, 90 percent of the world's fisheries are found in coastal zones and, from these, 70 percent of the catch is composed of organisms that live all or part of their lives in coastal lagoons (Flores-Verdugo, 1990; González-Farias et al., 1997). These areas are also very important from the ecological point of view, because they function as nursery grounds for a wide variety of organisms and provide food and shelter for diverse species including certain endangered species, e.g. crocodiles, marine turtles, osprey, jaguars, and manatees (Flores-Verdugo, 1990; González-Farias et al., 1997; Noreña-Barroso et al., 1998).

Mexico has around 150 coastal lagoons with a surface area of more than 1,162,000 ha (Table 11.4) and these lagoons provide economic contributions from tourism, fisheries, and aquaculture. Misuse of pesticides (mainly from agricultural uses) in adjacent hydrological basins has a negative impact on both biodiversity and on social and economic activities supported by coastal lagoons. With the data currently available, it is difficult to establish the degree of impact from pesticide runoff in the coastal zone of the different states. However, it is evident that the most affected areas are the coastal lagoons. A number of studies have demonstrated that the main source of pesticide residues in coastal lagoons is runoff from agricultural fields (Botello *et al.*, 1994a; Banderas, 1994; Benítez and Bárcenas, 1997; Carvalho *et al.*, 1996, Readman *et al.*, 1992). INEGI (1995) estimated approximately 265 m^3 s^{-1} of contaminated water from agricultural fields enter coastal lagoons or the sea.

Pesticide residues tend to increase over time due to the indiscriminate use of those chemicals. This, in turn, can promote the development of resistance by target species and cause an increase in the number of pesticide applications required per crop cycle. For example, the frequency of pesticide application to cotton has increased from 6 to 30 per crop cycle (INEGI, 1995), and, for tomato, chilli, and Cucurbitaceae crops, there are 27–30, 12–15, and 7–10 applications, respectively, in Sinaloa (Bernal Ruíz and Urías Morales, 1995). This increase in application frequency tends to raise the concentration of pesticides in agricultural drainage runoff. Other factors that enhance the levels of pesticides transported from agricultural drainage areas to coastal lagoons include the fact that in most parts of Mexico there is no specific and adequate area for disposal of pesticide rinse water and containers and that pilots involved in aerial crop-dusting and fumigation activities usually wash their tanks in the agricultural drainage ditches.

In Sinaloa, estimates of the quantity of pesticides required for the annual production of one hectare of horticultural or Graminaceae products, are 7–8 L of liquid and 23 kg of dry a.i. (Hernández Moreno and Valenzuela Rivera, 1995), which can be rounded to roughly 30 kg a.i. ha^{-1}. This estimate is quite high, as much as one order magnitude higher than that of Carvalho *et al.* (1996) at 3.3 kg a.i. ha^{-1} for the Culiacán River Valley in Sinaloa and 1.0 kg a.i. ha^{-1} for the Laguna de Términos system in Campeche (calculated from data of Benítez and Bárcenas, 1997). The magnitude of these differences indicates the need to establish, at state and national levels, data bases of agrochemicals both sold and applied in the different regions. The data collected should include specific information about both the formulation and the concentration of the a.i. used in order to obtain realistic figures for pesticide usage throughout the country.

Pesticide residues in the coastal zone also impact bird populations, aquaculture enterprises, and the health of inhabitants of the region. It is widely known that pesticides impact bird populations, but, unfortunately there are few studies examining their impacts in Mexico and no correlation between the collected data and pesticides used near the study site has been made (Albert, 1996). The effect on aquaculture has not been evaluated, but it is remarkable that those areas in which important agricultural centers are located are the places in which aquaculture is developing. This is the case for Sinaloa and Chiapas where extensive irrigated land exists (Figure 11.6) and where shrimp farms are being constructed. In 1995, there were 8,284 ha of shrimp ponds in Sinaloa, with production of more than 8,500 T of white shrimp *Penaeus vannamei* Boone (Decapoda: Penaeidae). In most shrimp farms, water is taken from coastal lagoons that receive agricultural drainage and, thus, there is the possibility of massive mortality from diseases promoted by the toxic effect of high pesticide residues. The impact of pesticide residues on the health of inhabitants of coastal regions ranges from pesticide bioconcentration by consumption of contaminated food to mortality (mainly field workers who apply pesticides). The Health Minister reported 1,576 deaths from pesticide intoxication for all of Mexico in 1993. Of these deaths, 998 occurred in coastal states, mainly Jalisco (112), Michoacan (115), and Nayarit (277) (INEGI, 1995). An increase in certain cancer rates is another potential impact of pesticides on public health. The highest children's leukemia incidence in the country occurs in the northern part of Sinaloa were huge quantities of pesticides are used (Berrelleza and Sosa Pérez, 1995; Antuna Medina, 1995).

Bioconcentration of pesticides in Mexico's human population is also a matter of serious concern. Even though there are few references on the subject (Waliszewski *et al.*, 1995; 2000; 2001; Elvia *et al.*, 2000), the study by Waliszewski *et al.* (1995) calls attention to this matter because of the high occurrence frequency and the concentration levels of OCs in human tissue samples. They collected samples of adipose tissue from 90 individuals and, of these, 100 percent were positive for p,p'-DDE and p,p'-DDT and 54 percent showed detectable levels of o,p'-DDT. The total DDT body burden reported for individuals from Mexico is 24.82 mg kg^{-1}, which is one order of magnitude higher than body burdens reported for other countries.

There have been several proposals for remediation of impacts caused by the excessive use of pesticides including: educational and training programs for farmers, authorities, and citizens of agricultural regions (Hernández Monge *et al.*, 1995); development of specific areas for disposal and treatment of pesticide containers (Miranda Castellanos, 1995); integrated pest management (rotation of crops, massive release of sterile pest organisms, use of natural predators, and pathogens, etc.) (Núñez Acosta, 1995); development of inland lagoons as contaminant traps to avoid the direct discharge of pesticides into coastal lagoons (González-Farias et al, 1997); and a general integrated coastal zone management plan (Carvalho *et al.*, 1996) that attempts to reconcile all human activities affecting the coastal zone with the environment.

ACKNOWLEDGMENT

The author expresses his gratitude to the International Atomic Energy Agency for support through contract number IAEA-79391R2.

REFERENCES

Albert, L.A. 1996. Persistent pesticides in Mexico. *Rev Environ Contam Toxicol.* 147:1–44.

Albert, L.A. and Aranda, E. 1986. La legislación mexicana sobre plaguicidas. Análisis y propuesta de modificaciones. *Folia Entomol Mex.* 68:75–87.

Albert, L.A. and Armienta, V.M. 1977. Contaminación por plaguicidas organoclorados en un sistema de drenaje agricola del estado de Sinaloa. *Protec Calidad del Agua.* 3:5–17.

Álvarez, A. and Gaitán, J. 1994. Lagunas costeras y el litoral mexicano: Geología. In: de la Lanza Espino, G. and Cáceres Martínez, C. (eds) *Lagunas Costeras y el Litoral Mexicano.* La Paz, BCS, México: Universidad Autónoma de Baja California Sur, pp. 13–71.

Antuna Medina, A. 1995. Impacto de los plaguicidas sobre salud en Sinaloa. In: Memorias simposium agroquímicos: Aplicación y efectos. Culiacán, Sin, México: Universidad Autónoma de Sinaloa, pp. 99–103.

Banderas, A.G. 1994. Impacto ambiental de los desarrollos hidroagrícolas sobre las lagunas costeras del noroeste de México. In: de la Lanza Espino, G. and Cáceres Martinez, C. (eds) *Lagunas Costeras y el Litoral Mexicano.* La Paz, BCS, México: Universidad Autonóma de Baja California Sur, pp. 471–95.

Benítez, J.A. and Bárcenas, C. 1997. Patrones de uso de los plaguicidas en la zona costera del Golfo de México. In: Botello, A.V., Rojas-Galaviz, J.L., Benítez, J.A. and Zárate-Lomelí, D. (eds) *Golfo de México, Contaminación e Impacto Ambiental: Diagnóstico y Tendencias.* EPOMEX Serie Científica 5. Campeche Cam, México: Universidad Autónoma de Campeche, pp. 155–67.

Bernal Ruíz, C.R. and Urías Morales, C. 1995. Sistemas de producción de hortalizas orgánicas en Sinaloa. In: *Memorias Simposium Agroquimícos: Aplicación y Efectos.* Culiacán, Sin, México: Universidad Autónoma de Sinaloa, pp. 39–50.

Berrelleza, T.J. and Sosa Perez, R. 1995. La bio-remediación, alternativa de mejoramiento en suelos contaminados con plaguicidas. In: *Memorias Simposium Agroquimícos: Aplicación y Efectos.* Culiacán, Sin, México: Universidad Autónoma de Sinaloa, pp. 126–30.

Botello, A.V., Diaz, G., Rueda, L. and Villanueva, S.F. 1994a. Organochlorine compounds in oysters and sediments from coastal lagoons of the Gulf of Mexico. *Bull Environ Contam Toxicol.* 53:238–45.

Botello, A.V., Ponce, G., Villanueva, S. and Rueda, L. 1994b. Contaminación. In: de la Lanza Espino, G. and Cáceres Martinez, C. (eds) *Lagunas Costeras y el Litoral Mexicano.* La Paz, BCS, México: Universidad Autonóma de Baja California Sur, pp. 445–70.

CAADES. 1987. *Sinaloa, Agricultura y Desarrollo.* Culiacán, Sin, México: Confederación de Asociaciones Agrícolas del Estado de Sinaloa.

Carvalho, F.P., Fowler, S.W., Gonzalez-Farias, F. and Mee, L.D. 1996. Agrochemical residues in the Altata-Ensenada del Pabellon coastal lagoon (Sinaloa, Mexico): a need for integrated coastal zone management. *Int J Environ Health Res.* 6:209–20.

Carvalho, F.P., Fowler, S.W., Villeneuve, J.P. and Horvat, M. 1997. Pesticide residues in the marine environment and analytical quality assurance of results. In: *Environmental Behavior of Crop Protection Chemicals.* Proc. Int. Symp. on Use of Nuclear and Related Techniques for Studying the Environmental Behavior of Crop Protection Chemicals. 1–5 July 1996. Vienna: IAEA/FAO, pp. 35–57.

CICOPLAFEST. 1988. *Catálogo Oficial de Plaguicidas.* México, Distrito Federal: SAGAR (Secretaría de Agricultura, GAnadería y Desarrollo Rural), SEMARNAP (SEcretaría de Medio Ambiente, Recursos NAturales y Pesca), SSA (Secretaría de Salubridad y Asistencia) y SECOFI (Secretaría de COmercio y Fomento Industrial).

CICOPLAFEST. 1996. *Catálogo Oficial de Plaguicidas*. México, Distrito Federal: SAGAR, SEMARNAP, SSA y SECOFI.

Contreras, F. 1988. *Las Lagunas Costeras Mexicanas*, 2nd edn. México, Distrito Federal: Centro de Ecodesarrollo.

de la Lanza, G. 1986. Calidad ambiental de la laguna de Mezcaltitán, Nayarit, México, durante el estiaje. *An Centro Cienc del Mar y Limnol, Univ Nal Autón México*. 13(2):315–28.

de Llasera, M.P.G. and Bernal-González, M. 2001. Presence of carbamate pesticides in environmental waters from the northwest of Mexico: determination by liquid chromatography. *Water Research*. 35(8):1933–40.

Diario Oficial de la Federación. 1971. Ley federal para prevenir y controlar la contaminación ambiental, 23 de marzo de 1971. México City, México.

Diario Oficial de la Federación. 1988. Ley general del equilibrio ecológico y la protección al ambiente, 28 de enero de 1988. México City, México.

Díaz Simental, V.M. 1995. Salud y medio ambiente. In: *Memorias Simposium Agroquimícos: Aplicación y Efectos*. Culiacán, Sin, México: Universidad Autónoma de Sinaloa, pp. 21–5.

Elvia, L.F., Sioban, H.D., Bernardo, H.P. and Constanza, S.C. 2000. Organochlorine pesticide exposure in rural and urban areas in Mexico. *J Expos Analy Environ Epidemiol*. 10(4):394–9.

Flores-Verdugo, F. 1990. Algunos aspectos sobre ecología, uso e importancia de los ecosistemas de manglar. In: de la Rosa-Vélez, J. and González-Farias, F. (eds) *Temas de oceanografía biológica en México*, Volume1. Ensenada, BC, México: Universidad Autónoma de Baja California, pp. 21–56.

Gold-Bouchot, G., Silva-Herrera, T. and Zapata-Pérez, O. 1993. Organochlorine pesticide residues in the Río Palizada, Campeche, Mexico. *Mar Poll Bull*. 26(11):648–50.

Gold-Bouchot, G., Silva-Herrera, T. and Zapata-Pérez, O. 1995. Organochlorine pesticides in biota and sediments from Río Palizada, Mexico. *Mar Poll Bull*. 54(11):554–61.

González-Farias, F., Carvalho, F.P., Fowler, S.W. and Mee, L.D. 1997. A tropical coastal lagoon affected by agricultural activities: the importance of radiolabeled pesticide studies. In: *Environmental Behavior of Crop Protection Chemicals*. Proc. Int. Symp. on Use of Nuclear and Related Techniques for Studying the Environmental Behavior of Crop Protection Chemicals. 1–5 July 1996. Vienna: IAEA/FAO, pp. 289–99.

Gutiérrez-Galindo, E., Sañudo-Wilhelmy, S.A. and Flores-Báez, B.P. 1983. Variación espacial y temporal de pesticidas organoclorados en el mejillón *Mytillus cailfornianus* (Conrad) de Baja California, Parte 1. *Cienc Mar (Méx)*. 9(1):7–18.

Gutiérrez-Galindo, E., Flores-Muñóz, G., Ortega, M.L. and Villaescusa-Celaya, Y.J.A. 1992. Pesticidas en las aguas costeras del Golfo de California: programa de vigilancia con mejillón 1987–1988. *Cien Mar (Méx.)*. 18(2):77–99.

Hernández Monge, J., Sauceda López, R. and Gómez Soto, M. 1995. Educación y capacitación en el uso y aplicación de agroquimícos. In: *Memorias Simposium Agroquimícos: Aplicación y Efectos*. Culiacán, Sin, México: Universidad Autónoma de Sinaloa, pp. 30–4.

Hernández Moreno, M.R. and Valenzuela Rivera, M.L. 1995. La leucemia y su relación con los plaguicidas en Sinaloa. In: *Memorias Simposium Agroquimícos: Aplicación y Efectos*. Culiacán, Sin, México: Universidad Autónoma de Sinaloa, pp. 134–5.

IMW. 1995. International Mussel Watch Project: Initial Implementation Phase, Final Report May 1995. Washington DC: NOAA Technical Memorandum National Ocean Service (NOS) Ocean Resources Conservation and Assessment (ORCA) Nr 95.

Instituto Nacional de Estadística, Geografía e Informática (INEGI). 1994a. *La Industria Química en México*, 1993 edn. Aguascalientes, Aguascalientes State, México: INEGI.

INEGI. 1994b. Estados Unidos Mexicanos. resultados definitivos: VII Censo agrícola-ganadero, Tomo I. Aguascalientes, Aguascalientes State, México: INEGI.

INEGI. 1995. Estadísticas del medio ambiente México, 1994. Aguascalientes, Aguascalientes State, México: INEGI.

López Pérez, J. 1995. Prácticas decadentes, actuales y emergentes en el uso de pesticidas en la zona centro-norte de Sinaloa. In: *Memorias Simposium Agroquimícos: Aplicación y Efectos*. Culiacán, Sin, México: Universidad Autónoma de Sinaloa, pp. 114–25.

Michel, M. and Gutiérrez-Galindo, E. 1989. Pesticides and PCBs in oysters from Mazatlan, Sinaloa, Mexico. *Mar Poll Bull.* 20(9):469–72.

Miranda Castellanos, R. 1995. Confinamiento y tratamiento de residuos, un servicio para la industria y la preservación de los ecosistemas. In: *Memorias Simposium Agroquimícos: Aplicación y Efectos*. Culiacán, Sin, México: Universidad Autónoma de Sinaloa, pp. 35–8.

Narro, J.G. 1979. El uso de plaguicidas en la agricultura Mexicana. In: Ondarza, R. (ed.) *Los Reguladores de las Plantas y los Insectos*. México, Distrito Federal: CONACYT, pp. 27–40.

Noreña-Barroso, E., Zapata-Peréz, O., Ceja-Moreno, V. and Gold-Bouchot, G. 1998. Hydrocarbon and organochlorine residue concentrations in sediments from Bay of Chetumal, Mexico. *Bull Environ Contam Toxicol.* 61(1):80–7.

Núñez Acosta, H.M. 1995. El manejo integrado de plagas, alternativa para disminuir el uso de agroquimícos en Sinaloa. In: *Memorias Simposium Agroquimícos: Aplicación y Efectos*. Culiacán, Sin, México: Universidad Autónoma de Sinaloa, pp. 75–8.

Núñez Cebreros, R.D. 1995. Alternativas para disminuir el uso de plaguicidas. In: *Memorias Simposium Agroquimícos: Aplicación y Efectos*. Culiacán, Sin, México: Universidad Autónoma de Sinaloa, pp. 148–50.

Readman, J.W., Liong Wee Kwong, L., Mee, L.D., Bartocci, J., Nilve, G., Rodríguez-Solano, J.A. and González-Farias, F. 1992. Persistent organophosphorus pesticides in tropical marine environments. *Mar Poll Bull.* 24(8):398–402.

Restrepo, I. 1988. *Naturaleza Muerta: Los Plaguicidas en México*. México, Distrito Federal: Ediciones Océano.

Reyes, G.G., Villagrana, C.L. and Alvarez, G.L. 1999. Environmental conditions and pesticide pollution of two coastal ecosystems in the Gulf of California, Mexico. *Ecotoxicology and Environmental Safety* 44(3):280–6.

Rosales, M.T.L. 1979. Sobre la dispersión de compuestos organoclorados en el medio ambiente marino: Nota científica. *An Centro Cienc del Mar y Limnol, Univ Nal Autón México.* 6(1):33–6.

Rosales, M.T.L. and Álvarez Leon, R. 1979. Niveles actuales de hidrocarburos organoclorados en sedimentos de lagunas costeras del Golfo de México. *An Centro Cienc del Mar y Limnol, Univ Nal Autón México* 6(2):1–6.

Rosales, M.T.L., Botello, A.V. and Mandelli, E.F. 1979. PCBs and organochlorine insecticides in oysters from coastal lagoons of the Gulf of Mexico, Mexico. *Bull Environ Contam Toxicol.* 21:652–6.

Rosales, M.T.L., Escalona, R.L., Alarcón, R.M. and Zamora, V. 1985. Organochlorine hydrocarbon residues in sediments of two different lagoons of northwest Mexico. *Bull Environ Contam Toxicol.* 35:322–30.

Rueda, L., Botello, A.V. and Díaz, G. 1997. Presencia de plaguicidas organoclorados en dos sistemas lagunares del estado de Chiapas, México. *Rev Int Contam Ambient* 13(2):55–61.

Saavedra-Vázquez, T.E. 1996. Normatividad en zonas costeras. In: Botello, A.V., Rojas-Galaviz, J.L., Benítez, J.A. and Zárate-Lomelí, D. (eds) *Golfo de México, Contaminación e Impacto Ambiental: Diagnóstico y Tendencias*. EPOMEX Serie Científica Nr 5. Campeche Cam, México: Universidad Autónoma de Campeche, pp. 605–40.

Simonich, S. and Hites, R.A. 1995. Global distribution of persistent organochlorine compounds. *Science*. 269:1851–4.

Waliszewski, S.M., Pardio Sedas, V.T., Infanzon, R.M. and Rivera, J. 1995. Determination of organochlorine pesticide residues in human adipose tissue: 1992 study in Mexico. *Bull Environ Contam Toxicol*. 55:43–9.

Waliszewski, S.M., Aguirre, A.A., Infanzon, R.M., Lopez-Carrillo, L. and Torres-Sanchez, L. 2000. Comparison of organochlorine pesticide levels in adipose tissue and blood serum from mothers living in Veracruz, Mexico. *Bull Environ Contam Toxicol*. 64(1):8–15.

Waliszewski, S.M., Aguirre, A.A., Infanzon, R.M., Silva, C.S. and Siliceo, J. 2001. Organochlorine pesticide levels in maternal adipose tissue, maternal blood serum, umbilical blood serum, and milk from inhabitants of Veracruz, Mexico. *Arch Environ Contam Toxicol*. 40(3):432–8.

Chapter 12

The use of pesticides in Costa Rica and their impact on coastal ecosystems

Elba M. de la Cruz and Luisa E. Castillo

INTRODUCTION

Costa Rica is the second smallest country in Central America, with an area of 51,100 km², and extends from approximately Lat. 8° to 11°N and between Long. 83° and 86°W. Costa Rica is bordered by Nicaragua on the north, Panama on the south, the Caribbean Sea on the east, and the Pacific Ocean on the west. It has an elongated form that stretches from northeast to southeast with a greatest length of 480 km on the northwest–southeast axis and a narrowest width between the Caribbean and the Pacific of only 118 km (Figure 12.1).

The highest regions of Costa Rica are in the center of the country; its lowlands are more extensive and flat on the Caribbean side and to the north than on the Pacific side. Costa Rican geology dates from 150 M years ago; the consolidation of its mountainous backbone was associated with a long history of volcanic activity. Sixty-eight volcanoes have been identified of which nine are considered to be active. Its mountainous backbone can be divided into two units, separated in the middle of the country by two valleys, those of the Rio Grande de Tárcoles and the Reventazón (Castillo-Muños, 1983; Trejos, 1991).

Costa Rica has a patrimonial sea area of 520,000 km², about 10 times its national territory, although the characteristics of the Pacific and the Caribbean coastal regions are quite different. The Caribbean Coast is straight and short, 212 km, while the Pacific Coast is very irregular with many peninsulas, capes, points, islands, and gulfs over a length of 1,328 km (Quesada, 1990).

Costa Rica's tropical location between two oceans with its complex mountain systems causes a great variety of climatic conditions. There are two defined rainfall regimes: one for the Caribbean side and another for the Pacific side. On the Caribbean side, including both the northern lowlands and the Caribbean coastal regions, there is not a defined dry season. In the coastal zone, there are relatively dry periods, one in March and April and another in September and October. On the Pacific side there are two distinct seasons: one rainy and one dry. The rainy season extends from May to the middle of December and the dry season runs from January to April.

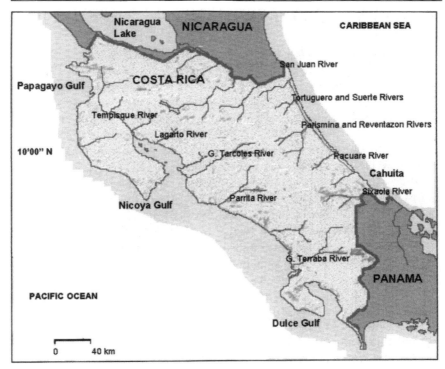

Figure 12.1 Costa Rica's location and primary river systems

A mountainous backbone, coupled with abundant rainfall, results in Costa Rica having an extensive hydrographic system. The Costa Rican Electricity Institute has divided this hydrographic system into 34 watersheds. The system comprises two versants, one toward the Pacific side and another toward the Caribbean side (Trejos, 1991). The later is usually subdivided into two parts. A northern subversant carries water toward Lake Nicaragua and the San Juan River and through it to the Caribbean. The Caribbean or Atlantic sub-versant, carries water directly to the Caribbean. Due to the narrowness of its territory, Costa Rica has relatively small watersheds (Trejos, 1991). The largest watersheds – those more than 2,000 km^2 – include Grande de Térraba (5,000 km^2); Tempisque (3,400 km^2); Reventazón-Parismina (3,000 km^2); Sixaola (2,700 km^2); San Carlos (2,650 km^2); Grande de Tárcoles (2,150 km^2); and the Sarapiquí (2,150 km^2).

Costa Rica is a democratic republic; its Constitution established that the government of the republic is to be exercised through three distinct and independent powers: the Legislative, Executive, and Judicial powers. Costa Rica's territory is divided into seven provinces including San José (the capital), Alajuela, Cartago, Heredia, Guanacaste, Puntarenas, and Limón. These provinces are subdivided into cantons and in each canton there is a municipality whose leaders are elected by the people to administer the community's interests (Trejos, 1991).

PESTICIDE REGULATIONS

Control of pesticide labeling and handling in Costa Rica is primarily the responsibility of the Ministries of Agriculture and Health. Pesticide use was first regulated
by a 1954 law requiring information on the physical properties, recommended
uses, and health risks of all locally produced and imported pesticides. Then in
1976, the Regulation on Pesticide Control was passed to implement procedures
for the registration and control of all pesticides with both agricultural and domestic
uses entering the country. Evaluation of a pesticide's toxicity is the duty of the
Ministry of Health. This ministry had, and still has, the authority to ban or restrict
registered uses of a compound if they consider it dangerous to human or animal
health (Castro, 1998). There are two primary laws and many related laws and
regulations controlling the major aspects of pesticide use in Costa Rica. The
Phytosanitation Protection Law of 1968 (revised in 1978 and 1997) is administered
by the Ministry of Agriculture and the General Law of Health 1973 (revised in
1975, 1980, 1982, and 1988) is administered by the Ministry of Health.

The Ministry of Agriculture has complete authority to regulate the use of all
agricultural crop protection chemicals including their environmental, wildlife, and
human health effects. They also have the right to determine when chemical control
must be replaced by biological control to reduce environmental pollution. Under
the authority of the General Law of Health, the Ministry of Health promulgates
rules for importing, handling, storing, transporting, marketing, distributing, and
applying pesticides. All pest control products not under the Phytosanitation Law
and capable of poisoning or causing serious damage to the health of humans or
non-target organisms must be registered and receive a permit from the Ministry
of Health before being used. This law allows health authorities to institute certain
preventive measures, e.g. to retain or remove products from the market, to destroy
or neutralize contaminated materials, and to confiscate damaged or suspicious
products. They also have the authority to close pesticide storehouses, formulating
plants and retail shops, and cancel pesticide permits or registrations (Castro, 1998).

In 1989, the National Pesticide Use Advisory Commission was reorganized –
the previous commission was created in 1972 – and tasked to evaluate the toxicology
of pesticides, to recommend banning dangerous substances, to re-examine approved
pesticide registrations, and to make suggestions or observations to the Ministries
of Agriculture and Health (Hilje *et al.*, 1992; Castro, 1998). It may re-evaluate a
registration acting upon a request from one of the Ministries (Castro, 1998). The
1989 Commission is composed of nine members with two representatives from
the agrochemical industry, two from the Ministry of Agriculture, and one each
from the Ministry of Health, the National Agronomist Association, the Ministry
of Work and Social Security, the Ministry of Environment and Energy, and the
National Center for Poison Control (Castro, 1998). The Commission is responsible
for coordinating its activities and developing consensus recommendations from
stakeholders among different pesticide-related interest groups. The law 'Registration, Use and Control of Agriculturally Used Pesticides and Related Products

(1995)' regulates a pesticide's commercial life including registration, labeling, unloading, manufacturing, formulation, packaging, commerce, storage, transport, use and management, destruction of empty pesticide packages, pesticide residues, unused pesticides, and spill cleanup.

In principle, all occurrences of acute, subacute, and chronic effects from either a voluntary or accidental pesticide poisoning must be reported to the Ministry of Health – charged with keeping a registry of these cases. Every person who handles or applies pesticides on a regular basis must have a pre-exposure medical checkup followed by an annual medical checkup. In special cases, medical checkups may be more frequent. The law does not allow persons less than 18 years old to work with or apply pesticides. Persons applying pesticides by aerial or ground application methods must inform the Ministry of Agriculture of the date, time, location, pesticide, and method of application at least 72 h in advance. The local Ministry official then notifies apiculturists 48 h in advance to protect their bees and beehives. Signs must be posted to warn people to keep themselves and their animals out of the application area. Controls and procedures following accidents are not well established in the law. Responsibility can rest with the landowner (for failure to inform the Ministry of Agriculture), the owner of the aerial (crop dusting) application company (for a malfunctioning airplane), the pilot (for spraying the wrong field), the professional agriculturist (for recommending the wrong products or application rates), or even the injured party (for failure to protect himself or his animals).

Water resources are protected under two laws, one promulgated in 1989 to protect important aquifers and the other in 1997 to regulate industrial wastewater and effluents. The initial use of water quality criteria in Costa Rica was to legislate pesticide residue concentrations in water bodies. The maximum allowable pesticide concentrations in waste or natural waters are 0.05 mg L^{-1} for $\sum$-OC pesticides and 0.1 mg L^{-1} for both $\sum$-OP and $\sum$-carbamate pesticides (Castro, 1998).

Agricultural and other pest control practices in Costa Rica are highly dependent on synthetic pesticide use. However, no official policy exists to reduce the quantity of pesticides used or to change from the more toxic and dangerous products to less toxic ones. A list of the a.i.(s) regulated or prohibited in Costa Rica is presented in Table 12.1. Pesticides classified as highly toxic are restricted for use and can only be sold with a professional prescription.

Even though there are laws and regulations governing pesticide use in Costa Rica, these are often transgressed, causing health and environmental problems. These problems include exposure and poisoning of workers and the general population (with some individuals being less than 18 years old) and exposure of aquatic organisms, domestic animals, and wildlife with fatal consequences occasionally resulting from the exposures. Improved laws, especially in environmental quality criteria, and improved implementation and enforcement is a must.

Table 12.1 Pesticides with prohibited or restricted use in Costa Rica[a]

Year	a.i.(s)	Legal status
1960	cianogas	prohibited
1960	mercurial	prohibited
1982	arsenic compounds[b]	prohibited except they could still be used to combat fungal diseases in coffee
1987	carbofuran, ethyl parathion, methyl parathion, phosphine, phorate, monocrotophos	restricted use; sold only with authorization and red strip label
1987	2,4,5-T	prohibited
1988	aldrin, DDT[c], dieldrin, toxaphene, chlordecone, chlordimeform, dibromochloropropane, ethylene dibromide, dinoseb, nitrofen	prohibited
1989	captafol	prohibited
1990	lead arsena(i)te[b], endrin, penta-chlorophenol, cihexatin	prohibited
1991	chlordane, heptachlor	all uses prohibited in 1998
1992	daminozide	restricted use; only for ornamental plants
1995	methyl bromide	restricted use
1996	captan	restricted use
1996	lindane and its isomers, ethephon[d]	prohibited
1998	declorane (mirex) and arsenic compounds[b]	prohibited

Notes:
a Source: adapted from Castro, 1998; UNA/IRET, 1999; Phytosanitary Department of the Ministry of Agriculture, Costa Rica.
b Lead arsen(i)ate was prohibited in 1990. However, in 1991, its use was again permitted. It was in 1998 that importation and use of all arsenic compounds was prohibited in Costa Rica.
c Law still allows the Ministry of Health to use DDT in exceptional situations for combating malaria-carrying mosquitoes when there is no alternative.
d Ethephon is banned only for coffee bean ripening.

PAST AND PRESENT PESTICIDE USE IN COSTA RICA

Pesticide imports

Environmental problems created by the misuse of pesticides in developing countries are extensive. Costa Rica, having an agriculture-based economy, has also been influenced by the production of these chemicals. Information regarding methods of pest control during pre colonial, colonial, and republican times is almost non-existent (Hilje *et al.*, 1989). The first chemical used to control pests in Costa Rica was called 'Tree Tanglefoot' – a product made from natural tree resins which have been polymerized with castor oil and further waterproofed with vegetable waxes; its mode of action is mechanical – and was introduced by 1916. In 1926, copper sulfate was used extensively to combat diseases in crops. During this time the convenience of using chemical substances to control any pest damaging a crop was already a common thought among farmers (Hilje *et al.*, 1989). Since then, pesticide importation and use in Costa Rica has slowly increased, reaching its highest quantities after 1992. Figure 12.2 shows the quantity of formulated pesticides imported from 1970 to 1996. Even though the available pesticide import data has always been deficient, in the 1980s and 1990s the nature of the data permitted separation of disinfectants, organic solvents, and other similar products and the calculation of kg of a.i. being imported. The quantity of formulated pesticides imported during 1970 amounted to 5.6 M kg and in 1996 surpassed 14 M kg (Figure 12.2). The imported quantity of pesticides had risen to 18 M kg in 1997 (de la Cruz *et al.*, 1998). There was a modest reduction in the quantity of pesticides imported during the late 1980s. The reasons for this may be attributed to improvements in the data registers, which permitted better separation of other products from pesticides, and decreases in the prices of Costa Rican export cash crops. The increase in the quantity of pesticides imported during the 1990s must be attributed to the expansion of cultivated area for highly pesticide dependent crops, e.g. banana, rice, melon, watermelon, pineapple, and ornamental plants, during the same period. In Costa Rica, the most important pest control method for insects, other invertebrates, vertebrates, weeds, and diseases is pesticides (Hilje *et al.*, 1992). Hilje (1984) has calculated that in 95 percent of the cases where an insect was the organism causing damage on a plantation, it was controlled with pesticides.

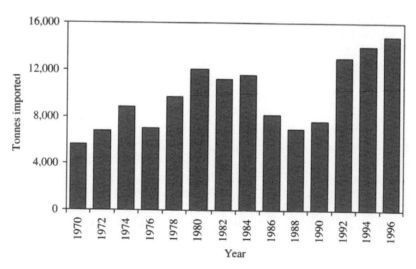

Figure 12.2 Tonnes of formulated pesticides imported by Costa Rica from 1970–96 (adapted from Castillo *et al.*, 1989; Hilje *et al.*, 1992; IRET database)

The type of pesticides used in Costa Rica has also changed with time. Nevertheless, about 13 products have dominated the import list for the last 12 years (1985 to 1997), among them mancozeb, 2,4-D, glyphosate, chlorothalonil, ethoprophos, paraquat, terbufos, cadusafos, methyl bromide, carbofuran, propanil, tridemorph, and fenamiphos (Table 12.2). During the 1990s, the nematicide cadusafos became a dominant imported pesticide. It is primarily used on banana plantations and was heavily imported at the beginning of the 1990s, but by 1995 the quantity of imports was reduced due to its high price. Then, traditional products such as terbufos, carbofuran, and ethoprophos again increased (Chaverri and Blanco 1995; IRET database).

Inorganic pesticides, e.g. copper and sulfur related compounds, constituted the bulk of pesticides used before the 1950s. OC insecticides predominated during the 1960s and 1970s. From 1977 to 1979, about 815 T of OCs including DDT, dieldrin, heptachlor, chlordane, aldrin, endrin, toxaphene, endosulfan, and lindane were imported. By the 1987 to 1989 period, the quantity of OCs had declined to 242 T and included declorane (mirex), lindane, chlordane, heptachlor, pentachlorophenol, and endosulfan. From 1995 to 1997, only about 183 T of the OCs endosulfan and chloroneb were imported (Vega *et al.*, 1983; Hidalgo, 1986; Castillo, *et al.*, 1989; Hilje *et al.*, 1989; Hilje *et al.*, 1992; Instituto Regional de Estudios en Sustancias Tóxicas (IRET) database). The dramatic reduction in the importation of OC pesticides can be attributed to prohibition and other restrictive regulations imposed on their use by Costa Rica. Other organic pesticides, e.g. dithiocarbamates, carbamates, phenoxyacetic acid, OPs, benzonitriles, morpholines, bipyridils, anilides, triazines, and pyrethroids, replaced the OCs during the 1970s to the 1990s. From 1970 to 1979, no detailed import records of a.i.(s) and quantities exist, but products including mancozeb, methyl bromide, aldicarb, 2,4-D, glyphosate, chlorothalonil, tridemorph, terbufos, paraquat, propanil, ethoprophos, cupric compounds, diuron, methamidophos, carbofuran, carbendazim, thiabendazole, and terbuthylazine were being imported (Vega *et al.*, 1983). DBCP, a well-known nematicide with negative effects on male reproduction, was also imported during this time.

Currently Costa Rica imports approximately 280 pesticides that are sold under more than 2,000 different brand names (IRET database). From 1992 to 1997, Costa Rica imported about 40.1 M kg of pesticide a.i.(s) (Figure 12.3 A) for approximately US$530 M (Figure 12.3 B) and the yearly quantity of pesticides imported continued to increase during the period. Total pesticide a.i. imports in 1997 (8,972 T) were 59 percent higher than in 1992 (5,656 T) (Figure 12.3 A) and an increase of 39 percent for formulated products occurred over the same period. The cost of these pesticides escalated from US$74.6 M to US$117 M for the same period (Figure 12.3 B). Of the a.i.(s) imported by Costa Rica between 1995 and 1997, 17 constituted 80 percent of the total quantity of pesticides imported (Table 12.3).

A comparison of the biocide groups imported during the periods 1977 to 1979, 1985 to 1987, and 1995 to 1997 is shown in Figure 12.4. During the 1970s,

Table 12.2 The 25 most imported pesticides in Costa from 1985–97[a]

Pesticide	Tons[b]	% of total	∑%	Pesticide	Tons[b]	% of total	∑%
mancozeb	27,265.8	20.3	20.3	Aldicarb	2,767.6	2.1	68.5
2,4-D	8,509.9	6.3	26.6	Carbaryl	2,562.3	1.9	70.4
Glyphosate	6,472.3	4.8	31.4	Foxim (phoxim)	2,456.7	1.8	72.2
Chlorothalonil	5,925.5	4.4	35.8	Diuron	1,710.0	1.2	73.5
Ethoprophos	5,847.2	4.3	40.2	2,4-D + Picloram	1,355.7	1.0	74.5
Paraquat	5,530.0	4.1	44.3	Maneb	1,310.8	1.0	75.5
Terbufos	5,136.7	3.8	48.1	Oxamyl	1,205.8	0.9	76.4
Cadusafos	4,736.7	3.5	51.6	Terbutryn	1,168.5	0.9	77.2
Methyl bromide	4,698.1	3.5	55.1	Methamidophos	1,149.4	0.8	78.1
Carbofuran	4,515.7	3.4	58.4	Pendimethalin	1,142.0	0.8	78.9
Propanil	4,319.6	3.2	61.7	Terbuthylazine	1,136.2	0.8	79.8
Tridemorph	3,380.3	2.5	64.2	Propiconazole	1,091.5	0.8	80.6
Fenamiphos	3,040.6	2.3	66.4				
				Total imports	134,545.2	100.0	

Notes:

a Source: adapted from IRET database.

b Includes only formulated products; except for 2,4-D + picloram, the a.i.(s) in mixtures are not included.

herbicides (35.3 percent), insecticides plus nematicides (30.1 percent), and fungicides (22.8 percent) were the most imported biocide groups. The later two periods differ from the 1977 to 1979 period, but were similar to each other. In them, the most imported biocide groups were the fungicides (45.9 percent and 47.1 percent, respectively), followed by the herbicides (28.0 percent and 26.5 percent, respectively), and the insecticides plus nematicides (23.1 percent and 16.3 percent, respectively). In the 1995 to 1997 period, fumigants, e.g. methyl bromide, constituted 10 percent of total pesticide imports. Generally, fungicides, herbicides, and

A. Quantity of pesticides imported

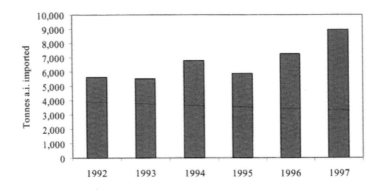

B. Cost of pesticides

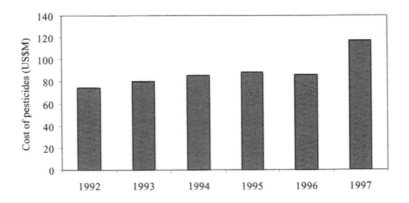

Figure 12.3 Tonnes of pesticide a.i.(s) imported by Costa Rica from 1992–97 and the cost of pesticide a.i.(s) imported from 1992–97 (adapted from Castillo, 1997; de la Cruz *et al.*, 1998; IRET, database)

Table 12.3 Major pesticide a.i. imported from 1995–97 in Costa Rica[a]

a.i.	Chemical family	Biocide action[b]	Imported (T)	Cumulative (T)	Cumulative %
Mancozeb	dithiocarbamate	fungicide	5,702.9	5,702.9	25.8
Methyl bromide	halogenated aliphatic	fumigant	2,168.3	7,871.2	35.6
2,4-D	phenoxyacetic acid	herbicide	1,973.4	9,844.6	44.5
Glyphosate	organophosphorus	herbicide	1,414.7	11,259.3	50.9
Chlorothalonil	chlorobenzonitrile	fungicide	1,223.2	12,482.5	56.4
Tridemorph	morfoline	fungicide	1,082.1	13,564.5	61.3
Terbufos	organothiophosphate	ins/nem	965.0	14,529.6	65.6
Paraquat	quaternary ammonium	herbicide	483.5	15,013.1	67.8
Propanil	anilide	herbicide	473.9	15,487.0	70.0
Ethoprophos	organothiophosphate	ins/nem	339.7	15,826.7	71.5
Cupric	inorganic	ins/acar	333.5	16,160.2	73.0
Methamidophos	phosphoramidothioate	ins/nem	315.1	16,475.3	74.4
Cadusafos	organothiophosphate	ins/nem	271.2	16,746.5	75.6
Diuron	phenylurea	herbicide	262.0	17,008.5	76.8
Propineb	dithiocarbamate	fungicide	261.1	17,269.6	78.0
Carbofuran	benzofuranyl methylcarbamate	ins/nem/acar	249.0	17,518.6	79.1
Terbuthylazine	chlorotriazine	herbicide	204.4	17,723.0	80.0
Total 90% a.i.				19,991.8	
Total 100% a.i.				22,134.9	

Notes:

a Source: adapted from IRET database and de la Cruz et al., 1998.

b Biocide abbreviations indicate: ins. for insecticides; acar. for acaricides; and nem. for nematicides.

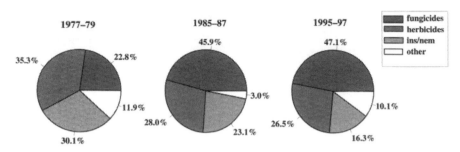

Figure 12.4 Biocides by class imported by Costa Rica during: 1977–79, 1985–87, and 1995–97 (adapted from Castillo *et al.*, 1989, Hilje *et al.*, 1992; de la Cruz et *al.*, 1998; IRET, database)

insecticides plus nematicides comprised about 91 percent of the pesticide a.i.(s) imported during these three periods.

Pesticide use

Pesticides are primarily employed in Costa Rica in agricultural activities and to a lesser degree in other activities such as forestry, husbandry, and public health. For many crops, pesticides are the primary method of pest control practiced. The number of kg of pesticide a.i.(s) imported per ha of cultivated area increased from 12.8 kg a.i. ha^{-1} in 1992 to 20.5 kg a.i. ha^{-1} in 1997 (Chaverri and Blanco, 1995; de la Cruz, 1998) and is similar to the value reported for the Netherlands in 1991 (20 kg a.i. ha^{-1}) (Teunissen-Ordelman and Scrap, 1997) but higher than that of Japan (10 kg a.i. ha^{-1}) (WHO, 1990). The quantity of formulated products increased from 29.5 kg ha^{-1} in 1992 to 41.3 kg ha^{-1} in 1997 (Table 12.4). Wesseling and Castillo (1989) and Chaverri and Blanco (1995) calculated that the total quantity of formulated pesticide products used per person in Costa Rica was equal to 4 kg, but by 1997, this figure had increased to almost 5.3 kg (approximately 2.6 kg a.i. per person) (Table 12.4).

By 1950, banana (about 30,000 ha), coffee (49,000 ha), and sugarcane (22,700 ha) were being planted as monocultures in Costa Rica and some insects, nematodes, diseases, and vertebrates had been designated as pests (Costa Rica, 1953; Araya, 1982). The intensive use of pesticides probably started at this time (Hilje *et al.*, 1989). By 1950 the chemical industry was already established in Costa Rica and was involved in pesticide production, formulation, and marketing. Before 1950, there were six established commercial pesticide companies and, between 1950 and 1960, 19 new companies were established. By 1983, there were 160 commercial pesticide companies in Costa Rica, not including small retailers (Hilje *et al.*, 1992). The introduction of some exotic pests and the proliferation of other native species as pests also occurred after 1950 and led to the increased use of synthetic pesticides (Hilje *et al.*, 1989; 1992).

Table 12.4 Total cultivated area (ha), quantity of a.i. and formulated pesticides utilized per ha (in kg), and quantity of a.i. (in kg) per person imported by Costa Rica from 1992–97

Year	1992	1993	1994	1995	1996	1997
hectares of cultivated land (M)	441.8	445.7	441.6	446.9	418.8	438.1
kg of a.i. per hectare	12.8	12.4	15.4	13.2	17.3	20.5
kg formulated pesticide per ha	29.5	28.1	31.7	30.9	34.6	41.3
kg formulated pesticide per person	4.2	4.0	4.3	4.2	4.3	5.3
kg a.i. per person	1.8	1.7	2.1	1.8	2.2	2.6

Source: Adapted from SEPSA, 1998; Chaverri and Blanco, 1995; de la Cruz *et al.*, 1998; Costa Rica, 1999; IRET database.

In the mid-1990s, the major crops grown in Costa Rica, based on both area cultivated and total production, included banana (52,165 ha), coffee (108,000 ha), rice (44,112 ha), vegetables (14,134 ha), fruits (51,043 ha), ornamentals (4,600 ha), root crops (13,253 ha), sugarcane (43,000 ha), other grains (74,556 ha), and pasturage (1,565,076.3 ha) (SEPSA, 1995; MAG, 1997). The geographic distribution of these crops is shown in Figure 12.5. Crops such as banana, rice, sugarcane, other grains, and some fruits are grown near coastal areas and consequently their agricultural practices may influence coastal ecosystems. Banana production requires more pesticide use per hectare (about 45 kg a.i. ha^{-1}) than any other Costa Rican crop, followed by fruits and vegetables (20 kg of a.i. ha^{-1}), and rice (10 kg a.i. ha^{-1}) while pasturage requires the lowest input of pesticides (0.25 kg a.i. ha^{-1}) (Table 12.5). Coffee production, which utilizes the greatest amount of land area, requires 6.5 kg a.i. ha^{-1} (Castillo *et al.*, 1997). From 1995 to 1997, almost 24 percent of total a.i. imports was directly related to banana production (Table 12.6).

Table 12.7 summarizes the various production stages; biocide functions, e.g. weed control, seed treatment, and insect control; formulation types; expected environmental transport mechanisms; type of pollutants; and potential ecosystems exposed for banana, root crops, date and oil palm, coffee, rice, sugarcane, and ornamental plant plantations (data compiled from field observations; Cortés, 1994; Subirós, 1995). Pesticides used in these agricultural activities influence coastal areas primarily through the extent of area cultivated for crops located near a coastal zone; number of pesticide application per crop; proximity of aquatic ecosystems to fields where pesticides are applied; direct application to aquatic ecosystems, either incidentally or from other practices; toxicity of the products used; intensity of precipitation periods characteristic of some Costa Rican regions, e.g. the Atlantic coastal zone receives >4,000 mm of rainfall annually, and the rainfall's temporal proximity to the pesticide application; and pesticide application technique. Aerial applications have a greater potential for wind drift – compared to manual applications – and can result in pollutants being carried away from target areas.

Table 12.5 Number of hectares cultivated for major crops harvested in Costa Rica, per ha pesticide a.i. use for each crop, and total quantity of a.i.(s) used per crop

	Crop						
	Banana	Vegetable/fruits	Rice	Other grains	Coffee	Sugar cane	Pastures
ha x 1,000	52.1	65.2	44.1	74.6	108.0	43.0	1,565.1
kg a.i. ha^{-1}	45	20	10	7.5	6.5	3.5	0.25
T a.i.	2,347	1,303	441	559	702	150	391

Source: adapted from SEPSA, 1995; Castillo et al., 1997; MAG, 1997.

Table 12.6 Quantity of pesticide a.i. (T) imported during 1995–97 by Costa Rica (CR), the Punta Morales formulating plant (PM), and by companies directly related to banana cultivation (BC)[a]

Year	Costa Rica	% by CR[b]	Punta Morales formulating plant	% by PM	Banana companies	% by BC
1995	5,899.1	25.8	316.1	5.3	1,423.8	24.1
1996	7,263.6	35.0	639.2	8.8	1,747.1	24.0
1997	8,971.9	39.2	1,309.4	14.6	2,060.2	23.0
Total	22,135.5	100.0	2,264.6	10.2	5,231.1	23.6

Notes:
a Source: adapted from IRET database; de la Cruz et al., 1998.
b Indicates the percent of total pesticide imports by Costa Rica for the years 1995–97.

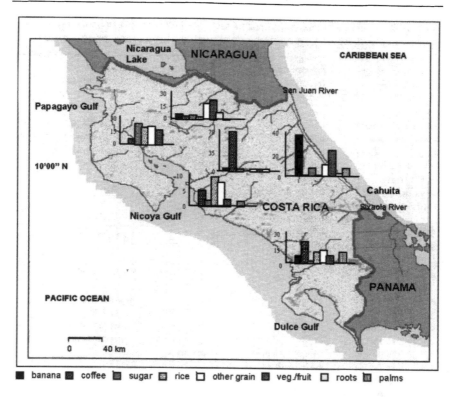

banana ■ coffee ▨ sugar ▤ rice ☐ other grain ▨ veg./fruit ☐ roots ▨ palms

Figure 12.5 Extension (ha × 1,000) and regional distribution of major crop production in Costa Rica

Additionally pesticides applied directly to plants and soils are subject to runoff and erosion forces, which may carry these pollutants through overflow or plantation drainage systems, into river systems, and eventually into estuarine and marine ecosystems.

Information on pesticide use in other business activities, e.g. aquaculture and salt production, is sparse. However, toxic compounds are used in these types of activities to control algae populations and to prevent boring organisms from damaging walls of shrimp ponds or water retention ponds (used for salt production).

Other uses of pesticides and pollution sources

About 10 percent of the a.i.(s) entering Costa Rica during the 1990s were imported by a single pesticide formulating plant located in the vicinity of the Nicoya Gulf on the Pacific Coast (Figure 12.1). Their imports in 1995 made up 5.3 percent of total imports, equivalent to 316.1 T a.i. and this increased to 14.6 percent of total pesticide imports in 1997, corresponding to 1,309.4 T (Table 12.6).

Table 12.7 Adverse environmental impacts associated with crop production cycles, emissions, and ecosystem exposure associated with the crop production life cycle[a]

Production stage/ Biocide action	Environmental transport	Pollution source	Exposed ecosystems
BANANA			
On the plantation control of weeds fungicides (aerial) nematicides herbicides	aerial drift erosion runoff leaching	non-point	takes place in lowlands at the lower end of the watershed near coastal areas
insecticides (impregnated bags)	evaporation from and contact with plastic bags	non-point	
Packing facilities water from washing process post-harvest application	effluents	point	
Pesticide storage facilities cellar, mixture preparations	_[b]	point	fresh waters coastal waters terrestrial and ground waters
Aerial application facilities mixture preparation, tank cleaning, storage, wastes	runoff, erosion aerial drift	non-point	
ROOT CROPS			
Seed preparation (nematicides, fungicides)	aerial drift	point	some root crops are planted near coastal areas
On the plantation weed control: pre-emergent herbicides insecticides	aerial drift runoff, erosion	non-point	
Packing facilities water from washing process post-harvest fungicide application	effluents	point	fresh waters coastal waters terrestrial and ground waters
Pesticide storage facilities cellars, mixture preparation	–	point	
DATE AND OIL PALM			
Pre-planting seeds: fungicides weeds: herbicides	aerial drift runoff, erosion	point	occurs in lowlands at lower end of watershed near coastal areas

continued...

Table 12.7 continued

Production stage/ Biocide action	Environmental transport	Pollution source	Exposed ecosystems
On the plantation herbicides insecticides	aerial drift runoff, erosion	non-point	fresh waters coastal waters terrestrial and ground waters
Packing facilities fungicides	effluents	point	

ORNAMENTAL AND FLOWER PRODUCTION

On the plantation herbicides, insecticides fungicides, acaricides nematicides, molluscicides	aerial drift runoff, erosion	non-point	occurs in the upper, middle, and lower parts of watersheds in all regions of the country may influence coastal areas
Pesticide storage facilities cellar, mixture preparations	–	point	fresh waters coastal waters terrestrial and ground waters

RICE

Seed preparation insecticides	aerial drift	point	takes place in lowlands at lower end of watershed near coastal areas
Pre-planting weed control: herbicides soil insects: insecticides	runoff, erosion	non-point	
Post-planting soil and stem insects: insecticides	aerial drift runoff, erosion	non-point	
Pre-emergence and Post-emergence herbicides soil and foliar insets: insecticides	aerial drift runoff, erosion	non-point	fresh waters coastal waters terrestrial and ground waters
Blooming foliar and spike insects: insecticides diseases: fungicides	aerial drift runoff, erosion	non-point	
Pesticide storage facilities cellar, mixture preparations	–	point	

continued...

Table 12.7 continued

Production stage/ Biocide action	Environmental transport	Pollution source	Exposed ecosystems
SUGAR CANE			
Seed treatment fungicides	runoff, erosion	point	planted in the middle and lower section of watersheds
Pre-planting land preparation: insecticides weeds:herbicides	aerial drift runoff, erosion	non-point	
Post-planting weeds: herbicides	aerial drift runoff, erosion	non-point	
Ripening acceleration: herbicides	aerial drift runoff, erosion	non-point	fresh waters coastal waters terrestrial and ground waters
Production fungicides, insecticides biological control, rodenticides	aerial drift runoff, erosion	non-point	
COFFEE			
Pre-planting weeds: herbicides	runoff, erosion		cultivated in the upper and middle sections of watersheds, but rapid river water flow may carry pesticides used in this crop to coastal areas
On the plantation weeds: herbicides insects: insecticides diseases: fungicides	runoff, erosion aerial drift		fresh waters terrestrial and ground waters

Notes:
a Source: adapted from Subirós, 1995; Cortés, 1994; and field inquires and observations by the authors.
b En dash (–) indicates no information available.

It is critical that environmental monitoring programs be established near formulating plants to detect spill events; to reduce the response time for dealing with accidental releases of toxic substances; and to protect natural ecosystems, especially aquatic ecosystems, which are highly vulnerable to pesticide pollution. Most pesticide a.i.(s) used in formulation plants are classified as extremely to highly toxic to fish and crustaceans, e.g. mancozeb and chlorothalonil (Table 12.8). Most of these chemicals are imported in large quantities and are handled as technical grade materials. Degradation products or metabolites can be more toxic and

Table 12.8 Pesticides imported by a formulation plant located on the Nicoya Gulf (northwestern Pacific Coast of Costa Rica) and their aquatic toxicity to laboratory organisms[a]

a.i.	Purity or grade[b] (%)	Quantity (T)	LC_{50} (96 hour) fish[c]	LC_{50} (24–48 hour) crustaceans[d]
2,4-D	technical grade	168.1	high[e]	–[f]
Benomyl	90–97	7.9	extreme	extreme
Captan	90	18.9	extreme	high
Carbendazim	90–99	43.5	extreme	extreme
Chlorothalonil	96–98	21.5	extreme	extreme
Phorate	85	30.8	extreme	extreme (96)
Isazofos	95	72.0	extreme	extreme
Malathion	96	36.7	extreme	extreme
Mancozeb	33–90	455.4	extreme	extreme
MCPA	98–99	7.9	moderate	light
Methamidophos	60–75	90.4	moderate	extreme
Methyl parathion	75–85	6.9	high	extreme
Propanil	95	119.5	high	high
Propiconazole	88	69.3	moderate	moderate to high
Tridemorph	75–84	751.8	high	high
Glyphosate	55–95	38.7	moderate	light
Ferbam	76–90	31.8	–	–
Paraquat	42	76.8	moderate	high
Pendimethalin	90	42.0	extreme	extreme
Terbufos	85	123.0	extreme	extreme
Other	5–100	51.7	–	–

Notes:
a Source: adapted from Tomlin, 1997; UNA/IRET, 1999.
b Technical grade pesticides are pesticide chemicals in a pure form (usually 95–100 percent a.i.), which are then formulated into pesticide products, e.g. wettable powders, dusts, emulsifiable concentrates, granules, etc.
c Rainbow trout.
d *Daphnia* sp.
e Toxicity classification in mg L^{-1}: extreme (<1), high (1–10), moderate (10–100), and light (>100).
f En dash (–) indicates no information available.

persistent than the parent compounds. Examples of this include chlorothalonil, phorate, malathion, mancozeb (ETU), methyl parathion, ferbam, and terbufos (UNA/IRET, 1999). Furthermore, available toxicity data are routinely obtained from laboratory species used in temperate countries and primarily for the parent compounds. These data account for the toxicity of primary and secondary metabolites, but not for the sensitivity of local species to either the parent compound or its metabolites. Special security measures at these facilities are needed to prevent accidental releases with subsequent damage to ecosystems. Environmental monitoring programs are quite useful for evaluating the impact of any accidental release and should be implemented. The proximity of a major pesticide formulator to the Nicoya Gulf, which is an important economic and recreational ecosystem for Costa

Rica, provides ample reason for maintaining an environmental monitoring program in this vicinity.

ENVIRONMENTAL RESIDUE STUDIES

Most studies of Costa Rican aquatic ecosystems in coastal areas conducted before 1998 focused on chemical residues, emphasizing OC residues (Table 12.9). We have identified 16 such studies in the literature, though there may be other unpublished studies for which we do not have data. Of these, seven studies are from the Pacific region and nine from the Caribbean Coast. The first pesticide residue study (1983 to 1984 on the Pacific Coast) examined near-shore aquatic ecosystem impacts of OC pesticides to eggs of eight species of aquatic birds (Hidalgo, 1986). The next (1987 to 1988) examined samples of water, sediment, and biota from areas of the Caribbean Coast influenced by banana plantation pesticide applications (von Düszeln, 1988).

Samples of larvae of the aquatic mayfly *Euthyplocia hecuba* Hagen (Ephemeroptera: Polymitarcyidae) collected from streams in the upper reached of Rio Tempisquito on the western slopes of Volcán Orosí and Cerro Cacao, which are in the Parque Nationale de Guanacaste in northern Costa Rica, had quantifiable levels of eight OC pesticides – levels of DDT, endrin, α-HCH, and γ-HCH were below detection limits (Standley and Sweeney, 1995). Mayflies are especially useful as indications of aquatic contamination because their larvae bioconcentrate the lipophilic OC pesticides, which are often too dilute in the stream water to detect. Their larvae can thus give a representation of material deposited into and transferred through a stream's catchment basin. It does not, however, reflect the total deposition in the basin because natural filtration occurs throughout the catchment. Larval contaminant loading will reflect the processes occurring including physico-chemical properties of the pesticides, loss mechanisms including microbial degradation and volatilization, and processing by the aquatic organisms including bioaccumulation, metabolism, and biomagnification. OC pesticides found (all units are in ng g^{-1} ww) in mayfly larval tissues included heptachlor epoxide (37), α-endosulfan (51), aldrin (54), DDE (67), dieldrin (100), β-endosulfan (150), endrin aldehyde (150), and endosulfan sulfate (2,000) (Standley and Sweeney, 1995).

AQUATIC ECOSYSTEM IMPACTS

The high diversity of the Costa Rican terrestrial (50,660 km^2), aquatic (440 km^2), and marine ecosystems (520,000 km^2) results in a highly diverse flora and fauna – many of which remain undescribed. It is believed that regions with extremely high biodiversity such as Costa Rica are preserves of germplasm diversity for future generations. With the rate of environmental destruction increasing, universities and research centers must strive to protect those ecosystems that are still unaffected,

Table 12.9 Summary of pesticide studies in aquatic ecosystems near shore in Costa Rica (only the maximum concentrations measured are shown)

Year/location	Substrates	Pesticide class	Results	Remarks
1983–84 Marine Nicoya Gulf (Pacific)	biota: eggs of eight species of aquatic birds (n = 34)	OC	*In biota:* OC: 4.16 mg kg^{-1} fw p,p'-DDE: 3.19 mg kg^{-1} fw	A positive correlation between eggshell thinning and DDE levels in birds was found.
1987–88 Freshwater streams in banana production area Arenal Lake and tributaries (Caribbean)	surface waters sediments biota	OC OP paraquat	*Water:* chlorothalonil: 11 μg L^{-1} chlorpyrifos: 0.18 μg L^{-1} paraquat: 5.6 μg L^{-1} *In biota:* OC: 58.3 mg kg^{-1} fw	OC and OP pesticides were detected in few water samples.
1989 Marine (coastal Caribbean)	sediment form shallow coastal waters	OP	*Sediments:* chlorpyrifos: 34 μg kg^{-1} dw parathion: 12 μg kg^{-1} dw	Sampling area was influenced by drainage from banana plantations.
1988–91 Marine Nicoya Gulf (estuarine) Pacific	biota: *Anadara tuberculosa*	OC	*In biota:* DDT: 134 μg kg^{-1} dw chlordane: 119 μg kg^{-1} dw lindane: 706 μg kg^{-1} dw heptachlor: 29.9 μg kg^{-1} dw mirex: 2.28 μg kg^{-1} dw	*Anadara tuberculosa* can be used as bioindicator in mangrove ecosystems. OC concentrations are higher in rainy season. Positive correlation between OC levels and lipid content. Positive correlation between OC levels and PCB content.
1991 Marine (coastal) Pacific and Caribbean	biota: several bivalve species	OC	*In biota:* DDT: 199.5 μg kg^{-1} dw chlordane: 16.0 μg kg^{-1} dw BHC: 2.82 μg kg^{-1} dw lindane: 4.2 μg kg^{-1} dw heptachlor: 1.75 μg kg^{-1} dw	OC levels were below national or international recommended action limits for human consumption.

continued…

Table 12.9 continued

Year/location	Substrates	Pesticide class	Results	Remarks
1992 Freshwater drainage channels, streams, river in banana plantation area Marine-coral reef Caribbean	surface water sediment biota: sea cucumber	various pesticides used in banana plantations	aldrin: 1.76 mg kg^{-1} dw dieldrin: 4.73 μg kg^{-1} dw endrin: 1.29 μg kg^{-1} dw mirex: 0.85 μg kg^{-1} dw *Water:* chlorothalonil: 8 μg L^{-1} *In sediment:* chlorothalonil: 40 μg kg^{-1} dw *In biota:* chlorpyrifos: 8 μg kg^{-1} dw	Chlorothalonil and chlorpyrifos used in banana production. Limited number of samples analyzed.
1992 Freshwater drainage channels, streams, rivers in banana plantations Caribbean	water sediments biota: fish and invertebrates	various pesticides used in banana production	*Water:* chlorothalonil: 0.9 μg L^{-1} propiconazole: 2.2 μg L^{-1} thiabendazole: 66.0 μg L^{-1}	Preliminary results of ongoing integrated study. Highest concentrations found in samples from drainage channels.
1992 Freshwater drainage channels and stream in rice paddy area Pacific	water sediments biota: fish and invertebrates	various pesticides used in rice production	*Water:* propanil: 5.1 μg L^{-1} cypermethrin: 6.6 μg L^{-1} oxadiazon: 0.6 μg L^{-1} edifenphos: 0.7 μg L^{-1} quinclorac: 790 μg L^{-1} methamidophos: 82 μg L^{-1}	Preliminary results of ongoing integrated study.
1993–96 Freshwater drainage canals and streams, Suerte River, and river junction with the Tortuguero Lagoon	water sediments	various pesticides used in banana production	*Water:* cadusafos 2 μg L^{-1} propiconazole 3.6 μg L^{-1} thiabendazole 17 μg L^{-1} imazalil 8.7 μg L^{-1} carbofuran 6.2 μg L^{-1}	

Year/location	Substrates	Pesticide class	Results	Remarks
1995–97 Freshwater rivers and coastal lagoons, marine river mouth, sea outlet, and coastal waters Caribbean	water sediment biota: fish, crabs, shrimp, and bivalves	various pesticides used in banana production	ametryn 1.7 μg L^{-1} chlorpyrifos < 0.1 μg L^{-1} *In sediment:* cadusafos 16 μg kg^{-1} dw imazalil 446 μg kg^{-1} dw propiconazole 33 μg kg^{-1} dw thiabendazole 435 μg kg^{-1} dw	Cadusafos detected in coastal waters. Many of the sampling sites are situated in protected areas of the coastal zone at the lagoon system. Frequently, multiple residues were present in the samples.
1998 Freshwater Tempisque River and Nicoya Gulf Estuary Pacific	water sediment biota: fish, crabs, shrimp, bivalves, and gastropods	various pesticides used in rice and technicals imported to formulate	*Water:* carbofuran: 6.27 μg L^{-1} propiconazole: 1.5 μg L^{-1} ethoprophos: 0.28 μg L^{-1} cadusafos: 0.07 μg L^{-1} diazinon: 0.31 μg L^{-1} fenamiphos 0.40 μg L^{-1} *Sediment:* propiconazole: 19 μg kg^{-1} dw *In biota:* DDE: 10 μg kg^{-1} dw *Water:* ametryn: 0.13 μg L^{-1}	Pesticide residues were detected only in one sampling day.
1998 Freshwater drainage system Arenal-Tempisque Pacific	water sediment biota: invertebrates	OC and various pesticides used in rice	*Water:* DDT: 0.5 μg L^{-1} lindane: 0.04 μg L^{-1}	Samples from drainage systems had higher incidence of pesticides.

Table 12.9 continued

continued...

Table 12.9 continued

Year/location	Substrates	Pesticide class	Results	Remarks
			propanil: 0.8 μg L^{-1} chlorpyrifos: 0.2 μg L^{-1} *In biota:* aquatic bird's eggs OC: 78 mg kg^{-1} fw *Shrimp:* HCB: 13 μg kg^{-1} fw lindane: 23 μg kg^{-1} fw aldrin: 10 μg kg^{-1} fw heptachlor: 3 μg kg^{-1} fw chlordane: 7 μg kg^{-1} fw $o,p + p,p$-DDT: 25 μg kg^{-1} fw	There was no relationship found between the season and pesticide content. A relationship between the type of crop and pesticides present in soil was proposed. OC concentrations in bird's eggs were lower than those of Hidalgo, 1986.
1998 Fresh water drainage system Arenal -Tempisque Pacific	water in rice paddies and sugar cane fields	Some pesticides used in rice and sugar cane	*Water:* ametryn: 1.0 μg L^{-1}	Samples for phenoxyacetic acids were not taken.

Source: adapted from Hidalgo, 1986; von Düszelen, 1988; Readman et al., 1992; Abarca and Ruepert, 1992; Castillo et al., 1994; de la Cruz, 1994; Farrington and Tripp, 1994; Castillo et al., 1995; Rodríguez, 1997; de la Cruz et al., 1998; and Castillo et al., 2000.

assess the degree of damage in those ecosystems already degraded, and establish preventive measures against future damage in those affected. Despite protection of 19 percent of Costa Rica's territorial area as biological preserves, these areas are not immune from the extensive use of agrochemicals. Special attention must focus on coastal and marine ecosystems because they receive and may be reservoirs for contaminants. Costa Rica has about 1,300 km of coastline, influenced by 34 river basins, and most of it is not protected. Coastal areas are vital to the country's economy and are an integral part of the everyday life of Costa Ricans. The Nicoya Gulf, located on the northwest Pacific coastline, and the Tortuguero lowland lagoons, located on the northeast Caribbean coastline, are two of the most important coastal areas (Figure 12.1).

Pacific coast

The Nicoya Gulf is, and has always been, a very important fishing ground and thus it has a large influence on the Costa Rican economy. More than 11 percent of Costa Rica's population lives in its vicinity (Costa Rica, 1987) and an even higher percentage utilizes the area for recreational and commercial purposes. The Nicoya Gulf's productivity has decreased over time. In 1976, 90 percent of Costa Rica's fish catch came from this area (Phillips, 1983) but this has declined to 50 percent even though fishermen have increased their fishing effort (Campos, 1987). The decreasing fish populations were explained by Campos in terms of overfishing and shifts in fishing areas but more comprehensive studies including fish population dynamics, ecological change analysis, and pollutant loading in the Gulf are needed to adequately explain the decline. Estuaries play a fundamental role in the life cycle of many marine organisms. Estuaries and especially mangrove forest areas within the estuaries are important nursery grounds for juveniles of many commercial and non-commercial fish species, invertebrates, and aquatic birds (Jaccarini and Martens, 1992). Early stages in the life cycle of a species are known to be the most vulnerable to the influence of chemicals (Castillo, 1987). Nicoya Gulf's estuary is a very important nursery ground for many species of crab, shrimp, fish, and mollusk (de Vries et al., 1983a; 1983b; Dittel et al., 1985). The mangrove forests coupled with islands located in the Gulf are the natural habitats of many marine and terrestrial birds. They also provide a natural refuge for migratory bird species during the Northern Hemisphere's winter season (Smith and Stiles, 1979). Nicoya Gulf provides many resources for Costa Rica and its pollutant load must be carefully studied, monitored, and controlled. Regular monitoring programs for chemicals used in the region are necessary and research to study cause and effect relationships for contaminants must be conducted to evaluate the risk from contaminants to the Gulf's populations and to the human population utilizing it. The proper use of resources within a sustainable development management strategy will improve the quality of life of both humans and other species inhabiting and sharing Nicoya Gulf. Studies conducted on pesticide loads in different compartments of Pacific Coast ecosystems are summarized in Table 12.9.

The influence of rice and sugarcane production on coastal aquatic ecosystems can be observed in Nicoya Gulf. Pesticides generally used on these crops, e.g. propanil, quinclorac, cypermethrin, oxadiazon, and ametryn, have been reported in surface water from the region (Table 12.9). Pesticide residues reported in biota were primarily OCs, including aldrin, lindane, DDT and its metabolites, and the OP chlorpyrifos (Figure 12.6) (Hidalgo, 1986; de la Cruz, 1994; Castillo *et al.*, 1995; de la Cruz *et al.*, 1998; Osorio, 1998; Rodríguez, 1997).

Hidalgo (1986) reported the presence of OC residues in eggs of eight different species of near-shore aquatic birds. Between 1983 and 1984, a total of 137 eggs was collected on an island nesting site located in the estuary formed by the Tempisque River and Nicoya Gulf. Residues of *p,p'*-DDE were found in all samples; the highest concentration was found in eggs of wood stork *Mycteria americana* L. (Ciconiiformes: Ciconiidae) and the lowest in eggs of white ibis *Eudocimus albus* L. (Ciconiiformes: Threskiornithidae). Heptachlor epoxide, HCB, *p,p'*-DDT, and endrin were present in a high percentage of the samples. For all except two species, a strong correlation was found between shell thickness and *p,p'*-DDE residues. In some eggs of *M. americana* having the highest DDE concentrations, the author

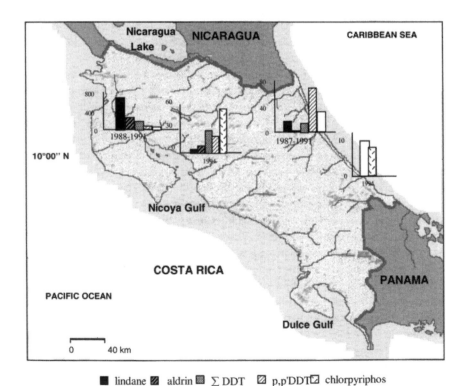

■ lindane ▨ aldrin ▧ ∑DDT ▨ p,p'DDT▨ chlorpyriphos

Figure 12.6 Pesticide residues (μg kg⁻¹ fw) reported for coastal and natural freshwater biota of Costa Rica.

observed cracks. In a 1996 study of OC content in eggs of aquatic birds from the Tempisque River estuary, the concentrations of HCB, lindane, aldrin, dieldrin, heptachlor epoxide, chlordane, and DDT were much lower that those reported by Hidalgo in 1986 (Rodríguez, 1997). According to the International Mussel Watch project, DDT and its metabolites are the most prevalent OCs found in biota (Farrington and Tripp, 1994). OC residues were also reported in the cockle *Anadara tuberculosa* Sowerby (Pelecypoda: Arcidae), inhabiting the mangrove ecosystems of the Nicoya Gulf (de la Cruz, 1994). The author reported a maximum concentration of 134 μg kg^{-1} dw of $\sum$ DDTs. DDT, lindane, and chlordane accounted for most of the total OC residues measured at different study sites from 137 cockles collected on the Pacific Coast of Costa Rica between 1988 and 1991. OC residues reported in water and biota samples by Rodríguez (1997) were much lower than reported in previous studies.

OP and carbamate insecticides, herbicides, and fungicides have also been studied in the Pacific aquatic ecosystems (Castillo *et al.*, 1995; de la Cruz *et al.*, 1998; Osorio, 1998; Rodríguez, 1997). Residues of the herbicides propanil, cypermethrin, oxadiazon, quinclorac, and ametryn were detected in water samples from this region. Residues of the fungicide edifenphos and the insecticides methamidophos and chlorpyrifos have also been reported. Some of these compounds are toxic to aquatic fauna and may cause effects from short or long-term exposure, e.g. methamidophos, edifenphos, cypermethrin, oxadiazon, and propanil (Tomlin, 1997; UNA/IRET, 1999).

Further south along the Pacific Coast lies Golfo Dulce, which was surveyed by Spongberg and Davis (1998) using sediment core samples collected from around the shores of the gulf to quantify persistent pesticide contamination while recording organic matter content and particle size distribution of the cores. They found that the gulf to date had only minimal accumulations of pesticides, although the Esquinas River sediments were found to contain numerous pesticide residues including all BHC isomers, heptachlor, endrin, dieldrin, endosulfans, and DDTs. The Coto-Colorado River sediments were coarser in texture with low organic content that may not retain pesticides efficiently. Thus, any residues transported into the gulf by this river system may be transported great distances within the gulf. They also found that the Golfito area had little pesticide contamination, although hydrocarbons were in great abundance in the sediments.

Caribbean coast

On the Caribbean Coast of Costa Rica, the Tortuguero Lagoons ecosystem is a very important part of the natural refuges and conservation zones. The long stretches of beach are used by marine turtles to lay their eggs. Along the southern Caribbean Coast important coral reef communities exist. Studies of the pollutant load in different compartments of the Costa Rican Caribbean Coast are summarized in Table 12.9. Most of the completed pesticide residue studies focusing on the aquatic ecosystems of the Caribbean Plains are related to banana culture techniques and consequently the influence of this crop is generally observed (Table 12.9).

The most frequently detected OC residues in fish collected from banana production areas and Lake Arenal during 1987 and 1988 were HCB, dieldrin, heptachlor, DDE, and lindane (von Düszeln, 1988). Also, residues of cadusafos, chlorpyrifos, propiconazole, thiabendazole, ethoprophos, carbofuran, fenamiphos, diazinon, lindane, parathion, chlorothalonil, 2,4-D, and paraquat have been reported from freshwater and marine ecosystems of this region (von Düszeln, 1988; Readman et al., 1989; Abarca and Ruepert, 1992; Castillo et al., 1995, Castillo, 1997; Castillo et al., 1997; de la Cruz et al., 1998; Castillo et al., 2000). Residues of chlorpyrifos, used on banana plantations, cotton fields, and on horticultural crops, have been reported in several studies. Von Düszeln (1988) reported 0.18 μg L^{-1} in surface water and Castillo et al. (1995) found a maximum concentration of 0.07 μg L^{-1} in surface water. The highest level of chlorpyrifos in sediment samples (34.2 μg kg^{-1} dw) was reported by Readman et al. (1992) from an area located downstream from a banana plantation, the Parismina River outlet to the Caribbean Sea. Castillo et al. (1995) found 161 μg kg^{-1} dw chlorpyrifos in sediments from the drainage channel of a banana packing plant facility. Castillo et al. (2000) detected chlorpyrifos primarily in effluents of packing plants but it was also detected in 9 percent of stream samples and in one water sample from the Tortuguero National conservation area (mean concentrations were <0.1 μg L^{-1}). In sediment samples, they detected chlorpyrifos in effluent channels of packing plants (mean concentration 112 μg kg^{-1} dw; range 23 to 320 μg kg^{-1} dw). Abarca and Ruepert (1992) measured 8 μg kg^{-1} dw in sea cucumber (*Holothuria* sp) collected from a coral reef in an area influenced by banana plantations.

The herbicide paraquat was detected in surface water and sediments near banana plantations at concentrations of 5.6 μg L^{-1} and 4.1 mg kg^{-1}, respectively, by von Düszeln (1988). The maximum concentration of the fungicide chlorothalonil in water was 11 μg L^{-1} (von Düszeln, 1988) and in sediments 40 μg kg^{-1} (Abarca and Ruepert, 1992). Another fungicide propiconazole is widely distributed in surface waters (maximum concentration of 2.2 μg L^{-1}) and sediments (maximum concentration of 19 μg kg^{-1}) of aquatic ecosystems associated with areas near banana plantations (Figure 12.7) (Castillo et al., 1995; Castillo 1997; Castillo et al., 1997; de la Cruz et al., 1998). Other compounds frequently found in surface waters are the fungicide thiabendazole (maximum concentration of 66.0 μg L^{-1} in effluents of a banana packing plant); the nematicides carbofuran (6.27 μg L^{-1}), ethoprophos (0.28 μg L^{-1}), cadusafos (0.07 μg L^{-1}), and fenamiphos (0.4 μg L^{-1}); and the insecticide diazinon (0.31 μg L^{-1}) (Castillo et al., 1995; de la Cruz et al., 1998). Residues of cadusafos were reported from the Caribbean Sea (0.05 μg L^{-1}) near the Tortuguero Lagoon outlet (de la Cruz et al., 1998). Some of the compounds evaluated pose a risk for acute or chronic toxicity to biota in the aquatic community based on field exposure levels and reference values from the literature (Teunissen-Ordelman and Schrap, 1997; Tomlin, 1997; de la Cruz and Castillo, 1999; UNA/IRET, 1999; Castillo et al., 2000).

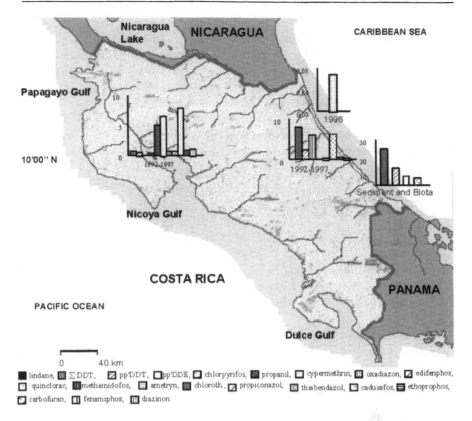

Figure 12.7 Pesticide residues (μg L^{-1}) in surface waters of Costa Rica from 1992–98. Quinclorac and methamidofos concentrations must be multiplied by 100 and thiabendazol by 10. Residues reported for marine sediments (μg kg^{-1} dw) and biota are also included (μg kg^{-1} fw).

Despite the quantities of pesticides used in Costa Rica, studies of their presence and impact on aquatic ecosystems, especially for coastal areas, are few. Data from before 1992 mainly provide information on OC pesticide residues. Recent data includes information on other pesticide chemical classes. More data are needed to adequately assess the frequency and biological impact of these compounds in marine coastal ecosystems of the Costa Rican Caribbean zone. A better understanding of their behavior in the freshwater and estuarine ecosystems of the region is imperative.

Because there are no national monitoring programs in place, research programs are the only source of information about pesticide residues in Costa Rica's aquatic ecosystems. Only a small number of the more common pesticides are included in these research programs. Among those not monitored, yet imported in large quantities, are the fungicide mancozeb, the herbicides glyphosate and terbuthylazine, the fumigant methyl bromide (known to damage the ozone layer), and copper compounds. Several heavily used pesticides have been analyzed in very few studies, including the herbicides paraquat and atrazine, the insecticides monocrotophos and methamidophos, and the nematicides fenamiphos and chlorothalonil.

ALTERNATIVE AGRICULTURAL PRACTICES

Biological control and integrated pest management

The first formal attempts at biological control in Costa Rica were a 1915 effort to control a locust pest with a pathogenic bacteria and a cooperative program from 1924 to 1935 with United States entomologists to control citrus black fly populations using a parasite (Hilje *et al.*, 1989). Currently IPM research is conducted at governmental universities, the Ministry of Agriculture (MAG), the Sugar Cane Office for Research and Extension (DIECA), the Phytoprotection Department of the Center for Tropical Agriculture for Research and Teaching (CATIE), and the Institute for Coffee Research (ICAFE). A Costa Rican Journal of Integrated Pest Management and an Integrated Pest Management Information Bulletin are regularly published by CATIE.

There are several examples of the successful implementation of biological control and IPM practices in Costa Rica. Sugarcane stalk-borer *Diatraea saccharalis* Fabricius, *D. tabernella*, and *D. guatemallela* (Lepidoptera: Pyralidae, Crambinae) have been controlled on sugarcane plantations utilizing brachonid wasp parasitoids *Cotesia flavipes* Cameron (Hymenoptera: Braconidae, a gregarious, larval endoparasitoid native to southeast Asia), from 1985 to 1991. The Mediterranean fruit fly *Ceratitis capitata* Wiedemann (Diptera: Tephritidae) has been controlled with pupal parasites and sterile males (1989). Also pheromones have been used to sample population levels of the potato tuber moths *Tecia solanivora* Povolny (Lepidoptera: Gelechiidae) and *Phthorimaea operculella* Zeller (Lepidoptera: Gelechiidae) to determine the threshold for insecticide applications (Rodríguez, 1989; Rodríguez *et al.*, 1989). By utilizing this program, the cost of potato production has been reduced by about 60 percent during the dry season and 40 percent during the rainy season. In potato culture, three different methods are used to schedule pesticide applications: calendar, observation, and sampling (Rodríguez *et al.*, 1988). Based on which method was used during two cycles of potato production, a quantitative difference in the number of pesticide applications was observed. The number of insecticide applications was reduced from 30 to 20 using the observation method and from 30 to 7 applications using a sampling method. Fungicide applications were reduced from 45 to 32 with both observation and sampling methods. Thus, the total use of pesticides was reduced by almost 50 percent using a sampling method instead of the calendar method.

Other examples include producing tomato seedlings free of the geminivirus transmitted by the sweetpotato whitefly *Bemisia tabaci* Gennadius (Homoptera:

Aleyrodidae) and the use of threshold criteria for the control of the fruitworm *Heliothis zea* Boddie (Lepidoptera: Noctuidae) on tomato (and corn) by counting the number of eggs, larvae, or perforated fruit to determine the timing of pesticide applications (CATIE, 1991; Pérez *et al.*, 1997; Hilje, 1998). Also, cultural practices and organic amendments to the soil have been used to reduce disease severity. Using compost as the growing medium reduces the severity of soil-borne fungi on seedlings. The soil fungi *Rhizoctonia solani* Kühn (anamorph or asexual growth stage of *Thanatephorus cucumeris*, see below) is well controlled in broccoli *Brassica oleracea* var *italica* L. using compost as fertilizer or soil amendment (Alvarado, 1997). Lime ($CaCO_3$) and phosphorus (P_2O_5) application to soil results in bean plants *Phaseolus vulgaris* L. being less susceptible to fungal attack by web blight *Thanatephorus cucumeris* (A.B. Frank) Donk (telemorph or sexual growth stage of *Rhizoctonia solani*). The reduction in susceptibility level was attributed to the importance of calcium as a component of calcium pectate in the middle lamella and of phosphorus as a component of phospholipids in cellular membranes (Chaverri *et al.*, 1993).

Biological control and IPM techniques are used on a number of crops in Costa Rica including cotton, cacao, coffee, sugarcane, pepper, coconuts, beans, fruits, maize, mango, apples, potatoes, plantain, cabbage, tobacco, tomato, and in forestry (reviewed by Garcia, 1999). The Center for Research in Molecular and Cellular Biology of the University of Costa Rica – dedicated to using biotechnology to enhance crop protection – has been working to produce whiteleaf virus-resistant rice plants. Also the Laboratorio de Cultivo de Tejidos y Células Vegetales (*In vitro* Tissue and Cell Culture Laboratory) of the Universidad Nacional has produced genetically improved rice varieties using radiation-induced gene mutation techniques. Their new rice varieties are cheaper to produce, more resistant to diseases, and have higher yields.

Alternative agriculture

Organic agriculture started in Costa Rica during the mid-1980s as the result of individual efforts – not a government-sponsored initiative – with the production of vegetables for the national market (Soto, 1998). This type of agriculture is becoming more important in Costa Rica because of a reduction in production costs and an increase in profitability per ha of cultivated land (Boyce *et al.*, 1994; Hitz, 1995; Garcia, 1999; Ramírez and Soto, 1999). In 1996, there were 3,000 ha cultivated organically in Costa Rica and by 1997 that had increased to 6,000 ha (Proyecto Estado de la Nación, 1998; Soto, 1998). Organic agriculture in Costa Rica is primarily used in cacao, banana, and coffee production, but other crops are being tried. The first organic product exported by Costa Rica was coffee and in 1997 the category of organic coffee for export was officially established. Organic bananas are produced, mainly in the Caribbean region, but are primarily used in the preparation of organic banana puree. Currently Costa Rica exports organic coffee, pineapple, banana puree and fresh bananas, cocoa, blackberries, sugar, orange juice, mango, ginger, vanilla, and other aromatic spices (Proyecto Estado de la Nación, 1998).

Since 1990, the University of Costa Rica and the National Institute for Learning (INA) have conducted a pilot project where regular training in organic agriculture for grain and vegetable production has been conducted emphasizing transfer of knowledge from farmer to farmer. The program has trained many groups of small farmers with special emphasis placed on organic compost preparation techniques. Currently most government universities have included courses in organic agriculture and agroecology in their bachelor, master, and doctoral programs (Proyecto Estado de la Nación, 1998).

The National Association of Organic Agriculture was established in 1992 with the primary objectives of collaborating in organic certification management and development of organic production legislation (Proyecto Estado de la Nación, 1998; Garcia, 1999). Organic farm certification was established to validate the effort of farmers growing crops organically, to assure consumers of product quality, and to help organic farmers compete with farmers falsely claiming to be organic farmers. At present, there are two certifying agencies in Costa Rica that provide inspection and certification services to organic farms and processors. They certify organic farms not only in Costa Rica but in other Latin American countries including Panama, Guatemala, Honduras, Colombia, Paraguay, and others. The government of Costa Rica accredits these certification agencies. One of these agencies, Eco-lógica, has a reciprocity program with other international agencies in the USA (Oregon Tilth Certified Organic, OTCO; Quality Assurance International, QAI; and Washington State Department of Agriculture, WSDA) and Europe (Ecocert) (G. Delgado personal communication in 1999; unreferenced).

For some organic products the rate of return for farmers has reached 46 percent. Because the quality of organic coffee for export is so high, Costa Rica has one of the highest price differentials (up to 35 percent) in the international coffee market (Proyecto Estado de la Nación, 1998). However, the commercialization of organic products in local markets is low, but increasing, with four supermarket chains in Costa Rica offering organic products.

SUMMARY AND RECOMMENDATIONS

Agricultural and other pest control practices in Costa Rica are highly pesticide dependent and no official government policy exists to reduce the quantity of pesticides or to switch from highly toxic and dangerous products to less toxic ones. Even though legislation regulating pesticide use in Costa Rica is in place, these compounds may in many cases reach non-target ecosystems with deleterious consequences to ecosystem health.

Pesticide import data are indicative of the great variety, quantity, and frequency of a.i.(s) used in Costa Rica's agricultural production. OCs were heavily used during the 1970s and 1980s, but their use decreased dramatically after most were prohibited in 1988. There has been an increase in the quantity of pesticides imported by Costa Rica since 1988 when more contemporary pesticides, e.g. mancozeb, 2-4-D, glyphosate, chlorothalonil, ethoprophos, paraquat, and terbufos, began to dominate the marketplace. The most heavily used biocide groups during this period were fungicides followed by herbicides and nematicides.

Monoculture-type agriculture for producing banana, coffee, sugarcane, rice, ornamentals, and fruits is one reason for the intense and predominant use of pesticides in Costa Rica. This method of farming depends heavily on agrochemical use, which has many negative consequences. These include pest resistance development, soil deterioration, aquatic ecosystem degradation, the emergence and proliferation of secondary pests, adverse health effects on the general and agricultural labor populations, and various other environmental effects from exposure to pesticides residues. Both governmental and non-governmental institutions are working together with farmers' associations to conduct research and facilitate technology transfers with a goal of more rational and sustainable agricultural practices. Examples of this can be seen in Costa Rica's Integrated Pest Management (IPM) program and its organic farming movement.

Studies have been conducted in both coastal and inland aquatic ecosystems of Costa Rica. Published data relate primarily to pesticide residue levels (mainly the more persistent OCs) in one or two environmental matrices. However, more recently the tendency has been to emphasize more integrated studies, relating exposures and effects. These studies demonstrate the potential for adverse impacts associated with pesticide use. Persistent compounds and more contemporary substances have been detected in various environmental components of aquatic ecosystems in Costa Rica. More research is needed to better understand the chemical fate, distribution, and effects of pesticides in the tropics and the consequences of their use and misuse on tropical aquatic ecosystems. Special emphasis should be given to studies examining the potential impact on tropical aquatic ecosystems of repeated and continual low-level exposures to mixtures of pesticides.

Contemporary pesticides, e.g. ametryn, cadusafos, chlorothalonil, chlorpyrifos, cypermethrin, endosulfan, paraquat, propiconazole, quinclorac, diazinon, and carbofuran, which have been found in aquatic ecosystems (some are highly toxic to aquatic organisms), should be regarded as priority substances for future studies. A major need will be the development of sensitive methods to monitor pesticide effects on ecosystems. Criteria related to general water quality and specific criteria for tropical aquatic ecosystems must be developed for the entire Central American region. The concept of acceptable risk levels should be assessed relative to protection of valuable tropical aquatic ecosystems. It will be necessary to develop our own perspective of water quality, acceptable risk, and methods for environmental evaluation. Studies to develop pesticide reduction strategies especially for the more toxic pesticides, e.g. some nematicides and fungicides, are greatly needed.

It is well known that the adoption of alternatives to reduce pesticide consumption is economically feasible. Cases of alternative pest control methods used commercially for horticultural crops on modern and traditional farms are well documented.

In many cases, Costa Rican farmers have some experience with pesticide alternatives but they need more information to improve and select the best of them. Now, both Costa Rican farmers and consumers alike need a strong awareness process. Specifically workshops, pilot demonstration parcels with farmers, and extension training projects need to be developed and implemented or adapted to be suitable to the Costa Rica's needs.

REFERENCES

Abarca, L. and Ruepert, C. 1992. Plaguicidas encontrados en el Valle de la Estrella: estudio preliminar. *Tecnología en Marcha*. 12:31–8.

Alvarado, G. 1997. Efecto de la aplicación de dos fuentes de compost y enmiendas sobre la incidencia de *Rhizoctonia solani* y el contenido de nutrientes en bróculi (*Brassica oleracea* var. *italica*) [MSc thesis]. Universidad de Costa Rica. San José, Costa Rica.

Araya, C.A. 1982. *Historia económica de Costa Rica 1821–1921*. San José, Costa Rica: Editorial Fernandez Arce.

Boyce, J.K., Fernández, G.A., Fürst, E. and Segura, B.O. 1994. *Café y Desarrollo Sostenible: del Cultivo de Agroquímicos a la Producción Orgánica en Costa Rica*. Heredia, Costa Rica: Fundación UNA.

Campos, J. 1987. Fisheries development in Costa Rica. In: *Conferencia Sobre Pesca Artesanal y el Desarrollo Económico*, Volume 2. Universidad de Quebec. Conference held Aug 1986, Rimcuski, Canada, pp. 699–705.

Castillo, L.E. 1987. Toxicidad aguda y crónica del herbicida propanil en juveniles de *Machrobrachium rosembergii* [MSc thesis]. Universidad de Costa Rica. San José, Costa Rica.

Castillo, L.E. 1997. Presencia de agroquímicos en canales de drenaje y riesgo ambiental. In: *Simposio Internacional Sobre Riego y Drenaje en Banano*. EARTH, 12–22 August 1997. Guácimo, Costa Rica.

Castillo, L.E., de la Cruz, E. and Ruepert, C. 1997. Ecotoxicology and pesticides in tropical aquatic ecosystems of Central America. *Environ Toxicol Chem*. 16(1):41–51.

Castillo, L.E., Ruepert, C., Solís, E. and Martínez, E. 1994. Environmental impact assessment. In: *Environment, Health and pesticides: A Costa Rican–Swedish Research Program*. Project Progress Report, July 1995. Heredia, Costa Rica: Programa de Plaguicidas, Universidad Nacional, p. 6.

Castillo, L.E., Ruepert, C., Solís, E. and Martínez, E. 1995. Environmental impact assessment. In: *Environment, Health and pesticides: A Costa Rican–Swedish Research Program*. Project Progress Report, July 1995. Heredia, Costa Rica: Programa de Plaguicidas, Universidad Nacional, p. 15.

Castillo, L.E., Ruepert, C., Solís, E. and Martínez, E. 1995. *Ecological Consequences from Pesticide Use*. GTZ PN 90.2136.1–03.103. Final Report. Tropenökologisches Begleitprogramm. Heredia, Costa Rica: Deutsche Gesellschaft für Technische Zusammenarbeit.

Castillo, L.E., Wesseling, C., Aguilar, H., Castillo, C. and de Vos, P. 1989. Uso e impacto de los plaguicidas en tres países centroamericanos. *Estudios Sociales Centroamericanos*. 49: 119–39.

Castillo, L.E., Ruepert, C. and Solís, E. 2000. Pesticide residues in the aquatic environment of banana plantation areas in the North Atlantic Zone of Costa Rica. *Environ Toxicol Chem.* 19(8):1942–50.

Castillo-Muñoz, R. 1983. Geology. In: Janzen, D.H. (ed.) Costa Rican Natural History. Chicago: University of Chicago Press, pp. 47–62.

Castro, R. 1998. *Compedium Sobre la Legislación de Plaguicidas en Costa Rica.* San José, Costa Rica: Ministerio de Salud, Organización Panamericana de la Salud.

CATIE. 1991. Guía para el manejo integrado de plagas del cultivo de tomate. Turrialba, Costa Rica. Serie Técnica, Informe Técnico. pp. 138, 151.

Chaverri, F. and Blanco, J. 1995. *Importación,Formulación y Uso de Plaguicidas en Costa Rica: Periodo 1992–1993.* Final report to Programa de Plaguicidas, Pan American Health Organization (PAHO), Universidad Nacional. Heredia, Costa Rica.

Chaverri, F., González, L.C. and Bertsch, F. 1993. Efecto de la aplicación de calcio y fósforo en ultisoles e inceptisoles, sobre el desarrollo de la telaraña (*Thanatephorus cucumeris*) en frijol común (*Phaseolus vulgaris*). *Agronomía Costarricense.* 17(2):77–86.

Cortés, G. 1994. Atlas agropecuario de Costa Rica. Universidad Estatal a Distancia (EUNED) (ed.) San José, Costa Rica.

Costa Rica. 1953. Censo Agropecuario de 1950. Dirección General de Estadística y Censos, Ministerio de Economía y Hacienda. San José, Costa Rica, p. 160.

Costa Rica. 1987. Censo de Población 1984: Tomo 2. Dirección General de Estadistica y Censos, Ministerio de Econonía Industria y Comercio. San José, Costa Rica.

Costa Rica. 1999. Principales indicadores demográficos 1941–1997. Area de Estadística y Censos. Ministerio de Economía Industria y Comercio. www.inec.go.cr/INEC2/demograficos.htm.

de la Cruz, E. 1994. Stable pollutants in the bivalve *Anadara tuberculosa*, from the Nicoya Gulf, Costa Rica [PhD dissertation]. Vrije Universiteit. Brussels, Belgium. 216 p.

de la Cruz, E., Castillo, L.E. and Ruepert, C. 1998. *Study of the Fate and Impact of Organic and Inorganic Pollutants in the Costa Rican Coastal Zones.* Final report of European Union Research Project ERB CI1*-CT94–0076. Herdia, Costa Rica: Central American Institute for Studies on Toxic Substances.

de la Cruz, E. and Castillo, L.E. 1999. Presencia de agroquímicos en ecosistemas acuáticos de zonas costeras y análisis preliminar del riesgo ambiental. *Uniciencia.* 15–16:93–103.

de Vries, M., Epifanio, C.E. and Dittel, A.I. 1983a. Lunar rhythms in the egg hatching of the subtidal crustacean: *Callinectes arcuatus* Ordway (Decapoda: Brachyura). *Estuarine Coastal Shelf Sci.* 17:717–24.

de Vries, M., Epifanio, C.E. and Dittel, A.I. 1983b. Reproductive periodicity of the tropical crab *Callinectes arcuatus* Ordway in Central America. *Estuarine Coastal Shelf Sci.* 17:709–16.

Dittel, A.I., Epifanio, C.E. and Chavarria, J.B. 1985. Population biology of the portunid crab *Callinectes arcuatus* Ordway in the Gulf of Nicoya, Costa Rica, Central America. *Estuarine Coastal Shelf Sci.* 20:593–602.

Farrington, J.W. and Tripp, B.W. 1994. *International Mussel Watch Project, Initial Implementation Phase.* IMW Final Report, Coastal Chemical Contaminant Monitoring using Bivalves. Woods Hole, MA, USA.

García, J.E. 1999. *La Agrícultura Orgánica en Costa Rica.* San José, Costa Rica: Universidad Estatal a Distancia (EUNED).

Hidalgo, C. 1986. Determinación de residuos de plaguicidas organochlorados en huevos de ocho especies de aves acuaticas que anidan in la isla Pájaros, Guanacaste, Costa Rica [MSc thesis]. Universidad de Costa Rica. San José, Costa Rica.

Hilje, L. 1984. Estado actual del combate de plagas agrícolas en Costa Rica. *Ciencias Ambientales* (Costa Rica). 5–6:115–24.

Hilje, L. 1998. Un modelo de colaboración agrícola internacional para el manejo de moscas blancas y geminivirus en América Latina y el Caribe. *Manejo Integrado de Plagas*. 49:1–9.

Hilje, L., Cartín, V. and March, E. 1989. El combate de plagas agrícolas dentro del contexto histórico costarricense. *Manejo Integrado de Plagas* 14:68–86.

Hilje, L., Castillo, L.E., Thrupp, L.A. and Wesseling, C. 1992. El uso de los plaguicidas en Costa Rica. Heliconia/Universidad Estatal a Distancia (UNED) (ed.) San José, Costa Rica.

Hitz, W. 1995. Sustainable biomass use in organic agriculture, Case study: The Tapezco Organic Farmers Association. In: *Micro-hydro and Organic Farming for Sustainable Rural Development in Costa Rica*. San José, Costa Rica: Cromo S.A.

Instituto Regional de Estudios en Sustancias Tóxicas (IRET) database. Universidad Nacional. Heredia, Costa Rica.

Jaccarini, V. and Martens, E. 1992. The ecology of mangrove and related ecosystems. *Hydrobiologia*. 247:1–3.

Ministerio de Agricultura y Ganadería (MAG). 1997. Informe anual. Estadistica sectorial. www.mag.go.cr./est60o.htm.

Osorio, F. 1998. La calidad del Agua de la cuenca y del sistema nacional de riego y avenamiento: su reutilización agrioindustrial en Taboga y su efecto potencial en humedales del Bajo Tempisque, Costa Rica. [MSc thesis]. Universidad Nacional. Heredia, Costa Rica.

Pérez, O., Ramírez, O., Hilje, L. and Arremans, J.K. 1997. Potencial de adopción de dos opciones tecnológicas de Manejo Integrado de Plagas (MIP), aplicando tres técnicas de extensión con productores de tomate en el Valle Central occidental, Costa Rica. *Manejo Integrado de Plagas* 43:19–30.

Phillips, P. 1983. Dial and monthly variation in abundance, diversity and composition of the littoral fish populations in the Gulf of Nicoya, Costa Rica. *Revista de Biologia Tropical*. 31:97–306.

Proyecto Estado de la Nación. 1998. Estado de la Nación en Desarrollo Humano Sostenible N°4. EDITORAMA, S.A. San José, Costa Rica.

Quesada, C. 1990. Estrategía de Conservación para el Desarrollo Sostenible de Costa Rica, ECODES. Ministerio de Recursos Naturales Energía y Minas. San José, Costa Rica. 180 p.

Ramírez, L. and Soto, G. 1999. Una experiencia de papa orgánica en la zona de Cartago, Costa Rica. www.infoagro.go.cr/tecnologia/papa/experiencia_papa_organica.htm.

Readman, J.W., Wee Kwong, L.L., Mee, L.D., Bartocci, J., Nilve, G., Rodríguez-Solano, J.A. and González-Farias, F. 1992. Persistent organophosphorus pesticides in tropical marine environments. *Mar Poll Bull.* 24(8):398–402.

Rodríguez, C.L. 1989. Problemática del combate de la polilla de la papa en Costa Rica. *Investigación Agrícola* (Costa Rica). 3:1–4.

Rodríguez, C.L., Rodríguez, M. and Camacho, H. 1988. La situación de los plaguicidas en las hortalizas de Costa Rica. In: Taller de expertos de manejo integrado de plagas Workshop in Costa Rica.

Rodríguez, C.L., Arce, A. and Brenes, F. 1989. Avances programa de transferencia de feromonas en polillas de la papa en Costa Rica. In: 8th Congreso de Agronomía Nacional, 1989 Workshop in Cartago, Costa Rica.

Rodríguez, S.M. 1997. Residuos de plaguicidas en la zona de influencia del proyecto de riego Arena-Tempisque. Informe final, Centro de Invetigación en Contaminación Ambiental, Universidad de Costa Rica. San José, Costa Rica.

SEPSA. 1995. Boletín Estadístico nr 6, Planificación Agropecuaria. SEPSA, Ministerio de Agricultura. San José, Costa Rica.

SEPSA. 1998. Boletín Estadístico nr 9, Planificación Agropecuaria. SEPSA, Ministerio de Agricultura. San José, Costa Rica.

Smith, S. and Stiles, G. 1979. Banding studies of migrant shorebirds in northwestern Costa Rica. *Studies in Avian Biology.* 2:41–7.

Soto, G. 1998. La agricultura orgánica. San José, Costa Rica: Documento elaborado para el Proyecto de Estado de la Nación.

Spongberg, A.L. and Davis, P. 1998. Organochlorinated pesticide contaminants in Golfo Dulce, Costa Rica. *Revista de Biologia Tropical.* 46(Suppl. 6):111–24.

Standley, L.J. and Sweeney, B.W. 1995. Organochlorine pesticides in stream mayflies and terrestrial vegetation of undisturbed tropical catchments exposed to long-range atmospheric transport. *Journal of the North American Benthological Society.* 14(1):38–49.

Subirós, F. 1995. El cultivo de la caña de azúcar. UNED ed. San José, Costa Rica. 448 p.

Teunissen-Ordelman, H.G.K. and Schrap, S.M. 1997. 1996 Aquatic outlook: an analysis of issues pertaining to aquatic environments: pesticides. Riza Policy document nr 97.038. Amsterdam, Netherlands.

Tomlin, C. 1997. *The Pesticide Manual*, 11th edn. Crop protection publication of British Crop Protection Council and The Royal Society of Chemistry. United Kingdom.

Trejos, A. (ed.) 1991. *Illustrated Geography of Costa Rica.* San José, Costa Rica: Trejos Hermanos Sucesores, S.A.

Universidad Nacional, Instituto Regional de Estudios en Sustancias Tóxicas (UNA/IRET). 1999. Manual de plaguicidas: Guía para América Central. In: Castillo, L., Chaverri, F., Ruepert, C., Astorga, Y., Monge, P. and Wesseling, P. (eds) *Manual de Plaguicidas.* Heredia, Costa Rica: Editoria Universitaria EUNA.

Vega, S., Zuñiga, C.M., García, R., Rodríguez, A., Solano, G. and Maroto, I. 1983. *Importación y Exportación de Plaguicidas en Costa Rica: Mercado, Ecología, Salud.* Heredia, Costa Rica: Universidad Nacional.

von Düszeln, J. 1988. Análisis de plaguicidas en Costa Rica, con especial énfasis en muestras de agua y peces. Deutsche Gessellschaft für Technishe Zusammenarbeit (GTZ) Technical Report. Bremen, Germany: GTZ. Report nr GTZ PN 85.2039-7.

Wesseling, C. and Castillo, L.E. 1989. Uso y registro de los plaguicidas en América Central en la década de los ochenta. *Memorias de la Jornadas de Toxicología* (1):19–20.

WHO. 1990. *Public Health Impact of Pesticides Used in Agriculture.* Geneva, Switzerland: WHO.

Chapter 13

Pesticides in Colombia
Their application, use and legislation

Luz Angela Castro

INTRODUCTION

Pesticides have been used in Colombia to control not only agricultural pests but also vectors of both medical and veterinary importance (Calderón, 1994). The vectors for malaria, Chagas' disease, dengue, yellow fever, and Venezuelan equine encephalitis have been controlled with pesticides, leaving no doubt as to the contribution of judicious pesticide applications to the public health. However, any use of pesticides implies risks for humans, domestic animals, and the environment with contamination of air, water, soil, and foodstuffs as potential unintended effects that should be considered when making the decision to use pesticides. The potential risks should also serve to guide the development of regulations and procedures for determining economic and public health thresholds for the safe use of pesticides. A risk to benefits-based approach can and should guide users when they try to determine the correct choice of pesticide, application equipment, timing, and optimal conditions for employing pesticides.

Pesticide selection based on cost, efficacy, or non-target toxicity has become a difficult task because of the development of insect resistance to some pesticides. Resistance development has prompted the search for new pesticides with greater efficacy, although often at the cost of greater mammalian toxicity or unevaluated environmental effects. The substitution for persistent OCs by the more toxic OPs has increased the risk of contamination for sprayers and also for individuals living in nearby communities. Other pesticides developed in recent years include phenoxy-acetics, oximes, bipyridils, carbamates, triazoles, dinitrophenols, and a myriad of other chemical types. At present, several government ministries in Colombia, namely the Agriculture, Public Health, and Environment Secretariats, have proposed programs to improve the management of and conditions for the use and handling of pesticides in an effort to diminish the dangerous effects of these products. This development has led the Colombian Government to pass new legislation on pesticides, specifically Decree 1843 of 1991, that adopted and integrated the recommendations of the FAO Code of Conduct and International Sanitary Rules. The implementation process has now begun through involvement by the various sectional offices around Columbia.

The Colombian Government is a party to a number of international agreements on environmental issues including the Antarctic Treaty, Biodiversity, Climate Change, Desertification, Endangered Species, Hazardous Wastes, Marine Life Conservation, Nuclear Test Ban, Ozone Layer Protection, Ship Pollution, Tropical Timber 83, and Tropical Timber 94. Colombia has signed but not yet ratified the Antarctic-Environmental Protocol, Law of the Sea, and Marine Dumping treaties (CIA, 2000).

OVERVIEW OF COLOMBIA

Description and location

The fourth largest country in South America, Colombia – lying well south of the Tropic of Cancer – occupies the northwest corner of South America bordering the Caribbean Sea between Panama and Venezuela and bordering the North Pacific Ocean between Ecuador and Panama (Figure 13.1). It is roughly equidistant from both extremes of the American continents, although its major portion lies within the Northern Hemisphere. Colombia has maritime ports on the Caribbean Sea, on a coastline that stretches across 1,760 km, and the Pacific Ocean – along a coastline covering 1,448 km. Thus, Colombia is a strategic crossroads in the network of communication between North, Central, and South America. The Panama

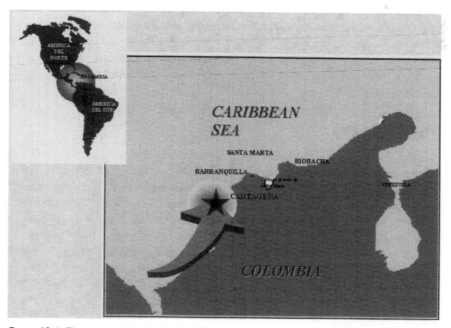

Figure 13.1 The geographic location of Colombia

Canal, with some of the most intensive maritime traffic in the world, is only a short distance away (350 to 450 km).

Colombia's maritime jurisdiction covers 100,210 km^2 and includes a territorial sea to 12 nautical miles offshore, the continental shelf to 200 m in depth or to the depth of exploitation, and an exclusive economic zone of 200 nautical miles offshore. Its maritime and terrestrial frontiers adjoin the Dominican Republic, Haiti, Jamaica, Honduras, Nicaragua, Costa Rica, Panama, Venezuela, Brazil, Peru, and Ecuador.

Colombia gained its independence from Spain in 1819 – earlier than most countries in the region. It established traditions of civilian government with regular, free elections. However, in recent years, assassinations, widespread guerrilla activities, and drug trafficking have severely disrupted normal public and private activities.

Extension and limits

The northernmost continental point in Colombia reaches Lat. 12°26'46"N at Punta Gallinas on the Guajira Peninsula in the Guajira Department; its southernmost point reaches Lat. 4°13'30"S in Quebrada San Antonio on the Amazon River in the Amazonas Department. The easternmost point is at the common boundary between Brasil, Venezuela, and Colombia at Long. 66°50'54"W in the Guainía Department and the westernmost reaches Long. 79°02'33"W in the Narino Department. Colombia's only important outlying possession is the San Andrés y Providencia Archipelago, located in the Caribbean Sea 160 km off the eastern coast of Nicaragua between Lat. 12° to 16°30'N and Long. 78° to 82°W. Colombia has a total national area of 1,138,910 km^2 with 1,038,700 km^2 of land and 100,210 km^2 of maritime area. Colombia has sovereignty over the Isla de Malpelo off the Pacific Coast and Roncador Cay, Serrana Bank, and Serranilla Bank in the Caribbean Sea.

Physiography

A short distance from its southern border with Ecuador, the Andes branch into three cordilleras, each lying roughly parallel in a south-north direction. These cordilleras help break the country into five natural units: the Andean region itself, the Pacific Lowlands, the Caribbean Coastal Plain, the eastern *Llanos* Plain, and the Amazonian basin lowlands. The Cordillera Occidental (western) and the Cordillera Oriental (eastern) are separated from the Cordillera Central by the Cauca and Magdalena valleys. In the Andean region, steep slopes alternate with long valleys and plateaus or high flat areas forming an extremely varied landscape. Elevation extremes range from sea-level on the coasts to 5,775 m on Pico Simon Bolívar and Pico Cristóbal Colón both located in the Sierra Nevada de Santa Marta massif.

West of the Cordillera Occidental, a narrow plain – composed of comparatively recent alluvial soils deposited by the San Juan River and numerous streams, which drain the western slopes of the Cordillera – faces the Pacific Ocean. One of the rainiest spots on Earth – with some locations receiving 990 cm annually – it is covered with marshlands and tropical rain forests. Low hills between the watersheds of the San Juan and Atrato rivers separate the Pacific and Caribbean coastal plains. The Caribbean Coastal Plain is a lowland formed by alluvium in the flood plains of the lower Magdalena, Cauca, Sinu, and Atrato valleys. Near the coast the plain is grassland subject to seasonal droughts, although its southern section, in contact with the northern spurs of the Andes, is forested marshland affected by seasonal flooding. East of the mouth of the Magdalena River, the coastal plain is interrupted by the heights of the Sierra Nevada de Santa Marta.

Colombia's eastern half is a vast inland plain, which extends well beyond the Venezuelan and Brazilian borders. The *Llanos* occupies the northern part while Amazonia lies south of the Guaviare River. The *Llanos* is a vast tropical savanna with a wet-and-dry tropical climate and poor soils. It is drained through the Arauca, Meta, and Vichada rivers to the Orinoco. Sparsely populated and undeveloped, the region traditionally has been used for open-range cattle ranching, although rich petroleum deposits were discovered in 1984 along the Andean Piedmont. Amazonia has a warm tropical climate and is almost as rainy as the Pacific Plain. Until the 1950s, this area was almost entirely covered with *selvas* (humid tropical forests). Large numbers of settlers along the upper Guaviare, Caquetá, and Putumayo rivers are rapidly turning the forest into cropland and pastures.

Colombian natural resources include petroleum, natural gas, coal, iron ore, nickel, gold, copper, and emeralds. Only 4 percent of Colombia's total land area is considered arable (1993 estimate) with 1 percent in permanent crops, 39 percent in permanent pastures, 48 percent in forests and woodlands, and the remaining 8 percent is unclassified (CIA, 2000).

Climate, natural regions, and agriculture

Situated at low latitudes, Colombia has a tropical climate along the coast, eastern plains, and Amazon Basin with cooler temperatures in the central highlands. Latitude and elevation within the cordilleras are the most important factors determining the climate. Weather conditions are controlled by the Inter-tropical Convergence Zone, a low-pressure belt that moves back and forth across the equator. The *Alisios*, or trade winds, flow from the north northeast and are strongest from December through May. Wet and dry seasons alternate in the Andes, the *Llanos*, and the Caribbean Plain. The Pacific region and Amazonia, which lie along the equator, have abundant rainfall evenly distributed throughout the year. On Guajira Peninsula semidesert conditions prevail with an annual precipitation of about 30 centimeters.

Tierra caliente, the warm lowlands, extend up to about 1,000 m followed by the *tierra templada*, or temperate zone, reaching upward to 2,000 m. Above this lies the *tierra fría*, or cool climate zone, extending up to 3,000 m. Above 4,000 m and below the permanent snow line at about 4,700 m is a zone of treeless *páramos*. The tropical plant species of the *selvas* give way gradually to subtropical woods in the *tierra templada* and to temperate species in the *tierra fría*. The *páramos* is covered with alpine meadows that give way to barren rock and snow.

Based on its topography, climate, vegetation, availability of natural resources, and degree of development and human population level, Colombia can be divided into six different regions: the Caribbean (Caribe), Pacific (Pacifico), Andean (Magdalena-Cauca), Amazonian (Amazonas), Oriental Plains (Orinoco), and the San Andrés y Providencia Archipelago (Instituto Geográfico Agustín Codazzi, 1996; Figure 13.2). The mountainous character of Colombia's territory, along with the attendant climatic variations of the different vertical zones, allows for the production of an unusually wide range of both tropical and temperate zone crops. The warm lowlands are devoted to rice, cotton, banana, sugarcane, cassava, and cacao while coffee, manioc, plantain, and citrus cultivation is typical of the intermediate *tierra templada*. Commercial cut flowers including carnations, roses, pompoms, and alstroemeria are grown on more than 4,200 ha in the *tierra fría*.

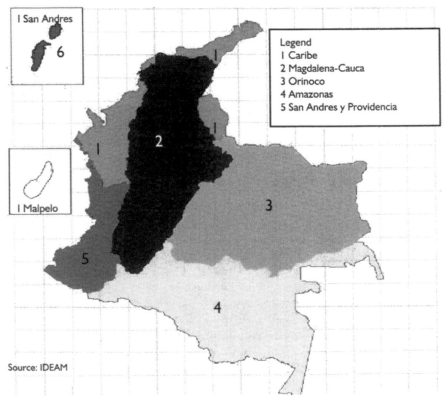

Figure 13.2 The natural regions of Colombia

Certain grains such as corn, wheat, barley, and oilseed grow well in the cool uplands, which also support the best dairylands. The *páramos* are well suited for potato farming with wheat production on its margins. Coffee constitutes the backbone of the Colombian economy, accounting for about half of all legal exports.

For a country with coasts on two seas, ocean fishing is little developed. River fish constitute the more abundant catch, although stocks have been reduced as a result of pollution and siltation. Shrimp farming has achieved some success in Colombia.

Hydrography

Colombia is rich in water resources with five primary hydrographic systems named for their principal river or destination including the Caribbean, Pacific, Amazon, Orinoco, and Catatumbo (Figure 13.3). The Caribbean hydrographic system drains

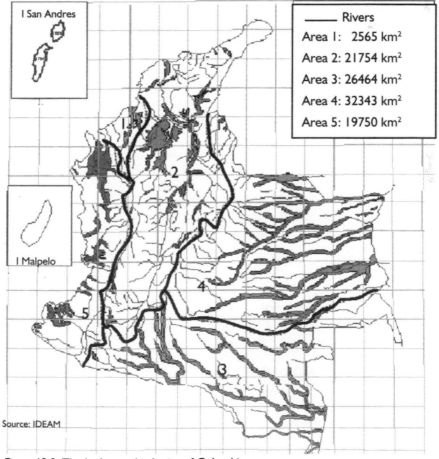

Figure 13.3 The hydrographic basins of Colombia

435,000 km². Its principal river is Magdalena, which is 1,600 km in length. With about 15 tributaries, it has a drainage area of 257,000 km². The Magdalena River system supports the primary farming and livestock region of Colombia. It has a mean flow of 6,700 m³ sec⁻¹ and carries 125 M m³ of sediments annually to the Caribbean Sea. The Pacific system is made up of numerous rivers of short length and high stream flow and covers about 90,000 km². These tributaries – rising in the Andes and flowing into the upper Amazon River that constitutes the border with Peru for 116 km – drain about 332,000 km². Ten rivers form Colombia's 263,000 km² drainage basin in the *Llanos* region and flow into the Orinoco River. The Catatumbo River drains 18,500 km² into the Maracaibo Lagoon. There are 14,300 km navigable by river boats along Colombia's many river systems.

Colombia also has a number of bays and harbors. The Caribbean ports of Cartagena, Barranquilla (Bocas de Ceniza), and Santa Marta have relatively deep water and are equipped with modern port facilities and services. However, silt deposited by the Magdalena River at its mouth requires constant dredging to maintain ship access to Barranquilla. Several coastal lagoons are located along the Caribbean and represent some of Colombia's most important estuarine ecosystems. They include Ciénaga de la Virgen, Ciénaga la Caimanera, and Ciénaga Grande de Santa Marta, which is the largest and one of the most important fishery grounds in Colombia. Other important estuarine ecosystems include Cartagena Bay, Barbacoas Bay, and Cispata Bay – the latter located in the Morrosquillo Gulf. The port of Buenaventura is located on the Pacific coastline and is situated on a mangrove-lined bay. Access to Buenaventura is easy for ships and there are also modern port facilities. Tumaco, to the south, has also been marked for development. Colombia's other major ports and harbors include Bahia de Portete, Puerto Bolivar, San Andrés, and Turbo in the Caribbean and Leticia on the Amazon River.

PESTICIDE REGULATION IN COLOMBIA

Pesticide production

Pesticides were first formulated in Colombia from imported a.i.(s) in 1962. In 1964, a.i.(s), e.g. maneb and mancozeb, were produced in Colombia from native or imported raw materials from DuPont, and Rohm and Haas in Barranquilla. These were followed by propanil and diuron a few years later. In 1985, Proficol el Carmen SA began production of propanil and in 1995 Fertilizantes y Plaguicidas del Valle produced mancozeb at its plant in Ginebra, Valle (ICA-ANDI, 1997).

The agrochemical sector has the capacity to produce 34,800 T of dry pesticides and 56 M L of liquid products annually. However, current production is estimated at only 60 percent of installed capacity. Of 40,000 T of formulated products sold annually, 84 percent is derived from local production and 16 percent from imports. Colombia exports approximately 17,000 T of dry and 4 M L of liquid pesticides (Ministerio de Salud, 1994). Currently 98 companies are registered to produce and sell pesticides in Colombia, of which 25 are part of multinational corporations and 73 are locally owned companies – including flower, tobacco, palm, and banana farmers who import pesticides directly (Ministerio de Salud, 1994).

There are 280 pesticide a.i.(s) registered in Colombia and these are used in 700 commercial formulations (230 insecticides, 250 fungicides, 150 herbicides, and 98 formulations in other classes). Total national production of a.i.(s) for 1994 was more than 28 M kg – equivalent to 53,234,581 kg or L of formulated products – of which 51.2 percent were fungicides, 36.3 percent herbicides, 11.1 percent insecticides, 1.2 percent coadjuvants, 0.2 percent pyrethroids, and less than 0.1 percent physiological regulators (Ministerio de Salud, 1994). During 1995, there were 20,259,240 kg a.i. of pesticides used in Colombia including 4,204,620 kg of insecticides; 7,280,340 kg of fungicides; 8,322,140 kg of herbicides (Table 13.1); 400,260 kg of coadjuvants; 35,900 kg of pyrethroids; and 15,990 kg of physiological regulators (ICA-ANDI, 1997). This represents an increase of 15.35 percent over the 1994 consumption of 17,146,930 kg.

Table 13.1 Sales of pesticides in Colombia, according to type of product 1975–95

Year	Total sales T a.i.	Insecticides[a]		Herbicides[b]		Fungicides	
		T a.i.	%	T a.i.	%	T a.i.	%
1975	16,551	7,519	45.4	4,553	27.5	4,479	27.1
1976	17,462	10,338	59.2	3,356	19.2	3,768	21.6
1977	23,656	15,223	64.4	3,532	14.9	4,901	20.7
1978	15,766	6,127	38.9	5,165	32.8	4,474	28.4
1979	12,927	4,138	32.0	488	37.8	3,909	30.2
1980	12,572	3,641	29.0	4,621	36.8	431	34.3
1981	12,211	3,159	25.9	4,599	37.7	4,453	36.5
1982	12,853	2,836	22.1	5,203	40.5	4,814	37.5
1983	14,354	3,237	22.6	5,672	39.5	5,445	37.9
1984	1,546	3,523	22.8	6,001	38.8	5,936	38.4
1985	17,046	3,912	22.9	6,113	35.9	7,021	41.2
1986	16,229	3,671	22.6	6,257	38.6	6,301	38.8
1987	23,379	3,912	16.7	653	27.9	12,937	55.3
1988	21,484	4,705	21.9	6,098	28.4	10,682	49.7
1989	19,968	3,694	18.5	7,207	36.1	9,067	45.4
1990	17,602	4,006	22.8	6,573	37.3	7,023	39.9
1991	16,396	3,507	21.4	6,368	38.8	6,521	39.8
1992	14,947	2,645	17.7	6,067	40.6	6,235	41.7
1993	14,718	225	15.3	5,719	38.9	6,749	45.8
1994	16,358	2,534	13.8	7,454	45.6	6,736	40.6
1995	19,806	4,204	21.2	8,322	42.0	728	36.8

Notes:
a Includes insecticides, soil disinfectants, molluscicides, acaridicides, nematicides and rodenticides.
b Includes defoliants.

Pesticide use and handling in Colombian agriculture

Farming is a major sector of the Colombian economy, accounting for 20 percent of 1995 GDP while providing employment for 40 percent of the country's workers and generating 50 percent of export revenues. In 1990, there were 3,855,000 ha under cultivation (Table 13.2), with 50 percent planted in 'technified' or technologically-enhanced crops (Ministerio de Agricultura, 1992). Note that 'technified' is a construction of the word 'technification', which is a back-formation from the Spanish *tecnificación*. The practice of technification in Latin America was spurred by the spread of coffee leaf rust – *Hemileia vastatrix*, a fungal disease of coffee known in Spanish as *la roya* – in the 1970s. Technification projects were assisted with funding from the United States Agency for International Development (USAID). In coffee production, technification goes beyond the intensive management of shade and shrubs to the application of agrochemical inputs and the introduction of higher-yielding, disease-resistant varieties of coffee that respond well to those inputs. Like the 'Green Revolution' that was expected to provide miraculous high-yield agriculture through new strains of rice, wheat, and corn, the sun coffee revolution has failed to fulfill its promise (Hull, 1999). The results have been increased erosion, increased runoff laden with pollutants, significant reduction in wildlife habitat, and increased exposure of workers to hazardous chemicals (Smithsonian Institution, 1996). The percentage of technified coffee acreage – based on 1993 estimates – ranges from 10 percent in El Salvador and Haiti to 40 percent in Costa Rica and nearly 70 percent in Columbia (FAO, 1996).

The intensive and indiscriminate use of chemical pesticides in agriculture has resulted in many problems, most of which have not yet been evaluated in Colombia. Up to 70 percent (11,477 T) of 1991 total pesticide use – estimated at 16,396 T – was applied to five crops: rice (22.9 percent), cotton (22.2 percent), potatoes (12 percent), pastures (7.8 percent), and horticultural (4.9 percent) (Table 13.3). For the eight principal crops grown in Colombia, pesticide use was 4.0 kg ha^{-1} of formulated product in 1991 in four regions of industrialized agricultural production. This is significantly higher than the 2.75 kg ha^{-1} estimate for the USA. The higher use rate may be partially attributed to the more rapid degradation from tropical temperatures and weather and also to the greater biodiversity found in the tropics

Table 13.2 Total area under cultivation in Colombia, 1995

Crop	Ha(s) (x 10^3)	Percent of total
Temporary crops	1,810.2	42.5
Permanent crops with coffee	1,411.2	33.1
Agriculture without coffee	3,221.4	75.7
Coffee	1,036.4	24.3
Agriculture with coffee	4,257.8	100.0

Source: Agroeconomics Actuality Nr 3 in Sociedad de Agricultores de Colombia (SAC), 1996.

Table 13.3 Market composition of pesticide groups by crop for 1991 in Colombia

Crop	% of Total	% Insecticides	% Herbicides	% Fungicides
Cotton	22.2	47.8	13.0	0.0
Rice	22.9	10.4	34.0	20.0
Potatoes	12.0	13.9	1.0	31.0
Horticultural	4.9	4.9	2.0	11.0
Grass	7.8	0.1	18.0	0.0
Flowers	4.8	6.8	0.0	11.0
Banana	4.6	0.1	3.0	14.0
Sorghum	4.3	5.9	4.0	3.0
Sugarcane	4.0	0.2	9.0	0.0
Coffee	3.0	1.3	5.0	3.0
Corn	1.3	1.0	2.0	0.0
Soybean	1.6	1.4	3.0	0.0
Fruits	1.2	1.3	1.0	2.0
Others	5.4	5.0	5.0	5.0
Totals	100.0	100.0	100.0	100.0

Source: ANDI, Committee for Agrochemical Industry, 1992. Taken from: Bustamante de Henao, 1994.

where there are more insect species and diseases. Further, the continuous reproduction of pests at higher rates in the tropics gives them greater opportunity to adapt to chemical controls. This implies greater risk of economic loss from pests for agricultural activities and increased inefficiency in the use of pesticides (Ardila, 1994).

Based on pesticide use for the period 1982 to 1984, Colombia was the Latin American country with the highest consumption of pesticides per person at 0.72 kg per person behind Costa Rica (Table 13.4). Recent data suggest that the per capita consumption of pesticides has increased in Colombia. Total national sales of pesticides – the sum of domestic production plus imports – in 1995 was 46,111 T (Table 13.5) and Colombia's population was 37 million (Departamento Administrativo Nacional de Estadística, 1995) giving a use rate of 1.25 kg per person. Some authors have found evidence that the actual rates of pesticide use are much higher than the optimum level for protection and growth of plants. This suggests that some farmers are maximizing production with pesticides instead of maximizing efficiency (Olson *et al.*, 1987).

That is the case in one 17,000 ha region of Colombia with 9,713 ha under cultivation. This region is planted with potatoes, corn, beans, tomatoes, and horticultural crops and experiences an unusually high use rate of pesticides. Here, farmers use 247.1 kg ha^{-1} of fungicides and 29.7 L ha^{-1} of insecticides per cropping cycle or six times the rates recommended by the Instituto Colombiano Agropecuario (ICA) – 37.1 kg ha^{-1} for fungicides and 9.9 L ha^{-1} for insecticides. Obviously this level of use has economic ramifications in addition to occupational health, population health, and environmental implications (Gonzalez, 1994).

Table 13.4 Use of pesticides in Colombia and other Latin American countries based on use data for the period 1982–84

Country	Population (M)	Cultivated area (km²)	Quantity of pesticides (T)	Kg per person	Kg km⁻²
Costa Rica	2.6	31,844	8,000	3.08	251
Guatemala	8.4	42,000	3,000	0.36	71
Colombia	29	310,000	21,000	0.72	68
Mexico	81	600,000	53,000	0.65	88
Brazil	136	1,200,000	42,000	0.31	35
Worldwide	4,000		2,000,000	0.5	

Source: Finkelman and Molina. 1988. As cited by the WHO in Ginebra, 1995. Taken from: Report Jesús Minambiente Conselor.

The top five crops for insecticide use in 1991 – requiring 85 percent of total insecticide consumption – were cotton (47.8 percent), potatoes (13.9 percent), rice (10.4 percent), flowers (6.8 percent), and sorghum (5.9 percent) (Table 13.3). The principal a.i.(s) used in Colombia during 1995 are listed in Table 13.6. Some authors justify the high usage rate of insecticides on cotton based on the complexity of pest species attacking the crop (Barriga, 1994). However, rice cultivation, which has a simpler pest situation, is a major consumer of all three groups of pesticides (insecticides, herbicides, and fungicides). Rice cultivation has benefitted most from research designed to reduce insecticide use thanks to the joint efforts of the ICA, CIAT (El Centro Internacional de Agricultura Tropical or International Center for Tropical Agriculture), and FEDEARROZ (Federacion Nacional de Arroceros de Colombia or National Federation of Rice Growers of Colombia). They have developed an integrated pest control system that reduces the number of insecticide applications from six to one per harvest, thus avoiding unnecessary costs while maintaining the quality of production (Vergara, 1990).

Herbicide use in 1991 was greatest for rice (34 percent) followed by pastures (18 percent), cotton (13 percent), sugar cane (9 percent) and coffee (5 percent) (Table 13.3). Applications to these five crops consumed 79 percent of all herbicides used nationally in 1991 (ICA-ANDI, 1996). The principal a.i.(s) in both herbicides and fungicides used in Colombia during 1995 are listed in Table 13.6. Table 13.7 lists the production, sales, imports, and exports for all different types of pesticide products used for crop protection in Colombia during 1995. Fungicides were used primarily for potatoes (31 percent), rice (20 percent), bananas (14 percent), horticultural (11 percent), and flowers (11 percent) and together represented 87 percent of all fungicide consumption (Table 13.3). Some authors (Barriga, 1994) justify the heavy use of fungicides on potatoes, coffee, and bananas because disease pressure from 'the drop' (late blight caused by *Phytophthora infestans* (Mont.) de Bary), '*la roya*' (coffee leaf rust caused by the fungal pathogen *Hemileia vastatrix* Berk. and Broome) and 'black sigatoka' (caused by *Mycosphaerella musicola* Mulder (anamorph: *Pseudocercospora musae* (Zimm.) Deighton)) can significantly impact crop value if the

Table 13.5 Pesticides: national sales, importation for domestic use and exports for 1988–95 in metric tonnes

Year	1988	1989	1990	1991	1992	1993	1994	1995
Insecticides[a]								
National sales	4,594	3,609	3,949	3,429	2,645	2,250	2,534	4,204
Domestic use imports	ND	66.5	138.9	114.0	205.9	ND[b]	3,486	3,517
Exports	ND	364.9	431.7	309.2	463.1	569.2	892.9	1,314
Pyrethroids								
National sales	110.5	84.4	56.9	77.4	37.0	30	34.42	35.9
Domestic use imports	ND	ND			ND	ND	91.12	93.92
Exports	ND	7.1	4.8	6.8	7.52	5.54	19.13	71.27
Herbicides[c]								
National sales	6,098	7,207	6,573	6,368	6,067	5,719	7,454	8,322
Domestic use imports	ND	46.2	145.4	107.1	110.4	ND	98.58	11.76
Exports	ND	2,342	2,051	2,254	2,502	2,153	2,361	4,279
Fungicides								
National sales	10,682	9,066	7,023	6,521	6,235	6,748	6,737	7,280
Domestic use imports	ND	13.4	319.5	406.8	114.0	ND	7,563	10,130
Exports	ND	3,777	4,784	6,569	7,765	7,651	7,697	9,364
Physiological regulators								
National sales	28.5	26.2	18.3	16.5	16.9	13.4	11.1	15.9
Domestic use imports	ND	0.0	73.4	17.3	0.8	ND	44.2	7.7
Exports	ND	0.0	0.0	0.0	0.0	0.05	0.09	0.07
Co-adjuvants								
National sales	605.7	647.0	612.4	580.4	408.9	268.1	375.7	400.2
Domestic use imports	ND	0.0	16.2	4.4	11.3	ND	242	35.14
Exports	ND	0.7	5.8	20.8	9.22	2.3	26.9	18.07
Totals:								
I. Sales national ind.	22,118	27,134	25,511	26,154	26,157	25,411	20,259	35,306
National sales	22,118	20,641	18,233	16,993	15,410	15,023	17,146	15,047
Exports	ND	6,493	7,278.5	9,160.5	10,747	10,382	10,998	15,047
II. Apparent national consumption	22,118	20,767	18,926	17,643	15,852	15,030	38,431	46,111
National sales	22,118	20,641	18,233	16,993	15,410	15,030	17,146	20,259
Domestic use imports	ND	126.2	693.4	649.6	442.4	ND	21,285	25,852

Source: DNP-UDA, based on ICA data tabulated by Agricultural Insumes Division for years 1988 to 1992. Taken from: Bustamante de Henao 1994. Data for 1992, 1993, 1994, and 1995 were taken directly from a document prepared by ICA-ANDI (8 January 1997).

Notes: a Includes insecticides, fungicides and soil disinfectants, molluscicides, acaridicides, nematicides and rodenticides. b ND indicates no data. c Includes defoliants.

Table 13.6 Principal active ingredients marketed in Colombia during 1995

Active ingredient	Production		Sales		Imports		Exports	
	T	%	T	%	T	%	T	%
Insecticides								
Trichogramma	1,446	26	1,150	27	–	–	–	–
Chlorpyrifos	877	16	291	7	288	8	246	19
Methamidophos	570	13	370	9	537	5	179	14
Monocrotophos	307	6	282	7	431	12	268	20
Malathion	278	5	–	–	337	10	–	–
Endosulfan	269	5	223	5	315	9	53	6
Carbaryl	230	4	194	5	–	–	–	–
Carbofuran	212	4	200	5	256	7	–	–
Dimethoate	–	–	145	4	–	–	–	–
Total	5,543		4,205		3,518		1,315	
Herbicides								
2,4-D	2,234	18	1,694	20	1,044	9	557	13
Picloram	1,767	15	1,057	13	–	–	–	–
Propanil	1,765	15	654	8	–	–	1,723	40
Glyphosate	1,437	12	1,400	17	–	–	–	–
Ametryn	490	4	–	–				
Paraquat	–	–	610	7	631	5	–	–
3,4-dichloranilin	–	–	–	–	1,556	13	–	–
Butachlor	–	–	–	–	643	6	–	–
Tipa 4	–	–	–	–	621	5	–	–
Diuron	–	–	–	–	–	–	550	13
Total	12,148		8,322		11,763		4,279	
Fungicides								
Mancozeb	13,168	81	5,014	69	972	10	8,413	90
Propineb	944	6	548	8	902	9	303	2
Sulphur	627	4	629	9	–	–	–	–
Copper oxychloride	301	2	181	3	–	–	211	2
Caustic soda	–	–	–	–	2,335	23	–	–
Calcium lignosulphate	–	–	–	–	677	7	–	–
Fenbuconazole	–	–	–	–	642	6	–	–
Total	16,193		7,280		10,134		9,364	

Source: ICA-ANDI (1997) Pesticide marketing 1994–95: imports, production, sales and exports.

diseases are not controlled (Barriga, 1994). However, there are published IPM programs for disease control that significantly reduce the need for fungicide applications – Corporación Colombiana de Investigación Agropecuaria (CORPOICA) produces written recommendations for potato cultivation.

Production of illegal crops, e.g. marihuana, opium poppy, and cocaine has been discouraged (and sometimes partly controlled) by means of herbicides, primarily paraquat and glyphosate. Beginning in 1979, some Colombian scientists were consulted by the US Government about the possibility of using herbicides for this purpose. Five herbicides were chosen as potential candidates – 2,4-D; 2,4,5-T;

Table 13.7 Production, sales, imports and exports of pesticides in Colombia during 1995 (kg × 10³)

Type of product	Production	Sales	Imports	Exports
Co-adjuvants	402.82	400.26	35.14	18.07
Pyrethroids	51.90	35.90	93.92	71.27
Insecticides	5,542.98	4,204.62	3,517.95	1,314.62
Fungicides	16,192.69	7,280.34	10,134.32	9,363.78
Herbicides	12,148.05	8,322.14	11,763.22	4,279.33
Physiological regulators	11.28	15.99	7.69	0.07
Total	34,349.72	20,259.24	25,552.24	15,047.14

Source: Tabulated by the Agricultural Insumes Division ICA and the Industrial Chamber for Crop Protection ANDI. 1995 and taken from the ICA's Marketing of pesticides: imports, production, sales, and exports, 1995.

paraquat; diquat; and glyphosate – although the final recommendation was not to employ any of them. Instead, they recommended initiating research into products with low-impact on human health and the environment. They further recommended utilizing mechanical destruction and controlled burning to control these illegal crops (Sánchez et al., 1994a). Presently in Colombia, glyphosate is utilized in aerial applications to illegal crops, although various groups including the Environmental Risk Control Deputy Directory and the Potentially Toxic Substances Division of the Ministerio de Salud (Ministry of Health) have objected to its continued use (Sánchez et al., 1994b).

Pesticide dispersal in the environment

Pesticides can leave agricultural landscapes through drift, water movement, and even through residues associated with agricultural products taken to market. This may lead to negative effects on people not involved in agricultural activities. Aerial application, crop irrigation, and certain weather-related events after application can increase the movement of pesticide residues and the severity of unintended effects from their usage. Both legal and illegal activities in conjunction with accidental spills or leaks of pesticides can also result in adverse effects on food and water supplies (Carlson and Wetzstein, 1993). Workers dealing directly with pesticides are most at risk from accidental exposures. Environmental economists can quantify these negative aspects of pesticide use. This information can be used as the basis for decision-makers to reach sound judgements about cancellation, registration, or use restriction of individual pesticides. For this purpose, health effects should be separated into short-term effects, e.g. illness or death caused by accidental exposure, and long-term effects, e.g. mutagenic, carcinogenic, and neurological effects, on public health (Carlson and Wetzstein, 1993).

Distribution of pesticide residues in marine ecosystems along the Caribbean coast

Several studies (Garay et al., 1994) have been conducted by the Oceanographic and Hydrographic Research Center of the Colombian Navy (Centro de Investigaciones Oceanográficas e Hidrográficas or CIOH) in conjunction with other international entities, namely IAEA, UNESCO, IOC (Intergovernmental Oceanographic Commission), and the University of Miami, in Cartagena Bay and La Virgen (a coastal lagoon located just north of Cartagena Bay). Sources, concentrations and distributions of pesticides were determined and evaluated, with special emphasis on OCs due to their long-term persistence in the environment (Tables 13.8 and 13.9). In 1980, CIOH and the University of Miami conducted a monitoring study of Cartagena Bay (Table 13.8) (Pagliardini et al., 1982). Water, sediment, and organism samples contained OC pesticides, viz. DDT and its metabolites, aldrin, heptachlor epoxide, dieldrin, and endrin, with a mean concentration of $0.32 \mu g L^{-1}$. An accidental spill of an unknown quantity of Lorsban® 4E occurred at the Dow Quimica facilities in Cartagena Bay on June 19, 1989 (El Universal,1989). Chlorpyrifos, the a. i. of this insecticide formulation, polluted the bay and killed thousands of fish. An evaluation of its environmental damage is still in progress. Cartagena Bay is also in communication with the Magdalena River through the Canal del Dique. The Magdalena River, which flows through Colombia's most important agricultural areas, receives agricultural pesticide runoff and carries this load to both Cartagena Bay and its mouth.

In 1992, CIOH with financial support from IOC/UNESCO carried out a study of La Virgen coastal lagoon (Table 13.9). They found OCs including aldrin, heptachlor epoxide, DDT and its metabolites, lindane, and PCBs in water, sediment, and fish samples. CIOH measured sediment concentrations of 2.08 ppb α-HCH and for 63.65 ng L^{-1} for HCB (Garay and Castro, 1993). Muscle tissue samples from commercial fish species also contained these same pesticides. Concentrations varied between 0.23 and 8.94 ng g^{-1} of aldrin and p,p'-DDE, respectively.

Based on the 1992 study, CIOH and IAEA initiated in 1996 an additional study of the lagoon but also included nearby rice fields, several open (raw) sewage inflows that introduce significant amounts of organic pollutants into the ecosystem, and the principal affluents to the lagoon. Samples were taken seasonally (dry, intermediate, and wet season). DDT, DDE, DDD, and lindane were found in sediment and water samples of the lagoon affluents in all seasons, demonstrating the persistence and low degradation rates for these OCs. Concentrations in sediment samples ranged from 0 to 1.426 ng g^{-1} dw during the dry season to 0 to 10.678 ng g^{-1} in the wet season. Sediment and water samples from the lagoon, also contained PCBs. Wet season PCB water sample concentrations of 173.4 ng L^{-1} were measured at the sampling station located near the open sewage inflow.

In 1975, three OCs (Table 13.8) were reported from the mouth of the Magdalena River near Barranquilla. In addition, OC pesticides were detected in samples of water, sediment, and organisms near Rada de Tumaco in the Pacific Ocean (Ramirez, 1988a).

Table 13.8 Concentration of OC pesticides detected in waters of Colombia's Caribbean Coast (ppb)

Compound detected	Magdalena River Mouth 1975	Cartagena Bay 1980	Cienaga Grande 1986	Cienaga de la Virgen 1996
Aldrin	10	0.13	0.2–1.1	NA
Heptachlor epoxide	10	0.30	NA	NA
DDT	140	0.18	0.1	24.486
Lindane	ND[a]	NA[b]	0.4–44.2	0.970
Heptachlor	ND	NA	28.2	0.875
Dieldrin	ND	0.02	0.2–1.9	NA
Endrin	ND	0.32	NA	NA

Notes:
a ND indicates compound not detected.
b NA indicates sample not analyzed for the compound.

Table 13.9 Concentration of OC pesticides detected in sediments and soil along Colombia's Caribbean coast (ppb)

Compound	Cienaga Grande 1985	Cienaga Virgen 1992	Chengue Bay 1993	Cienaga Virgen 1996	Zone adjacent to Cienaga Virgen 1996
Aldrin	0.2–1.1	0–0.39	0.09–0.45	NA[a]	NA
Heptachlor epoxide	NA	0–2.10	NA	NA	NA
DDT	0–0.1	0–0.06	0.04–20.81	0.274	4.736
Lindane	0.4–44.0	0–2.54	0.23–0.54	0.11	0.175
Heptachlor	NA	NA	0.45–1.40	NA	NA
Dieldrin	0.2–1.9	NA	NA	NA	NA

Notes:
a NA indicates sample not analyzed for the compound. Taken from Ramirez (1988b), Garay *et al.* (1994), and Espinosa, *et al.* (1995).

The Instituto de Investigaciones Marinas y Costeras (INVEMAR) reported in 1996 that lindane (0.4 to 44.2 μg g^{-1}), heptachlor (1.97 to 28.2 μg g^{-1}), aldrin (0.17 to 1.07 μg g^{-1}), dieldrin (0.19 to 1.91 μg g^{-1}), p,p'-DDT (0 to 0.08 μg g^{-1}), and other unidentified pesticides (38.226 μg g^{-1}) had been found in sediment samples from Ciénaga Grande de Santa Marta on the Colombian Caribbean (Casanova, 1995). Sediment samples taken the previous year from mangrove stands surrounding the cienaga contained the same pesticides but their concentrations varied slightly with the season.

Pesticide residue distribution in other ecosystems

Aerial application of pesticides constitutes an important source of air pollution from pesticides. The primary problems are the location of airstrips in or near towns and fumigation flight patterns occurring over agricultural areas just outside towns (INDERENA, 1991). Also, some commercial airfields simultaneously serve as storage, mixing, loading, and refilling sites for aerial applicators. Generally landing strips do not have disposal sites for empty pesticide containers or disposal facilities for unused pesticide or rinse water. Two special cases for Colombia occur in the city of Cartagena and in the department of Tolima. These areas have pesticide fumigation airfields situated less than 2 km from crowded neighborhoods and close to schools, hospitals, potable water sources, and irrigation channels. In Tolima Department, 52 percent of the airfields do not have disposal sites for empty pesticide containers or waste-disposal equipment for rinse water or unused spray mix. There have also been instances where empty containers were resold at the airfields for potable water storage for the local population.

The concentration of pesticides in the atmosphere around several agricultural towns has been measured. In a small city on the Colombian Pacific bordering Ecuador, aldicarb and carbofura – with well-known toxic effects inhibiting the respiration of soils (Typic Dystrandept and Typic Dystropet) and affecting organisms – are being applied in doses of 29 and 30 ppm (Burbano, 1981). In potato production soils, the harmful effects of aldicarb and carbofuran to micro-arthropod populations have been confirmed. Also, along the border with Brazil, toxic effects on 100 percent of the soybean harvest have been noted when the crop was planted in soils treated with paraquat more than six months earlier. Other studies carried out in cotton fields have shown that after application of diuron, atrazine, and 2,4,5-T at the recommended levels toxic concentrations of nitrites (as high as 46 ppm) led to phytotoxicity in cultivated areas (Vergara, 1990).

Monitoring of potable water from several Colombian towns showed contamination by chloroform, dichlorobromomethane, dibromochloromethane, bromoform, and several pesticides (Chemical Lab for Environmental Monitoring, 1996).

Botero et al. (1996) compared seasonal concentrations of 12 OC pesticides in samples of breast muscle, associated skin, and subcutaneous fat of blue-winged teal Anas discors L. (Anseriformes: Anatidae) collected in Ciénaga Grande De Santa Marta, Colombia in 1987 and 1988. This region is surrounded by important agricultural lands and is fed by the Magdalena River, the watershed which includes the majority of Colombia's industrial and agricultural regions. Previous preliminary studies by Plata et al. (1993) had shown low levels of OCs in sediments and in several fish species. No OCs were found to be at or above concentrations known to affect reproduction in waterfowl. The mean p,p'-DDE concentrations in blue-winged teal samples did not differ between the spring (0.037 μg g^{-1} ww; range, ND–1.10; percent fat, 13.14 percent) and the fall (0.039 μg g^{-1} ww; range, ND–0.47; percent fat, 6.87 percent) and no PCBs were detected in samples from Colombia. For p,p'-DDD and p,p'-DDT only one spring sample was positive with a concentration of 0.026 and 0.029 μg g^{-1} ww.

Pesticide residues in food

Direct application of pesticides to crops in the late production (pre-harvest) stage can result in residue levels that may pose a risk to consumers. In Colombia several cases of food contamination have been reported. These incidents may have been caused by inadequate preparation, improper storage, or improper transport, e.g. seeds being treated with organic mercury compounds and hexachlorobenzene during storage as treatments against rodents, insects, etc.

PESTICIDE LEGISLATION IN COLOMBIA

International framework

International guidelines developed by various United Nations' Organizations have been used as benchmarks for Colombia's pesticide regulatory legislation. The FAO Code of Conduct was introduced in Colombia via decree 1843 of 1991 and regulates the use and handling of pesticides. Other instruments and international procedures such as the London Directives and Prior Informed Consent procedures have been adopted by Colombian national authorities to encourage the exchange of data and the introduction of concepts that encourage the environmentally safe management of chemical products and pesticides.

Colombian regulations

Pesticide regulation in Colombia has focused on direct government action using the Command and Control paradigm, following the regulatory example of nations, e.g. the US, but without the depth of regulation and control employed elsewhere.

The basic Colombian regulations related to the use, handling, and disposal of pesticides are: Decree 843 of 1969, National Sanitary Code of 1979, Natural Resources Code of 1974, Decree 1843 of 1991, and Resolution 992 of 1992. The scope and significance of each specific regulation, as regards certain activities with pesticides and specific products, are discussed below.

Decree 843 of 1969 regulates agricultural consumables and describes the area of action and responsibility for the agricultural and health sectors. This decree established that anyone, e.g. individuals, small businesses, or large corporations, who participates in the production, importation, or use of fertilizers, pesticides, and drugs for agricultural purposes must register with the ICA.

The National Sanitary Code of 1979 establishes in a general way that the import, manufacture, storage, transport, commerce, handling, or disposal of dangerous substances should be handled with the necessary precautions to avoid damage to human, animal, or environmental health. Protection procedures, based on current regulations, must be adopted to protect humans from any risk due to pesticide related activities. The Natural Resources Code of 1974 outlines the general criteria for protection of renewable natural resources and the environment.

Decree 1843 of July 1991, in conjunction with Resolution 992 of the ICA, constitutes the basic rules for epidemiological surveillance and integrated management – by officials of the agriculture, public health, and environmental ministries – of the registration, use, and handling of pesticides. It attempts to maintain communication between the different sectors mentioned above, pesticide users, and industry representatives, with the aim of promoting studies that develop solutions to specific problems associated with pesticides. It specifies controls for agriculture, health, and environmental activities related to pesticides and establishes the concept of providing toxicological data as a requirement for obtaining trade (sale) licenses from the ICA. ICA Resolution 992 of 1992 specifically establishes regulations for trading (sales) and production of pesticides for agricultural purposes.

There are a number of other laws and regulations related to pesticides that have been passed and promulgated through the Departamento de Sanidad Vegetal y Asistencia Técnica in Colombia. A chronological listing follows with a brief description of their purpose(s) (Morales, 1978).

1938	Established regulations for inspection procedures, trade controls, insecticide and fungicide application methods, and application equipment.
1950	Set fines for violation of sanitary regulations.
1953	Regulated uses of 2,4-D and similar herbicides (superseded by Resolution 482 of 1968).
1956	Regulations on the import, manufacture, commerce, use, and application of pesticides.
1957	Set registration procedures for pesticides used in agriculture and established rules for packaging.
1963	Established rules for sales and application of pesticides.
1964	Regulated cotton production and instructed farmers to use agronomist engineers to supervise pest control.
1965	Set dates for sowing cotton and required licenses issued by the Ministerio de Agricultura (Ministry of Agriculture) to sell or apply pesticides.
1966	Regulated technical assistance and required written permission from an agronomist engineer to apply pesticides.
1967	Set requirement for a Certificate of Efficacy for pesticides. Precise definitions for active ingredient, technical material, additives, and toxicological requirements published by Ministerio de Salud.
1968	Agriculture sector was restructured by Decree 2420 that designated ICA to have control of all agricultural and cattle consumables.

Prohibitions of and restrictions to pesticide use in Colombia

The ICA and other government agencies have prohibited the production, importation, trade, or use of 21 specific pesticides (Table 13.10), as described chronologically below. In 1974, ICA canceled the registration of fungicides based on mercurial

Table 13.10 Chronology of Colombian legislation to prohibit the production, importation, trade, or use of 21 specific pesticides

Legislation	Pesticide	Directive
Resolution ICA 1158 of 1985	EDB	Prohibit import, production, sale for agricultural use
Resolution ICA 1042 of 1977	Phosvel	Cancels registration
Resolution 209 of 1978 Ministry of Agriculture	OCs, e.g. DDT, BHC, Lindane,	Prohibit use in coffee culture
Resolution ICA 749 of 1979	2,4,5-T and 2,4,5-TP	Cancels registration
Resolution ICA 1849 of 1985	Endrin	Prohibit import, production, sale for agricultural use
Decree 704 of 1986	DDT and its derivatives	Prohibits use in agriculture
Resolution ICA 891 of 1986	DDT and its derivatives	Cancels licenses of agricultural products containing this pesticide
Resolution ICA 930 of 1987	Dinoseb	Prohibit import, production, sale for agricultural use
Resolution 19408 of 1987	Chlordimeform and its salts	Prohibits use and handling
Resolution 366 of 1987, 531, 723, 874 and 724 of 1988 of the ICA	OCs aldrin, heptachlor, dieldrin, chlordane, camphechlor	Cancels licenses of products with these a.i.(s)
Decree 305 of 1988	OCs aldrin, heptachlor, dieldrin, chlordane, camphechlor	Prohibit import, production, sale
Resolution ICA 47 of 1988	Chlordimeform	Cancels licenses of products with this a.i.
Resolution ICA 5053 of 1989	Captafol	Prohibit import, production, sale for agricultural use
Resolution ICA 3028 of 1989	Paraquat	Prohibits aerial application
Resolution ICA 2308 of 1990	Tebuconazol	Prohibit import, production, sale for agricultural use
Resolution 2156, 2157, 2158, 2159 and 2857 of 1991 of the ICA	Lindane EC and WP formulations	Cancels licenses of products containing this a.i.
Resolution ICA 2471 of 1991	Parathion	Restricted use: cotton and technified pastures
	Methyl parathion	Restricted use: cotton and technified rice
Resolution ICA 243 of 1982	DBCP	Prohibit import, production, sale
Resolution ICA 29 of 1992	Fonofos	Prohibits agricultural use
Resolution 9913 of 6 Dec 1993, Ministry of Health	Maneb, zineb	Prohibits import, production, formulation, commercialization, use, handling, application
Resolution 10255 of 9 Dec 1993, Ministry of Health	Dieldrin, chlordane, mirex, PCP, DDT, BHC, lindane, heptachlor, dicofol	Prohibits import, production, formulation, commercialization, use, handling[b]

continued...

Table 13.10 continued

Legislation	Pesticide	Directive
Resolution 447 of 1974 Ministry of Agriculture	OCs	Prohibits use and sale for tobacco production
Resolution ICA 2180 of 1974	Mercury containing fungicides	Cancels agricultural use registration
Resolution 00138 of 17 Jan 1996 Ministry of Health	Pesticides containing methyl bromide	Prohibits import, manufacture commercialization, use
Resolution 02152/96 Ministry of Health	Methyl bromide	Restricted use: to treat exotic insect infestations on imported fresh vegetables[c]

Notes:
a Adapted from *Productos prohibidos en Colombia*, División Insumos Agricolas-ICA, 1999.
b Temporary exceptions include human use of lindane for ectoparasites – until the Ministry of Health determines that there are effective substitutes for this application – and endosulfan until a substitute of comparable effectiveness is available for use against the coffee berry borer *Hypothenemus hampei* Ferrari (Coleoptera: Scolytidae).
c Application in hermetically sealed containers with complete recovery of the methyl bromide after treatment – with the Division of Plant Health of the ICA supervising. This exception is valid only until an effective substitute is available.

compounds and the Ministerio de Agricultura forbade the use or trade of OC pesticides on tobacco crops. The ICA in 1977 (by Resolución ICA 1042 of 1977) canceled registration of pesticides containing leptophos (Phosvel) and the following year the Ministerio de Agricultura outlawed the employment of OC pesticides in coffee culture. ICA then canceled, in 1979, the registration of herbicides based on 2,4,5-T and 2,4,5-TP (Silvex). Following that, in 1982, the ICA banned the import, production, and trade of formulated pesticides based on dibromochloropropane (or DBCP) used to control soil pests.

In 1983, a court ruling (Codigo Sanitario Nacional) established that Title II, Law 09 of 1979 regulated water purification and FAO/WHO maximum permitted values for pesticides in potable water would be adopted. Further, the decree established that a desirable concentration of all pesticides in potable water was 'no detectable level' and that the Σ of all pesticide residues in potable water should not exceed 0.1 mg L^{-1} nor should individual tolerances for any particular pesticide be exceeded.

In 1985, the ICA stopped the import, production, and trade of pesticides based on ethylene dibromide (EDB) and endrin. Then, in 1986, the Ministerio de Salud and the Ministerio de Agricultura canceled the licenses of DDT, its derivatives and compounds, although temporary use was authorized for public health and sanitation programs and campaigns conducted by the Ministerio de Salud until 1994. The ICA followed suit by canceling all trade licenses for insecticides that contained DDT.

The next year, the ICA forbade the import, production, and trade of herbicides that contained the a.i. dinoseb and the Ministerio de Salud banned products based on chlordimeform and its salts. During both 1987 and 1988, the ICA canceled trade licenses of OC pesticides containing aldrin, heptachlor, dieldrin, chlordane, and camphechlor or toxaphene. In 1988, the Health and Agriculture ministries stopped their importation, production, formulation, and trade. However, they temporarily authorized the use of dieldrin and chlordane in wood products and camphechlor in the mixture of toxaphene and methyl parathion. In 1989, the ICA banned aerial fumigation with herbicides containing paraquat and the importation, production, and trade of fungicides containing captafol and canceled its trade licenses.

In 1990, the ICA stopped the import, production, trade, and application of fungicides that contained tebuconazole and, the following year, restricted the use of parathion to cotton and pastureland insect control and the use of methyl parathion to cotton and technified rice crops. In 1991, they canceled trade in lindane for certain uses. The following year they forbade insecticides based on fonofos. Also in 1992, the Ministerio de Salud received partial regulatory power from Chapter III of decree 1843 of 1991 to redefine toxicity categories from three to four – Category I, Extremely toxic; Category II, Highly toxic; Category III, Moderately toxic; and Category IV, Slightly toxic – and set toxicological classification criteria for pesticides. Then in 1993, the Ministerio de Salud stopped all import, production, formulation, trade, handling, and use of dieldrin, chlordane, declorane or mirex, pentachlorophenol, dicofol, DDT, BHC (HCH), heptachlor, lindane, and their related compounds. However, the ministry temporarily authorized the use of lindane to control ectoparasites for human health purposes and endosulfan until a comparable substitute was found against the coffee berry borer *Hypothenemus hampei* Ferrari (Coleoptera: Scolytidae). The Ministerio de Salud also authorized another year of use for DDT in anti-disease vector public health campaigns. They also stopped import, production, formulation, trade, handling, use, and application for the fungicides maneb and zineb and most related compounds. Only mancozeb is still in production in Colombia for both national consumption and for export. In 1994, mancozeb accounted for 80 percent (Table 13.6) of total national fungicide production, 70 percent of sales, 10 percent of fungicide imports and 88 percent of fungicide exports.

In 1995, the Ministerio de Salud attempted to ban endosulfan but failed because the rule was not published in the Official Government Diary. The following year the Ministerio de Salud banned the use of pesticides based on methyl bromide. However, a recent regulation allows use of this pesticide for quarantines established for sanitary control of imported products such as fresh fruits. Under this regulation, methyl bromide must be used in a hermetically sealed container or space and following treatment the pesticide must be completely recovered.

CONCLUSIONS

The Colombian government needs to implement more stringent control of pesticides through its Agriculture, Public Health, and Environment Ministries because existing controls are hampered by a lack of human, technical, and training resources. Analytical laboratories are needed in areas where pesticides are used the most. Because very little information is provided to government ministries by companies producing pesticides, enforceable regulations requiring such information is needed.

In Colombia during 1995, herbicides were the most used type of pesticide, specifically the OP herbicide glyphosate and the quaternary ammonium herbicide paraquat. These were used against illegal crops like cocaine, despite no technical recommendation for such use appearing on the label. However, their use has yet to be forbidden by law.

Data collected from the various Colombian Ministries lead to the conclusion that farmers in Colombia use more pesticides than they really need to use, due to lack of knowledge on the part of technicians who advise users of pesticides. Another obvious conclusion is that poor management of pesticides, either intentionally or accidentally, has lead to high levels of pesticides, especially OC pesticides, in coastal ecosystems of the Colombian Pacific Ocean and the Caribbean Sea. These high levels of pesticides can lead to a deterioration in the quality of life. Persistent pesticides are washed off crops and out of the soil by rain and eventually are carried downstream to rivers to be deposited in coastal ecosystems. These ecosystems continue to suffer from the impact of pesticides, due to their persistence. Although Colombia has been recognized on a regional basis as a country well advanced in the implementation of laws and regulations to control pesticides, there is still a long way to go to fully achieve total correspondence between words and actions.

Problems with pesticides must be approached through their different facets: agricultural, economic, political, and public health. Only in this way can priorities be set to adequately protect the quality of life, health, and food supply while achieving an environment for all Colombians that is less contaminated.

REFERENCES

Ardila, S. 1994. Agrochemicals and their relationship with the environment (Los agroquímicos y la relación agricultura medio ambiente. Mimeo). Los Andes University. Santa Fe de Bogotá, Colombia.

Asociación Nacional de Industriales (ANDI), Committee for the Agrochemical Industry. 1992. Pesticide sector: Proposal for business with Mexico. Santa Fe de Bogotá, Colombia: ANDI. 15 October 1992.

Barriga, J.M. 1994. Agrochemicals in Colombia, CENSAT (Centro National Salud Ambiente y Trabajo). In: *Los Plaguicidas en America Latina*, segunda edición. Memorias del seminario – taller internacional, held 21–25 April 1992 in Santa Fe de Bogotá. Santa Fe de Bogotá, Colombia: Ministerio de Salud.

Botero, J.E., Meyer, M.W., Hurley, S.S. and Rusch, D.H. 1996. Residues of organochlorines in mallards and blue-winged teal collected in Colombia and Wisconsin, 1984–1989. *Archives of Environmental Contamination and Toxicology* 31(2):225–31.

Burbano, A.A. 1981. Taken from: Vergara, R. 1990. Use and abuse of pesticides in Colombia. In: *Plaguicidas, ambiente y salud humana*: 1) Simposio internacional y 2) nacional, held 13–17 November 1990 in Palmira Valle, Colombia.

Bustamante de Henao, R. 1994. Consumables and agricultural equipment. In: Gonzalez, C. and Jaramillo, C.F. (eds) *Competitive Without Poverty: Studies for Country Development in Colombia*. Santa fé de Bogotá, Colombia: DNP-FONADE (Departamento Nacional de Planeación-Fondo Financiero de Proyectos de Desarrollo).

Calderón Carlos, E. 1994. Introducción a la primera edición. In: *Los Plaguicidas en America Latina*. Memorias del seminario – taller internacional. La problemática de los plaguicidas en la región de la Americas segunda edición. 21–25 April 1992. Santa Fé de Bogotá, Colombia.

Carlson, G.A. and Wetzstein, M.E. 1993. Pesticides and pest management. In: Carlson, G.A., Zilberman, D. and Miranowski, J.A. (eds) *Agricultural and Environmental Resource Economics*. New York: Oxford University Press.

Casanova, R. 1995. Estudio de la contaminación por compuestos organoclorados en la Costa Pacífica Colombiana. *Boletín Científico, Centro Control Contaminación del Pacífico* 5:15–23. Colombian Navy. San Andres de Tumaco, Colombia.

Chemical Laboratory for Environmental Monitoring. 1996. Study conducted 15 November 1995–15 January 1996. Santa fé de Bogotá, Colombia.

Central Intelligence Agency (CIA). 2000. *The World Factbook 2000*. http://www.odci.gov/cia/publications/factbook/index.html

Departamento Administrativo Nacional de Estadística (DANE). 1995. Santa fé de Bogotá, Colombia.

División Insumos Agricolas-ICA. 1999. Productos prohíbidos en Colombia. http://www.presidencia.gov.co/ica/nuestroinstituto/areaagricola/insumosagricolas/prohibidos.htm.

El Universal.19 June 1989. Cartagena, Colombia.

Espinosa, L., Ramirez, G. and Campos, N.H. 1995. Análisis de residuos de OC en los sedimentos de zonas de Manglar en la Ciénaga Grande de Santa Marta y la Bahía de Chengué. Punta Betín, Santa Marta, Colombia. Anales INVEMAR. 24:79–94.

FAO. 1996. Production yearbook Vol. 49:171–73. Rome: FAO.

Finkelman, J. and Molina, G. 1988. As cited by the WHO in Ginebra. 1995. Sanitary consequences of pesticide employment in agriculture.

Garay, J.A. and Castro, L.A. 1993. Final report of impact evaluation on marine ecosystems generated by the use of pesticides, Cienaga de la Virgen. Santa Fé de Bogotá, Colombia: Ciencias del Mar, Comisión Colombiana de Oceanografia (CCO).

Garay, J. Castro, L.A. and Pión, A. 1994. Impacto de los plaguicidas organoclorados en los ecosistemas de la Ciénaga de la Virgen, Municipio de Cartagena, Caribe Colombiano. Memorias IX. Seminario Nacional de Ciencia y Tecnología del Mar y Congreso Latinoamericano. Santa Fé de Bogotá, Colombia: Ciencias del Mar, Comisión Colombiana de Oceanografia (CCO).

Gonzalez, C. 1994. Inauguración del seminario taller sobre la problemática de plaguicidas en las región de la Américas. In: Los plaguicidas en America Latina. Memorias del seminario-taller internacional. La problemática de los plaguicidas en la región de la Americas segunda edición. 21–25 April 1992. Santa Fé de Bogotá, Colombia.

Hull, J.B. 1999. Can coffee drinkers save the rain forest? The Atlantic Monthly 284(2):19–21.

Instituto Colombiano Agropecuario-Asociación Nacional de Industriales (ICA-ANDI). 1996. Pesticides Marketing. Santa fé de Bogotá, Colombia: ICA.

ICA-ANDI. 1997. Pesticide Marketing 1994–1995: Imports, Production, Sales and Exports. (Comercialización de plaguicidas: Importación, producción, ventas, exportación 1994–1995, primera edición, producción editorial). Santa fé de Bogotá, Colombia: ICA.

Instituto Geográfico Agustín Codazzi. 1996. Atlas de Colombia Tercera Edición. Santa fé de Bogotá, Colombia. 620p.

Instituto Nacional de los Recursos Naturales Renovables y del Ambiente (INDERENA). 1991. Use and Abuse of Pesticides in Colombia. Santa fé de Bogotá, Colombia.

Ministerio de Agricultura. 1992. Santa fé de Bogotá, Colombia: Ministerio de Agricultura.

Ministerio de Salud. 1994. Pesticides in Latin America. Collection: Health, Environment, and Development. Volume 2. Santa fé de Bogotá, Colombia: Ministerio de Salud.

Morales, C. 1978. Regulations for imports, distribution, and trade of pesticides. Pesticide management and environmental protection seminar. Ministerio de Agricultura, ICA, Ministerio de Salud, University of California and USAID, held 13–17 February 1978. Santa fé de Bogotá, Colombia.

Olson, R., Frank, K., Grabouski, P. and Rehm, G. 1987. Economic and agronomic impacts of varied philosophies of soil testing. Agronomy J. 74:492–9. As cited by Daberkow, S. and Reichelderfer, H. 1988. Low input agriculture: trends, goals and prospects for input use. Am J Agricul Econ. 70(5):1159–66.

Pagliardini, J.L. et al. 1982. Síntesis del proyecto Bahía de Cartagena. CIOH Scientific Bulletin Nr 4. Cartagena, Colombia: CIOH.

Plata, J., Campos, N.H. and Ramírez, G. 1993. Flujo de compuestos organoclorados en las cadenas tróficas de la Ciénaga Grande de Santa Marta. *Caldasia* 17:199–204.

Ramirez, G. 1988a. Niveles de contaminación por plaguicidas organoclorados en sedimentos de la Cienaga Grande de Santa Marta. *INVEMAR Boletín Científico* 16:127–33.

Ramirez, G. 1988b. Estudio de plaguicidas organoclorados en sedimentos de la Ciénaga Grande de Santa Marta. Punta Betín, Santa Marta, Colombia. *Anales INVEMAR* 18: 127–33.

Sánchez, J., Romero, F. and Calderón, C. 1994a. Los plaguicidas: su impacto en el medio ambiente, la salud y el desarrollo. In: *Informe del Ministerio de Salud a propósito del uso del glifosato en la erradicación de cultivos de amapola, 1992*. Taken from: *Los Plaguicidas en America Latina*, segunda edición. Memorias del seminario-taller internacional, held 21–25 April 1992 in Santa Fe de Bogotá. Santa Fe de Bogotá, Colombia: Ministerio de Salud.

Sanchez, J., Romero, F. and Calderón, C. 1994b. Los plaguicidas: su impacto en el medio ambiente, la salud y el desarrollo. In: *Los Plaguicidas en America Latina*, segunda edición. Memorias del seminario-taller internacional, held 21–25 April 1992 in Santa Fe de Bogotá. Santa Fe de Bogotá, Colombia: Ministerio de Salud.

Smithsonian Institution. 1996. Biological conservation newsletter no. 161 November. Washington, DC: Department of Botany, National Museum of Natural History http://www.mnh.si.edu/botany/bcn/issue/161.html.

Sociedad de Agricultores de Colombia (SAC). 1996. *Agroeconomics Actuality* Nr 3.

United States Central Intelligence Agency (CIA). 2000. *The World Fact Book*. www.odci.gov/cia/publications/factbook/index.html.

Vergara, R. 1990. La problemática del uso y abuso de plaguicidas en Colombia. Cita del CIAT 1989. In: Plaguicidas, ambiente y salud humana: 1) Simposio internacional y 2) nacional, held 13–17 November 1990 in Palmira Valle, Colombia.

Chapter 14

Pesticide use in Cuban agriculture, present and prospects

*Gonzalo Dierksmeier, Pura Moreno,
R. Hernández and K. Martinez*

OVERVIEW

Geographical location and topography

With two major islands and many small keys, the Cuban archipelago is situated in the tropical Caribbean Sea between Lat. 19°47'36" to 23°17'09"N, Long. 80°53'55" to 84°57'54"W. It has a land mass area of 114,524 km² (National Geographic Society, 1981) and the two main islands are predominantly flat. Only 21 percent of Cuba's land area is mountainous and this is concentrated in three areas in the eastern, central, and western provinces. The mountain's heights vary between 200 and nearly 2,000 meters. The three mountainous regions, covered by dense forest, are the source of many of the watersheds and rivers of Cuba. They are economically important for the valuable timber and other useful plants, e.g. fruit trees and medicinal shrubs, found there. Coffee and some other minor crops, e.g. banana (small-scale production only) are planted in some areas of the mountains.

Primarily Cuba is one large savannah with the exception of a few small wetlands covering about 4 percent of the total land mass and located almost exclusively in the southern portion of the two main islands. The largest of these two wetlands Ciénaga de Zapata (southwest central part of the largest island) is a protected region because of its biodiversity. The greater part of the savannah has fertile soils and is primarily agricultural (Academia de Ciencias de Cuba, 1992).

Geology

The Cuban archipelago formed at the end of the Eocene period and its present shape was determined by tectonic plate movement. In general, the region has a low level of seismic activity; only in the eastern portion of the main island does sporadic seismic activity occur (Atlas Nacional de Cuba, 1970).

Climate

The Cuban climate is typical of Caribbean islands in that it is hot and humid with only two well-defined seasons, summer and winter. In summer, the daily average solar illumination is >8.5 h out of a total of 10–14 h of solar radiation. The maximum solar flux is approximately 1.2 cal cm^{-2} min^{-1} at midday. Average wind velocity is low during summer with a maximum velocity of 7–10 km h^{-1} occurring during daylight hours. Winds generally decrease at night. However, hurricanes may develop during the summer. They annually threaten, and sometimes desolate, the region. The average air temperature is approximately 30°C with minor exceptions for microclimates in hilly regions, where the temperature is slightly lower, and in the eastern province of Santiago de Cuba, where it is higher. Humidity is high throughout the year, but reaches its highest values in summer when it consistently exceeds 95 percent.

In winter, air temperatures are lower, especially in the western part of Cuba due to the influence of frequent incursions of cold Arctic air masses. Winter temperatures average 20°C with minimums well below the average after passage of an Arctic cold front. Average solar illumination in winter is shorter, averaging seven hours while maximum solar flux decreases to <0.8 cal cm^{-2} min^{-1} at midday. Wind velocity in winter is greater with a maximum velocity of 20 km h^{-1} during daylight hours. Hurricanes do not develop during this season and humidity levels are lower than in summer, but rarely drop below 70 percent.

Rainfall

The average yearly rainfall in Cuba is 1,345 mm, depending on the season and, to a lesser extent, on the region. The normal rainy season is from May to October when 80 percent of the total yearly rainfall occurs. Regionally, rainfall is unevenly distributed with less rainfall along the northern coast. In the mountainous regions rainfall is heavier, especially during the summer months. Cuba normally experiences typical tropical-type rainfall, i.e. heavy short-duration downfalls. Ecologically this causes intense soil erosion which consequently increases the risk of contamination of lakes, ponds, rivers, and, ultimately, coastal zones from both agrochemicals and sediments loaded with organic matter. There are no deserts in Cuba, although several locations receive far below the yearly average rainfall (Atlas climático de Cuba, 1987).

Watersheds

Cuba has 563 watersheds, whose rivers, because of topography, flow from the main island's center to the north or south – 236 flow north and 327 south. Some of these 'rivers' are wet weather streams, flowing only during the rainy season, while others, especially those that have their origins in the mountains, flow year-round. Because the main island is narrow, the average river is only 40 km long and few rivers are deep enough to be navigable. Likewise in Cuba, there are few natural lakes and lagoons. To retain part of the year's rainfall for the summer growing season, more than 200 artificial dams and lagoons have been constructed in or near regions with high agricultural water demand. More than 70 percent of the impounded water is used for that purpose (Academia de Ciencias de Cuba, 1992).

The groundwater table in most of Cuba is very shallow, especially in the south where in many places groundwater is near the soil surface. However, in the north it is quite different for there it is generally necessary to drill through bedrock to reach water. For these reasons, the risk of agrochemicals leaching into groundwater is higher in the south, especially where light, sandy soils predominate. Groundwater contamination from agrochemicals is a concern in some areas of Pinar del Rio Province and on the second largest island, The Isle of Youth (formerly the Isle of Pines), located south of Havana Province. Both regions are important citrus production areas.

Principal economic crops

The most important crops, listed by area under cultivation or economic contribution, are sugarcane, banana, citrus, rice, tobacco, legumes, and vegetables. Sugarcane is the most important by both measures. Fields are evenly distributed across the flatlands, and occasionally are located very near the coast. However, only herbicides and plant growth regulators are used in sugarcane production resulting in low risk of contamination to aquatic ecosystems. Cultivation practices result in the same herbicides being applied each year and this may actually enhance the degradation of these compounds and consequently reduce the risk to the environment.

Ecologically rice cultivation is more important not only because it requires a diversity of pesticides (Table 14.1) but also because the necessary application equipment for applying these compounds over large areas (primarily by aerial application) increases the likelihood of drift (Bossan et al., 1995) or accidental application to non-target aquatic ecosystems. Additionally, rice cultivation is ecologically important because of the proximity of most large rice producing regions to the coast (Figure 14.1). The last two factors contribute most to the potential for environmental contamination. Furthermore, drainage from rice paddy fields treated with pesticides may also impact local aquatic wildlife.

Most of Cuba's banana production is treated with fungicides to protect against disease. However, pesticides are applied only when the disease (or pest) reaches the EIL. This technique allows a reduction in the number of treatments and, thus, a lower impact on the environment. Banana plantations have experienced a reduction of about 50 percent in the annual number of applications compared to other banana growers in the Caribbean region.

Similar management practices in citrus, potato, and tobacco are in place but these crops are generally planted in sandy soil. The sandy soils contribute to the leaching hazard and consequent groundwater contamination from pesticide applications. Managing this risk requires the implementation of a monitoring program to evaluate leachable residue movement of highly persistent pesticide in

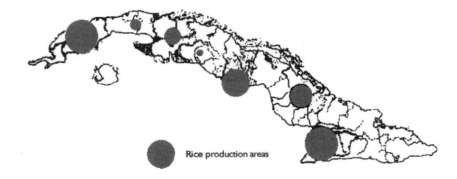

Figure 14.1 Map of the Cuban Archipelago showing the major rice production areas

Table 14.1 Pesticides authorized for use in rice cultivation

Pesticide	Rate (kg a.i. ha⁻¹)	Pesticide	Rate (kg a.i. ha⁻¹)
Benfuracarb	20	Dalapon	4.4–13.5
Benomyl	2	Deltamethrin	0.0125
Bentazone	1.2	Dimethoate	0.4–0.6
β-Cyfluthrin	0.0125	Edifenphos	0.5–0.7
Buprofezin (a chitin synthesis inhibitor)	0.25	Thiobencarb (a thiocarbamate herbicide)	1.5–5
Carbaryl	1.7–2.5	Fenpropathrin	0.1–0.2
Carbofuran	1.0	Fenthion	0.5–0.75
Chlorpyrifos	0.48–0.72	Fenvalerate	0.2
Cyfluthrin	0.0037–0.05	Fenitrothion	0.3
2,4-D (isopropyl ester)	0.6–1.2	Iprobenphos	0.5–0.7
λ-Cyhalothrin	0.006–0.01	Isoprothiolane (fungicide/insecticide)	0.5
Malathion	1.14–1.4	Methyl parathion	1–1.5
Molinate (a thiocarbamate herbicide)	1.8–2.5	Phosphamidon	0.5
Methamidophos	0.4–0.6	Propanil	1.08–3.5
Oxadiazon	1–2	Tebuconazol (a conazole fungicide)	0.25

Source: Lista Oficial de Plaguicidas Autorizados, 1995/1996 (Ministry of Agriculture, 1996).

order to restrict or ban those found in well water. Another problem in these sandy soil lands is increased contamination of nearby waterways from runoff and erosion, each bearing substantial quantities of pesticides. Table 14.2 presents results from an experiment designed to measure desorption of pesticides from two typical agricultural soils of Cuba. Desorption was higher from the sandy soil due to its low organic matter content.

Table 14.2 Desorption of pesticides in two Cuban soil types

Pesticide	Red ferralitic soil type total desorption (%)	Brown plastic soil type total desorption (%)
Simazine	73.34	68.30
Atrazine	62.30	67.93
Ametryn	60.14	75.45
Carbofuran	56.42	62.25
Bromacil	51.98	67.00
Prometryn	45.17	71.13
Pirimiphos-methyl	22.09	20.56

Source: Dierksmeier, forthcoming.

PESTICIDE USE AND DISTRIBUTION

Past and current use patterns

Pesticides have been used in Cuba since the early 1950s but the pesticide class and use patterns have changed over the years. The first organic pesticides used in Cuban agriculture were the herbicide 2,4-D, some OC insecticides, and dithiocarbamate fungicides. In this respect Cuba followed world trends in pesticide development and trade. During the 1960s, the triazine herbicides and some others were introduced for sugarcane weed control. Also, OP and carbamate insecticides were gradually substituted for more persistent OC insecticides. The introduction of synthetic pyrethroids in the late 1980s then contributed to a ban of all OCs. Currently no OC or other highly toxic or persistent pesticides are permitted to be used in agriculture. The present trend is to introduce and use only the less toxic and less persistent pesticides available on world markets.

Cuba's pesticide use pattern has undergone significant change from an initial use pattern of spraying according to a fixed schedule, independent of the presence or absence of the target pest. This placed an unnecessary chemical deposit on crops, increasing both the risk of undesirable environmental effects and significantly adding to production costs. In the late 1970s, a radical change was introduced whereby a pesticide was applied only if the pest density, disease prevalence, or weed density surpassed a threshold level that would result in economically significant damage to the final crop. This reduced the number of treatments, the unit cost, and the potential environmental impact.

In recent years an important new trend has developed with the introduction for agricultural use of a new generation of pesticides that further reduces residue concentrations on or in crops and the potential environmental impact from their use. This is because application rates for these new pesticides are very low (ranging from 15 to 500 g a.i. ha^{-1}) and they are generally easily biodegraded in the environment. Examples of this trend include the introduction of pyrethroids and other biorational insecticides, e.g. *Bacillus thuringiensis* Berliner, which was substituted for some carbamate and OP insecticides in the late 1970s, and the use of triazole fungicides in place of dithiocarbamates. More recently the introduction of sulfonyl urea herbicides in rice culture has reinforced this trend. This, coupled with new measures of biological control, has further reduced the demand for synthetic pesticides and stand as one of the significant advancements of the last decade. The trends that have taken place in plant protection in Cuban agriculture are similar to other countries (Farm Chemicals International, 1996) and follow world concerns about the use of and reliance on agrochemicals.

Location of applications with respect to the marine environment

Most pesticides used on Cuba's primary agricultural crops are applied by aerial or large ground-based mechanical sprayers. In some cases, pastures are also treated by aerial applications, which are highly subject to drift. Because of the prevailing wind direction during daylight hours (almost all applications occur after daybreak), the risk of direct contamination of north-coast coastal waters is extremely low. On the contrary, along the south coast, particularly near some large rice fields (Figure 14.1), the risk of direct contamination is much greater. However, for technical and economic reasons, aerial applications are allowed only when the wind velocity is very low (<3 m sec^{-1}). If this regulation is followed, drift and its concomitant contamination is effectively reduced. Another form of coastal contamination may occur when pesticides are adsorbed onto eroded soils in runoff and with drainage waters from rice culture areas. This form of environmental contamination is the most important, because rice fields are located along waterways or very near the coast.

TECHNICAL APPROACHES

Integrated pest management

One of the most important achievements in agriculture worldwide in the last twenty years is undoubtedly the development of the concept of IPM and the subsequent introduction of ideas associated with it into agricultural practice (Verreet, 1995). The most significant ideas from the standpoint of environmental impact are the selection of insect and disease resistant crop varieties, adequate soil preparation for the specific crop, planting at the optimal time, rational use of chemical pesticides to minimize their detrimental effects on natural enemies of noxious insects, use of biological pest controls when economical and feasible, and the use of pesticides only after a pest reaches an economic injury threshold population level. IPM guidelines effectively reduce the need for applying pesticides and thus minimize their introduction into the environment.

IPM has been implemented for several of Cuba's major crops with excellent results. IPM programs are in place for rice, coffee, citrus, tobacco, and banana with research ongoing to establish IPM guidelines for other crops, e.g. potato and some vegetables. To implement IPM, several steps are necessary. Basic and applied research is conducted by research institutes and universities, where extension education programs and large-scale demonstration projects are used to present the results of their work. Concurrently an intensive grass roots education and outreach program is conducted at the farmer level by the National Center of Plant Health, Agriculture Ministry. This includes training courses, workshops, and distribution of technical information using various media to present basic knowledge of IPM techniques for a specific crop. Another measure implemented to reduce pesticide residues in food crops is periodically checking for compliance with established pre-harvest intervals (Ministry of Agriculture, 1996). This work is done by the 14 provincial residue laboratories in Cuba. A substantial reduction in pesticide use is one of the results of IPM implementation. In some crops, such as banana, the actual use of pesticides is approximately 50 percent below the level needed without this new approach.

Pesticide regulations

Before 1987, the importation, distribution, storage, use, and waste disposal of pesticides were regulated by a patchwork of separate legal statutes, that considered in isolation were sufficient for each specific issue for which they were designed but overall lacked coordination and integration. For example, the statute on pesticide storage regulated all aspects related to this activity, e.g. storage building characteristics, ventilation facilities, and accident (fires, spills, etc.) procedures. However, it made no mention of pesticide quality assurance, maximum storage times, or other important considerations such as preventive health care for pesticide workers.

Finally in 1987, Cuba enacted a law that created the National Pesticide Registration Office (Gaceta Oficial de la República de Cuba Año 1987, 1989) which now strictly regulates the importation, transport, uses, storage, waste disposal, and other important aspects of pesticides. Before importation, all pesticides for agricultural or other uses must be registered. To register a new pesticide a.i. or a new formulated product of a known active material, it is mandatory that the producer or seller submit to the Registration Office all data required by legislation. These data include the chemical composition of the formulated product (a.i., impurities, solvents, co-adjuvants, inert materials, etc.); its biological effectiveness against targeted pests on crops for which the product will be used; the analytical methods used to obtain required data; toxicological evaluation data; ecotoxicology; and environmental behavior, fate, and transport. The law encourages submission of other aspects such as safe handling procedures for the formulated product.

The multidisciplinary scientific staff of the Registration Office evaluate all data submitted and decides which data needs independent verification, though verification of some pesticide product data is mandatory according to the registration law. Mandatory verification includes checking the physiochemical parameters of the product; its biological effectiveness against target pests, effectiveness in the crops proposed, and effectiveness under the climatic conditions and agricultural practices of Cuba; establishing residue levels in the proposed crops; and establishing the appropriate pre-harvest interval. The staff may require that the environmental behavior of the new formulated product, e.g. soil degradation, leaching potential, and water-sediment distribution (and degradation) be experimentally checked. In special cases, a product's effect on honey bees, earthworms, or fish is considered based on its toxicological properties and its possible uses. In all cases, the Registration Office verifies the required parameters through contracts with research institutes in Cuba.

If the new pesticide fulfills all requirements, the Registration Office grants a permit which is valid for importation and selling the new formulated product in Cuba for five years. If the new product fails to meet all requirements, the permit is refused and the importation of the compound is banned. Upon approval, the Registration Office will list the new formulated product in the 'Cuban Official Pesticide Authorized List' which it publishes yearly. Generally only those pesticides are allowed to be registered that do not present an excessive health hazard to consumers of agricultural products, wildlife, or the environment.

The quality of all imported pesticides and those formulated in Cuba is checked periodically by the Pesticide Chemistry Laboratories of the Ministry of Agriculture (there is one in each Cuban province) and the Plant Protection Research Institute (INISAV). Violation of permitted parameters results in the Pesticide Registration Office canceling the offending pesticide's permit. The Pesticide Registration Office has banned some persistent and health endangering pesticides from all uses in Cuba (Table 14.3).

ENVIRONMENTAL IMPACT OF PESTICIDES

Behavior in soil and water

Pesticide residues in crops and the environment are generally quite low due to the tropics' favorable climatic conditions for pesticide degradation and Cuba's strict regulations on their use. High solar radiation, air temperatures, soil temperatures, and moisture levels favor high dissipation rates for pesticides through photolysis, volatilization, and degradation (especially soil degradation from enhanced microbial activity) (Malbury *et al.*, 1996). However, several moderately persistent pesticides may be found in the environment but generally at low concentrations. These pesticides are almost exclusively OCs which are now banned in Cuba and thus their environmental concentrations should continue to decline.

INISAV studies the environmental behavior of pesticides with the goal of reducing the concentration of pesticide residues in crops, soils, and waters by systematically conducting laboratory and field experiments and monitoring programs with newly introduced pesticides. Specific adsorption constants (K values) for Cuban soil types and agricultural pesticides are developed which predict the

Table 14.3 Pesticides banned from use in Cuba

Aldrin	Heptachlor
Dieldrin	Leptophos
Camphechlor (Toxaphene)	Sodium fluoroacetate (Compound 1080)
Chlordimeform	Thallium salts (rodenticide)
Chlorobenzilate	2,4,5-T
Inorganic arsenic compounds	Dinoseb
DDT	Hexachlorocyclohexane (Lindane)
Dibromochloropropane (Nemagon)	Nitrofen (herbicide)
Inorganic mercurial compounds	Fluoracetamide (rodenticide)
Organic mercurial compounds	Cyhexatin
Endrin	Ethylene dibromide

Source: Dierksmeier, 1996.

Table 14.4 Specific adsorption constants (K) for selected pesticides in two important Cuban agricultural soils

	$K (\mu g\ g^{-1})^a$	
Pesticide	Red ferralitic soil	Brown plastic soil
Simazine	2.4	12.7
Atrazine	0.41	1.5
Ametryn	4.04	7.71
Carbofuran	0.29	1.54
Bromacil	2.89	23.9
Prometryn	4.47	29.98
Pirimiphos-ethyl	29.3	54.9

Source: Dierksmeier, forthcoming.

Note:
a Determined according to Freundlich's law.

movement of those pesticides in various ecosystems and in water-sediment systems. Table 14.4 presents K values for selected pesticides in two major Cuban soil types. Lower K values predict higher rates of leaching and runoff for the pesticide.

Leaching and upward capillary movement of pesticides in soils are two opposing phenomena, taking place simultaneously with upward capillary movement reducing the risk of water table contamination by the pesticide. Table 14.5 gives the leaching behavior of several pesticides commonly used in agriculture (Dierksmeier, forthcoming). Concentrations beyond the arable layer (25 cm depth) are low even under severe laboratory conditions. Table 14.6 shows the upward capillary movement of selected herbicides in a red ferralitic soil based on laboratory and field experiments. This demonstrates the opposing effect to leaching (Dierksmeier, 1986).

Theoretically dissipation and degradation of pesticides in tropical soils should occur at higher rates than in temperate zones. This is the situation, in part due to favorable weather conditions throughout the year that enhance the development

Table 14.5 Leaching of selected pesticides in two Cuban soil types

Pesticide	Soil type	Soil layer (cm)	Simulated annual rainfall (mm)		
			100	200	400
			Leaching (% of total pesticide found in the column)		
Ametryn	Red ferralitic	0–5	69.05	20.93	49.47
		5–10	12.18	12.67	31.92
		10–15	9.30	17.62	12.25
		15–20	7.75	19.70	3.63
		20–25	4.84	29.04	1.81
Atrazine	Red ferralitic	0–5	67.62	20.60	34.64
		5–10	14.16	18.04	34.75
		10–15	9.80	19.18	18.82
		15–20	5.84	16.83	7.85
		20–25	2.70	16.33	3.92
Propachlor	Red ferralitic	0–10	100	100	–[a]
		10–20	–	–	–
		20–30	–	–	–
	Brown plastic	0–10	42.7	60	–
		10–20	38.2	24	–
		20–30	19.1	16	–
Propiconazole	Red ferralitic	0–10	100	98.1	97.1
		10–20	–	1.9	2.9
		20–30	–	–	–
Metolachlor	Red ferralitic	0–10	49.3	29.3	18.8
		10–20	28.4	35.8	43.1
		20–30	22.3	34.8	38.1
	Brown plastic	0–10	52.5	44.8	47.0
		10–20	38.7	46.2	46.7
		20–30	8.8	9.0	6.3

Source: Dierksmeier, forthcoming.

Note:
a   En dash (–) indicates no data.

of a microflora, which contributes to the degradation of the pesticides in soil (Pemberton, 1981). Other factors like photolysis, volatilization (due to high soil temperature in summer), and runoff contribute to rapid dissipation of these compounds in soil (Laskowski *et al.*, 1983). This is illustrated in Table 14.7 for some triazine herbicides, which are commonly and extensively used in many crops including sugarcane.

The residues after harvest of several soil-applied pesticides are shown in Table 14.8. These results are from under field conditions. Other factors including root uptake of part of the applied pesticide may have been responsible for some of the residues. In some cases, concentrations in the soil are sufficiently high to injure crops grown in rotation (Stougard *et al.*, 1990).

Repeated, long-term application of a pesticide to the same crop species over many years causes selection and development of a specific microflora. This micro-

Table 14.6 Upward capillary movement of selected herbicides in a Cuban red ferralitic soil

Herbicide	Time post-application (d)	% Herbicide found in soil layers (cm depth)[a]		
		0–6	6–12	12–18
Ametryn	2	0.8	22	77.1
	4	3.04	17	79.8
	7	4.83	22.4	72.6
Prometryn	2	0.9	20.2	78.9
	4	1.8	21.4	76.8
	7	5.1	25.3	69.6
Simazine	2	0.57	15	84.3
	4	2.42	17	80.5
	7	8.11	28.8	62.9
Bromacil	2	9.78	42.1	62.9
	4	56.0	23.8	48.0
	7	48.1	42.5	20.0
Atrazine	2	17.7	40.5	41.8
	4	25.7	33.5	40.8
	7	48.3	22.2	31.95
Atrazine	7	1.69	5.20	93.11
(Field data)	14	1.26	2.47	96.27
	24	0.87	1.18	97.84

Source: Dierksmeier, 1986.

Note:
a The soil depth, 0 cm, indicates the soil surface.

Table 14.7 Dissipation of triazine herbicides in a Cuban red ferralitic soil under field conditions[a]

Herbicide	Residue (mg kg^{-1}) remaining after time (d)						
	0	4	17	30	46	67	120
Ametryn	1.45	1.01	1.08	–[b]	0.57	0.47	0.15
Atrazine	1.73	1.54	1.04	–	0.59	0.47	0.20
Simazine	1.55	1.50	1.02	0.87	–	0.76	0.44
Prometryn	1.20	0.92	0.68	0.63	0.46	0.32	0.25
Terbumeton	1.77	1.65	1.74	1.08	0.96	0.89	0.38

Notes:
a Sisinno et al., 1990.
b En dash (–) indicates no sample taken.

flora can degrade the pesticide at higher rates compared to the microflora in a pristine soil after first application of the pesticide. Figure 14.2 shows the degradation of carbofuran (a rice crop soil insecticide) after application to pristine soil and soil subject to long-term application of the pesticide. Degradation in the soil subject to long-term application of carbofuran is very rapid. In pristine soil of the same soil

Table 14.8 Pesticide residues in soil after application or harvest

Pesticide	Residue (mg kg⁻¹)	Detection limit (mg kg⁻¹)	Days post-application	Crops
Trifluralin	0.30	0.02	164	tomato
Nitrofen	ND[a]	0.02	120	vegetable
2,4-D	ND	0.05	22	rice
Dalapon	ND	0.30	10	rice
Ametryn	0.15	0.02	120	potato
Simazine	0.20	0.02	120	maize
Prometryn	0.25	0.02	120	potato
Desmetryn (a methylthiotriazine herbicide)	0.09	0.02	120	maize
Terbumeton	0.33	0.02	120	citrus
Diuron	1.96	0.05	70	pineapple
Bromacil	0.13	0.02	61	pineapple

Source: Dierksmeier, 1990.

Note:
a ND indicates below detection limit.

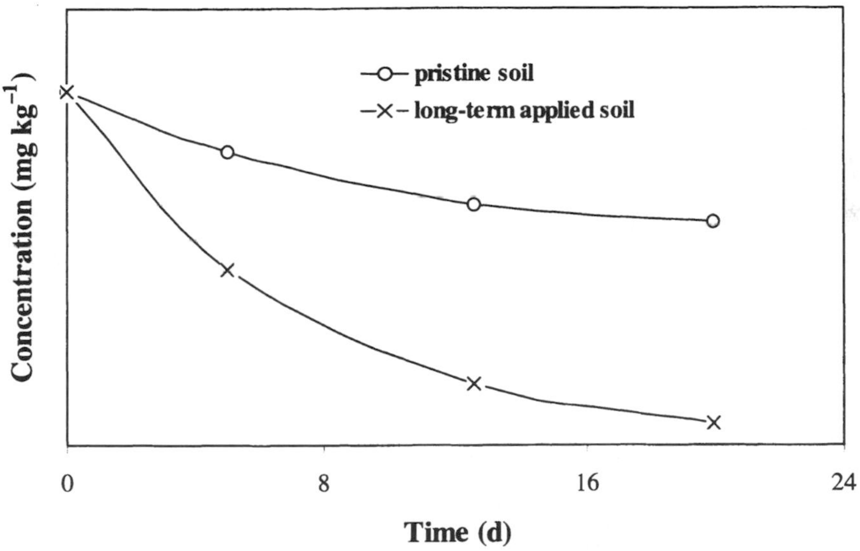

Figure 14.2 Degradation of carbofuran in a typical rice soil

type, the dissipation rate is much lower suggesting a microflora that is less efficient at degrading carbofuran. Ecologically this phenomenon is advantageous, reducing the time this chemical is in the environment. However, from the agricultural viewpoint, this reduces the chemical's effectiveness and results in increased application rates to adequately protect the crop. The same phenomenon may be taking place

Table 14.9 Dissipation of DDT and dieldrin under field conditions in a Cuban red ferralitic soil

Insecticide	Rate of application (mg a.i. kg^{-1})	Residues (mg kg^{-1}) remaining after time (months)								
		0	1	2	4	6	7	10	12	19
Dieldrin	3.33	1.91	1.61	1.34	1.55	1.03	0.92	0.78	0.61	0.38
DDT	7.50	4.44	4.33	4.31	–^{a}	3.11	1.95	0.96	0.43	–

Source: Dierksmeier, 1989.

Note:
a En dash (–) indicates no sample taken.

in other crops to which the same pesticide has been applied for many years including sugarcane, citrus, and banana. Furthermore, this process partly explains the low residue concentrations found in Cuba's agricultural soils.

The OC pesticides are a special case in that they show a high persistence (compared to other classes of pesticides under similar conditions) that may be due to a lack of soil microorganisms capable of transforming and degrading them. The soil behavior of DDT and dieldrin in a red ferralitic soil was studied in a field experiment in which both insecticides were applied to the soil, mixed with the top 10 cm layer of soil, and allowed to weather under environmental conditions (Table 14.9). Samples taken over 19 months were analyzed by gas chromatography yielding an estimated half-life of 180 d for both insecticides (Dierksmeier, 1989), compared to a half-life of 10 to 20 d for OP and carbamate insecticides and 40 d for triazine and diuron herbicides.

There has been increasing worldwide concern about water quality in part due to many reports of watershed contamination, coastal zone pollution, and pollutant effects on endangered aquatic species from industrial waste, domestic waste, and agrochemicals. In Cuba, most watersheds, lagoons, and dams are within or in close proximity to cultivated lands where pesticide applications are common. Thus, there is high risk in Cuba for agrochemical contamination of aquatic ecosystems. Several laboratory-based research projects have assessed the impact of pesticides in water and water-sediment systems (Dierksmeier *et al.*, 1994) under natural conditions (Table 14.10). Though it is impossible to extrapolate from these data to the complex tropical environment, they undoubtedly serve to throw some light on this complex problem. Some of the results were expected, based on reports from other counties with similar climatic conditions and the physiochemical properties of the pesticides considered. However, results for the synthetic pyrethroids are surprising and raise concern about their fate because of their present widespread use in Cuban agriculture. The synthetic pyrethroids are highly persistent, especially in sediments, and are highly toxic to fish (Tomlin, 1994). Further studies are needed to adequately assess the risk these compounds pose to Cuba's aquatic ecosystems.

Table 14.10 Behavior of pesticides in water-sediment systems

Pesticide		Concentration in water (mg L⁻¹) or sediment (mg kg⁻¹)							
Endosulfan	time (d)	0	5	8	16	21	63	78	140
	water	1.9	0.23	0.16	0.03	0.02	0.01	ND^a	ND
	sediment	5.2	206	–^{b}	35	25	–	–	1.3
Lindane	time (d)	0	5	8	13	16	21	63	127
	water	6	1.3	1.0	2.2	1.2	–	0.004	0.001
	sediment	1.6	4.8	–	1.9	0.17	0.13	–	ND
Permethrin	time (d)	0	5	11	19	45	50	136	170
	water	13.6	7.9	5.0	0.27	0.05	ND	–	–
	sediment	169	177	–	434	395	–	270	367
Cypermethrin	time (d)	0	5	11	19	45	50	136	170
	water	16.8	6.95	4.76	0.25	0.06	ND	–	–
	sediment	161	175	–	424	263	–	187	165
Dichlorvos	time (d)	0	5	11	20	45	50		
	water	2.5	2.4	1.5	0.04	0.01	ND		
	sediment	ND	ND	–	–	–	–		
Fenthion	time (d)	0	3	6	14	22	35	48	60
	water	10	4.4	2	1.8	0.24	0.10	0.02	ND
	sediment	35	–	22.3	2.5	1.4	1.13	0.12	0.10
DDT	time (d)	0	12	25	38	61	96	120	150
	water	0.96	0.08	0.05	0.003	0.008	0.006	–	0.003
	sediment	20.8	44.8	49.08	–	–	44.7	42.4	22.3

Source: Dierksmeier *et al.*, 1994.

Notes:
a ND indicates below detection limit.
b En dash (–) indicates no sample taken.

Over the last ten years, Cuba's program for monitoring drinking water shows no major problems with the water supply as all residues found were below the levels set by national and international regulations (GIFAP, 1989). More than one thousand samples from wells in agricultural regions of Havana Province (the most agricultural province with the highest pesticide consumption in Cuba) revealed no residues in excess of regulatory limits. This trend should continue because of past efforts to limit groundwater contamination and the future establishment of a comprehensive monitoring program for surface and ground waters.

Residues in the coastal zone

Currently no systematic monitoring of Cuba's coastal zone is in place for pesticide residues. However, International Mussel Watch has sampled offshore bivalves south of Havana Province as part of a comprehensive Latin American study (UNESCO *et al.*, 1994). The Centro de Investigación de Ingenería y Medio Ambiente (CIMAB) has monitored the contamination, including that from pesticides and hydrocarbons, of Cuba's principal bays. Recently research evaluating pesticide residue levels in sediment and biota near an extensive rice producing area has begun in the coastal zone near Los Palacios, south Pinar del Rio Province (see Figure 14.1) as part of a comprehensive international project focusing on the distribution, fate, and effects of pesticides on biota in the tropical marine environment using radiotracer technology. This project is sponsored by the International Atomic Energy Agency with technical assistance from the Marine Environmental Laboratory in Monaco and focuses on the evaluation of OC and OP pesticide residue levels in sediment and biota along Dayanigua Beach between the Carraguao and San Diego rivers (Figure 14.3).

Sampling sites included the mouths of the San Diego and Carraguao rivers and a segment of a tidal mangrove coast east of San Diego River and west of Carraguao River. Samples of the predominant species of bivalve the Flat Tree or Mangrove oyster *Isognomon alatus* Gmelin (Bivalva: Isognomonidae) were taken along a kilometer of coastline between the rivers (UNEP *et al.*, 1991). Table 14.11 (Dierksmeier *et al.*, 1996) summarizes the results of analyses for OC residues in the field samples. Only DDT residues were found in sediment and biota and no other OCs, pyrethroids, or PCBs were detected. The concentration of $\sum$ DDT (the sum of the parent compound and its metabolites) in sediments remained fairly stable during the dry season but declined after the start of the rainy season, perhaps due to intense runoff caused by strong tropical storms during that October and November. However, $\sum$ DDT concentration in biota remained nearly unchanged for all sample dates. The absence of residues from pesticides actually used in the rice fields near

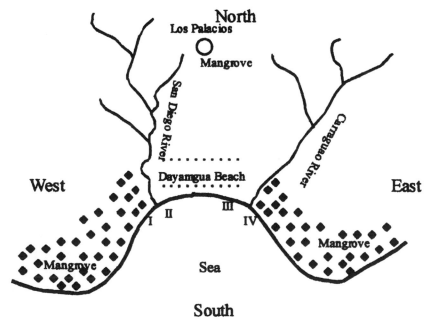

Figure 14.3 Sampling sites at Dayamgua Beach, Cuba

the sampling sites, e.g. endosulfan and synthetic pyrethroids, was surprising and in part explained by the results presented in Table 14.10 which show a strong adsorption of these pesticides in the sediment phase and a high dissipation rate from the aqueous phase (Dierksmeier *et al.*, 1994).

Table 14.12 summarizes the results for OP residues in the same field samples. No residues were found in sediment or biota samples above the detection limit. This may be due to the rapid degradation of OPs in aquatic systems or to the adsorption of these pesticides by sediments in drainage channels (Readman *et al.*, 1992). Each mechanism results in minimal quantities reaching the coastal zone (see Table 14.13) (Dierksmeier *et al.*, 1998). Experiments, carried out in rice field drainage channels (see Figure 14.4), demonstrated a very high reduction in the concentrations of OPs and other pesticides, especially endosulfan and cypermethrin (Dierksmeier *et al.*, 1998).

Table 14.11 Total $\sum$ DDT and its metabolites found in sediment and biota (μg kg^{-1})

Date			Sampling site			
	I	*II*		*III*		*IV*
	Sediment	*Sediment*	*Biota*	*Sediment*	*Biota*	*Sediment*
9 December 1994	2.39	–[a]	–	–	–	–
18 February 1995	11.63	14.46	12.96	13.51	19.45	12.44
25 April 1995	4.62	23.15	11.20	17.65	23.80	21.00
25 July 1995	ND[b]	ND	12.90	ND	14.74	ND
17 November 1995	ND	ND	–	ND	–	ND

Source: Dierksmeier *et al.*, 1996.

Notes:
a En dash (–) indicates no sample taken.
b ND indicates below detection limit of 0.25 μg kg^{-1} for OCs.

Table 14.12 OP residues in field samples

Sampling sites	Date	Type of sample (S = sediment; B = biota)	Pesticide residue (μg kg^{-1})
I–IV	8 December 1994	S	ND[a]
		B	ND
I–IV	18 February 1995	S	ND
		B	ND
I–IV	23 April 1995	S	ND
		B	ND
I–IV	25 July 1995	S	ND
		B	ND
I–IV	17 November 1995	S	ND
		B	ND

Source: Dierksmeier *et al.*, 1996.

Note:
a ND indicates below detection limit of 5 μg kg^{-1}.

Table 14.13 Dynamic adsorption of pesticides under field conditions

Pesticide	Experiment	Initial concentration (mg L⁻¹)	Concentration found at the indicated points (mg L⁻¹)			
			200 m	600 m	1,000 m	1,200 m
Dimethoate	1	0.6	0.40	0.01	ND[a]	ND
	2	0.3	0.01	ND	–[b]	ND
Iprobenfos	1	0.7	0.30	0.006	0.003	0.006
	2	0.35	0.27	0.02	–	0.003
Methyl parathion	1	1.5	0.40	0.009	0.003	0.008
	2	0.75	0.53	0.04	–	0.06
Malathion	1	1.4	0.50	0.008	0.007	0.001
	2	0.7	0.54	0.1	–	0.07
Chlorpyrifos	1	0.72	0.12	0.005	0.004	0.002
	2	0.36	0.11	0.003	–	0.001
Cypermethrin	1	0.05	0.02	ND	ND	ND
	2	0.025	0.02	ND	–	ND
α+β Endosulfan	1	1.0	0.14	0.006	0.002	0.001
	2	0.5	0.02	0.002	–	0.002
Carbofuran	2	0.35	0.085	0.053	–	ND

Source: Dierksmeier et al., 1998.

Notes:
a ND indicates below detection limit.
b En dash (–) indicates no sample taken.

MANAGEMENT OF PESTICIDE RESIDUES

Industrial pesticide residues

In Cuba there are no major pesticide manufacturers but there are several formulation facilities whose waste waters pose a risk of environmental contamination. However, strict laws and regulations for pesticide formulators compel them to decontaminate waste water before its release into the environment. The Ministry of Health conducts systematic monitoring of wells and surface waters around these plants to aid enforcement of regulatory measures. They have recently installed vapor and dust traps to reduce local effects caused by some formulation facilities, e.g. the formulation plant for the fanes herbicide, 2,4-D. Currently all formulation facilities meet regulatory standards and no major problems are associated with their activities.

Agricultural pesticide residues

Many agricultural activities may result, directly or indirectly, in environmental contamination by pesticides but the risk can be reduced through proper management techniques. Obviously activities resulting in direct environmental contamination are of greatest concern and these include weed control in water channels

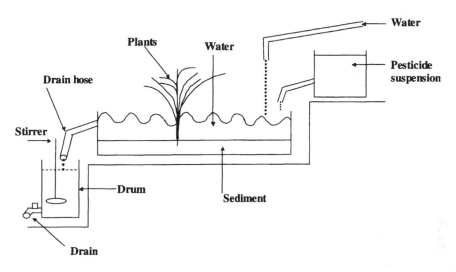

Figure 14.4 Channel system for the study of adsorption behavior of pesticides in a water/sediment system

and control of human and domestic animal disease vectors. Spraying to control either weeds or mosquito larvae results in high levels of pesticides in surface waters which may harm non-target organisms. These treatments are justified only if the benefit to risk ratio is sufficiently high to compensate for secondary effects to the environment (Hurlbert, 1975). There are no specific measures or regulations in place in Cuba to reduce or eliminate these practices. However, in some cases, mechanical weeding of drainage channels and the use of the mosquito fish *Gambusia affinis* Baird and Girad (Cyprinodontiformes: Poeciliidae) have given good results.

The indirect causes of pesticide contamination are numerous and while some are avoidable, others are inherent in pesticide use, and some are due to human failure. One of the most important indirect sources of pesticide contamination in the environment is from washing empty pesticide drums, washing plastic pesticide containers, and discarding the plastic bags in which some pesticides are packaged. Cleaning application equipment also creates problems because there are not always special places in the field to receive and treat the rinsing water adequately. In some cases, ignorant or irresponsible workers discard these wastes in violation of pertinent regulations.

Worldwide more than a billion empty drums are recycled each year. The cleaning and rinsing water contain approximately 500,000 kg of formulated pesticides (Munnecke, 1979) for disposal, which is generally discarded. Other Caribbean countries report similar problems with the discarding of empty containers (Espinosa Gonzales, 1996). Disposal of pesticide containers by dumping, burning, or burial is not always carried out correctly. Plastic pesticide containers are sometimes reused for storing and transporting fuels or, in some cases, for domestic uses. Good management practice requires thoroughly cleaning containers and adding the rinsing water to the spray mixture, thus saving money and protecting the environment.

Recently the Plant Protection Research Institute developed a simple and inexpensive procedure to receive, treat, and ultimately dispose of the products of this treatment as harmless wastes. The procedure was designed to reduce the quantities of pesticide residues from farms and small pesticide storehouses by flocculating the wastewater with ferric salts and calcium hydroxide. The resulting sediment is separated and treated with calcium oxide while the water phase is aerated and drained through a cement channel where heterotropic algae are grown. The result of this procedure is a harmless solid waste and clean water.

Contamination by air drift

Almost 35 percent of the pesticides produced annually worldwide are applied by aerial application (Ware *et al.*, 1970). The productivity increases and the economic benefits of this application technique are undeniable. Pesticide drift associated with aerial application is the primary concern. Drift is affected by a number of factors including wind velocity, air temperature, type of pesticide formulation, and aircraft altitude at application. Under Cuba's environmental conditions only 85 percent of the applied pesticide actually reached the targeted crop following aerial application (Garcia Perez, 1981). This is both economically and ecologically unacceptable. However, technically it is not advisable to restrict or ban the aerial application of pesticides, because of the trend toward merging small farms into large plantations and the concentration of some crop's production into areas where they grow satisfactorily. This has happened for sugarcane, rice, and banana in Cuba; corn, wheat, and soybeans in the United States; wheat and sunflowers in Argentina; and soybeans in Brazil (Groner, 1985). These trends will continue to make aerial application of pesticides necessary.

To reduce the drift of pesticides from aerial application, applicators should use ultra low volume (ULV) formulations specifically designed for the job and begin at daybreak or early morning when wind velocity and air temperature are lowest. Furthermore, they should avoid flying on hot days to reduce droplet evaporation and, flying as low as possible, use the maximum droplet size permitted, balancing a lower drift rate against spotty pesticide distribution and poorer pest control for larger droplet diameters (Akesson and Yates, 1964). Arriving at a satisfactory compromise between low drift and adequate biological effectiveness is essential to economically and ecologically cost-effective aerial application of pesticides. An adequate compromise between these factors will result in more efficient crop protection, savings in pesticide costs, and increase crop production with fewer environmental consequences. However, no matter what compromise is used, applicators cannot completely avoid damage to the flora and fauna on field edges. Fortunately unintended effects due to aerial application of pesticides in Cuba are limited due to strict regulations and mandatory monetary compensation when damage occurs from aerial application.

Volatilization of pesticides from crops and soils

Volatilization of pesticides applied to crops and soil is another source of contamination that can be reduced through proper management. Especially in the tropics, the physical process of volatilization is very rapid immediately after application of the pesticide but slows thereafter, and continues for long periods. Volatilization rates depend on several factors including the prevailing environmental conditions (wind velocity and air temperature), the vapor pressure of the pesticide, and its behavior on the plant (contact or systemic pesticide) (Kersting and Kuck, 1992). Over the long term, pesticide loss by volatilization is greater than the loss by drift. Lloyd-Jones (1971) found that DDT applied to soil with ambient temperatures between 20 and 30°C will volatilize at a rate of 0.9 to 4 kg a.i. ha^{-1} yr^{-1}. While there are no data available specific to Cuban environmental conditions, volatilization loss can be managed through the use of systemic pesticides which are rapidly absorbed into plant tissues and thus less subject to volatilize, though their use is not always feasible.

Contamination of the environment through leaching and runoff has been mentioned previously and can be reduced by increasing the soil's organic content to adsorb and hold pesticides until they are degraded by soil microorganisms and through application of various agricultural practices. Increasing soil organic matter content to reduce leaching is effective but expensive. Runoff may be reduced by proper ploughing techniques and application of foliar herbicides. Both methods are under consideration in Cuba to prevent further soil erosion and leaching, though additional research and development programs are necessary to optimize the cost–benefit ratio.

Management of pesticide residues in crops

Maintaining foodstuff residue levels below nationally and internationally acceptable MRLs is of utmost importance to protect consumer health and remain in compliance with international trade regulations. To accomplish this, Cuba examines each combination of pesticide and crop following FAO guidelines (FAO, 1990) and good agricultural practices. The dissipation of residues is followed analytically to establish pre-harvest intervals for each pesticide and crop based on MRL guidelines. The Cuban National Pesticide Registration Office issues yearly the list of experimentally determined pre-harvest intervals.

Despite these precautions, pesticide residue levels are sometimes above the MRLs. This may be due, among other reasons, to a farmer's lack of knowledge or violation of the pre-harvest interval. However, the number of such violations is relatively small as shown in Table 14.14, which presents monitoring results for several common crops. Only 1.6 percent of the samples had concentrations beyond the MRLs, which indicates that pesticide residue management in Cuba's crops is effective.

CONCLUSIONS AND PROSPECTS

Conclusions

Pesticide contamination in Cuba is very low and is due to the prevailing favorable weather conditions, the prudent use of these agrochemicals, and strict regulations concerning all aspects of pesticide importation, transport, storage, use, and waste disposal. For these reasons, there are no pesticide residues in drinking water beyond the MRLs while in crops, most pesticide residues are found in compliance with national and international standards.

In agricultural soils and in inland waters, pesticide dissipation is very rapid. However, some residues may be present in soil after crops with short growing seasons. Other residues may persist for long periods in river or reservoir sediments and these will require further research to thoroughly assess their behavior in the environment. In the coastal zone, only DDT and its metabolites were found in sediment and biota, but at very low concentrations and it is extremely unlikely that these will present future problems.

Prospects

Cuban agriculture, like that in other developing countries, faces certain common problems. There is a continuous reduction in the amount of useful agricultural land due to growing cities, recreational areas, industrialization, and park preserves. Furthermore, every year valuable soils are lost to erosion from the deforestation that began more than 100 years ago but continues today. An increasing population, which demands higher quantities and better qualities of agroproducts, and a need for expanded agricultural exports to contribute to national economic development pose a tremendous challenge to Cuban agriculture. It must produce more and better crops using less land and this necessarily implies an increase in productivity, achievable only through correct and timely application of science-based agricultural knowledge, including the prudent use of pesticides.

Despite worldwide efforts to find substitutes for agrochemicals, world food production will depend on the use of these chemicals for the foreseeable future (FAO, 1994). To relieve projected food shortages, Africa and Latin America are expected to increase their use of agrochemicals. In the future it will be necessary to increase research and residue monitoring of crops and the environment to preserve the environment and contribute to sustainable agriculture. This will require investment in a scientific infrastructure and periodically upgrading the knowledge

Table 14.14 Pesticide residues found in samples of some important crops in Cuba[a]

Pesticide		Cabbage	Tomato	Cucumber	Onion	Pepper	Citrus	Sweet potato	Beans	Potato	Carrot
											Crop
Zineb	No. of samples	–[b]	12	8	6	16	14	–	–	–	–
	Residue Levels[c]	–	7.2–3.6	ND[d]	6.5–ND	0.5–ND	ND	–	–	–	–
	MRL	–	0.5	0.5	7	2	1	–	–	–	–
Methyl parathion	No. of samples	36	36	4	8	24	14	8	–	–	6
	Residue levels	ND	0.1–ND	ND	ND	0.03–ND	ND	ND	–	–	ND
	MRL	0.1	0.05	0.1	0.05	0.05	0.1	0.1	–	–	0.05
Carbaryl	No. of samples	–	34	–	–	12	18	2	4	–	–
	Residue levels	–	0.2–ND	–	–	0.8–0.1	0.5–0.3	ND	ND	–	–
	MRL	2	1	–	–	3	2.5	1	0.5	–	–
Dimethoate	No. of samples	2	30	18	6	16	12	8	–	4	4
	Residue levels	ND	ND	ND	ND	ND	ND	ND	–	ND	ND
	MRL	0.2	0.25	1	0.5	0.2	0.2	0.05	–	0.1	0.1
Methamidophos	No. of samples	28	10	–	4	10	–	–	45	16	–
	Residue levels	ND	0.7–ND	–	0.02–0.01	0.2–0.07	–	–	ND	ND	–
	MRL	0.5	0.1	–	0.1	0.2	–	–	0.05	0.05	–
Endosulfan	No. of samples	6	8	–	6	–	–	8	–	–	–
	Residue levels	ND	ND	–	ND	–	–	ND	–	–	–
	MRL	2.0	0.5	–	0.2	–	–	–	–	–	–
Trichlorfon	No. of samples	–	8	12	4	12	–	–	–	–	–
	Residue levels	–	ND	ND	ND	0.8–0.12	–	–	–	–	–
	MRL	–	–	–	0.5	–	–	–	–	–	–
Malathion	No. of samples	36	26	4	4	25	26	8	–	2	4
	Residue levels	ND	ND	ND	ND	ND	ND	ND	–	ND	ND
	MRL	0.5	0.5	0.1	0.05	0.5	0.5	0.1	–	0.1	0.1

Notes:
a FAO, 1994. b En dash (–) indicates no data. c Residue levels in $\mu g\ kg^{-1}$. d ND indicates below detection limit.

and skill levels of Cuba's environmental and agricultural scientists. These goals will be achieved with the recent formation of the Ministry of Science Technology and the Environment and in cooperation with the Ministry of Agriculture. More support for basic and applied research will be available and enforcement of environmental regulations will be easier with an end result of better protection for Cuba's unique and diverse environment.

REFERENCES

Academia de Ciencias de Cuba. 1992. Comarna, Informe Nacional a la Conferencia de Naciones Unidas sobre Medio Ambiente y Desarrollo. Rio de Janeiro, Brazil.

Akesson, M.B. and Yates, W.E. 1964. Problems related to application of agricultural chemicals and resulting drift residues. *Ann Rev Entomol.* 9:285.

Atlas Climático de Cuba. 1987. Habana, Cuba: Editado por Instituto Cubano de Geodesia y Cartografía.

Atlas Nacional de Cuba. 1970. Habana, Cuba: Academia de Ciencias de Cuba.

Bossan, D., Worthano, H. and Masclet, P. 1995. Atmospheric transport of pesticides adsorbed on aerosols: 1, Photodegradation in simulated atmosphere. *Chemosphere.* 30(1): 21–9.

Dierksmeier, G. 1986. Ascenso Capilar de herbicidas en el suelo. In: Proc. 1st Jornada Científico Técnica de Sanidad Vegetal. Sancti Spiritus, Cuba, pp. 137–41.

Dierksmeier, G. 1989. Comportamiento del DDT y el dieldrin en suelos. *Ciencia y Técnica en la Agricultura, Serie Protección de Plantas* 12(4):21–4.

Dierksmeier G. 1990. Movimiento y persistencia de plaguicidas en el suelo. In: Proc. 10th Congress of the Latin American Association on Weeds. Habana, Cuba, pp. 17–26.

Dierksmeier, G. 1996. Pesticide contamination in the Cuban agricultural environment. *Trends Anal Chem.* 15(5):154–9.

Dierksmeier, G. (Forthcoming). *Plaguicidas. Residuos Efectos y Presencia en el Medio.* Habana, Cuba: Editorial Científico-Técnica.

Dierksmeier, G., Moreno, P.L., Hernández, R. and Sissino, A. 1994. Behavior of pesticide in water and sediment: Confirmation of residues by relative retention times using GC and HPLC and by relative Rf in TLC. In: Proc. 8th Int. IUPAC Congress of Pesticide Chemistry, held 5–9 July 1994 in Washington, DC, USA.

Dierksmeier, G., Hernández, R., Moreno, P.L., Martinez, K. and Ricardo, C. 1996. Organochlorine pesticides in sediment and biota in the coastal region to the south of the Pinar del Rio Province, Cuba. In: *Environmental Behavior of Crop Protection Chemicals.* Proc. Int. Symp. on the Use of Nuclear and Related Techniques for Studying Environmental Behavior of Crop Protection Chemicals, 1–5 July 1996. Vienna: IAEA/ FAO, pp. 343–7.

Dierksmeier, G., Moreno, P.L., Martínez, K., Hernández, R. and Linares, C. 1998. Environmental behavior of pesticides in rice field drainage water: impact on the coastal zone. In: *Extended Synopses.* Int. Symp. on Marine Pollution, held 5–9 October 1998 in Monaco, pp. 386–7.

Espinosa Gonzalez, J. 1996. Fate of pesticides under tropical field conditions: Implications and research needs in a developing country. In: *Environmental Behavior of Crop Protection*

Chemicals. Proc. Int. Symp. on the Use of Nuclear and Related Techniques for Studying Environmental Behavior of Crop Protection Chemicals, 1–5 July 1996. Vienna: IAEA/FAO, pp. 93–110.

FAO. 1990. Guidelines on producing residues data from supervised trials. Rome: FAO.

FAO. 1994. Amuario Estadístico. Rome: FAO.

Farm Chemicals International. 1996. Facing for the future. Willoughby, Ohio: Meister Publ. November 1996.

Gaceta Oficial de la República de Cuba Año 1987. 1989. Habana, Cuba: Government of Cuba, September 1989, pp. 69–70.

Garcia Perez, A. 1981. Evaluación de la uniformidad y recobrado de las aplicaciones aéreas de DDT PH 75 percent en el cultivo del arroz. In: Proc. 1st Scientific Meeting of Plant Protection, held 26–27 May 1981 in Villa Clara, Cuba.

GIFAP. 1989. National and international health based 'standards' for agricultural chemicals in drinking water. Agrupación Internacional de las Asociaciones Nacionales de Fabricantes de Productos Agroquímicos. GIFAP C/89/211, 21 March 1989. Brucelas, Bélgica: GIFAP.

Groner, H. 1985. *Técnica de Aplicación de Productos Fitosanitarios.* Revista BASF Reportes Agrícolas, Edición Especial Alemania.

Hurlbert, S.H. 1975. Secondary effects of pesticides on aquatic ecosystems. *Residue Rev.* 57:81–144.

Kersting, E. and Kuck, K.H. 1992. Volatilization behaviour of pesticides in field trials. *Brighton Crop Protect Conf Proc.* 2:829–34.

Laskowski, D.A., Swann, R.L., Mac Call, P.J. and Bidlack, H.P. 1983. Soil degradation studies. *Residue Rev.* 85:140–7.

Lloyd-Jones, C.P. 1971. Evaporation of DDT. *Nature.* 229:65.

Malbury, S., Cox, J. and Crosby, D. 1996. Environmental fate of rice pesticides in California. *Rev Environ Contam Toxicol.* 147:71–108.

Ministry of Agriculture. 1996. *Lista Oficial de Plaguicidas Autorizados: 1995/1996.* Habana, Cuba: Ministry of Agriculture, República de Cuba.

Munnecke, D.M. 1979. Chemical, physical and biological methods for disposal and detoxification of pesticides. *Residue Rev.* 70:1–26.

National Geographic Society. 1981. *National Geographic Atlas of the World,* 5th edn. Washington, DC: National Geographic Society, p. 112.

Pemberton, J.M. 1981. Genetic engineering and biological detoxification of environmental pollutants. *Residue Rev.* 78:1–11.

Readman, L., Lion Wee Kwong, Mee, L.D., Bartocci, J., Nilve, G., Rodriguez Solano, J.A. and Gonzalez Farias, F. 1992. Persistent organophosphorus pesticides in tropical marine environment. *Mar Poll Bull.* 24(8):398–402.

Sisinno, A., Hernández, R. and Merlo, M.E. 1990. Cinética de degradación de los herbicidas atrazina, ametrina, prometrina, simazina, terbutrina y desmetrina en suelo. In: 2nd Seminario Científico de Sanidad Vegetal. Habana, Cuba, pp. 57–63.

Stougard, R.N., Shea, P.J. and Martin, A.R. 1990. Effect of soil type and pH on adsorption, mobility, and efficacy of imazaquin and imazethapyr. *Weed Sci.* 38:67–73.

Tomlin, C. (ed.) 1994. *The Pesticide Manual: A World Compendium,* 10th edn. London: Crop Protection Publishers.

UNEP, FAO, IAEA, IOC. 1991. Sampling of selected marine organisms and sample preparation for the analysis of chlorinated hydrocarbons. In: *Reference Methods for Marine Pollution Studies* Nr 12 Rev 2. Nairobi: UNEP.

UNESCO, UNEP, NOAA. 1994. International Mussel Watch Project, Initial Implementation Phase. Final Report. Boston, Massachusetts: UNESCO/UNEP/NOAA. October 1994.

Verreet, J.A. 1995. Principles of integrated pest management: the IPM model. *PflanzenschutzNachrichten Bayer* 48:1.

Ware, G.W., Cahill, W.P., Gerhardt, P.D. and Witt, J.M. 1970. Pesticide drift IV: on target deposits from aerial application of pesticides. *J Econ Entomol.* 63:1982–89.

Chapter 15

Use, fate, and ecotoxicity of pesticides in Jamaica and the Commonwealth Caribbean

Ajai Mansingh, Dwight E. Robinson and Kathy M. Dalip

INTRODUCTION

Since their introduction into the Caribbean in 1945, synthetic organic pesticides have been used injudiciously in the region, without any appreciation or concern about the ecological and environmental consequences. The history and current status of research and data on the management of pests and pesticides, including establishment of the economic injury levels for pests; the efficacy of individual pesticides and alternate methods of pest management; legislative management of pesticides; the fate, persistence, and ecotoxicity of pesticides; and the environmental contamination by pesticide residues in the Commonwealth Caribbean are reviewed in this chapter.

OVERVIEW

The Commonwealth Caribbean

The English-speaking Commonwealth Caribbean community comprises two thinly populated mainland countries (Guyana in South America and Belize in Central America) and a chain of islands in the Caribbean Sea that are grouped into ten independent countries and five British territories (Figure 15.1). These islands start with Trinidad and Tobago in the south and Barbados in the east, extend northwest in an arc made up of the Windward (Grenada, St Vincent and the Grenadines, St Lucia, and Dominica) and the Leeward (Montserrat, St Kitts and Nevis, Anguilla, and Antigua and Barbuda) islands and the British Virgin Islands to Jamaica and the Cayman islands just south of Cuba, and include the Turks and Caicos islands and the Bahamas to the north of Cuba. The more than 4.5 million people who live on these islands depend primarily upon agriculture, fishing, mining, and tourism for their livelihoods. Only Trinidad, with limited oil but enormous gas reserves has developed a strong industrial economy.

The islands are volcanic in origin and the land is composed of white limestone, metamorphic rocks, and alluvium. Except for Trinidad, Barbados, and Antigua,

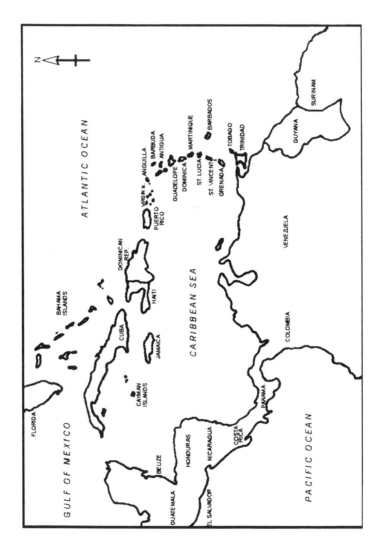

Figure 15.1 Map of the Caribbean Basin showing major islands and island groups

which are fairly flat, the other islands have rugged central mountain ranges, which slope toward the coastal plains. Rivers originate in the mountains and drain the valleys and plains into the sea. Annual rainfall ranges between 1,000 and 5,000 mm and temperatures between 25° and 35°C. Antigua receives significantly less rainfall than the other islands.

Jamaica, the largest island in the Commonwealth (area 1,140,480 ha; 235 km × 82 km), is situated between Lat. 17°30' to 18°30'N and Long. 76°30' to 78°30'W. The bird-shaped island is characterized by a central spine of rugged mountain ranges, which extend from east (highest peak, 2,300 m) to west (300 to 900 m high) and slope into valleys and the coastal plains in the north and south (Figure 15.2). Almost half of Jamaica is 300 m above sea level. Sixty percent of the land is composed of white limestone while the rest is made up of metamorphic rocks and alluvium. The island has twenty watersheds that are drained by nineteen major rivers, ten flowing generally north, eight south, one east and one west (Figure 15.3). Most watersheds experience at least twice-weekly rainfall although there are two defined rainy seasons, a minor one from May to June and a major one from September to early November. The annual rainfall ranges from 1,200 to 5,500 mm and the temperatures range between 23° and 33°C in the plains.

Land use

Since the early days of European colonization, agriculture has been the mainstay of the Caribbean economy, although only about 15 to 25 percent of arable land in the different islands is cultivated. In the mainland countries of Guyana and Belize,

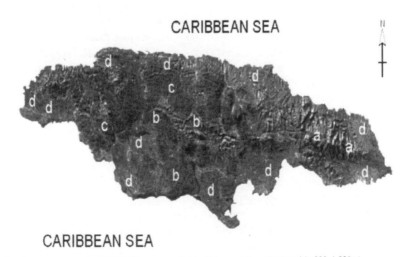

a – Blue Mountain ranges, alt. 1,500–2,135m surrounded by high mountains and valleys (alt. 900–1,500m)
b – mountain ranges, alt. 600–900m
c – cockpit country, alt. 300–600m, limestone hills
d – coastal plains, alt. 0–150m

Figure 15.2 Topography of Jamaica

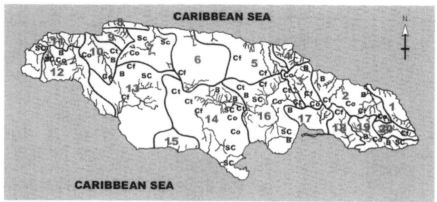

B – bananas
SC – sugar cane
Ct – citrus fruit
Co – coco
Cf – coffee

Figure 15.3 The watersheds of Jamaica and geographic distribution of the major
agricultural crops of Jamaica

cultivated land is only 1 percent and 2 percent of arable land, respectively (Table 15.1). Sugarcane is the major crop for the entire region except in Guyana where 120,000 ha are under rice and only 44,000 ha are under sugarcane (Higman, 1975). A century after its introduction in the region in 1872, bananas have become a major crop in many islands, particularly in the Windwards. Coconut, cocoa, citrus, vegetables, beans, coffee, cotton, peanuts, ornamentals, and a variety of root and other tropical crops are grown on a small scale on different islands.

Jamaica has diversified its agriculture from sugarcane to bananas (in the 1870s), coconuts (around 1910), citrus and vegetables (in the 1920s), and to mangos and ornamentals in the 1980s (Table 15.2; Figure 15.3). Although coffee plantations were developed in the middle of the 1700s, their fortunes fluctuated until the 1970s, when massive renewal and expansion of the crop were initiated.

Agronomic practices introduced by the Europeans in the coastal plains have remained essentially unchanged, although agriculture has since been extended to hillsides where slopes of up to 70° are cultivated. In many areas land is still being cleared by cutting trees and burning brush. There is no terracing of land or management of water flow. Soil erosion on the different islands is undocumented but is probably not very different from Jamaica where the estimated loss of top soil is about 13,000 T km^{-2} year^{-1} (Eyre, 1990).

Table 15.1 Agricultural use of land in different Commonwealth Caribbean countries

Country	Area (km²)	Land use (%)				Coastline (km)	Area (ha) under crops	Area (ha) under pasture
		Arable land	Permanent crops	Permanent pastures	Forests			
Antigua and Barbuda	442	18	0	9	11	153	25,920	2,835
The Bahamas	13,940	1	0	0	32	3,542	14,985	810
Barbados	430	37	0	5	12	97	25,920	4,050
Belize	22,960	2	1	2	92	386	46,980	17,010
Dominica	754	9	13	3	67	148	17,010	2,025
Grenada	340	15	18	3	9	121	16,200	810
Guyana	214,970	2	0	3	84	459	833,895	2,430,000
Jamaica	10,990	14	6	24	17	1,022	241,380	247,050
Montserrat	100	20	0	10	40	40	2,025	810
St Kitts-Nevis-Anguilla	352	16	13	2	13	196	16,200	4,050
St Lucia	620	8	21	5	13	158	21,060	2,835
St Vincent and the Grenadines	389	10	18	5	36	84	18,225	810
Trinidad and Tobago	5,128	15	9	2	46	362	139,320	6,075

Source: Adapted from Higman, 1975 with additional information from *The CIA World Factbook, 2000.*

Table 15.2 Area cultivated and annual consumption of pesticides in Jamaica by major crops based on a survey of farmers by the authors and data from each Commodity Board or Association

Product	Area (ha)	Major pests	Pesticides used[a]	Mean application rate (kg or L a.i.)	Pesticide load (kg or L a.i. year⁻¹)
Banana	34,000	Banana borer, thrips, nematodes	I/N: ethoprophos, isazofos, chlorpyrifos and other OPs	1.5 kg ha⁻¹; 3 times per y	15 kg ha⁻¹ y⁻¹; 510,000 kg y⁻¹
		Weeds	H: paraquat, ametryn, glyphosate	5 L ha⁻¹; 3 times per y	4.5 L ha⁻¹ y⁻¹; 153,000 L y⁻¹
		Sigatoka disease	F: hexaconazole, chlorothalonil, tridemorph and others	0.5 L ha⁻¹; 6 times per y	3 L ha⁻¹ y⁻¹; 102,000 L y⁻¹
Cattle	1,500,000	Ticks, screw worm	I/A: amitraz	0.003 L per animal per spray; 26 sprays per y	0.078 L per animal per y; 117,000 L y⁻¹
Citrus	12,000	Citrus root weevil, ants	I: carbaryl	9 kg ha⁻¹; 2 times per y	18 kg ha⁻¹ y⁻¹; 216,000 kg y⁻¹
		Leaf miner, aphids, scale insects	I: malathion, dimethoate, diazinon	2.8 L ha⁻¹; 3 times per y	8.4 L ha⁻¹ y⁻¹; 100,800 L y⁻¹
		Gummosis, scab, foot rot	F: benomyl, fosetyl, copper hydroxide	3.4 kg ha⁻¹ and 2.6 L ha⁻¹; 2 times per y	6.8 kg ha⁻¹ y⁻¹; 81,600 kg y⁻¹ & 5.2 L ha⁻¹ y⁻¹; 62,400 L y⁻¹
		Weeds	H: paraquat, glyphosate	0.3 L ha⁻¹; 2 times per y	0.6 L ha⁻¹ y⁻¹; 7,200 L y⁻¹
Coffee	10,112	Coffee berry borer	I: endosulfan	0.3 L ha⁻¹; 2 times per y	0.6 L ha⁻¹ y⁻¹; 6,070 L y⁻¹
		Coffee leaf miner	I: dimethoate, diazinon, carbofuran	0.5 L ha⁻¹; 2–4 times per y	1.5 L ha⁻¹ y⁻¹; 15,168 L y⁻¹
		Coffee leaf rust, anthracnose and brown eye spot	F: copper oxychloride	1.1 kg ha⁻¹; 2 times per y	2.2 kg ha⁻¹ y⁻¹; 22,246 kg y⁻¹
Ornamentals	300	Weeds	H: paraquat, glyphosate	0.6 L ha⁻¹; 2–4 times per y	1.5 L ha⁻¹ y⁻¹; 13,651 L y⁻¹
		Mites	I: various OPs	0.35 L ha⁻¹; 20 times per y	7 L ha⁻¹ y⁻¹; 2,100 L y⁻¹
		Rust	F: copper-based	0.5 kg ha⁻¹ week⁻¹	26 kg ha⁻¹ y⁻¹; 7,800 kg y⁻¹

Product	Area (ha)	Major pests	Pesticides used[a]	Mean application rate (kg or L a.i.)	Pesticide load (kg or L a.i. year^{-1})
Sugarcane[2]	41,000	West Indian canefly, Weeds	I: fenitrothion, malathion H: 2,4-D, ametryn/amitraz, diuron	0.1–0.2 L ha^{-1}; occasionally 1.5 & 6 L ha^{-1}; 1.5 times per y 4.3 kg ha^{-1}; 1.5 times per y	40.5–80.9 L ha^{-1} y^{-1} 2.3 L ha^{-1} y^{-1}; 94,300 L y^{-1} 6.5 kg ha^{-1} y^{-1}; 266,500 kg y^{-1}
Vegetables	2,500	Mites, aphids, army-worms, semi loopers, diamondback moth, whiteflies, cucumber beetles	I: deltamethrin, λ- cyhalothrin, malathion, profenofos, other OPs	1 L ha^{-1}; 54 times per y	54 L ha^{-1} y^{-1}; 108,000 L y^{-1}

Note:
a Initials indicate: I-insecticide; N-nematicide; H-herbicide; F-fungicide; / indicates multiple use.

AGRICULTURAL MANAGEMENT, RESEARCH, AND TRAINING

Management

Large plantations owned by the British were the norm until the early twentieth century when departments or ministries assumed greater responsibility for managing agriculture. In many Commonwealth countries, semi-autonomous boards were set up to address the needs of farmers for major crops such as rice in Guyana, banana in the Windwards, and sugarcane, coffee, banana, coconut, cocoa, and citrus in Jamaica. These needs include supplying planting material, agricultural extension services, and marketing assistance.

Research

From plantation days when naturalist Hans Sloane first recorded the Jamaican citrus root weevil *Exophthalmus vittatus* L. (Coleoptera: Curculionidae) in 1725 to the 1970s, most research on plant protection in the Caribbean was restricted to recording and describing crop pests and their outbreaks. The notable exceptions to this trend have been the excellent work done on sugarcane pests by the Caroni Sugarcane Research Institute in Trinidad and the Sugarcane Research Institute in Jamaica. The Imperial College of Tropical Agriculture, founded in Trinidad in 1921, did not have much impact on insect pest control research in the region. This trend continued even when the college became the Faculty of Agriculture of the newly founded University of the West Indies (UWI) in 1962. At about the same time, a laboratory of the Commonwealth Institute of Biological Control, London, England was set up in Trinidad and two regional organizations – the Caribbean Agricultural Research and Development Institute (CARDI) and the Inter-American Institute for Cooperation in Agriculture (IICA) funded by various international agencies – became active in different countries of the Commonwealth.

In spite of the infrastructure, research on plant protection continued to be of marginal value. The introduction of modern synthetic organic pesticides in the region in 1945 further exacerbated the problems of local entomologists. Pesticides provided an excellent cost–benefit ratio, much less dependence on a usually unreliable labor force, and the euphoria of being current with contemporary technology. The practice of chemical pesticide reliance created a 'mutant culture' within agriculture, the 'pesticide subculture', which has become deeply ingrained and difficult to reverse even in agricultural policy makers.

Until the 1970s, almost no data existed on any crop pest that could be used for developing even short-term strategies for its control. To develop this type of data, A. Mansingh established an Insect Toxicology and Physiology Laboratory in 1974, which in 1985 became an interdisciplinary Pesticide and Pest Research Group (PPRG) in the Faculty of Pure and Applied Sciences at the UWI, Mona, Jamaica. The group embarked upon the relevant research as outlined in Figure 15.4.

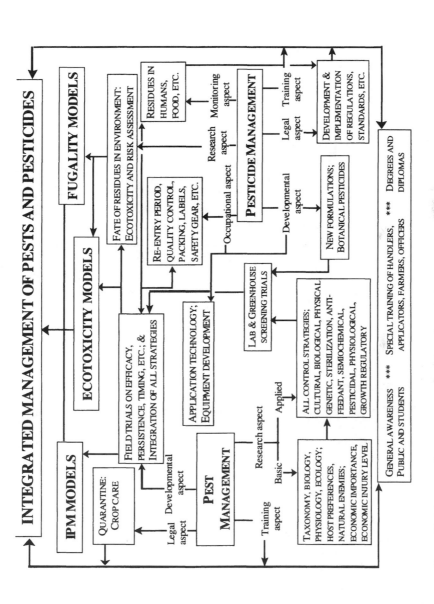

Figure 15.4 Integrated management of pests and pesticides in tropical island ecosystems: a prospectus

Training

It is unfortunate that training in pesticide use has traditionally been neglected by the UWI, agricultural schools, and Ministries of Agriculture of the Commonwealth Caribbean. Users have little appreciation of the occupational and environmental hazards of pesticides. Most receive training from their peers, which perpetuates the misuse and abuse of pesticides. Very few farmers can read or understand the manufacturer's instructions, which are usually printed in fine print. Even most extension and training officers have little knowledge of the three 'W's (why?, what?, and when?) and three 'H's (how much?, how?, and how often?) of pesticide use. Occasionally the UWI, CARDI, and the Caribbean Conservation Association (CCA) hold training programs on pesticide application for a few farmers, depending upon availability of international funding.

The College of Agriculture in Jamaica offers a two-year diploma and three-year associate degree program after grade 10 of high school, which superficially cover plant protection and pesticides. The Faculty of Agriculture, UWI offers a general BSc degree in Agriculture and MSc and PhD degrees in different disciplines, including plant protection, but the program is weak. The PPRG of the UWI, Jamaica offers quite extensive courses on insect taxonomy, ecology, physiology, and integrated management of pests and pesticides to final year undergraduate students. Its graduate school trains at least three students per year, offering masters and doctorate degrees in various fields.

USE OF PESTICIDES

Consumption

It is difficult to establish the trend in pesticide use over the decades or calculate a pesticide load for the Commonwealth Caribbean countries as import data were never recorded until the 1970s. Even now, most countries record only the quantities of formulations imported – each may contain from 5 to 80 percent a.i. Data in Table 15.3 suggest that the relative consumption of different pesticide groups varies with crop and country. Consumption by group in Barbados and Trinidad is herbicides > insecticides > fungicides while in Jamaica and other islands the relative ranking is insecticides > herbicides > fungicides. However, when the quantity of a.i.(s) is considered, the use of herbicides in Jamaica is about 2- and 2.8-fold more than fungicides and insecticides, respectively, while in Guyana, these differences are 2.5- and 33.2-fold, respectively. There may be a similar trend for a.i. consumption in other Commonwealth countries depending on the mix of crops grown and local pest problems, although Belize currently uses more fungicides than herbicides or insecticides.

The greatest quantity of pesticides per hectare of cultivated land is utilized in the cultivated fields of Barbados, followed by St Vincent > Dominica > St Lucia > others (Table 15.3). Pesticide loads (kg a.i. ha^{-1} cultivated) in Belize (0.1), Guyana

Table 15.3 Amount (kg) of pesticides imported into Commonwealth Caribbean countries

Country	Insecticides[a]		Herbicides		Fungicides		Rodenticides		Others		Total	kg ha⁻¹ of cultivated land
					Quantity (kg) (and % of total) of pesticides imported							
Barbados[b]	125,943	(18.5)	362,183	(53.2)	28,315	(4.2)	–[c]		164,208	(24.1)	680,649	53.08
Belize[d]	8,770	(13.4)	15,500	(23.6)	19,800	(30.1)	–		21,610	(32.9)	65,680	0.10
Dominica[b]	196,865	(56.9)	134,686	(38.9)	14,077	(4.1)	–		186	(0.05)	345,814	27.32
Grenada[b]	72,010	(70.6)	21,876	(21.4)	5,876	(5.8)	1,576	(1.5)	694	(0.7)	102,032	3.29
Guyana[d]	87,458	(26.6)	220,948	(67.3)	6,654	(2.0)	–		13,147	(4.0)	328,207	0.39
St Lucia[b]	309,294	(62.3)	151,059	(30.4)	18,905	(3.8)	4,011	(0.8)	14,567	(2.9)	496,659	22.97
St Vincent[b]	387,644	(87.9)	43,011	(9.7)	10,102	(2.3)	1	(0.0002)	949	(0.2)	441,194	37.48
Trinidad[b]	171,280	(24.3)	333,984	(47.4)	27,443	(3.9)	–		172,413	(24.5)	705,120	2.98
Jamaica[e1]	781,836	(45.2)	715,375	(41.4)	232,201	(13.4)	–		–		1,729,412	7.16
Jamaica[e2]	98,751	(19.2)	275,807	(53.6)	140,314	(27.2)	–		–		514,872	2.13

Source: PCA, 2000.

Notes:
a Quantity includes insecticides, nematicides, and acaricides.
b Data are on formulations with 5–80% active ingredient; Source: deGeorges, 1989; CCA/IRF, 1991.
c En dash (–) indicates no data available.
d Data in kg of active ingredient; Source: Gooding, 1980.
e Data in kg of (1) formulations and (2) active ingredient

(0.39) and Jamaica (2.13) are much lower than that for the United Kingdom (3.9) in 1977 (Gooding, 1980).

Estimates of the annual consumption (kg and L) of pesticides for specific crops are available only for Jamaica (Table 15.2). The use order by crop is bananas (807,000) > citrus (130,000) > livestock (117,000) > vegetables (108,000) > sugarcane (50,000) > coffee (43,484) > ornamentals (9,900). The ranked pesticide load (kg or L ha^{-1} year^{-1}) by crop type is vegetable fields (54) > ornamental (33) > banana (23.7) > citrus (12.5) > coffee (4.35) > sugarcane (1.2) > livestock (0.078 L per head).

Environmentally however, the use of endosulfan and other pesticides in coffee may be regarded as the most dangerous, because this crop is grown mainly in mountainous watersheds and highlands, where pesticide runoff into rivers is more frequent. Similarly, a high pesticide load in banana fields – particularly that of isazofos, which is highly toxic to aquatic fauna – could contaminate rivers and coastal waters. With the banning of dieldrin in 1989, entry of the last persistent OC into the Jamaican environment was stopped.

Importation and storage

Most of the pesticides imported into the region are formulations that may be sold as is or diluted by reformulation at facilities in Jamaica. Until the 1990s, there were no regular or formal government inspections at ports of entry and storage in both warehouses and retail stores was unregulated. At home, small farmers usually store chemicals inside their houses, with the consequence of more than occasional food contamination and poisoning.

Transport and retailing

Transport of pesticides along with foodstuffs is common. Retail sale to small farmers has always created problems because retailers sell pesticides in paper bags similar to those used for sugar and flour, or in empty drink bottles, a practice that has poisoned and killed a few people.

Dilution and disposal

The proverb 'medicine can't harm' is the guiding principle used for diluting pesticides. Even when farmers follow the manufacturer's recommendations, they add a little extra as insurance to the usual amount (the maximum listed rate), raising the upper limit of application concentration. Often, farmers measure by measuring cup, without regard for the percentage of a.i. in their formulation.

Distributors and end-users of pesticides do not consider management of spills necessary. For example, in 1987 a multinational dealer dumped more than 500 gallons of phosdrin (a.i. is mevinphos) in a municipal dump just outside Kingston, poisoning several people and killing dozens of pigs. Unused spray mixtures may be disposed anywhere, including in rivers. Empty pesticide containers and bags are routinely disposed of as litter. Disposal of chlorpyrifos-coated plastic bags used for covering developing banana bunches is largely indiscriminate, particularly in the Windwards, where pesticides are eventually carried into rivers and streams by wind or torrential downpours. This situation is improving to some extent on a few plantations in Jamaica where the bags are now collected for burial or exported to Costa Rica for recycling.

Application

Pesticides are applied to control crop pests, termites, domestic pests, and mosquitoes. The US Drug Enforcement Agency (US DEA) sprays herbicides for marijuana eradication and many 'smart' individuals throw insecticides into rivers for fishing. The timing and frequency of pesticide application are still based on routine calendar recommendations and not on EIL or even sighting the pest. For instance, coffee farmers spray endosulfan twice a year and crucifers are sprayed twice a week without ever checking for the presence of a pest (Table 15.2).

All small and medium-sized farmers (<30 ha) use knapsack sprayers, although larger farmers may have motorized sprayers. Selection of an appropriate nozzle or calibration of spray equipment has never been taught or practiced. Aerial spraying in the Commonwealth Caribbean is restricted to large areas of sugarcane in Trinidad, rice in Guyana, bananas in the Windwards and Jamaica, drug eradication by the US DEA, and mosquito control by the Ministry of Health.

Few farmers wear any protective gear or consider the wind velocity and direction, the prospects for rain, the nearness to a river or people's homes, the local topography, or the height of target plants before spraying operations commence. In the Windwards, aerial spraying is regularly carried out without any warning being given to people living in homes within and on the periphery of plantations. The authors have observed – on more than one occasion – that children, laundry, cooking ware, livestock, and home gardens were exposed directly to settling from aerial pesticide sprays. Also, drift and settling of aerial sprays occurs in various drains and streams running through the plantations and discharging ultimately into the sea.

LEGISLATIVE MANAGEMENT OF PESTICIDES

Until the 1970s, pesticides were handled by the pharmaceutical division in the Ministry of Health for the various islands. Jamaica passed a Pesticide Act in 1975 that regulated registration, importation, transport, storage, retailing, and manufacturing of pesticide formulations. All highly toxic pesticides and OCs, except dieldrin and chlordane, were banned in 1973. A list of restricted use pesticides was also prepared. In the 1980s, chlordane and dieldrin were banned while endosulfan and isazofos – two environmentally toxic compounds – continue to be listed among the restricted use pesticides. Unfortunately, the registration of pesticides and issuance of import permits were handled in a half-hearted manner while other regulations were never enforced.

The existence of regulations 'on the books' did not satisfy international monetary agencies who restricted funds for agricultural development to effect legal 'active' control of pesticides. In 1990, the World Bank invited A. Mansingh to serve as a consultant for developing policies and strategies for legislative control of pesticides in Jamaica. The German development agency, GTZ, later provided financial and technical assistance for the World Bank plan and in 1993, the Pesticide Control Authority of Jamaica (PCA) was formally established.

The PCA is housed in the Ministry of Health and is managed by a board of directors, which has members from the UWI and the Ministries of Agriculture, Environment, Health, Justice, and Commerce and Industry. A registrar and two deputy registrars execute the PCA mandate. In three years, the PCA streamlined and updated procedures for pesticide registration and issuance of import licences; prepared lists of banned, restricted use, unrestricted sale, and unused and unwanted stocks of pesticides; initiated registration of storage and manufacturing sites; and resumed programs for promoting pesticide awareness. Efforts are currently underway to train and license pest control operators. Legislation covering the various phases of a pesticide's life cycle is constantly being drafted. A major milestone was achieved by the PCA in 1998 when all the identified stocks of unwanted pesticides in the island were exported to the USA for incineration.

Full training of pest control operators and enforcement of most regulations is unlikely to be achieved in the near future. Having voted down a proposal for a levy on the sale of pesticides, the Government of Jamaica has restrained the activities of the PCA by forcing it to be dependent upon meager budgetary allocations. Notwithstanding these constraints, the PCA, in collaboration with the Natural Resources Conservation Authority and the Customs Department and with technical assistance and advice from the US EPA, has brought Jamaica to the forefront of African and Caribbean countries in the legislative management of pesticides.

The chronology and status of pesticide management in Trinidad and Barbados have been very similar to Jamaica. However, the eastern Caribbean states (the Windward and Leeward islands) with funding from the US AID have taken more than five years to develop their common legislation, which has yet to be promulgated and enforced, except in the case of the registration and importation of pesticides.

ISLAND ECOSYSTEM AND THE NEED FOR INTEGRATED MANAGEMENT OF PESTS AND PESTICIDES

Introduction

Pesticide application technology, developed for the continental land masses of Europe and North America, was adopted by the islands without the modifications needed to address the particular requirements of their topography, size, climate, and agricultural practices. Several natural features and aspects of human activity render an island's ecosystem prone to pesticide contamination (Mansingh, 1993). These include a thin (0.1 to 1.5 m) soil cover that favors leaching and runoff; crop cultivation and the use of agrochemicals on high mountainous slopes where most rivers originate; tilling steep slopes with poor agronomic practices, which encourages soil erosion; small holdings and a mixed crop system in which different pest complexes may require different pesticides; the vagaries of wind currents on slopes and in valleys, which promote aerial drift of spray particles; a pattern of frequent and intense rainfall, which facilitates the regular runoff of chemicals; complex dynamics of water-flow in unusually small and short rivers, which can transport residues to coastal waters quickly; and the close proximity of farms, rivers, homes, and seacoast, which makes them all vulnerable to residue exposure.

To reduce the ecological and environmental consequences of pesticide use, IPM is practiced by which insecticide applications are based upon an EIL. Many countries outside the Commonwealth Caribbean – with decades of research data – have established EILs for various pests and thus have reduced pesticide consumption on many crops, e.g. cotton, corn, and vegetables, by 50 to 70 percent. However, pesticides remain as the major element in their IPM programs.

For tropical countries in general – and the Commonwealth Caribbean in particular – there remains a paucity of the basic data needed for establishing EILs for individual pests and for initiating IPM strategies against them. Even international data on cosmopolitan pests must be validated under local conditions. The problem is further exacerbated by the reluctance of small farmers to risk their meager income by adopting recommended alternate strategies. In any case, partial or almost total reliance on organic pesticides is likely to continue globally, at least until botanical pesticides are popularized, their cost per application lowered, and new products developed for pests not currently targeted.

So long as synthetic organic pesticides continue to be even a minor element of IPM, equal or greater emphasis must be given to the management of pesticides, which includes their selection, application, fate, persistence, transport and runoff, ecotoxicity, and the environmental risk assessment of their residues. Therefore, it is proposed that rather than practicing IPM, a system, called Integrated Management of Pests and Pesticides (IMPP) and outlined in Figure 15.4, be adopted, practiced, and promoted in developing countries, particularly in island ecosystems (Mansingh, 1993).

IMPP in the Caribbean

Recognizing the need for IMPP in the Commonwealth Caribbean and guided by the philosophy that 'the developing countries must focus their attention on developing an intermediate technology for sustainable agricultural production, by integrating the practice of "risk reduction" and "safe use" of pesticides, with indigenous technology, which is economical, *effective and environmentally friendly'* (Mansingh, 1993), the PPRG laboratories have generated significant data for IMPP in the region since 1974.

Pest management

In the 1970s, regional sugarcane entomologists achieved success in biological control of the sugarcane moth borer *Diatraea saccharalis* Fabricus (Lepidoptera: Crambinae) and eliminated the use of insecticides for this pest. However, control of the frog-hopper *Aeneolamaia varia saccharina* Fabricus (Heteroptera: Cercopidae) in Trinidad is still dependent upon insecticides. In the 1980s, Dr Gene Pollard initiated pest management projects at the UWI, Trinidad. CARDI also has a few pest management projects, which are in their infancy.

Data on the life cycle, distribution, economic importance, alternate host plants, the natural enemy complex of major pests, and the laboratory and field efficacy of pesticides against them are essential for developing IPM and IMPP strategies. Research on the management of cattle ticks including *Boophilus microplus* Canestrini, *Amblyomma cajennennense* Fabricus, *A. variegatum* Fabricus, and *Dermacentor (Anocentor) nitens* Neumann (Acari: Ixodidae) (Rawlins, 1977) and screwworm fly *Cochliomyia hominivorax* Coquerel (Diptera: Calliphoridae) changed the acaricidal usage pattern in Jamaica and enabled investigators to propose strategies for managing these parasites (Rawlins and Mansingh, 1987). The IAEA, FAO, and the Ministry of Agriculture, Jamaica with technical involvement of the PPRG launched a five-year sterile-insect release program for screwworm eradication in Jamaica in late 1998.

The coffee berry borer *Hypothenemus hampei* Ferrari (Coleoptera: Scolytidae) threatened Jamaica's coffee industry after accidental introduction to the island in the mid-1970s. Data from other countries on the biology of this pest were suspect, resulting in total dependence on the indiscriminate and excessive use of insecticides (Mansingh, 1991). Studies of its life cycle (Johanneson, 1983), infestation pattern, insecticidal susceptibility (Rhodes, 1987), boring behavior (Boothe, 1987), economic importance, the field efficacy of insecticides, and cultural control practices (Reid and Mansingh, 1985; Reid, 1987) led to demonstration of a nutritional diapause in the pest (Mansingh, 1991). This led to development of the first ever IPM model for the suppression of its population (Reid, 1987). This model reduced the use of pesticides in coffee culture by 30 percent. An expert computer system, developed by Mansingh and Reichgelt (1997) in collaboration with the Pesticide and Pest Research Group, would further reduce insecticide use through computer-based evaluation and recommendations to farmers.

The citrus root weevil (CRW) complex *Exophthalmus vittatus* L. and *Pachnaeus citri* Marshall (Coleoptera: Curculionidae) has been the major pest of citrus crops in the region. Dieldrin, without justification, had been used for control of the pest between 1958 and 1989. Studies on susceptibility of the pest to various insecticides, the persistence of dieldrin on citrus plantations (Biggs-Allen, 1990), the potential of an entomopathogenic nematode against the pest (Myers, 1996), the role of alternate host plants and egg and larval parasites (Clark-Harris, 1998), and the efficacy of botanical formulations (Robinson and Mansingh, 1999, unpublished data) have forced the discontinuation of dieldrin use. This has led to the development of an IPM strategy that encompasses the use of alternate hosts as a trap-crop and treatment with an experimental botanical anti-feedant formulation, code named 'Ashima', developed by the PPRG.

Traditionally vegetable farmers have used the greatest quantity of insecticides per unit of land cultivated, yet little had been done by the region's scientists to alleviate their problems. Under the guidance of pesticide salesmen, vegetable farmers fought pest resistance by increasing the dose of an insecticide before finally switching to a different chemical, which then provided only a temporary respite. The PPRG alleviated this problem by demonstrating the efficacy of two OP pesticides and one *Bacillus thuringiensis* Berliner formulation (Forbes, 1995). Data on population fluctuations of the diamondback moth (DBM) *Plutella xylostella* L. (Lepidoptera: Plutellidae) and the cabbage looper *Trichoplusia ni* Hubner (Lepidoptera: Noctuidae), their natural enemies (parasitoids and predators, including spiders), EILs (Alam, 1996), the laboratory and field efficacy of insecticides (Forbes, 1995) and plant extracts (Wilson, 1993), the planting of tobacco within and mustard around cabbage plots (Mansingh and Napier, 1998, unpublished data), and spraying with three experimental botanical formulations code-named 'Ashima, Abhijai, and Dejoun' have provided enough information to allow cabbage cultivation without the use of synthetic organic pesticides.

Coconut mites *Eriophyes guerreronis* Keifer (Acari: Eriophyidae) have been damaging the quality and quantity of nuts produced in the Caribbean for over two decades. Vamidothion had been recommended for mite control but its partitioning between coconut meat and milk and the persistence of its residues inside the nut (Dasgupta *et al.*, 1998) aborted plans to use this or any other insecticide against the mite. Recently McDonald (1999) has quantified the economic loss inflicted by the pest in Jamaica but found no effective natural enemy that could be utilized for suppressing its population.

Weed management in the region has benefitted greatly from the work of Dr Richard Braithwaite of the Faculty of Agriculture, UWI, Trinidad. His colleague David Hutton in Jamaica has contributed significantly to nematode management in crops, planting materials, and ornamentals with recommendations on nematicides and alternate control techniques. Research on the field management of plant diseases has not been a priority of Jamaican scientists, although the use of fungicides in the region is quite high.

Ecological consequences

The ecological consequences of pesticide use have largely remained undocumented in the Caribbean except for the three 'R's – resistance, resurgence of pest population, and replacement by secondary pests. Various pests, ectoparasites, and disease vectors in the Caribbean have acquired resistance to pesticides (Rawlins and Mansingh, 1978; Biggs-Allen, 1990; Forbes, 1995; Sookhai-Mahadeo, 1997; Witter and Mansingh, 1997). Jamaican populations of cattle ticks have developed 15- to 67-fold resistance to carbaryl and a few OP insecticides (Rawlins, 1977; Rawlins and Mansingh, 1978). The CRW complex is extremely tolerant to dieldrin and other insecticides (Biggs-Allen, 1990) and the DBM has developed several thousand-fold resistance to most insecticides as compared to the susceptible Chinese strain (Forbes, 1995). The coffee berry borer is 300- to 700-fold resistant to endosulfan (Witter and Mansingh, 1997) and Trinidad populations of the mosquito *Aedes aegypti* L. (Diptera: Culicidae) – the vector for yellow fever – have also developed resistance to several OP insecticides (Sookhai-Mahadeo, 1997).

The resurgence of DBM populations and outbreaks of the cabbage looper in the 1990s in areas of heavy insecticide use around Jamaica have been attributed to deep declines in the populations of their natural enemies (Alam, 1996). Likewise, regular calendar-based spraying of endosulfan in the 1980s against the coffee berry borer has caused a minor and occasional pest of coffee, the leaf miner *Perileucoptera coffeella* Guérin-Méneville (Lepidoptera: Lyonetiidae), to become a regular and often serious pest in the 1990s (Dalip and Mansingh, 1995).

Environmental consequences

The consequences of pesticide use are difficult to assess because baseline data on biodiversity before the pesticide era are unavailable. The ecotoxicity of pesticide residues on selected terrestrial and aquatic fauna in Jamaica have been documented and will be discussed later in this chapter.

Pesticide management

Pesticide management requires a collaborative approach by entomologists, ecologists, biologists, chemists, physiologists, mechanical engineers, soil scientists, and lawyers for investigating and developing legislative management initiatives covering all phases of the pesticide life cycle. Further, collaboration by various stakeholders is required for developing more effective and environmentally friendly formulations; improving the efficiency of application equipment; establishing guidelines for occupational safety; determining the impact on non-target species; investigating the fate, persistence, and partitioning of residues in different environmental matrices; monitoring environmental contamination from pesticide residues; developing residue management techniques; assessing occupational and environmental risk from the residues; and integrating these findings within IPM models for specific pests. This should lead to the development of IMPP models and strategies (Figure 15.4). Further promotion of awareness programs for the general public, universal training of pesticide users, and improved efficiency of extension services would ensure implementation of IMPP.

Legislative management

Scientists have a large share of responsibility for the legislative management of pesticides in the Commonwealth Caribbean. A. Mansingh has played a key role in developing policy and strategy for the establishment of the PCA in Jamaica and as a USAID consultant reviewing pesticide regulations and developing policy on environmental health risks for the eastern Caribbean states. Members of the PCA board regularly discharge responsibilities including the assessment of pesticides for registration, categorization of a pesticide as restricted use or banned for a certain crop or certain geographical areas, and providing technical advice to legislators for drafting new legislation.

Regrettably, the Pesticide Boards in the Commonwealth Caribbean still do not pay due attention to labeling practices for pesticides, which ought to focus more on instructing the user rather than simply satisfying registration requirements. The print size on labels must be large enough to be easily readable by a normal person, color coded to indicate toxicity, instruct farmers not to use the produce until the EIL has been reached, and warn applicators of environmental hazards (Mansingh, 1993).

Public awareness

The PPRG can justifiably claim credit for initiating and implementing pesticide awareness programs in the Commonwealth Caribbean during the 1970s and 1980s through printed and electronic media, seminars, and direct training of farmers. This responsibility now rests with the Pesticide Control Boards or Authorities in the various islands and countries.

New formulations

A major need of IMPP in the Caribbean islands is development of formulations that quickly penetrate plant tissues or adhere sufficiently to plant surfaces so as not to be washed off by the frequent rain showers of the region. The ultimate goal of IMPP is to replace chemical pesticides with botanical pesticides suited to the local economy and environment (Williams and Mansingh, 1996). The PPRG has responded to the challenge by exploring the pesticide potential of tropical plants in Jamaica (Williams, 1991; Wilson, 1993; Williams and Mansingh, 1993; Mansingh and Williams, 1998). Partial chemical characterization of the active compounds has been accomplished for some of them (Williams and Mansingh, 1995) and some formulations have been tested successfully in the field (Wilson, 1993; Williams and Mansingh, 1993). Extracts of several plants, including *Artocarpus altilis* Park, *Azadirachta indica* A. Juss. and *Hibiscus rosa-sinensis* L., show great promise as acaricides (Blair *et al.*, 1995; Mansingh *et al.*, unpublished data).

Equipment

For the Caribbean islands, appropriate spray and agronomic equipment are essential. Sprayers that ensure restricted coverage with a minimum of drift are an ecological, environmental, and economic necessity for small holdings with multiple crop plots. Simple tools for weeding and terracing the slopes and making drain channels are needed to prevent soil erosion and pesticide runoff.

Occupational and public health

Occupational hazards of pesticides on any of the Caribbean islands have not been documented. Pesticide handlers in formulation plants, major retail stores, and on a few coffee plantations in Jamaica are tested occasionally for cholinesterase activity. A survey by the authors on different islands revealed that more than 70 percent of individuals involved in handling pesticides experience headache, vomiting, diarrhea, or skin rashes at least once a year, which they attribute to exposure to pesticides (Mansingh, 1993). However, few bothered seeking medical advice. There are no poison control centers in any of the Commonwealth Caribbean countries. The PCA in Jamaica is trying to establish a poison control center at the UWI hospital. In the meantime, however, only severe cases requiring hospitalization are to be found in hospital records, making it difficult to track the number of pesticide poisoning cases throughout the Commonwealth Caribbean. Pesticide symptoms and treatment are not taught in the region's medical schools. While there have been many cases of pesticide poisoning due to food contamination, there are no reliable statistics available.

FATE AND PERSISTENCE OF PESTICIDES

Behavior of pesticides in the tropical environment has not received sufficient attention, primarily because of a shortage of expertise and the paucity of research funds. In the Commonwealth Caribbean, only the interdisciplinary PPRG has investigated the dissipation and degradation of dieldrin and endosulfan (Singh, 1985; Singh *et al.*, 1991), chlorpyrifos (Morris, 1991), copper (II) hydroxide (Kocide), and copper oxychloride (Nelson, 1993) under laboratory conditions that simulate the Jamaican environment (Table 15.4). Similar studies under Jamaican field conditions were conducted on endosulfan, ethoprophos (or ethoprop) (Robinson, 1997; Robinson *et al.*, 1997; 1999), and isazofos (Robinson, 1997, unpublished data). An interesting finding was that the photolysis of dieldrin in the Jamaican environment was faster than in temperate countries. The half-life of this OC when incorporated in plantation soils was also quite short (Gayle, 1989; Biggs-Allen, 1990).

The persistence of bioactivity of twelve insecticides in as many Jamaican soil types was much less than the corresponding persistence in temperate countries (Anderson, 1987; Gayle, 1989). Chlordane and dieldrin had the longest persistence,

Table 15.4 Laboratory and field data on the dissipation and degradation of selected pesticides in the Jamaican environment

Insecticide	Process	Condition	$t_{1/2}$ (d)	Reference
Chlordane	Dissipation from soil	Field	579	Gayle, 1989
Chlorpyrifos	Volatilization	Laboratory	13.9	Morris, 1991
		Field	13.6–14.2	Morris, 1991
	Hydrolysis	Laboratory	0.11–48.1	Morris, 1991
	Photolysis	Laboratory	0.06–0.10	Morris, 1991
	Dissipation from soil	Field	5.8	Dalip, 1998
	Dissipation from coffee leaves	Field	1.1	Dalip, 1998
Deltamethrin	Dissipation from soil	Field	1.8	Dalip, 1998
	Dissipation from coffee leaves	Field	5.9–6.8	Dalip, 1998
Diazinon	Dissipation from soil	Field	7.3	Dalip, 1998
	Dissipation from coffee leaves	Field	1.4–1.8	Dalip, 1998
Dieldrin	Volatilization	Laboratory	4.40	Singh et al., 1991
	Hydrolysis	Laboratory	88.4–103	Singh et al., 1991
	Photolysis	Laboratory	1.7	Singh et al., 1991
		Field	20.7	Singh et al., 1991
	Dissipation from soil	Field	357–592	Gayle, 1989
			46.7–59.8	Biggs-Allen, 1990
Dimethoate	Dissipation from soil	Field	3.2	Dalip, 1998
	Dissipation from coffee leaves	Field	1.8–2.1	Dalip, 1998
Endosulfan (α)	Volatilization	Laboratory	3.92	Singh et al., 1991
	Hydrolysis	Laboratory	27.5–93.2	Singh et al., 1991
		Field	104.9–303.2	Robinson et al., 1997
	Photolysis	Laboratory	47.8	Singh et al., 1991
		Field	8.7–40.3	Robinson et al., 1997
	Microbial	Laboratory	28.5	Robinson et al., 1997
Endosulfan (β)	Volatilization	Laboratory	2.06	Singh et al., 1991
	Hydrolysis	Laboratory	23.5–87.7	Singh et al., 1991
		Field	86.9–547.5	Robinson et al., 1997
	Photolysis	Laboratory	32.9	Singh et al., 1991
		Field	8.7–32.6	Robinson et al., 1997
	Microbial	Laboratory	24.5	Robinson et al., 1997
Ethoprophos	Volatilization	Laboratory	4.2–64.8	Robinson, 1997
	Hydrolysis	Laboratory	24.7	Robinson, 1997
		Field	64.9–132.8	Robinson, 1997
Ethoprophos	Photolysis	Laboratory	14.4	Robinson, 1997
		Field	4.7	Robinson, 1997
	Microbial	Laboratory	10.9	Robinson, 1997
Isazofos	Hydrolysis	Laboratory	60.5–104.3	Robinson, unpub.
	Dissipation from soil	Field	17.7–175.1	Robinson, unpub.
			3.1	Dalip, 1998
	Dissipation from coffee leaves	Field	3.7–20.7	Robinson, unpub.
			1.3–1.7	Dalip, 1998
λ-cyhalothrin	Dissipation from soil	Field	6.5	Dalip, 1998
	Dissipation from coffee leaves	Field	11.7–12.7	Dalip, 1998

continued…

Table 15.4 continued

Insecticide	Process	Condition	$t_{1/2}$ (d)	Reference
Triadimefon	Volatilization	Laboratory	5.7–9.3	Nelson, 1993
	Hydrolysis	Laboratory	93.4–250.2	Nelson, 1993
	Photolysis	Laboratory	33.8–68.6	Nelson, 1993
		Field	8.9–9.6	Nelson, 1993
	Dissipation from coffee leaves	Field	1.7–7.8	Nelson, 1993
Copper oxychloride	Dissipation from coffee leaves	Field	40.1	Nelson, 1993
Copper hydroxide	Dissipation from coffee leaves	Field	32.9	Nelson, 1993

followed by carbamates > pyrethroids > OPs (Table 15.5). Generally persistence was greater in clay followed by sandy loam > loamy sand > and sandy soils at 10–20 percent moisture levels.

Often leaching and runoff from soils may be the major route of residue loss. A laboratory study revealed that 92 percent of endosulfan and 93 percent of dieldrin remained in the top 0 to 5 cm layer of a sandy loam soil column (Singh, 1985). In a similar study, Robinson (1997) found that leaching was significantly greater in sandy clay than in clay soil columns; percolation of ethoprophos was 33 percent and 28 percent and of endosulfan 9 percent and 3 percent in the two soil types, respectively.

Presence of vegetative cover favors leaching versus runoff of insecticides, particularly on gentle slopes. On a Jamaican Blue Mountain coffee plantation, leaching of soil-applied ethoprophos was 2 percent and 1.9 percent in the top 10 to 15 cm layer at 23° and 38° slopes, respectively, on weeded (without vegetative cover) plots and 2.8 percent and 2.2 percent, respectively, on unweeded (vegetated) plots. Leaching of soil-applied endosulfan on the two slopes was 1.7 percent and 0.8 percent, respectively, in the weeded and 2.7 percent and 1.5 percent, respectively, on the unweeded slopes (Robinson *et al.*, 1997; 1999). Thus, soil vegetative cover plays an important role in the persistence of residues, even on different slopes. Dissipation of ethoprophos was not significantly different in the weeded and unweeded plots on the 23° slope (93.6 percent and 89.5 percent, respectively) and on the 38° slope (92.4 and 91.2 percent, respectively). However, the dissipation of endosulfan was significantly ($P < 0.01$) different between the weeded (60.5 percent and 60 percent, respectively) and unweeded plots (54 percent and 57 percent, respectively) at the two slopes. The runoff of ethoprophos in the weeded and unweeded plots on a 5° slope (1.3 percent and 1.5 percent, respectively) was significantly less than on the 23° (5.3 percent and 4.2 percent, respectively) or the 38° (47.3 percent and 6.4 percent, respectively) slopes (Robinson *et al.*, 1999). Similarly runoff of endosulfan on the 5° slope in the weeded (0.21 percent) and unweeded (0.15 percent) plots was significantly lower ($P < 0.01$) than the runoff

Table 15.5 Persistence of various insecticides in different Jamaican soil types at 15 percent moisture level as determined by laboratory bioassay

Insecticide	LT_{50} (weeks) in different soil types[a]												Ref.[b]
	CCL	MSL	SCL	ELC	YS	YSL	CLS	VGLS	KC	SAC	HDLC	CLC	
Pyrethroids													
Decis	3.2	3.8	11.8	–[c]	–	–	–	–	–	–	–	–	2
Permethrin	3.9	4.1	3.1	–	–	–	–	–	–	–	–	–	2
Phenothrin	2.6	3.1	2.3	–	–	–	–	–	–	–	–	–	2
Organochlorines													
Chlordane	25	169.3	–	43.9	9.9	8.4	15	9.4	103.8	98.1	64.6	98.1	1, 2
Dieldrin	31	48.8	–	49.2	17.5	27.9	18.1	22.1	29.6	34.3	52.1	35.2	1, 2
Organophosphates													
Chlorpyrifos	–	2.2	–	0.67	3.2	0.7	0.9	1.3	2.6	0.9	0.6	1.3	1
Diazinon	–	1.7	–	3.5	0.9	0.4	0.9	0.6	2.2	1.1	1.1	3.2	1
Diethyl bromophos	–	3.6	–	2.8	2.4	3.6	3.3	2.4	2.1	2.9	1.1	4.0	1
Dimethyl bromophos	–	4.9	–	2.1	2.6	1.7	2.1	1.2	3.8	2.3	2.9	4.1	1
Malathion	–	2.6	–	13	8.4	14.4	3.6	1.7	8.3	3.4	6.9	7.4	1
Mevinphos	–	0.3	–	0.3	0.2	0.2	0.4	0.1	1.3	0.2	0.4	0.3	1
Carbamates													
Carbaryl	–	79.4	–	59	26	97.7	22.1	100	67	30.1	11.4	16.3	1

Notes:
a Soil types abbreviated as follows: CCL–Chudleigh Clay Loam; MSL–Marvelly Sandy Loam; SCL–Syndenhan Clay Loam; ELC–Ewarton Linstead Clay; YS–Yallahs Sand; YSL–Yallahs Sandy Loam; CLS–Caymanas Loamy Sand; VGLS–Valda Gravelly Loamy Sand; KC–Killancholly Clay; SAC–Saint Ann Clay; HDCL–Halls Delight Channery Loam; CLC–Charlton Linstead Clay.
b References are (1) Anderson, 1987 and (2) Gayle, 1989.
c En dash (–) indicates no data.

on the 23° slope (8.4 and 5.2 percent, respectively) and the 38° slope (9.1 and 4.5 percent, respectively).

Persistence ($t_{1/2}$ values in days) of various insecticides on coffee leaves and in soil (soil data in parentheses) on a Jamaican coffee plantation was chlorpyrifos, 1.1 (5.8); isazofos, 1.3 (3.1); diazinon, 1.8 (7.3); dimethoate, 2.1 (3.2); deltamethrin, 5.9 (1.8); and cyhalothrin, 12.7 (6.5) (Table 15.4; Dalip, 1999).

CONTAMINATION OF THE ENVIRONMENT: SOIL, SURFACE WATERS, GROUND WATERS, COASTAL WATERS AND AQUATIC FAUNA

Soils

A 1985 soil survey revealed the presence of dieldrin (92 μg kg^{-1} to 1224 μg kg^{-1}) and endosulfan (6.5 μg kg^{-1} to 400 μg kg^{-1}) on all citrus, coffee, banana, vegetable, and sugarcane plantations studied. No other OC insecticides were detected, although DDT, chlordane, and aldrin had been used in many of these fields until the late 1970s. Obviously their residues have been flushed out by heavy rains from the frequent storms prevalent in tropical climes (Mansingh, 1987). In coffee plantation soils the usual level of α-endosulfan may be as high as 300 μg kg^{-1} soon after foliar application of the pesticide but four weeks later α-endosulfan residues have dropped to 0.2 to 37.8 μg kg^{-1}, while β-endosulfan and endosulfan sulphate residues range from 6.4 to 13.8 μg kg^{-1} and 9.3 to 38.8 μg kg^{-1}, respectively (Robinson and Mansingh, 1999).

Rivers

Most Jamaican rivers that have been monitored over the years are contaminated with insecticides, mainly the endosulfans (Henry, 1984, Lawrence, 1984; Mansingh et al., 1995, 1997, 2000; Robinson and Mansingh, 1999; Witter et al., 1999; Table 15.6). In many instances, residue levels in the water and sediments of these rivers were higher than the LC$_{50}$ values for many aquatic species from around the world (Portman and Wilson, 1971; Worthing, 1987). In a rapid survey conducted in 1994, residues were detected in many other rivers (Mansingh et al., 1997; Table 15.7). Chlorpyrifos residues were detected in all rivers draining banana plantations, and diazinon and endrin in some of them (Witter et al., 1999).

Ground waters

Although runoff of pesticide residues is the primary transport mechanism on hillside farms in Jamaica, leaching does occur and has contaminated three of the six natural springs and nine of thirteen water-supply wells that were monitored. Endosulfans were the most significant residues in all the contaminated wells, although

Table 15.6 Residues of OC and OP insecticides detected in rivers of some Jamaican watersheds

Watershed	River	Year	Matrix	Residues in water ($\mu g\ L^{-1}$) and sediment and fauna (ng g^{-1} ww)					
				α-Endosulfan	β-Endosulfan	Endosulfan sulphate	Dieldrin	DDT/DDE	Diazinon
Hope River	Hope	1989–91	Water	0.006–7.23	0.024–6.0	0.010–1.06	0.013–0.019	—[a]	8.00–42.3
			Sediment	0.024–483	0.037–1.58	0.055–1.43	0.23–0.82	—	0.115–33.0
	Mammee	1989–91	Water	0.001–31.8	0.003–7.20	0.004–0.481	0.004–0.412	—	0.011–8.13
			Sediment	0.016–715	0.008–36.0	0.325–65.0	0.041–1.66	—	0.213–28.7
	Hog Hole	1989–91	Water	0.031–0.625	0.064–15.0	0.002–14.5	0.008–7.85	—	0.029–134
			Sediment	0.046–325	0.003–1.56	0.470–0.57	0.064–2.21	—	0.006–0.905
	Salt	1989–91	Water	0.024–1.79	0.170–0.246	0.317–8.49	0.013	—	40.1
			Sediment	0.046–950	235	0.385–0.82	0.001–1.01	—	3.03–23.6
Portland	Spanish	1991–92	Water	0.245–6.25	0.03–2.40	0.09–0.143	0.21	—	—
			Sediment	3.81	—	1.52–11.6	0.02	—	—
			Fauna	10.3–21.5	3.40–16.3	3.80–21.2	—	—	—
	Swift	1991–92	Water	1.00–2.11	0.80–3.19	0.029–6.90	0.76	0.92–5.26	—
			Sediment	0.777–94.3	0.30–1.2	0.011–4.94	—	0.11	—
			Fauna	4.01–16.1	2.40–28.8	1.20–6.50	—	—	—
Yallahs	Origin	1989–91	Water	0.19–0.55	0.11–0.77	0.01–24.3	0.12–0.29	—	0.03–1.69
			Sediment	2.00–5,961	26.3–1,972	—	19.0–22.6	—	13.0–132.6
Yallahs	Lower	1989–91	Water	0.60–1.70	0.40–1.00	0.01–10.50	0.03–0.36	—	0.35–1.24
			Sediment	2.00–4,394	0.40–2,690	1.00–27.90	2.50–19.90	—	3.50–135.0
Black River	Hector's	1989–92	Water	0.02–0.48	0.02–0.73	0.10–0.12	0.08–0.52	—	2.50–23.6
			Sediment	2.00–26.3	1.50	31.3–162.0	0.04–1.01	—	3.50–89.3
	One Eye	1989–92	Water	0.02–0.13	0.02–0.31	0.01–0.03	0.03–0.29	—	0.04–1.27
			Sediment	2.00–21.80	1.54	23.10–209	2.50–23.5	—	3.50–74.2
	Black	1989–92	Water	0.02–0.27	0.02–0.33	0.01–0.23	0.03–0.60	—	0.04–1.00
			Sediment	2.00–76.90	1.50–236	1.00–148	2.50–28.40	—	3.50–298

continued…

Table 15.6 continued

Watershed	River	Year	Matrix	Residues in water (μg L^{-1}) and sediment and fauna (ng g^{-1} ww)					
				α-Endosulfan	β-Endosulfan	Endosulfan sulphate	Dieldrin	DDT/DDE	Diazinon
Rio Cobre	Rio Pedro	1982–83	Water	–	–	–	0.002–0.08	0.01–0.2	–
			Sediment	4.0	1.7–3.7	–	1.00–2.00	2.20–8.30	–
			Fauna	0.08	1.4	–	1.0–9.2	7.2	–
	Thomas	1982–83	Water	–	–	–	0.01–0.07	0.01–0.10	–
			Sediment	0.43–4.7	–	–	1.00–2.00	2.0–11.10	–
			Fauna	–	–	–	1.3–3.3	10.0–13.1	–
	Rio Cobre	1982–83	Water	–	6.0	–	0.02–0.08	0.09–0.20	–
			Sediment	1.16–2.75	–	–	1.0–2.9	3.4–14.7	–
		1994	Fauna	8.2–14.6	7.9–10.3	–	0.65–0.92	1.3–1.6	–

Source: Adapted from Mansingh et al., 1997.

Notes:
a En dash (–) indicates below detectable limit.

Table 15.7 Residues of OC and OP insecticides detected in selected Jamaican rivers sampled during a rapid survey in 1994

| River | Matrix | Residues in water ($\mu g\ L^{-1}$) and sediment and fauna (ng g^{-1} ww)[a] | | | |
		α-Endosulfan	β-Endosulfan	Endosulfan sulphate	Chlorpyrifos
Morant	Water	0.024	0.09	0.003	–[b]
	Sediment	5.92	–	–	0.519
Buff Bay	Water	0.347	0.306	0.185	–
	Sediment	25.19	31.91	–	–
Wagwater	Water	–	–	–	–
	Sediment	–	–	–	–
Rio Magno	Water	–	–	–	–
	Sediment	33.48	–	–	–
Rio Minho	Water	0.354	0.291	–	–
	Sediment	0.913–21.42	17.78	–	2.48–135.2
Cabarita River	Water	–	–	–	0.022
	Sediment	41.28	49.35	–	0.069
South Negril	Water	0.024	0.27	–	–
	Sediment	–	–	–	–
Great River	Water	0.03	–	–	–
	Sediment	55.28	38.47	–	–
Martha Brae	Water	0.02	0.051	–	–
	Sediment	–	–	–	0.501

Source: Adapted from Mansingh *et al.*, 1997.

Notes:
a Also detected in sediment were ethoprophos (1.38 $\mu g\ kg^{-1}$) in Spanish River, diazinon (0.774–1.08 $\mu g\ kg^{-1}$) in Rio Minho and Alley, and endrin (0.006 $\mu g\ kg^{-1}$) in Martha Brae River and in water, chlorpyrifos (0.001 $\mu g\ L^{-1}$) in Spanish River, endosulfan sulphate (0.003–0.244 $\mu g\ L^{-1}$) in Morant River, Buff Bay River and Rio Cobre.
b En dash (–) indicates below detection limit.

ethoprophos, profenofos, diazinon, and chlorpyrifos were also detected in a few wells (Witter *et al.*, 1999).

Coastal waters

Coastal waters in Jamaica (Mansingh and Wilson, 1995; Mansingh *et al.*, 1997) and St Lucia (Shim, 1985; Ramsammy *et al.*, 1985; deGeorges, 1989) are contaminated with pesticide residues. The mean residue levels detected in water ($\mu g\ L^{-1}$) and sediment ($\mu g\ kg^{-1}$) samples collected from seven stations in Kingston Harbor, Jamaica during July and August of 1993 were: α-endosulfan, 2.2 and 0.52; β-endosulfan, 7.86 and 0.38; endosulfan sulphate, 0.0003 and 0; p,p'-DDT, 7.02 and 0.35; dieldrin, 1.88 and 0.001; aldrin, 0 and 9.18; endrin, 0.26 and 0.006; lindane, 0 and 0.51; HCB, 0 and 1.01; and diazinon, 0.05 and 0.005, respectively (Mansingh and Wilson, 1995). A 1995 to1996 study confirmed the presence of the endosulfans but also detected high levels of chlorpyrifos in water (0.02 to 2.35 μg L⁻¹) and sediments (0.12 to 24.64 μg kg⁻¹) (Mansingh *et al.*, 2000). Obviously these residues were transported into the bay by the Rio Cobre during the past few decades.

On the Portland Coast of Jamaica, where the Rio Grande, Swift, and Spanish rivers discharge, the mean level of residues in water (μg L⁻¹) and sediment (μg kg⁻¹), respectively, were: α-endosulfan, 0.42 and 1.8; β-endosulfan, 0 and 6.14; endosulfan sulphate, 0 and 3.9; dieldrin, 0 and 0.1; and p,p'-DDE, 0.84 and 6.14 (Robinson and Mansingh, 1999). Aldrin, heptachlor, endrin, total (Σ) DDT, and lindane were detected in fairly high concentrations in the coastal waters of St Lucia in the 1980s (Table 15.8).

Aquatic fauna and flora

Fish, shrimp, and oysters in Jamaican rivers and coastal waters accumulate pesticide residues from surrounding waters (Mansingh *et al.*, 1997). In the period 1982 to 1983, when endosulfan was little used, the most commonly detected residues and range of concentrations (in μg kg⁻¹ fresh wt) in shrimp from different rivers in the Rio Cobre basin were dieldrin (0.7 to 9.2) and DDE (1.0 to 7.2). However by 1995, the most frequent residues (in μg kg⁻¹ fresh wt.) were those of α-endosulfan (8.2 to 14.6), β-endosulfan (7.9 to 10.3), dieldrin (0.65 to 0.92), DDE (1.3 to 1.6), and chlorpyrifos (2.5 to 5.7) (Mansingh *et al.*, 1997; 2000). Almost 60 percent of shrimp and fish samples – collected monthly from the rivers of the Portland watershed – had accumulated α- and β-endosulfan and endosulfan sulphate to concentrations ranging from 4 to 21.5 ng g⁻¹ ww, 2.4 to 28.8 ng g⁻¹ ww, and 1.2 to 21.4 ng g⁻¹ ww, respectively (Robinson and Mansingh, 1999).

Deonarine (1980) has demonstrated selective partitioning of residues between flora and fauna in the brackish waters of the Caroni Swamp of Trinidad. Detritus contained only dieldrin and DDT residues while algae contained these in addition to heptachlor residues. Angiosperms had accumulated dieldrin, heptachlor, and endrin while crustacea and fish had accumulated all three (Table 15.8). Aldrin, dieldrin, DDT, and chlordane were also detected in fish samples from St Lucia's coastline (Shim, 1985).

The accumulation of residues in coastal fauna varied with the species. On the Portland Coast of Jamaica, mean levels of α- and β-endosulfan and endosulfan sulphate were found to be 86.1, 30.9, and 24 ng g⁻¹ fresh weight, respectively (Robinson and Mansingh, 1999). In Kingston Harbor, only one of four benthic fish sampled contained diazinon (0.31 ng g⁻¹ ww) while the oyster *Isognomon alatus* Gmelin had accumulated 0.04, 0.26, and 2.24 ng g⁻¹ ww of diazinon, α-endosulfan, and aldrin, respectively (Mansingh and Wilson, 1995).

A 1994 International Mussel Watch survey of mussels from Kingston Harbor and the Bowden Coast of Jamaica revealed the presence of Σ DDT at levels of 5.45 to 24.02 ng g⁻¹ dw (units are the same for all residues), methoxychlor (2.94), aldrin (0.1), dieldrin (0.13), endrin (0.18 to 0.34), oxychlordane (0 to 0.64), γ-chlordane (0.05 to 1.4), α-chlordane (0 to 1.25), α-HCH (0 to 0.72), β-HCH (0.05

Table 15.8 Residues of OC insecticides detected in water, flora and fauna in the Caroni Swamp in Trinidad[a] and coastal waters of St Lucia[b]

Site	Year	Matrix	Aldrin	Dieldrin	Heptachlor/Heptachlor epoxide	Endrin	Σ-DDT	Chlordane	Lindane
					Residues in water ($\mu g\ L^{-1}$) and sediment and fauna ($ng\ g^{-1}$)				
Trinidad									
Caroni Swamp	1978	Detritus	–[c]	0.038	–	–	0.039–11.5	–	–
		Algae	–	0.02	0.01–1.4	–	0.3–0.5	–	–
		Angiosperm	–	0.01–0.2	0.21–14.16	0.4	–	–	–
		Crustacea	–	0.49–2.71	0.75–19.00	11.62–70.15	7.00–70.15	–	–
		Fish	–	0.01–1.20	0.19–8.90	6.76	0.28–58.20	–	–
St Lucia									
Castries Harbour	1984	Coastal water	6.9	–	6.5	13.2	61.2	–	7.0
		Fish	0.001	0.005	–	–	0.012	0.058	–
Vigie Hoc		Coastal water	–	–	8.2	–	29.2	–	10.8
Cas en Bas		Coastal water	–	–	–	–	31.2	–	3.5

Notes:
a Data from Deonarine, 1980.
b Data from Shim, 1985 and Ramsammy et al., 1985.
c En dash (–) indicates pesticide not detected.

to 1.14), lindane HCH (0 to 0.49), and mirex (0.15 to 0.16) (Serricano, 1995). All residues except mirex had been detected previously in Kingston Harbor water and sediment by Mansingh and Wilson (1995). Bioaccumulation of residues by fish along St Lucia's coastline did not reflect the contamination of the surrounding water as these fish contained only aldrin, dieldrin, DDT, and chlordane residues (Table 15.8).

Ecotoxicity

Toxicity from pesticide residues to non-target terrestrial and aquatic organisms can be expected in the Caribbean islands, particularly because residues of acutely toxic OPs can persist for a few weeks (Table 15.5). Anderson (1987) found that the residual toxicity of several insecticides to 15 species of nematodes varied by species in different Jamaican soil types. Generally unidentified species of the genera *Dorilaimus* and *Pratylenchus* were more susceptible than others and diazinon and dieldrin had a longer-term effect than malathion. In a loamy sand soil type, the populations of most species were reduced by 70 to 90 percent within two months and did not recover significantly even seven months after treatment. Similar adverse effects were recorded for *Apholenchoides* spp., *Hoplolaimus* spp., and *Xiphinema* spp. in diazinon and dieldrin-treated soil and for *Pratylenchus* spp. and *Helicotylenchus* spp. in malathion-treated sandy loam plots. Most nematode species were affected to varying degrees in sandy soil treated with these chemicals. However, in fields with sandy loam soil, populations of all nematode species, except *Dorilaimus* spp., increased in treated plots, probably due to near extinction of their natural enemies by the pesticides in that particular soil type.

The impact of dieldrin residues on mite populations was much more severe than that of diazinon and malathion residues (Anderson, 1987). After three months, populations of mites in a loamy sand soil and a sandy loam soil were reduced by approximately 70 percent, 85 percent, and 95 percent in malathion, diazinon, and dieldrin-treated plots, respectively, but recovered by approximately 80 percent in the OP-treated plots during the following three months. In dieldrin-treated plots, recovery was approximately 25 percent in loamy sand soil during the following nine months but, in the sandy loam soil, almost total recovery was achieved within 10 to 12 months post treatment. In another sandy loam soil, mite populations recovered – after an initial reduction of 70 to 80 percent three months post-treatment – within an additional three months in malathion, seven months in diazinon, and eight months in dieldrin-treated plots. In a sandy soil, however, populations were reduced by 70 to 90 percent within a month following treatment with the three insecticides, but recovery was almost complete within four to five months in the malathion and diazinon-treated plots. In dieldrin-treated plots recovery was only about 25 percent one year after treatment (Anderson, 1987).

One week after isazofos application to coffee plants, arthropod (Diptera, Coleoptera, Arachnida, etc.) fauna on the plants remained suppressed by 30 to 70 percent, but spray drift and runoff from leaves had no effect on the soil fauna (earthworms, nematodes, spiders, etc.) under the tree canopy (Dalip and Robinson, 1997, unpublished data). Spray drift and runoff of chlorpyrifos, isazofos, diazinon, dimethoate, deltamethrin, and cyhalothrin from coffee leaves had no significant effect on the soil arachnids and insects on a Jamaican highland plantation (Dalip, 1999).

Acute and chronic toxicity values

Acute and chronic toxicity values of a toxicant are determined for an arbitrary period and are commonly expressed as one or more of the following: a NOEC or LOEC and a LC_{50} or LC_{90} (Cooney, 1995). However, three factors may be operative at these levels. These factors are first, toxicity may not be observable but detectable at concentrations below NOEC and visible and chronic or lethal effects may develop beyond the time-frame of the test; second, the ability of individuals at the LOEC and sub-lethal toxic concentrations to completely recover from the initial adverse effects; and third, the chronic effects in the survivors of LOEC and sub-lethal toxic doses, which may result in the retardation and inhibition, development, and reproduction and even survival of the F_1 generation. For instance, exposure of the Jamaican shrimp *Macrobrachium amazonicum* Heller (Crustacea: Palaemonidae) to the NOEC of dieldrin (0.0002 ppb) for seven days reduced the ventilatory rate by 21 percent; increased the respiratory reversal frequency, a measure of stress, by 6.7-fold; and reduced the resting and active oxygen consumption rates by 43 percent and 70 percent, respectively (Lawrence *et al.*, 1986a). Similarly after four days of exposure to the LOEC (0.01 ppb) of dieldrin, the ventilatory and cardiac rate of another Jamaican shrimp *Macrobrachium fastinum* de Sassure was reduced by 43 percent and 14.4 percent, respectively, while the active oxygen consumption rate was increased by 48 percent and the resting consumption rate was reduced by 13 percent (Lawrence *et al.*, 1986b). The authors concluded that 'the aerobic metabolic scope for activity decreased' in both species, which 'have been forced by the stressor into a metabolic zone of collapse'.

Some plant extracts induced no observable toxic effects on the cattle tick *Boophilus microplus* Canestrini (Acari: Ixodidae) but inhibited oogenesis in the treated ticks, embryogenesis in the eggs laid by them, and survival of F_1 larvae. Other extracts inflicted some mortality associated with severe adverse effects on these processes (Mansingh and Williams, 1998). Similar effects were manifested by sub-lethal concentrations of several synthetic organic insecticides (Mansingh and Rawlins, 1979).

The authors suggest that selected physiological and visible toxicity parameters be combined and examined in immature, developing, and sexually mature stages of aquatic organisms, e.g. fish, shrimp, and oysters, to assist in evaluating the effects of toxic chemicals on survival (including escape from natural enemies), feeding, growth, development, reproduction of individuals, and survival of offspring before evaluating the environmental risk potential of a chemical. The visible symptoms of toxicity must also be quantified in terms of the percentage of test individuals affected with mild or pronounced symptoms and recoverability within a specific time frame. Thus, at NOEC no individual exposed to a toxicant must show any toxic symptoms and at LOEC less than 50 percent of the individuals must display recoverable and mildly toxic symptoms. A new category must be created for a higher concentration at which more than 50 percent of the individuals show pronounced toxic symptoms from which almost all recover within the time frame, as this concentration is likely to create chronic and terminal lesions in many test animals (Mansingh *et al.*, forthcoming) (Table 15.9).

Although NOEC and LOEC should be determined for exposure durations exceeding at least a week, data based on one to 28 days exposure periods of the shrimp *M. faustinum* and the red Jamaican tilapia hybrid (*Tilapia mossambica* (Peters) × *Tilapia nilotica, Tilapia aurea*) to several insecticides are presented in Table 15.9 (Robinson, 1997; Walker, 1997; Mansingh *et al.*, Forthcoming). The toxicity ranking of the insecticides to the fish was chlorpyrifos > endosulfan > profenofos > chlorpyrifos. The maximum lethal effect on the fish in descending order of ranking was exerted by endosulfan > isazofos > profenofos > ethoprophos (Robinson, 1997; Walker, 1997) (Table 15.10). It is pertinent to note that the low slope values and range of LC_{10}, LC_{50}, and LC_{90} values for endosulfan and isazofos compared with those of ethoprophos and profenofos may be of significance in their environmental risk assessment. Likewise, the toxicity of dieldrin to shrimp and its slope value was higher than the LC_{50} of endosulfan for red Jamaican tilapia. The lower slope values indicate less heterogeneity in the population for tolerance to the particular chemical and greater environmental risk from its residues.

Bioaccumulation and degradation of residues

The limited data available on Caribbean species under static laboratory conditions suggest rapid partitioning of residues from water to organic phases. The shrimp *M. faustinum* exposed to the NOEC (0.001 ppb) of dieldrin, accumulated approximately 5.5 ppb dieldrin in 10 h, and 15 ppb at a steady rate in 100 h before reaching a plateau, which was maintained for a further 80 h of exposure. The pattern of bioconcentration was almost the same when the shrimp were exposed to the LOEC of 0.01 ppb but the quantity of bioaccumulation was about threefold more than at the NOEC (Mansingh *et al.*, Forthcoming). When placed in clean water after a 24 h exposure to 0.01 ppb dieldrin, the shrimp eliminated approximately 45 percent of the 10 ppb body burden in 5 h after which a steady state was maintained.

Similarly exposure of sexually mature red Jamaican tilapia to sub-lethal concentrations of diazinon (17.6 ppm), endosulfan (0.0075 ppm), ethoprophos (1.0 ppm), and isazofos (0.15 ppm) resulted in a maximum accumulation of 56.6, 1.0, 3.4, and 68.2 $\mu g\ g^{-1}$ of the respective insecticides after 4, 1–4, 12, and 8 hours, respectively (Robinson, 1997; Walker, 1997). The rate of uptake decreased and fluctuated around the lower level during the next three days. Metabolites of diazinon and ethoprofos were not detected but metabolites of isazofos and endosulfan

Table 15.9 Toxic symptoms displayed by the shrimp *Macrobrachium faustinum* and the fish red *Tilapia* hybrid exposed to sub-lethal concentrations of different insecticides

Insecticide[a] and concentration	Time of exposure (d)	Test organism, symptoms and percent affected	Toxicity level
Dieldrin[1]	14	*Macrobrachium faustinum*	NOEC
0.001 μg L^{-1}		No effect on any	
0.005 μg L^{-1}	3–8	42% hyperactive but recovered	LOEC
	9–14	All normal	
0.01 μg L^{-1}	2–5	> 70% with pronounced symptoms	LOEC?
0.015 μg L^{-1}	6–14	All recovered	
0.02 μg L^{-1}	14	All with pronounced symptoms; 20% dead; 50% recovered	?
Endosulfan[2]	3	*Tilapia*	NOEC
0.01 μg L^{-1}		No effect on any	
0.015 μg L^{-1}	3	30% hyperactive and side swimming, but recovered	LOEC
Ethoprophos[3]			
1 mg L^{-1}	3	No effect on any	NOEC
2 mg L^{-1}	3	25% hyperactive, but recovered	LOEC
3–4 mg L^{-1}	3	40–50% with mild to pronounced symptoms	LOEC?
Isazofos[4]			
0.15 μg L^{-1}	1	No effect on any	NOEC
0.2 μg L^{-1}	1	40% with hyperactivity, but recovered	LOEC
Profenofos[4]			
0.5 μg L^{-1}	1	No effect on any	NOEC
1–2 μg L^{-1}	1	40% hyperactive or side swimming, but recovered	LOEC
Chlorpyrifos[5]			
0.005 mg L^{-1}	28	No effect on any	NOEC
0.01 mg L^{-1}	28	7% with hyperactivity, but recovered	LOEC
0.05 mg L^{-1}	28	58% hyperactive and uncoordinated	LOEC?
0.1 mg L^{-1}	28	100% with pronounced symptoms, but all recovered	LOEC?

Note:
a Source of data as follows: (1) Mansingh *et al.*, Forthcoming; (2) Robinson, 1997; (3) Robinson and Mansingh, 1998; (4) Walker, 1997; and (5) Thomas, 1998 (unpub.).

(endosulfan lactone and sulfate) began to appear shortly after maximum uptake of the parent compounds. Their concentrations fluctuated after an initial peak (Robinson, 1997; Walker, 1997). When the contaminated fish were exposed to clean water, elimination of diazinon was steady and complete within three days, while only about 9 percent of isazofos was lost during the first 4 h and a further 1 percent during the next 68 h (Walker, 1997). Elimination of α- and β-endosulfan began almost immediately but at a slow rate, reaching a maximum of about 15 percent and 25 percent, respectively, after 12 h and about 26 and 39 percent, respectively, after 72 h. Metabolites of endosulfan (lactone and sulfate) increased

Table 15.10 LC$_{50}$ levels of various insecticides for the shrimp *Macrobrachium faustinum* and the fish red Jamaican *Tilapia* hybrid determined under static laboratory conditions

Insecticide[a]	Organism	Duration of exposure (h)	LC$_{50}$ (mg L^{-1})	Fiducial limits (mg L^{-1})	Slope
Dieldrin[1]	M. faustinum	24	1.23×10^{-4}	$(0.65–2.57) \times 10^{-4}$	2.063 ± 0.150
Endosulfan[2]	Tilapia	24	0.031	0.027–0.035	8.185 ± 1.598
Ethoprophos[3]	Tilapia	24	8.41	7.21–10.23	4.141 ± 0.934
Isazofos[4]	Tilapia	24	0.36	0.34–0.40	5.982 ± 4.493
Profenofos[4]	Tilapia	24	4.69	3.43–6.67	2.912 ± 0.606

Note:
a Source of data as follows: (1) Mansingh *et al.*, forthcoming; (2) Robinson, 1997; (3) Robinson and Mansingh, 1998; and (4) Walker, 1997.

by approximately 80 to 86 percent during the first 8 h and fluctuated around slightly lower levels for the next 64 h. The fish eliminated about 83 percent of accumulated ethoprophos from their body during the first 12 h and a further 3 percent during the next 60 h (Robinson, 1997).

Partitioning of dieldrin (ppb) residues in the shrimp *M. faustinum* occurred in the order hepatopancreas (0.1) > gonads (0.045) > gills (0.02) > large claws > muscles > exoskeleton (Mansingh *et al.*, Forthcoming). Similar results were obtained for ethoprophos and endosulfan residues (μg g^{-1} ww) in tilapia, for which the order of partitioning was gonads (95 and 61, respectively) > liver (62 and 16, respectively) > gut (31 and 10, respectively) > gills (19 and 10, respectively), and skin/muscle/ bone (9 and 2.5, respectively) (Robinson and Mansingh, 1998).

Residue management

Environmentally the major emphasis in IMPP ought to be on the management of residues. As demonstrated by Robinson (1997), 30 to 60 cm grass bands grown in experimental trenches at 25° slope reduced runoff of ethoprophos by 10 to 40 percent and endosulfan by 78 to 81 percent, respectively. Some environmentally aware coffee farmers are implementing strategies that involve growing vegetation across slopes and between rows of crops or trees and trenching the bottom of slopes and allowing vegetation to grow (personal observations by the authors). Further, encouraging results from the use of certain plant leaves to promote the rapid degradation of pesticide residues may lead to their use as mulch beneath coffee trees and in recommended field drains and trenches (Mansingh *et al.*, 1998, unpublished data).

ENVIRONMENTAL STANDARDS AND RISK ASSESSMENT

Constrained by the paucity of data and resources, Caribbean countries have adopted the FAO/WHO and EPA guidelines for residue standards in foodstuffs and environmental matrices. However, data generated by local researchers are considered by regulators in the pesticide registration, use approval, and licensing process. In 1996 for instance, Robinson's (unpublished) data convinced the PCA board to deny the request of a powerful lobby of coffee farmers to allow the use of isazofos in the Blue Mountain watersheds (a 1997 personal communication between A. Mansingh and the PCA; unreferenced). Based upon the author's data on chronic and acute toxicity, we recommend that, in addition to requiring LC_{50-95} data on fish for registration purposes, regression slope values should be required to better develop environmental risk assessments, particularly for ecosystems characterized by torrential rainfalls, which can cause sudden influxes of pesticide residues into the aquatic environment.

CONCLUSIONS

The success of pest and pesticide management in different member states of the Commonwealth Caribbean is directly related to a state of awakening and a general appreciation for environmental problems developed among policy makers and the general public. Import licenses and registration of pesticides are current practice in all the countries and other stages of a pesticide's life cycle, e.g. storage, packing, labeling, transport, retailing, aerial spraying, and applicator training, are being addressed and should be regulated within five years.

IPM implementation and education is still in its infancy in the Caribbean and progress has been painfully slow. Meanwhile, the abuse and misuse of pesticides, contamination of ecosystems, and effects on non-target organisms continues. National funding to develop alternate strategies is scarce and a mechanism for transferring new technology to farmers is non-existent. The activities of scientists and technologists in the region need to be directed, streamlined, coordinated, networked, and adequately funded if the Commonwealth Caribbean is to develop and implement workable strategies for the integrated management of pests and pesticides.

ACKNOWLEDGMENT

Writing of this chapter was jointly financed by the University of the West Indies, Mona, Jamaica and the IAEA (Contract No. 8053).

REFERENCES

Alam, M.M. 1996. Fluctuation in the populations of *Plutella xylostella, Trichoplusia ni, Bemisia tabaci* and their natural enemies [PhD dissertation]. University of the West Indies (UWI). Mona, Jamaica.

Anderson, M.A. 1987. Efficacy and persistence of several insecticides in different types of Jamaican soils and their effect on soil fauna [PhD dissertation]. University of the West Indies. Mona, Jamaica.

Biggs-Allen, G. 1990. Investigations on the population fluctuations, economic importance and chemical control of citrus root weevils *Exophthalmus vittatus* L. and *Pachnaeus citri* Marshall (Coleoptera: Curculionidae) [MSc thesis]. University of the West Indies. Mona, Jamaica.

Blair, M., Mansingh, A. and Roberts, E.V. 1995. Pesticidal potential of tropical plants: the acaricidal activity of crude extracts and fractions of five Jamaican plants. In: Proc. 2nd Conf., Fac. Nat. Sci., University of the West Indies, held 7–9 March 1995 in Mona, Jamaica, pp. 22–3.

Boothe, R.A. 1987. Preferences of *Hypothenemus hampei* Ferr. for three varieties of *Coffea arabica* L. [MSc thesis]. University of the West Indies. Mona, Jamaica.

Caribbean Conservation Association/Island Resources Foundation (CCA/IRF). 1991. Dominica country environmental profile. St Michael, Barbados: CCA/IRF.

Clarke-Harris, D. 1998. Potentials of egg parasites and alternative host plants in the integrated management of citrus root weevils in Jamaica [MSc thesis]. University of the West Indies. Mona, Jamaica.

Cooney, J.D. 1995. Freshwater tests, Chapter 2. In: Rand, G.M. (ed.) *Fundamentals of Aquatic Toxicology: Effects, Environmental Fate, and Risk Assessment*, 2nd edn. Washington, DC: Taylor & Francis, pp. 71–102.

Dalip, K.M. 1999. Management of the coffee leaf miner, *Perileucoptera coffeella* (Guérin-Ménèville, 1842), in Jamaica: epidemiolgy, economic importance, parasitoid complex and effect of selected insecticides and their ecological impact [PhD dissertation]. University of the West Indies. Mona, Jamaica.

Dalip, K.M. and Mansingh, A. 1995. Epidemiology and management of the coffee leaf miner, *Perileucoptera coffeella* (Guérin-Ménèville). In: Proc. 2nd Conf. Fac. Nat. Sci., University of the West Indies, held 7–9 March 1995 at Mona, Jamaica, pp. 25–6.

Dasgupta, T.P., Mansingh, A., Roberts, E.V. and Singh, N.C. 1998. Persistence of vamidithion in coconuts of different stages of development. *Jamaica J Sci Tech.* 7/8:38–44.

deGeorges, P.A. 1989. *Pesticides and Environmental Monitoring in the Eastern Caribbean: Current Settings and Needs*, Volume 1. Bridgetown, Barbados: USAID/RDO/C (United States Agency for International Development/Regional Development Office/Caribbean).

Deonarine, G.I. 1980. Studies on the biomagnification of some chlorinated hydrocarbons in a neotropical mangrove swamp [MSc thesis]. University of the West Indies, St Augustine, Trinidad.

Eyre, L.A. 1990. Forestry and watershed management. Ministry of Development, Planning and Production. National Conservation Strategy Consultation Workshop, held 25–27 April 1990 in Kingston, Jamaica.

Forbes, R. 1995. Potentials of insecticides and natural enemies in the management of the diamondback moth *Plutella xylostella* L. in Jamaica [MSc thesis]. University of the West Indies, Mona, Jamaica.

Gayle, D.A. 1989. Bioefficacy and persistence of certain insecticides in three Jamaican soils [MSc thesis]. University of the West Indies, Mona, Jamaica.

Gooding, E.G.B. (ed.) 1980. Pest and pesticide management in the Caribbean. In: Proc. Seminar and Workshop, Volume III, Country Papers, 3–7 November 1980. Christ Church, Barbados: Consortium for International Crop Protection (CICP)/USAID.

Henry, C. 1984. Organochlorine residues in a Jamaican river and uptake and elimination of dieldrin by the shrimp *Macrobrachium faustinum* de Saussure [MSc thesis]. University of the West Indies, Mona, Jamaica.

Higman, B. 1975. Symposium on the Caribbean today. Kingston, Jamaica: The Social Development Commission.

Johanneson, N.E. 1983. The external morphology and life cycle of *Hypothenemus hampei* (Ferrari) in Jamaica [MSc thesis]. University of the West Indies, Mona, Jamaica.

Lawrence, V. 1984. Organochlorine residues in the Rio Cobre, Jamaica, and the effect of dieldrin on the physiology of shrimp [MSc thesis]. University of the West Indies, Mona, Jamaica.

Lawrence, V., Young, R.E. and Mansingh, A. 1986a. The effect of sub-lethal doses of dieldrin on the ventilatory and cardiac activity in two species of shrimps. *J Comp Biochem Physiol.* 85C:177–81.

Lawrence, V., Young, R.E. and Mansingh, A. 1986b. The effect of sublethal doses of dieldrin on resting and active metabolism in two species of shrimp. *J Comp Biochem Physiol.* 85C:182–6.

Mansingh, A. 1987. Pesticide residues in the environment. In: Leslie, K. (ed.) *Pesticides and Food.* Kingston, Jamaica: Caribbean Food and Nutrition Institute (WHO/PAHO), pp. 18–29.

Mansingh, A. 1991. Limitations of insecticides in the management of the coffee berry borer *Hypothenemus hampei* Ferrari. *J Coffee Res.* 21:67–98.

Mansingh, A. 1993. Pesticide management – a global dilemma. In: *Pesticide Management: Policies and Practice.* Proc. UWI-OUSEC-USAID-DT Workshop, held 26–29 May 1993 in St Vincent and the Grenadines.

Mansingh, A. and Rawlins, S.C. 1979. Inhibition of oviposition in the cattle tick *B. microplus* by certain acaricides. *Pestic Sci.* 10:485–94.

Mansingh, G. and Reichgelt, H. 1997. CPEST: an expert system for integrated pest management for coffee. In: Proc. 3rd Annual Conf., Fac. Pure Appl. Sci., University of the West Indies, held 14–17 January 1997 in Mona, Jamaica.

Mansingh, A. and Williams, L.A.D. 1998. Pesticidal potential of tropical plants. II. Acaricidal activity of crude extracts of several Jamaican plants. *Insect Sci Appl.* 18:149–55.

Mansingh, A. and Wilson, A. 1995. Insecticide contamination of Jamaican environment. III. Baseline studies on the status of insecticidal pollution of Kingston Harbour. *Mar Poll Bull.* 30(10):640–5.

Mansingh, A., Robinson, D.E. and Wilson, A. 1995. Insecticide contamination of Jamaican environment. I. Pattern of fluctuations in residue levels in the rivers of Hope watershed during 1989–1991. *Jamaican J Sci Tech.* 6:52–67.

Mansingh, A., Robinson, D.E. and Dalip, K.M. 1997. Insecticide contamination of Jamaican environment. *Trends Anal Chem.* 16(3):115–23.

Mansingh, A., Henry, C. and Robinson, D.E. Forthcoming. Ecotoxicity of insecticides in Jamaican environment. II. Toxicity, bioaccumulation and tissue partitioning of dieldrin by the shrimp *Microbrachium faustinum* de Sassure. *Aquatic Toxicol.* Forthcoming.

Mansingh, A., Robinson, D.E., Henry, C. and Lawrence, V. 2000. Pesticide contamination of Jamaican environment. II. Insecticide residues in the rivers and shrimps of Rio Cobre basin, 1982–1996. *Environmental Monitoring and Assessment.* 63(3):459–80.

McDonald, S.A. 1999. Infestation pattern, economic importance and population fluctuations of the coconut mite, *Eriophyes guerreronis*, in Jamaica [MSc thesis]. University of the West Indies, Mona, Jamaica.

Morris, S.V. 1991. Dynamic studies of the degradation of chlorpyrifos under tropical conditions [MSc thesis]. University of the West Indies, Mona, Jamaica.

Myers, J. 1996. Biocontrol potential of the nematode *Steinernema carpocapsae* in the management of citrus and sweet potato weevils in Jamaica [MSc thesis]. University of the West Indies, Mona, Jamaica.

Nelson, O. 1993. Studies of degradation of the fungicide triadimefon under tropical conditions. MSc thesis. University of the West Indies. Mona, Jamaica.

Pesticides Control Authority (PCA). 2000. Register of pesticides approved by the Pesticides Control Authority. Kingston, Jamaica: Ministry of Health.

Portman, J.E. and Wilson, K.W. 1971. *The Toxicity of 140 Substances to the Brown Shrimp and Other Marine Animals.* Ministry of Agriculture, Fisheries, and Food. UK Shellfish Information Leaflet Nr 22.

Ramsammy, J.R., Shim, D.J. and Ward, R.W. 1985. Technical report on pollution monitoring component 1982–1984: protection of the marine and coastal environment of the Caribbean islands. Castries, St Lucia: Caribbean Environmental Health Institute.

Rawlins, S.C. 1977. Toxicological and biological studies on Jamaican and other Caribbean populations of the cattle tick *Boophilus microplus* (Canestrini) (Acarina: Ixodidae) [PhD dissertation]. University of the West Indies, Mona, Jamaica.

Rawlins, S.C. and Mansingh, A. 1978. Patterns of resistance to various insecticides in some Jamaican populations of *Boophilus microplus. J Econ Entomol.* 71:956–60.

Rawlins, S.C. and Mansingh, A. 1987. A review of ticks and screwworms affecting livestock in the Caribbean. *Insect Sci Appl.* 8:259–67.

Reid, J.C. 1987. Economic status and integrated management of the coffee berry borer *Hypothenemus hampei* (Ferr.) in Jamaica [PhD dissertation]. University of the West Indies, Mona, Jamaica.

Reid, J.C. and Mansingh, A. 1985. Economic losses due to *Hypothenemus hampei* (Ferr.) during processing of coffee berries in Jamaica. *Trop Pest Manag.* 31:55–9.

Rhodes, L. 1987. Infestation pattern and insecticidal susceptibility of *Hypothenemus hampei* (Ferrari) (Coleoptera: Scolytidae) [MSc thesis]. University of the West Indies, Mona, Jamaica.

Robinson, D.E. 1997. The fate, biological impact and management of ethoprophos and endosulfan in the Jamaican environment [PhD dissertation]. University of the West Indies, Mona, Jamaica.

Robinson, D.E. and Mansingh, A. 1998. Ecotoxicity studies in Jamaican environment. I. Toxicity, bioaccumulation, elimination and tissue partitioning of ethoprophos by the fish *Tilapia* in brackish water microcosm. In: Proc Int Symp on Marine Pollution held 5–9 Oct 1998 in Monaco. Extended synopses, IAEA Publ. pp. 139–40.

Robinson, D.E. and Mansingh, A. 1999. Insecticide contamination of the Jamaican environment. IV. Transport of residues from coffee plantations in the Blue Mountains to coastal waters in eastern Jamaica. *Environmental Monitoring and Assessment.* 54(2):125–41.

Robinson, D.E., Mansingh, A. and Dasgupta, T. 1997. Fate of endosulfan in soil and in river and coastal waters of Jamaica. In: *Environmental Behavior of Crop Protection Chemicals.* Proc. Int. Symp. on Use of Nuclear and Related Techniques for Studying Environmental Behavior of Crop Protection Chemicals, 1–5 July 1996. Vienna: IAEA/FAO, pp. 301–11.

Robinson, D.E., Mansingh, A. and Dasgupta, T.P. 1999. Fate and transport of ethoprophos in the Jamaican environment. *Sci Total Environ.* 237/238:373–8.

Serricano, J.L. 1995. Annual report, International Mussel Watch Program. Silver Spring, MD: NOAA.

Shim, D.J. 1985. Marine and coastal pollution monitoring. In: Proc. Workshop on Pollution Monitoring, held 19 July 1985 in Castries, St Lucia. Carribean Environmental Health Institute.

Singh, N. 1985. Physiochemical studies on selected pesticides [PhD dissertation]. University of the West Indies, Mona, Jamaica.

Singh, N.C., Dasgupta, T.P., Roberts, E.V. and Mansingh, A. 1991. Dynamics of pesticides in tropical conditions: 1 Kinetic studies of volatilization, hydrolysis, and photolysis of diledrin and α- and β-endosulfan. *J Agric Food Chem.* 39:575–9.

Sookhai-Mahadeo, S. 1997. The status of organophosphate insecticide resistance in strains of *Aedes aegypti* (Linn.) (Diptera: Culicidae) from Trinidad and Tobago [MSc thesis]. University of the West Indies, St Augustine, Trinidad.

US Central Intelligence Agency. 2000. *The CIA World Factbook 2000.* Washington DC: CIA. http://www.odci.gov/cia/publications/factbook/.

Walker, N. 1997. Toxic effects of selected insecticides on Jamaican red hybrid tilapia [MSc thesis]. University of the West Indies, Mona, Jamaica.

Williams, L.A.D. 1991. Biological activity in the leaf extracts of *Artocarpus altilis* Park and other Jamaican plants [PhD dissertation]. University of the West Indies. Mona, Jamaica.

Williams, L.A.D. and Mansingh, A. 1993. Pesticidal potential of tropical plants: 1 Insecticidal activity in leaf extracts of sixty plants. *Insect Sci.* 14(5/6):697–700.

Williams, L.A.D. and Mansingh, A. 1995. Insecticidally active terpenes from *Artocarpus altillis* Park. *Philippines J Sci.* 124:345–57.

Williams, L.A.D. and Mansingh, A. 1996. A review of insecticidal and acaricidal action of compounds from *Azadirachta indica* (A. Juss) and their use in tropical pest management. *Integrated Pest Manage Rev.* 1:133–45.

Wilson, A. 1993. Pesticidal properties of selected plant extracts and their formulations [MSc thesis]. University of the West Indies, Mona, Jamaica.

Witter, D. and Mansingh, A. 1997. Resistance in the coffee berry borer, *Hypothenemus hampei*, to endosulfan in three Jamaican populations. In: Proc, 3rd Annual Conf, Fac, Pure Appl, Sci, University of the West Indies, held 14–17 January 1997 in Mona, Jamaica, pp. 23–4.

Witter, J.V., Robinson, D.E., Mansingh, A. and Dalip, K.M. 1999. Insecticide contamination of Jamaican environment. V. Island-wide rapid survey of residues in surface and ground water. *Environmental Monitoring and Assessment.* 56(3):257–67.

Worthing, C.R. (ed.) 1987. *The Pesticide Manual: A World Compendium.* Heath, UK: The British Crop Protection Program.

Chapter 16

Coastal watershed-based ecological risk assessment – Gulf of Mexico

Stephen J. Klaine and Foster L. Mayer

INTRODUCTION

Rationale

The goal of USEPA's ecosystem protection program is to improve the Agency's overall ability to protect, maintain, and restore the ecological integrity of the nation's lands and waters by moving toward a place-driven or more site-specific focus. This goal represents a refocusing of the Agency's effort to improve and protect environmental quality in a more holistic manner. By moving toward a place-driven approach, the Agency can develop the paradigm necessary to address ecological, economic, and social needs indigenous to the particular geography or ecosystem. This new integrated ecosystem management will restore and maintain the health, sustainability, and biological diversity of ecosystems while supporting sustainable economies and communities.

The fundamental unit for this place-driven approach is a watershed. Observing, assessing and managing ecosystems on a watershed basis has many advantages. The watershed has predefined geographic boundaries and no one individual or group decides who is included or excluded from the program. These boundaries are easy for everyone to visualize based on the hydrology of the watershed and the simple concept that water runs downhill. Hydrology also provides continuity within the watershed and serves as the basis for understanding how each portion of the watershed may impact other portions (a tremendous educational tool for all levels of players).

By moving away from a simple point-source, permit-driven approach, the watershed approach allows consideration of particular socioeconomic factors that may dictate the adoption of certain scenarios in some watersheds that would be totally unacceptable elsewhere. This approach requires that all government (federal, state, local, or tribal) and private sectors inhabiting the watershed be represented in the development of the watershed assessment plan. These stakeholders need to be assembled early, educated on the process, allowed to provide significant input into plan development, and kept informed on the progress of the assessment. It is this group, ultimately, who must participate in risk management with buy-in and adoption of the watershed management plan that is the ultimate product of this exercise.

From a political standpoint, a watershed can be either a useful or a potentially divisive unit. If the watershed is entirely within a single political jurisdiction (e.g. city, county, or state), then the politics may be straightforward. If, however, the watershed falls between two or more jurisdictions, then significant groundwork is needed to make sure the jurisdictions communicate and work together. Encouraging such cooperation may be a very useful role of the Agency.

From scientific considerations, a watershed approach is extremely useful. This facilitates the deterministic characterization of several causal relationships. Establishment of cause and effect relationships between land use and water quality, between upper and lower portions of the watershed, and between physical modifications of the hydrological conduit and aquatic resources will produce management scenarios that maximize both economic productivity and environmental quality within the watershed. The only potential problem, scientifically, from this geographic unit, is that many terrestrial animal resources span more than one watershed. After consideration of all sociological, political, economic, and scientific information, the watershed is amenable to many different management scenarios. Among these are several that have been attempted previously with only limited success. They include Total Maximum Daily Loading, Pollutant Trading, and Land Use Trading.

Finally the use of the watershed as the principal assessment and management unit allows flexibility with scale; watersheds may be as small or as large as needed given other practical considerations. This facilitates a smooth transition and extrapolation into the regional and national issues addressed by Thornton *et al.* (1994). In this chapter we illustrate some of the principles of watershed-based ecological risk assessment using the Gulf of Mexico region as an example.

Aquatic ecosystems are subject to a complex and dynamic array of physical, chemical, and biological interactions. Ecologists do not have a complete understanding of how toxic chemicals and other stressors influence these interactions and how perturbations at one level of organization are expressed at other levels. Regulators and Environmental Managers depend on information provided by scientists to protect and improve our aquatic resources.

Approach

It is critical at this point to stress the approach that the Integrated Ecosystem Protection Research Program should take, not only for the Gulf of Mexico and the southeastern coast, but for the rest of the country as well. While the approach may appear similar to that laid out formally in the Framework for Ecological Risk Assessment (USEPA, 1992), there are several critical differences. The first is in problem formulation; this stage will require many group meetings during which 'turf' issues (real or perceived questions of who has the responsibility for or authority over the issue in discussion) need to be resolved and a focus is produced that centers on critical ecological resources as well as watershed uses. This is much more intricate than the more familiar 'site-specific' risk assessment. For this reason, we have called this phase 'Watershed Assessment Plan Development'.

The ultimate goal of this phase is to develop a conceptual model of how the watershed works. That is, how land use within the watershed contribute stressors to the aquatic system, how that aquatic system processes and transports these stressors, and how these stressors interact with critical ecological resources within the watershed. This conceptual model contains hypotheses regarding stressor exposure scenarios and the impact on ecological components of the watershed. These hypotheses will be tested during the analysis phase. Much of the analysis phase of risk assessments is similar as long as careful consideration is paid to characterization of watershed processes. The approach may be similar but the scale may be different. It also follows that risk characterization would also be similar.

While the analysis and risk characterization phases may be accomplished through integrated, interdisciplinary teams of government, university, and private sector scientists, risk management and the development of watershed management scenarios must be coordinated with previously identified stakeholders. This requires a continuous educational dialogue between scientists and stakeholders via a computer network. Education will flow both ways as stakeholders become more aware of good data, causal relationships, and the assessment process while scientists become more aware of the needs of the stakeholders to maintain socio-political harmony in an economically stable watershed.

The purpose of the following sections of this chapter is to facilitate the selection and accomplishment of watershed-based risk assessments on the Gulf of Mexico.

THE GULF OF MEXICO

Description, economic, and ecological resources

The Gulf of Mexico is a semi-enclosed, subtropical sea with an area of approximately $1.6 \, M \, km^2$. The Gulf is geographically unique, having two entrances, the Straits of Florida and the Yucatan Strait. Along the 26,000 km Gulf coastline, some 30 major estuaries are located on the US coast. The Gulf shoreline is enhanced with one of the most extensive barrier island systems in the world. The freshwater input to the Gulf Basin from precipitation and estuarine inflow from its 33 river systems and 207 estuaries influences the mixing and distribution of nutrients and pollutants in the Gulf and the health or condition of estuarine and near-shore waters. Sixty-one counties of five states (Florida, Alabama, Mississippi, Louisiana, and Texas) form the US Gulf of Mexico shoreline.

The Gulf of Mexico has many important economic and ecological resources: it is home to one-sixth of the US population; its shipping ports handle 45 percent of all US imports and exports; it provides habitats for migratory waterfowl and threatened and endangered species; and it contains half the US's total wetlands. Catches of fish, shrimp, and shellfish in US Territorial Waters in the Gulf of Mexico are greater than those taken from coastal waters of New England, mid- and South Atlantic, and Chesapeake Bay regions combined (USEPA, 1990). The Gulf of Mexico provides approximately 40 percent of the commercial fish landings and one-third of the recreational fishing activities in the continental US. Tourist-related travel expenditures to Gulf Coast areas amount to more than $US20 billion per year. Most of these expenditures are associated with outdoor recreational activity focused at the Gulf of Mexico shorefront's accessible beach areas and barrier islands. These beaches are a major inducement for coastal tourism, as well as a primary resource for resident recreational activity.

Many different habitat types are found along the Gulf Coast including man-groves, unforested wetlands (fresh, brackish, and saline marshes), sandy beaches, and forested wetlands (bottomland hardwoods, cypress-tupelo gum swamps, and tropical hardwood hummocks). Wetlands have an important role in buffering anthropogenic insults to Gulf waters and also a nurturing role for Gulf wildlife including fish, shrimp, shellfish, and the animals that eat them. Marshes and mangroves form an interface between marine and terrestrial habitats, while forested wetlands occur inland from marsh areas. Wetland habitats may occupy narrow bands or vast expanses and can consist of sharply delineated zones of different species, monotonous stands of a single species, or mixed plant species communities. The importance of coastal wetlands to the coastal environment has been well documented (La Roe *et al.*, 1995). They are characterized by high organic productivity, high detritus production, and efficient nutrient recycling. They also provide habitats for a great number and wide diversity of invertebrates, fish, reptiles, birds, and mammals, and are particularly important as nursery grounds for juvenile forms of many important fish species. The state of Louisiana (LA) contains most of the Gulf's coastal estuaries and marshes (approximately 60 percent).

The deterioration of these coastal wetlands is an issue of concern (USEPA, 1999). Wetland changes observed in Texas (TX) during the past several decades appear to be driven by subsidence and sea level increases. Open water areas are appearing in wetlands along their seaward margins, while new wetlands are encroaching onto previously non-wetland habitat along the landward margin of wetland areas on the mainland, the backside of barrier islands, and on dredging spoil banks. In addition, wetlands are being affected by human activities including canal dredging operations, impoundments, and accelerated subsidence caused by fluid withdrawals. The magnitudes of these wetland acreage changes in most of Texas have not been determined at the present time. In the Freeport, TX area along the Louisiana border, wetland loss is occurring at rates similar to those that occur in adjacent parts of the Louisiana coast. For example, in the Sabine Basin area of coastal Texas, 20,548 ha of wetlands were lost between 1952 and 1974 (Gosselink *et al.*, 1979).

In Louisiana, the annual rate of wetlands loss has been measured at 130 km^2 for the period 1955–78 (Turner and Cahoon, 1987). Several factors contribute to wetlands loss in coastal Louisiana, including sediment deprivation (a result of a 50 percent decrease in the suspended sediment load of the Mississippi River since the 1950s and the channelization of the river, which has prevented over bank sediment deposition), subsidence and sea level rise, and the construction of pipeline and navigation canals through the wetlands (Turner and Cahoon, 1987).

Wetland losses from sediment deprivation and submergence occur in such complex and synergistic ways that it is difficult to quantify their magnitude with great precision. Further, indirect impacts also contribute to wetlands losses and interact with sediment deprivation and submergence factors. Turner and Cahoon (1987) concluded that indirect impacts account for 20–60 percent of the total amount of wetlands loss. The report cautions, however, that these numbers are based on the interpretation of limited data and should be used with caution to indicate only the relative magnitudes of possible impacts.

In Mississippi (MS), the mainland marshes behind Mississippi Sound occur as discontinuous wetlands associated with estuarine environments. The most extensive wetland areas in Mississippi occur east of the Pearl River delta near the western border of the state and in the Pascagoula River delta near the eastern border. The wetlands of Mississippi are more stable than those of Louisiana, reflecting the more stable substrate and more active sedimentation per unit of wetland area.

Most of the wetlands in Alabama (AL) occur on the Mobile River delta or along the northern Mississippi Sound. Between 1955 and 1979, freshwater marshes and estuarine marshes declined by 69 percent and 29 percent, respectively, in these areas (La Roe et al., 1995). On a percentage basis, wetland loss has occurred more rapidly in Alabama during these years than it did in Louisiana. Major causes of losses of non-freshwater wetlands were industrial development, residential and commercial development, navigation maintenance, natural succession, and erosion or subsidence (La Roe et al., 1995). Loss of freshwater marshlands was mainly attributable to commercial and residential development and silviculture (Roach, 1987).

For the most part, coastal marsh habitat in Florida (FL) occurs north of Tampa Bay. To the south, because of milder winter temperatures, mangrove swamp predominates. Emergent wetlands are rare in the Florida panhandle. The limited areas of wetlands that do occur occupy narrow, often discontinuous bands in saline and brackish zones behind barrier islands and spits, near river mouths, and along some embayments. The most extensive and continuous expanse of coastal marshland in the state occurs along the coastline between Cape San Blas and Pasco County, just north of Tampa Bay. These wetlands are dominated by needle rush *Juncus roemerianus* Scheele. This stretch of coast is exposed to very low wave energy because of the broad and very shallow offshore bank. This low-energy environment allows the marsh edge to grow out directly into Gulf waters.

There are an estimated 189,945 ha of mangroves existing within Florida. About 90 percent of Florida's mangroves occur along the Gulf Coast from the Caloosahatchee River to the southernmost tip of the Florida Peninsula. Mangrove ecosystems function much like coastal marshes, providing shoreline stabilization, storm protection, and habitat for wildlife and fisheries and enhancing aesthetic and economic values. Destruction of mangrove forests in Florida has occurred in various ways including land filling, diking, and loss from pollution damage. Any process, natural or man-induced, that coats the aerial roots with fine sediments or covers them with water for extended periods has the potential to cause mangrove destruction (Odum et al., 1982).

In general, along the coastline of the Gulf of Mexico, luxuriant growth of seagrasses and the concomitant high diversity of associated marine species are found within a band roughly parallel to the coastline in waters less than 20 m deep in the eastern Gulf and in the more protected waters of the northern and western Gulf. A great deal of work documenting the ecological importance of seagrass meadows and the roles they serve in coastal ecosystems has been accomplished. A few of the more comprehensive (including regional) works are those of Zieman (1982), Thayer et al. (1984), Lewis et al. (1985), Iverson and Bittaker (1986), Stevenson (1988), Durako (1988), and Kenworthy et al. (1988).

Major sources of contaminants that could threaten Gulf coastal ecosystems include the petrochemical industry, hazardous waste sites and disposal facilities, agriculture and livestock farming, manufacturing industry activities, fossil fuel and nuclear power plant operations, pulp and paper mill plants, municipal wastewater treatment facilities, and the large amount of maritime shipping activities. Commercial and recreational fishing must also be considered to avoid disruption of ecosystem stability through over-harvesting or alteration of natural predator–prey relationships. A few reports have demonstrated the types and quantities of chemical contaminants entering the Gulf of Mexico (Brecken-Folse et al., 1993; USEPA, 2000). Restricted usage and closures due to poor water quality illustrate the potential for adverse effects from these stressors and serve to remind citizens and regulators that there are limits to the resiliency of the Gulf of Mexico.

Protected and endangered species resources

Two US federal laws address protected and endangered species: the Marine Mammal Protection Act of 1972 and the Endangered Species Act of 1973. The Marine Mammal Protection Act seeks to protect and conserve marine mammals and their habitats so as not to allow species or population stocks to diminish or decline beyond the point at which they cease to be a significant, functioning ecosystem element or allow species to go below optimum sustainable population levels. The Endangered Species Act provides for listing of threatened or endangered plant or animal species in support of protection and conservation of these species and their ecosystems or habitats. The US Fish and Wildlife Service (FWS) in the Department of the Interior and the National Marine Fisheries Service (NMFS) in the Department of Commerce have authority for administering these laws. The FWS and NMFS have noted that the following protected species occurring in the Gulf of Mexico near-coastal environments could potentially be impacted: endangered – green turtle, hawksbill turtle, Kemp's ridley turtle, leatherback turtle, olive ridley turtle, West Indian manatee, and sperm whale; threatened – Gulf sturgeon and loggerhead turtle; proposed – smalltooth sawfish.

Coastal and marine mammals

Twenty-nine species of cetaceans, one sirenian, and one exotic pinniped have been sighted in the northern Gulf of Mexico. Seven species of baleen whales and 22 species of toothed whales and dolphins have also been reported. Of these, seven species of whales and the sirenian are endangered.

Alabama, Choctawhatchee, and Perdido Key beach mice live in the coastal sand dunes of Florida and Alabama. Portions of these coastal dune areas have been designated as critical habitat for these endangered species (USDI, 2000) (Table 16.1).

Marine turtles

Green turtle nesting sites have been documented on Santa Rosa Island, FL and the Yucatan Peninsula, but are isolated and infrequent. Leatherback turtles nest on coarse-grained beaches in tropical latitudes, but occasionally enter shallow waters with rare occurrences in the Florida panhandle. The Atlantic hawksbill sea turtle is the least commonly reported turtle in the Gulf of Mexico, with reports of stranded turtles coming only from Texas. This turtle occurs more frequently in the tropical Gulf of Mexico and Carribean feeding on reef invertebrates. The loggerhead sea turtle is found and nests on all Gulf of Mexico shorelines. The Kemp's ridley sea turtle is the most endangered of the world's sea turtles; nesting female turtles have dwindled to less than 1,000 individuals.

Coastal and marine birds

The piping plover is a threatened migratory species, wintering mostly on the Texas and Louisiana intertidal flats and beaches. The whooping crane population (<200 individuals) winters along the Texas coastal marshlands from November to April. The Arctic peregrine falcon winters on the beaches and barrier islands of the US and Mexico. Bald Eagles nest in upland and wetland areas of the Gulf States. Coastal nesting has been documented in Mississippi and in Florida at St Vincent, St Marks, Lower Suwannee National Wildlife Refuges and Hondo Bay. Brown pelicans have been removed from the endangered species list in Alabama and Florida but remain on the list in Mississippi, Louisiana and Texas. The Eskimo curlew is one of the rarest native North American birds; only 18 birds were reported between 1983 and 1987. Most sightings occurred in coastal Texas, where they feed on invertebrates and crowberries on their migratory route from Canada to South America.

Table 16.1 Gulf of Mexico resources: endangered, threatened, and proposed species

Marine mammals	Marine turtles	Fish and reptiles	Coastal and marine birds	Field mice Rodentia: Cricetidae
Baleen whales (Cetacea: Balaenopteridae) Blue whale (Balaenoptera musculus L.) Fin whale (Balaenoptera physalus L.) Humpback whale (Megaptera novaeangliae Borowski) Northern right whale (Eubalaena glacialis Muller) Sei whale (Balaenoptera borealis Lesson) Toothed whales (Cetacea: :Physeteridae) Great sperm whale (Physeter catodon L.) Sirenians (Sirenia: Trichechidae) West Indian manatee (Trichechus manatus L.) Pinnipeds (Pinnipedia: Otariidae) California sea lion (rare, introduced) (Zolophus californianus Lesson)	Green sea turtle (Chelonia mydas L.) Testudines: Cheloniidae Atlantic hawksbill sea turtle (Eretmochelys imbricata L.) Testudines: Cheloniidae Kemp's ridley sea turtle (Lepidochelys kempii Garman) Testudines: Cheloniidae Leatherback sea turtle (Dermochelys coriacea L.) Testudines: Dermochelyidae Loggerhead sea turtle (Caretta caretta L.) Testudines: Cheloniidae Olive ridley sea turtle (Lepidochelys olivacea Eschscholtz) Testudines: Cheloniidae	Gulf sturgeon (Acipenser oxyrinchus desotoi Vladykov) (Acipenseriformes: Acipenseridae) American crocodile (Crocodylus acutus Cuvier) Crocodylia: Crocodylidae Smalltooth sawfish (Pristis pectinata Latham) Pristiformes: Pristidae	Arctic peregrine falcon (Falco peregrinus tundrius White) Falconiformes: Falconidae Bald eagle (Haliaeetus leucocephalus L.) Falconiformes: Accipitridae Brown pelican (Pelecanus occidentalis L.) Pelecaniformes: Pelecanidae Eskimo cerlew (Numenius borealis Forster) Charadriiformes: Scolopacidae Piping plover (Charadrius melodus Ord) Charadriiformes: Charadriidae Whooping crane (Grus americana L.) Gruiformes: Gruidae	Alabama beach mouse (Peromyscus polionotus ammobates Bowen) Choctawhatchee beach mouse (Peromyscus polionotus allophrys) Perdido Key beach mouse (Peromyscus polionotus trissyllepsis Bowen)

Fish and reptiles

The Gulf sturgeon is listed as a threatened species and found in major river systems and marine waters of the central and eastern Gulf of Mexico. It is an anadromous species (fish that live most of their lives in the ocean but migrate into freshwater streams to spawn), where adults migrate to freshwater during warmer months, but do not always spawn. The endangered American crocodile also occupies southern Florida coastal wetlands.

Many of these protected and endangered species have suffered population declines from over-harvesting, habitat destruction, unbalanced predation, reproductive interruption from pesticides and other chemicals, and direct toxicity or reduced immune responses from chemical contaminants. These examples help to illustrate the diverse and delicate balance of nature played out on Gulf of Mexico coastal environments an underscore the importance of vigilant monitoring and characterization programs.

Commercial and recreational fish resources

In 2000, the Gulf of Mexico provided 19.7 percent by weight but 27.3 percent in dollar terms of the commercial fish landings at ports within the 50 states. Commercial landings of all fisheries in the Gulf during 2000 totaled 813,515.1 T, valued at US$994.2 million (National Marine Fisheries Service (NMFS), 2001). Four Gulf ports ranked among the top 10 US ports in quantity of landings; Cameron (Cameron Parish), LA ranked number two behind Dutch Harbor, AK while Empire-Venice (Plaquemines Parish), LA was the number three port, Intracoastal City (Vermilion Parish), LA was number five, and Pascagoula-Moss Point, MS ranked number nine (NMFS, 2001). Four Gulf ports also ranked among the top 10 US ports ranked by dollar value of landings; Dulac-Chauvin (Terrebonne Parish), LA ranked number four behind New Bedford, MA followed by Empire-Venice, LA at number five, and Key West, FL, Port Arthur, TX, and Bayou La Batre (Mobile County), AL ranked eight through ten, respectively (NMFS, 2001).

Menhaden represents the most important Gulf species in quantity landed during 2000 (Table 16.2) while shrimp represents the most important Gulf species in value landed during 2000 (NMFS, 2001). Other significant Gulf commercial fisheries include crabs, oysters, spiny lobsters, grouper and scamp, striped mullet, yellowfin tuna, red snapper, and an assortment of finfish (see Table 16.3 for common names and phylogenetic information).

Commercial harvest of both oysters and crabs takes place in Gulf bays and estuaries. The Gulf oyster fishery accounted for 64 percent of the national total with landings of 11,133.6 T of meats, valued at US$50,173,841. The Gulf blue crab fishery accounted for 37.3 percent of the national total with landings of 31,191.8T, valued at US$51,015,611 (NMFS, 2001).

Excluding menhaden, catch of at least 10 species of commercial finfish reached a minimum of US$2 million each in value landed from the Gulf of Mexico during 2000. In decreasing order of value, they were groupers and scamp, striped mullet,

Table 16.2 Gulf of Mexico commercial fish resources: commercial landings from Gulf waters for 2000 listed in order of commercial value[a]

Commercial catch classification	Metric tonnes	Value (US$)
Shrimp (all species)	130,877.9	655,483,824
Menhaden	591,436.0	80,671,578
Blue crab (all forms)	31,191.8	51,015,611
Oysters, eastern	11,133.6	50,173,841
Stone crab	3,073.5	28,366,615
Spiny lobster	2,339.7	25,248,076
Grouper (all) and scamp	4,224.5	18,938,114
Striped mullet	7,440.2	11,406,589
Yellowfin tuna	1,954.1	10,557,379
Red snapper	2,193.0	10,264,018
Gag	990.9	5,387,166
Black drum	2,629.7	4,186,629
Yellowtail snapper	655.5	2,868,078
Vermilion snapper	641.1	2,779,039
Mackerel (all species)	1,425.6	2,722,440
Swordfish	468.1	2,108,551
Sharks (all species)	1,288.5	1,970,487
Pompano (all species)	210.0	1,405,250
Flatfish	315.1	1,199,273
Total	813,515.1	994,239,081

Note:
a Summary of data from the National Marine Fisheries Service (NMFS), Fisheries Statistics and Economics Division, Silver Spring, MD, USA, 2000 http://www.st.nmfs.gov/.

yellowfin tuna, red snapper, gag, black drum, yellowtail snapper, vermilion snapper, king and cero mackerel, and swordfish (Table 16.2). Commercial fishing resources may generate at least three times the dockside value as the fishery's product moves through processing stages and wholesale and retail markets (Mager and Ruebsamen, 1989).

Both the commercial and recreational fisheries in the Gulf of Mexico are dominated by estuary-dependent species. Approximately 46 percent of the southeastern United States wetlands and estuaries important to fisheries are located within the Gulf of Mexico (Mager and Ruebsamen, 1989). The life history of estuary-dependent species involves spawning on the continental shelf; transport of eggs, larvae, or juveniles to the estuarine nursery grounds; growth and maturation in the estuary; and migrations of the young adults back to the shelf for spawning. After spawning, the adults generally remain on the continental shelf. Most estuary-related species of importance, such as menhaden and shrimp, live for 18–36 months, but some sciaenids, such as croaker and black drum, may live for several years (Darnell, 1988).

Finfish and shellfish are sensitive to petroleum hydrocarbons in varying degrees during all their life stages (Anderson, 1985). Two of the most important species in the Gulf of Mexico, menhaden and shrimp, are both considered continental shelf

Table 16.3 Species data for commercial and recreational mollusc, crustacean, and fish resources mentioned in the chapter

Group	Common name	Scientific name (authority)	Order: Family
Mollusks	American oyster	Crassostrea virginica (Gmelin)	Mollusca: Bivalvia
Crustaceans	Blue crab	Callinectes sapidus (Rathbun)	Decapoda: Portunidae
	Gulf stone crab	Menippe adina	Decapoda: Menippidae
	Stone crab	Menippe mercenaria (Say)	Decapoda: Menippidae
	Brown shrimp	Penaeus aztecus (Ives)	Decapoda: Penaeidae
	Pink shrimp	P. duorarum (Burkenroad)	Decapoda: Penaeidae
	Royal red shrimp	Hymenopenaeus (Pleoticus) robustus (Smith)	Decapoda: Penaeidae
	Rock shrimp	Sicyonia brevirostris (Stimpson)	Decapoda: Penaeidae
	Seabob shrimp	Xiphopenaeus kroyeri (Heller)	Decapoda: Penaeidae
	White shrimp	P. setiferus (L.)	Decapoda: Penaeidae
	Spiny lobster	Panulirus argus (Latreille)	Decapoda: Palinuridae
Fish groups	Sharks		Carchardinidae, Lamnidae, Alopiidae
	Anchovies	Anchoa spp.	Engraulidae: Clupeiformes
	Drum	Including Larimus fasciatus (Holbrook), Pogonias cromis (L.), and Sciaenops ocellatus (L.)	Perciformes: Sciaenidae
	Groupers	various spp.	Perciformes: Serranidae
	Mackerels	Including Scomberomorus maculatus (Mitchill), Seriola zonata (Mitchill), and Sarda sarda (Bloch)	Perciformes: Scombridae
	Mullets	Mugil spp.	Perciformes: Mugilidae
	Snappers		Perciformes: Lutjanidae
Fish species	Bonefish	Albula vulpes (L.)	Albuliformes: Albulidae
	Gulf menhaden	Brevoortia patronus (Goode)	Clupeiformes: Clupeidae
	Shoal flounder	Syacium gunteri (Ginsburg)	Pleuronectiformes: Paralichthyidae
	Amberjack	Seriola dumerili (Risso)	Perciformes: Carangidae
	Pompano	Trachinotus carolinus (L.), and Alectis ciliaris Bloch	Perciformes: Carangidae
	Snook	Centropomus undecimalis (Bloch)	Perciformes: Centropomidae
	Dolphin fish (Dorado)	Coryphaena hippurus (L.)	Perciformes: Coryphaenidae
	Black marlin	Makaira nigricans (Lacepède)	Perciformes: Istiophoridae
	White marlin	Tetrapturus albidus (Poey)	Perciformes: Istiophoridae
	Sailfish	Istiophorus albicans (Latreille)	Perciformes: Istiophoridae

Group	Common name	Scientific name (authority)	Order: Family
	Red snapper	*Lutjanus campechanus* (Poey)	Perciformes: Lutjanidae
	Vermilion snapper	*Rhomboplites atrorubens* (Cuvier)	Perciformes: Lutjanidae
	Yellowtail snapper	*Ocyurus chrysurus* Bloch	Perciformes: Lutjanidae
	Striped mullet	*Mullus barbatus* (L.)	Perciformes: Mullidae
	Black mullet	*Mugil cephalus* (L.)	Perciformes: Mugilidae
	Atlantic threadfin	*Polydactylus octonemus* (Girard)	Perciformes: Polynemidae
	Bluefish	*Pomatomus saltatrix* (L.)	Perciformes: Pomatomidae
	Atlantic croaker	*Micropogonias undulatus* (L.)	Perciformes: Sciaenidae
	Black drum	*Pogonias cromis* (L.)	Perciformes: Sciaenidae
	Southern kingfish	*Menticirrhus americanus* (L.)	Perciformes:Sciaenidae
	Sand seatrout	*Cynoscion arenarius* (Ginsburg)	Perciformes: Sciaenidae
	Silver seatrout	*Cynoscion nothus* (Holbrook)	Perciformes: Sciaenidae
	Spotted seatrout	*Cynoscion nebulosus* (Cuvier)	Perciformes: Sciaenidae
	Spot	*Leiostomus xanthurus* (Lacepède)	Perciformes: Sciaenidae
	Whiting	*Menticirrhus americanus* (L.)	Perciformes: Sciaenidae
	Bluefin tuna	*Thunnus thynnus thynnus* (L.)	Perciformes: Scombridae
	Yellowfin tuna	*Thunnus albacares* (Bonnaterre)	Perciformes: Scombridae
	Little tunny	*Euthynnus alletteratus* (Rafinesque)	Perciformes: Scombridae
	Wahoo	*Acanthocybium solandri* (Cuvier)	Perciformes: Scombridae
	Scamp	*Mycteroperca phenax* (Jordan & Swain)	Perciformes: Serranidae
	Seabass	*Epinephelus itajara* (Lichtenstein)	Perciformes: Serranidae
	Longspine porgies	*Stenotomus caprinus* (Jordan & Gilbert)	Perciformes: Sparidae
	Atlantic sheepshead	*Archosargus probatocephalus* (Walbaum)	Perciformes: Sparidae
	Cobia	*Rachycentron canadum* (L.)	Perciformes: Rachycentridae
	Cutlassfish	*Trichiurus lepturus* (L.)	Perciformes: Trichiuridae
	Gag	*Mycteroperca microlepis* Goode & Bean	Perciformes: Serranidae
	Swordfish	*Xiphias gladius* (L.)	Perciformes: Xiphiidae
	Blackfin searobin	*Prionotus* sp.	Scorpaeniformes: Triglidae
	Mexican searobin	*Prionotus paralatus* (Ginsburg)	Scorpaeniformes: Triglidae
	Sea catfish	*Bagre marinus* (Mitchill)	Siluriformes: Ariidae

species, but they are estuary dependent. This means that spawning times and places are synchronized with hydrography such that eggs, larvae, and young arrive at estuary nursery areas during the appropriate season. Menhaden spawn near the water surface in a localized area of the middle continental shelf proximate to the Mississippi River delta during the winter and early spring. During late spring young menhaden enter the nursery areas from the delta region to as far west as Galveston Bay, TX. The two major shrimp species, brown (*Penaeus aztecus* Ives) and white (*P. setiferus* L.), spawn in a widely distributed area of the middle continental shelf during non-overlapping short periods in spring and summer. Their larvae become dispersed throughout the water column and along a broad east-west band on the inner shelf before tides, currents, and wind bring the postlarvae into estuary nursery areas in early spring. In this manner, brown and white shrimp utilize extensive nursery areas from the west coast of Florida to the east coast of Mexico.

Darnell *et al.* (1983) and Darnell and Kleypas (1987) found that the density distribution of total fish and penaeid shrimp catch in the northwestern Gulf was highest near-shore off Louisiana. For all seasons the greatest abundance of fish and penaeids occurred between Galveston Bay, TX and the Mississippi River. This may be directly attributable to the extensive estuary nursery areas of Louisiana. The density distribution of total fish catch for the eastern Gulf was patchy. High densities were associated with particular habitat types (e.g. east Mississippi Delta area, Florida Big Bend seagrass beds, near-shore areas off the Everglades, southwest Florida mid to outer shelf, and the DeSoto Canyon area). In addition, estuaries of the central and western Gulf are known to export considerable quantities of organic material, thereby enriching the adjacent continental shelf areas (Darnel and Soniat, 1979).

Approximately 30 percent of the other numerically significant fish species from the continental shelf are also estuary dependent species. Low salinity estuaries are dominated by oysters, crabs, sciaenids (drums and croakers), anchovies, and mullets. Populations from the inshore shelf zone (7 to 14 m) are dominated seasonally by Atlantic croaker, spot, drum, silver seatrout, southern kingfish, and Atlantic thread-fin. Populations from the middle shelf zone (27 to 46 m) include sciaenids, but are dominated by longspine porgies. The blackfin searobin, Mexican searobin, and shoal flounder are dominant on the outer shelf zone (64 to 110 m). Natural reefs and banks, located mainly between the middle and outer shelf zones, and the numerous offshore platforms acting as artificial reefs support such fishes as snapper, groupers, mackerels, and seabass. Oceanic fish such as tuna and swordfish inhabit deep ocean waters south of the central and western Gulf.

A total of nine species of penaeid shrimp contributes to the Gulf of Mexico commercial shrimp fishery. Brown, white, pink, royal red, and rock shrimp constitute the bulk of the harvest. Brown shrimp are centered in the northwestern Gulf. White shrimp are centered on the mud and sand bottoms in the central Gulf. Royal red shrimp are located in the deep waters of the Mississippi and DeSoto canyons. Pink shrimp have an almost continuous distribution throughout the Gulf, but most commercial pink and rock shrimp catches are made on the shell, coral sand, and coral silt bottoms off southern Florida.

Gulf menhaden occur in the inshore waters of the north-central Gulf from eastern Florida to eastern Texas with approximately 93 percent of the harvest occurring within 16 km of shore. This industrial bottom fisher also occurs in the near-shore waters of the north-central Gulf, taking advantage of the seemingly inexhaustible numbers of sciaenids. Atlantic croaker constitutes the largest component of the catch, which includes spot, sand seatrout, silver seatrout, cutlass fish, sea catfish, and longspine porgy.

Data analysis by the NMFS (1989) indicates a major change in characteristics of the finfishery occurred during the interval from 1981 through 1987. The number of commercial species landed from the Gulf of Mexico increased significantly from 27 in 1981 to 82 in 1987. In addition, the number of species with a value over US$1 million has tripled from three in 1981 to nine in 1987 (Linton, 1988; NMFS, 1989). Some of the more valuable species include snappers, groupers, mackerels, black drum, spotted and silver seatrout, pompano, yellowfin and bluefin tuna, shark, swordfish, amberjack, and sheepshead. The majority of this catch is harvested from the north-central and northwestern Gulf, where a hard substrate aided by numerous offshore petroleum platforms is thought to be a positive contributing factor for several species (Linton, 1988).

Recreational activities and associated tourism resources

The Gulf of Mexico provides over 25 percent of the recreational fishing activities in the continental US. The northern Gulf of Mexico's coastal zone is one of the major recreational regions of the United States, particularly in connection with marine fishing and beach-related activities. The shorefronts along the Gulf coasts of Florida, Alabama, Mississippi, Louisiana, and Texas offer a diversity of natural and developed landscapes and seascapes. The coastal beaches, barrier islands, estuarine bays and sounds, river deltas, and tidal marshes are extensively and intensively utilized for recreational activity by residents of the Gulf South and tourists from throughout the nation, as well as from foreign countries. Publicly-owned and administered areas, such as national seashores, parks, beaches, and wildlife lands, as well as specially designated preservation areas, such as historic and natural sites and landmarks, wilderness areas, wildlife sanctuaries, and scenic rivers, attract residents and visitors throughout the year. Commercial and private recreational facilities and establishments, such as resorts, marinas, amusement parks, and ornamental gardens, also serve as primary interest areas and support services for people who seek enjoyment from the recreational resources associated with the Gulf of Mexico. According to the National Oceanic and Atmospheric Agency (NOAA, 1988a), there are approximately 1,078 km of public recreational beaches along the Gulf Coast. These beaches are a major inducement for coastal tourism, as well as being a primary resource for resident recreational activity.

Approximately 40 million residents of the Gulf Coast states have a major interest in water-related and water-enhanced recreational activity, with approximately two-thirds of the Gulf shorefront composed of beaches; also, there is one water craft for fewer than every 20 people living in the Gulf region. Thirty-five percent of the Gulf Coast states' population lives in coastal counties/parishes or the area most directly affected by Gulf activity and about one-third of registered water craft are likely candidates for use in association with marine recreational activity.

Between 1984 and 1987, NOAA's Office of Strategic Assessment inventoried public recreation areas in coastal areas throughout the US. Their final report and data atlas (NOAA, 1988a) indicates 308 public agencies (289 local, 14 state, and 5 federal) owned or managed outdoor recreation areas and facilities in coastal areas of the Gulf of Mexico region. Public agencies managed 4,137 recreation sites greater than one acre in size in the Gulf's coastal zone. According to NOAA's report, 601 of these public recreation sites provide access to tidally influenced water and 215 provide access to the open waters of the Gulf of Mexico. The atlas provides extensive data on public recreation lands and waters, as well as the number of boat ramps, boating slips, docks, fishing piers, campsites, artificial reefs, and beach miles associated with every coastal county/parish associated with the Gulf of Mexico region.

The Sports Fishing Institute (1988) estimated resident and tourist sport fishermen in the five Gulf States spent an estimated US$6.5 billion, generating a total economic output exceeding US$10 billion and accounting for 188,000 person years of employment in 1985 (Table 16.4). Schmied and Burgess (1987) estimated the sales generated from saltwater fishing alone in 1985 in the five Gulf states at US$1.4 billion, accounting for 17,120 person years of employment. Conclusively sports fishing in the marine environment of the Gulf region is a major industry that is important to the economic viability of each Gulf state. Furthermore, it is estimated (NMFS, 1988a) that 31 percent of American saltwater fishermen did some deep-sea fishing (more than 5 km offshore). A NMFS (2001) recreational fisheries survey estimated that in 2000 the Gulf states excluding Texas accounted for 25 percent of marine recreational angling participants (over 2.6 million), 27

Table 16.4 Gulf of Mexico recreational fish resources: economic impact

US state	Resident and tourist sport fishermen		
	Expenditures[a]	Economic output[a]	Employment
Alabama	519.1	804.4	16,754
Florida	3,100.0	4,200.0	85,584
Louisiana	538.5	893.4	15,104
Mississippi	428.0	806.7	16,160
Texas	1,900.0	3,300.0	53,089
Total	6,485.6	10,004.5	186,691

Note:
a Expenditures and economic output in US$M.

percent of trips (20.4 million), and 35 percent of the catch (over 149 million fish) (NMFS, 2001). Trips originated primarily from west Florida (72 percent), followed by Louisiana (18 percent), and Alabama and Mississippi at 5 percent each. Spotted seatrout were the most common species caught (19 percent) followed by sand seatrout, red drum, white grunt, Atlantic croaker, red snapper, and gray snapper.

According to NOAA (Meade and Leeworthy, 1986) in FY 1982, US$525 million in public funds were spent for outdoor recreation in coastal counties of the Gulf of Mexico region, an average of US$44 per resident. Total public recreation expenditures and expenditures per capita are less in the Gulf Coast region than any other coastal region in the United States (the Pacific, South Atlantic, and North Atlantic). Notwithstanding the Gulf's low ranking, three south Florida coastal counties (Dade, Palm Beach, and Broward) and one Texas coastal county (Harris) ranked among the top 20 coastal counties in the nation (1,339 counties) for public recreation expenditures in FY 1982. This is not surprising considering tourism is Florida's number one industry and local governments spend considerable sums maintaining beaches. This study (Meade and Leeworthy, 1986) supports that recreational resources, activities, participation, and expenditures are not constant along the Gulf of Mexico shorefront, but are focused in major areas primarily where public beaches and major urban centers are closely related.

POLLUTION STRESS IN THE GULF OF MEXICO

Despite the large size of the Gulf of Mexico, evidence indicates that pollution and overuse are adversely affecting Gulf waters and biota (USEPA, 1999). Evidence includes fish kills, red and brown tides, permanent or conditional closures of shellfish producing areas (NOAA, 1991), discovery of the 'dead zone' off the Texas and Louisiana coastlines (Rabalais et al., 1991), and the tons of chemicals that are discharged into Gulf waters (Brecken-Folse et al., 1993; USEPA, 2000). Long-term monitoring studies using improved methods are beginning to provide a better understanding of conditions and limitations of the vast resources of the Gulf of Mexico.

Indicators used to characterize near-shore and coastal water quality include salinity, biochemical oxygen demand (BOD), chemical oxygen demand (COD), petroleum hydrocarbons (oil and grease), total suspended solids, total dissolved solids, nutrient loadings, heavy and trace metal concentrations, coliform bacteria, synthetic organics, tributyltins from marine paints, toxic substances and hazardous wastes, radionuclides, toxins, and solid wastes (trash). These parameters must be assessed in the context of the natural components of the water system such as substrate, suspended particulates, and current patterns and circulation. For instance, water quality in an estuary is a function of freshwater flow, the quantities of nutrients brought with this flow, and the degree of mixing in the estuary.

Salinity is an important parameter because it affects the distribution of marsh vegetation and benthic macroinvertebrates. Saltwater intrusion is defined as the introduction, accumulation, or formation of saline water in a water body or aquifer of lesser salinity (Texas Water Commission, 1986). The states of Texas and Louisiana have identified saltwater intrusion as a major water quality concern (Texas Water Commission, 1986; Louisiana Department of Environmental Quality, 1988).

Heavy metals normally enter the marine environment through the weathering of the earth's crust. When this background level is increased by inputs from man's activities, heavy metals become toxic to most organisms, including humans. Once ingested, they exert harmful toxic effects through inhibiting enzymes and biochemical processes. Thirteen heavy metals are listed by the USEPA on its priority pollutant list: antimony, arsenic, beryllium, cadmium, chromium, copper, lead, mercury, nickel, selenium, silver, thallium, and zinc. Coastal counties/parishes identified by NOAA as receiving the greatest heavy metal discharges include Ascension, LA and Calhoun, Harris, Orange, and Jefferson, TX. Counties or parishes, where lead discharges are the greatest, include the above counties and Mobile, AL and St Charles and Iberville, LA (NOAA, 1985).

Synthetic organic compounds, chlorinated hydrocarbons in particular, are recognized as carcinogens. They accumulate in the fats and tissues of animals, and many are known to bioconcentrate in the food chain. Major human health concerns relate to ingestion of both contaminated water and organisms contaminated with these compounds. Synthetic organic compounds are known to affect the central nervous system, liver, kidney, and reproductive system of both humans and other organisms. When discharged into marine waters, these compounds adhere to particulate matter and become deposited in bottom sediments. One-hundred-and-twelve of the 126 priority pollutants identified by the USEPA are synthetic organic chemicals. Sixty-seven of these are chlorinated hydrocarbons.

Sources of pollutants in the Gulf of Mexico

Monitoring the concentrations of these constituents is important in determining the health of the Gulf's coastal waters. However, prevention of the causes of water quality degradation usually focuses on identifying and regulating the sources of the pollutants. Sources of pollutants in the Gulf's coastal waters include point, non-point, and upland sources. The following geographical information about sources of pollutants is primarily derived from NOAA's Strategic Assessment Data Atlases that provide selected water quality characteristics along the Gulf Coast (NOAA, 1985) and USEPA's Toxic Release Inventory Database (TRI).

Major point sources of contaminants occurring along the Gulf Coast include the petrochemical industry, hazardous waste sites and disposal facilities, agricultural and livestock farming, manufacturing industry activities, fossil fuel and nuclear power plant operations, pulp and paper mill plants, commercial and recreational fishing, municipal wastewater treatment, and the large amount of maritime shipping activities.

A major economic activity that occurs primarily in the central and western Gulf of Mexico and that generates and discharges pollutants is the petrochemical industry. Within this industry are oil and gas exploration, development, and production activities; pipeline transport; tanker movement of both imported and domestic petroleum products and crude oil; and petroleum and petrochemcial refinery operations. Discharges of particular concern relating to oil development operations include produced waters and drilling muds and cuttings discharges into the coastal zone. Oil spills and leakage from pipelines result in increased hydrocarbon levels in Gulf waters. Canals and navigation channels cut to support the oil industry have increased turbidity and saltwater intrusion within the coastal zone. Pollutants discharged from refining operations include high levels of BOD, total suspended solids, chemical oxygen demand, total Kjeldahl nitrogen (a measurement of total organic nitrogen and ammonia), and petroleum hydrocarbons. The petrochemical industry contributes almost one-fourth of all petroleum hydrocarbons entering Gulf of Mexico waters (NOAA, 1985). The NOAA Data Atlas identifies 90 petrochemical plants within the Gulf Coast region, primarily in the Houston/Galveston, TX area and the Baton Rouge/New Orleans, LA region along the Mississippi River. One of these plants in Norco, LA, produces more than 113,500 m^3 of wastewater per day.

Wastewater discharges include relatively large amounts of toxic chemicals, petroleum hydrocarbons, phosphorus, and nitrogen. Wastewater discharges from industrial, municipal, and other activities for the five states along the Gulf Coast were approximately 32,427 T in 1999 and accounted for 28 percent of the nation's total wastewater discharges (USEPA, 2000). The chemical and allied products industry discharges more than 68 percent of all industrial wastewater entering the Gulf region. Coastal counties or parishes for which industrial wastewater discharge is the greatest include Brazoria and Jefferson, TX, and Calcasieu, LA. Cooling water discharges from power plants account for almost half the wastewater discharged from point and non-point sources in the Gulf. Coastal counties where cooling water is discharged in the greatest amounts from power plants include Hillsborough and Citrus, FL and Harris, Galveston, Gambers, and Jefferson, TX.

Along with wastewater, industrial point sources in coastal counties also discharge large amounts of chlorinated hydrocarbons and phosphorus. NOAA estimated that in 1980, industrial sources of chlorinated hydrocarbons were approximately 15 M T per year and phosphorus loadings were about 16 M T per year (60 percent of the total amount of phosphorus discharged). The chemical and allied products industry accounts for almost 60 percent of the discharges of chlorinated hydrocarbons and 45 percent of phosphorus discharges into the Gulf. Other industries discharging chlorinated hydrocarbons include petroleum refining, primary metals production, and pulp and paper products. Coastal counties/parishes where direct discharges of chlorinated hydrocarbons are the greatest include Orange, Jefferson, and Brazoria, TX, and Calcasieu, LA. Coastal counties/parishes where direct discharges of PCBs are the greatest include Harris, and Brazoria, TX; St Charles, Iberia, and St. James, LA; and Hillsborough, FL (NOAA, 1985).

Municipal wastewater treatment plants occurring along the Gulf Coast have been identified as important sources of natural organics and nutrients, pathogens, metals, and toxic organic compounds (all found in sludge produced as a by-product of treatment activities). These plants are the single largest direct point source dischargers of total Kjeldahl nitrogen and fecal coliform bacteria. Sludge disposal is a problem for many communities along the Gulf Coast. Most municipalities use landfills, but available land is becoming scarce. Sludge production was projected for the year 2000 for each Gulf county/parish by NOAA's Strategic Assessment Program. All counties/parishes are expected to have increases, with the exception of St. John the Baptist and West Baton Rouge, Louisiana parishes. Pinellas County, FL and Harris County, TX will have the greatest production by the year 2000 (NOAA, 1985).

The Gulf Coast also contains a large number of hazardous waste disposal facilities that serve as potential sites for releases of hazardous substances into coastal waters. In 1985, NOAA mapped and identified these facilities in their strategic assessment data atlases of the Gulf Coast (NOAA, 1985). Thirty-five National Priority Sites in the Gulf coastal area were identified by USEPA for cleanup under the Superfund program in 1985. Harris County, TX, which borders Galveston Bay, contained 210 uncontrolled and potentially hazardous waste sites. Regulated hazardous waste disposal facilities include hazardous waste landfills or land treatment sites, surface impoundments and injection wells, and incinerators. These facilities are primarily located in the Galveston Bay and Mississippi River chemical corridor (south of Baton Rouge, LA) areas. The NOAA report identified 119 such facilities along the Mississippi River corridor.

Hydrologic modification operations for land development result in both point and non-point source discharges. Dredged material is a potential source of heavy metals, petroleum hydrocarbons, PCBs, and other pollutants that are concentrated in harbors and shipping channels. Disposal sites established to receive large loads of dredged materials are located near harbors and major rivers where frequent dredging is required to maintain navigable waterways. There are 37 Dredged Material Disposal Sites under USEPA interim designation in the near-shore waters of the Gulf.

Non-point source pollution, as defined by the Clean Water Act 'is caused by diffuse sources that are not regulated as point sources and normally is associated with agricultural, silvicultural, and urban runoff, runoff from construction activities, etc. ... non-point source pollution does not result from a discharge at a specific, single location but generally results from land runoff, precipitation, atmospheric deposition, or percolation'. Stormwater runoff from urban areas results from paving and other alterations of natural surfaces that reduce soil permeability. Untreated urban runoff is a major source of non-point pollution. Urban stormwater runoff is estimated at 20 B L per day along the Gulf Coast. Approximately 18,000 T of BOD and 13,600 T of petroleum hydrocarbons (particularly from improperly disposed crankcase oil) enter the Gulf of Mexico from urban runoff. About 90 percent of fecal coliform loading comes from urban runoff. Other pollutants include Kjeldahl nitrogen, phosphorus, heavy metals, and chlorinated hydrocarbons. Orleans Parish, Louisiana, has the highest rate of runoff per square mile (~259 ha). The next highest are Pinellas County, FL, and Harris County, TX (NOAA, 1985).

Agricultural activities are another important source of non-point source pollution, particularly nutrient loading and toxic chemical discharges. Seven coastal counties are primarily farmland: Jackson, Kleberg, Refugio, San Patricio, and Wharton, TX; and Glades and Osceola, FL. Texas has the greatest concentration of farms. Livestock are the dominant farming activity in all states except Alabama. Various farming activities have different effects on water quality. Livestock can cause overgrazing, which results in erosion and runoff. Agricultural runoff usually contains fertilizers and pesticides that deteriorate coastal water quality. Runoff from heavily fertilized agricultural lands usually increases the availability of nutrients and intensifies eutrophication problems. Non-urban runoff accounts for almost 50 percent of the total discharges of Kjeldahl nitrogen into coastal Gulf waters and 36 percent of the total discharge of phosphorus (90 percent of all non-point source phosphorus discharges). Coastal counties/parishes with the greatest phosphorus discharge from non-point sources include Nueces, Victoria, and Jackson, TX; and Acadia, Vermilion, and Lafourche, LA (NOAA, 1985).

The impacts of pesticide runoff can include fish kills, uptake and/or accumulation in biota, and changes in estuarine population structure. The development and use of newer pesticides with shorter half-lives and lower bioconcentration potentials make it likely that impacts will occur primarily near sites of application, and coastal zones may be at high risk for aquatic toxicity (Pait *et al.*, 1992).

'Agricultural Pesticides in Coastal Areas: A National Summary' (Pait *et al.*, 1992) reports the Gulf of Mexico as having the highest applications of agricultural pesticides of US coastal watersheds, and to be at highest risk. The report assessed the use of 35 commonly applied pesticides in US estuarine watersheds; selection of the 35 was based on use and potential for the pesticide to impact the coastal aquatic environment. Pesticide usage was based on 1987 data. Pesticide use estimates were taken from the National Pesticide Use Database-Resources for the Future (RFF), and supplied the types and amounts of pesticides used agriculturally by county, and further prorated to determine usage for estuarine drainage areas (Pait *et al.*, 1992). Pait developed a hazard rating system to provide a means of ranking the pesticides in terms of their potential to impact the estuarine environment. These coefficients were multiplied by the quantities of pesticides used in 1989/90 to rank pesticides of concern based on usage and toxicity. The pesticides of most concern are: trifluralin, methyl parathion, chlorothalonil, parathion, phorate, carbofuran, carbaryl, thiobencarb, propanil, and alachlor.

One of the major sources of contaminants to the coastal/near-shore Gulf waters is upstream waters found in the drainage of the Mississippi River and Atchafalaya River systems. The Mississippi River is a major source of freshwater, with a drainage basin of almost 3,240,000 km^2 (41 percent of the land area of the lower 48 states) (NOAA, 1985). The Mississippi and the Atchafalaya rivers, along with 20 other major river systems along the Gulf Coast (with discharge rates >30 m^3 per second), are major sources of nutrients, heavy metals, pathogens, and organic and inorganic pollutants entering coastal and estuarine waters from upstream industrial, municipal, and agricultural sources. Tidal activity within estuaries then determines the residence time of the substances prior to being discharged into the open Gulf waters. Eighty percent of heavy metals (arsenic, cadmium, chromium, copper, lead, mercury, and zinc) are brought to Gulf coastal areas by rivers and streams, primarily the Mississippi and Atchafalaya river systems. Almost 12 times more wastewater enters the Gulf's coastal waters from upstream sources than from all (industrial, municipal, and other) coastal activities. Approximately 80 percent of all organic material, 75 percent of Kjeldahl nitrogen, and 80 percent of all phosphorus entering the Gulf is carried there by rivers and streams (NOAA, 1985).

Frequency of the most toxic releases to Gulf estuarine drainage systems (EDSs)

A report by Brecken-Folse *et al.* (1993) presents data on toxic releases and frequency of release for each estuarine drainage system (EDS) across the Gulf. This report examines quantities of chemicals released as reported in TRI (USEPA Toxic Release Inventory) and PCS (USEPA Permit Compliance System) databases and calculated relative toxicities based on literature values for chemicals. Ammonia was found in 17 EDSs, chlorine in 11, copper or copper compounds in 9, zinc or zinc compounds in 9, ethylbenzene in 7, ammonium sulfate and cyanide compounds in 6, chloroform, glycol ethers, molybdenum trioxide, and nickel or nickel compounds in 5, ammonium nitrate, catechol, cobalt or cobalt compounds, methanol, phenol, and sulfuric acid in 4, with remaining chemicals listed in Table 16.5. Ammonia releases (in 1989) entered near-shore Gulf waters through 68 percent (17 of 25 EDSs) of Gulf drainage systems.

Ranking the top ten Gulf-wide toxic releases (1989)

Toxicity Indices were calculated and totaled for each chemical shown in Table 16.5 to compare the total impact each chemical may have had Gulf-wide, and which EDSs were at greatest risk, by release of that compound. It should be noted that the most frequently released chemicals did not necessarily correspond with the top ten Gulf-wide toxic releases. Three chemicals made the top ten toxic release list with frequencies less than four (chromium, hydrazine, and sulfuric acid).

Contaminated areas in the Gulf of Mexico

Understanding the geographical distribution of contaminant concentrations along the Gulf Coast provides a description of the general health of coastal water quality of the northern Gulf of Mexico. Areas identified as receiving large amounts of contaminants may not be the final receptacles for the compounds. Contaminants may be transformed biologically or chemically, broken down into new substances;

Table 16.5 Significant Gulf of Mexico chemical stressors

Chemical released	Occurrences	CAS registry number
1,1,1-Trichloroethane	1	# 000071556
1,2,4-Trimethylbenzene	1	# 000095636
1,2-Dichloroethane	1	# 000107062
4,6-Dinitro-o-cresol	1	# 000534521
Acetone	5	# 000067641
Aluminum (fume or dust)	1	# 007429905
Ammonia[a]	17	# 007664417
Ammonium nitrate (solutions)	4	# 006484522
Ammonium sulfate (solutions)[a]	6	# 007783202
Antimony or antimony compounds	1	# 007440360
Benzene	1	# 000071432
Biphenyl	2	# 000092524
Catechol	4	# 000120809
Chlorine[a]	11	# 007782505
Chloroform	5	# 000067663
Chromium or chromium compounds[a]	3	# 007440473
Cobalt or cobalt compounds	4	# 007440484
Copper or copper compounds[a]	9	# 007440508
Cyanide compounds[a]	6	Cyanide C
Cyclohexane	2	# 000110827
Dichloromethane	1	# 000075092
Epichlorohydrin	1	# 000106898
Ethylbenzene[a]	7	# 000100414
Formaldehyde	2	# 000050000
Glycol ethers	5	various
Hexachloro-1,3-butadiene	1	# 000087683
Hexachlorobenzene	1	# 000118741
Hydrazine[a]	1	# 000302012
Hydrochloric acid	3	# 007647010
Hydrogen cyanide	2	# 000074908
Hydrogen fluoride	1	# 007664393
Lead compounds	1	# 007439921
Manganese or manganese compounds	1	# 007439965
Mercury or mercury compounds	2	# 007439976
Methanol	4	# 000067561
Molybdenum trioxide	5	# 001313275
Naphthalene	1	# 000091203
Nickel or nickel compounds	5	# 007440020
Nitroglycerin	1	# 000055630
Phenol	4	# 0001088952
Phenyl mixture	1	various
Styrene	1	# 000100425
Sulfuric acid[a]	4	# 007664939
Tetrachloroethylene	2	# 000127184
Toluene	3	# 000108883
Trichloroethylene	1	# 000079016
Vinyl chloride	1	# 000075014
Xylene (mixed isomers)	5	# 001330207
Zinc or zinc compounds[a]	9	

Note:
a Indicates chemicals with high release rates and high toxicities based on quantities released.

they may be mixed vertically and horizontally in the water and translocated by currents and tides.

At present, only NOAA's Status and Trends Program and EPA's EMAP Program provide comprehensive field measurements quantifying the health of the coastal water quality of the entire Gulf Coast (NOAA, 1988b; Summers et al., 1993; USEPA, 1999). This program has made systematic measurements of the concentrations of synthetic organics and other compounds in bivalves, bottomfish, and sediments at sites along the Gulf Coast since 1984. Sites were deliberately selected away from major point sources of contamination in order to determine where the sources (described previously) are influencing regional water quality. Sixty-four sites were sampled along the Gulf Coast, usually in embayments and behind barrier islands. Except for some sites near the Florida cities of Jacksonville, Tampa, Panama City, and Ft Walton Beach, levels of contamination within the Gulf of Mexico coastal areas were relatively low when compared to the rest of the United States coastal areas. These four areas in Florida contained contaminants that ranked within the highest 20 concentrations found in US coastal waters. Tampa Bay contained high levels of lead, polychlorinated biphenyls, chlorinated pesticides, and total organic carbon. St Andrew Bay and Choctawhatchee Bay, FL contained high levels of the same compounds as Tampa Bay, as well as DDT, polyaromatic hydrocarbons, and arsenic. Mississippi Sound, below Mississippi, showed high levels of polyaromatic hydrocarbons and chlorinated pesticides. It is evident that the highest concentrations of contamination of the Gulf Coast occur in coastal areas that are in proximity to urban development existing close to the shoreline.

It appears that stresses and effects on ecosystems from chemical contaminants are localized, but data availability and the methods used in data collection (large-scale random sampling or small-scale intensive sampling) may mask larger trends in environmental degradation. However, contaminants are continuously entering the estuarine drainage systems and the potential for ecosystem effect exists; indeed, some large-scale system effects may have already occurred. Sources, environmental reservoirs and effects or potential effects of many toxic substances released into the Gulf must be identified and their impacts studied to mitigate further degradation of Gulf of Mexico environments.

Summary

Relatively large amounts (in the order of 278,572 T per year) of toxic substances were discharged from industrial and municipal sites into estuarine drainage areas of the Gulf of Mexico in 1999 (USEPA, 2000). Other contaminant sources include produced waters (13.6 M kg) from near-shore oil and gas platforms and pesticide runoff from agricultural fields (2.3 M kg in 1989). These and other sources have contributed to elevated levels of these contaminants in near-shore waters of the Gulf (NOAA, 1988a).

Approximately 19.5 M kg of toxics were discharged into municipal treatment systems from coastal counties of which 450,000 kg were released into Gulf near-shore waters after treatment. Approximately 5.4 M kg of toxics were discharged into surface waters resulting in a total of 5.9 M kg reaching Gulf waters (Brecken-Folse *et al.*, 1993; USEPA, 2000).

Site specific mostly short-term adverse or potentially adverse effects due to toxic contamination have occurred in the coastal waters of the five states bordering the Gulf of Mexico. This is reflected in many seafood advisories and listings under section 304(l) of the Water Quality Act. Potential sources of contaminants include industrial and municipal discharges, pesticide runoff from agricultural activities, produced water from near-shore oil and gas platforms, urban runoff, and atmospheric deposition. Potential ecosystem effects are suggested when such large amounts of toxic substances enter the near-shore waters, in many instances on an annual basis and there is documented evidence of adverse short-term effects.

Approximately 4.5 M kg of pesticides were applied to agricultural fields in Gulf coastal counties in 1987 and 2.3 M kg in 1989. According to a rating index developed by NOAA in 1987, potential contamination of the Laguna Madre estuarine drainage system (south Texas) was large, followed by Tampa Bay, FL and Charlotte Harbor, FL. The index applied to the 1989 data showed Laguna Madre also to be great, but also included Atchafalaya and Vermilion bays, LA and Matagorda Bay, TX.

Results of federally and state-funded reports emphasize the need for standardized monitoring methods and programs that provide comparable data sets of known quality that can help determine changes in the near-shore environment. Data standardization suggests potential benefits from examining and implementing appropriate pollution prevention strategies that can reduce or eliminate contamination problems from manufacturing, agricultural, and urban and suburban citizen usage, in addition to current end-of-pipe controls.

WATERSHED DISCRIMINATION

Land-use

Before watershed-based risk assessments can be undertaken, it is important to understand the usefulness of the products. Each watershed-based risk assessment is an expensive, long-term venture that will be extremely useful for the stakeholders within that watershed. It is unreasonable, however, to conduct a definitive study in every watershed along the Gulf of Mexico and Southeast US coasts. Therefore, before conducting the first project in this area, a thorough understanding is needed in the differences and similarities between watersheds of this region. This, in itself, is no small task. However, resources are available to expedite this task. First, the EMAP-Louisianian Province program has thematic mapping images that, once classified, could facilitate a better understanding of land use in this region. One critical causal relationship needed for a watershed-based risk assessment is the influence of various land uses on water quality. Thus, it can be assumed that one factor that would discriminate between watersheds would be land use.

Hydrology

It follows, also, that hydrology would be another factor to aid in the discrimination between watersheds. Other governmental agencies, especially the US Geological Survey (USGS) could be a major source of hydrological information. The National Water Quality Assessment Program has several initiatives within the Gulf Coast and the Southeast region. These include the Trinity River Basin in Texas, Chicot–Evangeline Basin in Louisiana, Mobile River Basin in Alabama, Apalachicola–Chattahoochee Basin in Georgia and Florida, Georgia–Florida Coastal Plain in Georgia and Florida, South Florida (including the Everglades) in Florida, Santee Basin and Coastal Drainage in North and South Carolina, and the Albermarle–Pamlico Drainage in Virginia and North Carolina. Concerted efforts along these lines have begun over the last 10 years. These, along with other efforts including EMAP, may contribute significantly to the characterization of the hydrological network in this region.

Natural resources

The remaining discriminator between watersheds is natural resources. Many of these are summarized in a previous section of this chapter. If these are inventoried on a watershed basis and integrated with land use and hydrology data the degree of similarity and dissimilarity among watersheds can be quantified. The assembly and manipulation of these databases should be accomplished using Geographic Information Systems (GIS). This would not only aid in developing the science but also serve as a visualization technique for education and communication.

WATERSHED-BASED RISK ASSESSMENT

Introduction

The purpose of this section is to expand the theory presented in 'A Framework for Ecological Risk Assessment' (USEPA, 1992) to include the more holistic and robust case of entire watersheds. The concise outline presented in the framework document provides a logical starting point for the evolution of a watershed-based risk assessment paradigm as well as the further development of a protocol for conducting these risk assessments. The entire framework is summarized in Figure 16.1. The same general approach can be taken for watersheds, whatever the size, except that scale plays a critical role in how assessors and managers visualize, conduct, and conclude the assessment.

In the Framework document, the assessment is divided into three phases: Problem Formulation, Analysis, and Risk Characterization. Before Problem Formulation extensive interaction occurs between the risk manager and the risk assessor. In addition to setting time and funding limits, these discussions address policy issues that may affect the selection of assessment endpoints. These discussions

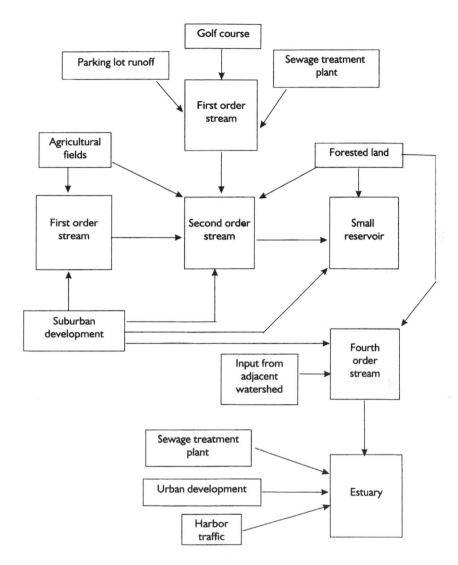

Figure 16.1 Relationship between land use and aquatic ecosystems in small, developed suburban watersheds. Arrows represent water movement

are critical to the success of the ecological risk assessment since they drive the Problem Formulation phase.

It includes a preliminary characterization of exposure and effects, site history, and various surveys and studies. This information is used to determine assessment and measurement endpoints. A review of all existing data on the watershed and the stressors identified is necessary. Working hypotheses are formulated on how the stressor(s) may affect ecological components.

The Analysis phase consists of two interrelated paths that occur simultaneously: exposure characterization and ecological effects characterization. The spatial and temporal distribution of the chemicals of concern and their interaction with the ecological system are addressed. In addition, the impact of the chemicals on individuals, populations and communities is quantified. Data requirements should be part of the work plan for inclusion in the remedial investigation. As a required component of the work plan, it is necessary to address data gaps and data quality objectives (USEPA, 1992).

Risk Characterization uses input from the Analysis phase to determine the likelihood of chemical exposure resulting in adverse ecological effects. Various issues including cause and effect, strength ('robustness') of the data, and scientific uncertainties are used to judge the ecological significance of the risk. The results from this process are interpreted by the risk assessor in an understandable and usable format. This information should be included in the work plan for inclusion in the remedial investigation.

Based on the Framework document, a watershed-based ecological risk assessment should have the following phases: Assessment Plan Development, Analysis, and Risk Characterization. Note that the last two phases carry the same name as those in the Framework document. This is intended to stress that the approach in these sections can and should be similar to that in the Framework document. The first phase carries a different name because the approach will be very different: the assessment considers an entire watershed.

Watershed assessment plan development

Like the Problem Formulation phase of the Framework document, the Watershed Assessment Plan Development is the first phase of watershed ecological risk assessment and establishes the goals, breadth, and focus of the assessment. It is a systematic planning step that identifies the major factors to be considered in a particular assessment and is linked to the regulatory and policy context of the assessment. The Framework document points out that the dialogue between the risk assessor and risk manager is critical to the success of the risk assessment. By bringing the management perspective to the discussion, risk managers charged with protecting societal values can ensure that the risk assessment will provide relevant information to making decisions on the issue under consideration. By bringing scientific knowledge to the discussion, the ecological risk assessor ensures that the assessment addresses all important ecological concerns (USEPA, 1992).

Risk assessment team

The initial step in the Watershed Assessment Plan Development is to bring together a much larger group than a risk assessor and a risk manager. It is essential to get input from all interest groups within the watershed. These include all major land use groups such as industries; farmers; urban and suburban developers; homeowner associations; non-governmental organizations with vested interests in specific land uses for the watershed, e.g. environmental groups, hunt clubs, and fishing groups; and government agencies that manage land within the watershed. In addition, major users of the natural resources need to be assembled. The challenge to these groups, or stakeholders, is to begin to define critical resources within the watershed as well as critical land uses within the watershed. This exercise develops a watershed-specific set of goals for resource use and protection. In essence, this exercise allows the stakeholders within the watershed to redefine the goal of integrated ecosystem protection unique to their watershed. For example, in agrarian watersheds a critical resource might be enough water for irrigation while the critical use might be economically viable agriculture. In another watershed a critical land use might be suburban development while the critical resource might be natural resources for recreation.

Stakeholders

It is worth stressing, at this point, that it is important to bring all stakeholders to the process in the beginning. Land users, developers, water management districts, environmental action groups, local, state, and federal government agencies are among those that need to be represented in the process. It is critical to the goal of ecosystem protection that all stakeholders be familiar with the process and adopt the approach that they have had a part in developing. When the risk assessment is finished and watershed management options discussed, these same stakeholders will have to 'buy in' to management scenarios.

Just as the Framework document stresses that the outcome of Problem Formulation is a conceptual model, the outcome of Watershed Assessment Plan Development is a conceptual model that describes how a given stressor might affect the ecological components in the environment. The difference is that there are many different stressors to consider as well as many different ecological components to maintain. In addition, there are many economic and social factors to consider in these models.

Identify critical resources

As stated earlier, a major challenge to the stakeholders is to adequately identify not only the fragile natural resources within the watershed that need to be protected and maintained but also identify the critical land use activities within the watershed that are critical to the economic survival of the watershed human community. This exercise may very well be the most complex portion of the watershed-based risk assessment and the utility of the assessment may hinge on how well this task is performed. Ultimately this exercise should provide a clear set of hypotheses amenable to scientific inquiry around which to structure the risk assessment.

The framework document identifies three components in this portion of Problem Formulation: stressor characteristics, ecosystems potentially at risk, and ecological effects. All of these interrelate and help determine endpoint selection and, ultimately, the conceptual model.

The 'ecosystem potentially at risk' portion must be expanded within a watershed risk assessment for several reasons. In a watershed there are many diverse types of ecosystems to consider. This phase must also consider critical land uses as well as natural resources. Furthermore, watersheds may differ by virtue of the differences in rationale of the stakeholders. That is, critical land uses for one watershed may be banned in another; preserved natural resources in one watershed may be harvested in another).

Since this phase is a subtle blend of economic needs and environmental quality, it is important to have stakeholders or ad hoc members who can act as resources without needing to represent a certain interest within the watershed. Economic data and environmental inventories must be acquired. In a watershed it is important to consider not only the ecosystems potentially at risk but all natural resources since management and land use scenarios may very likely change which ecosystem is at risk.

Identifying natural resources and critical land uses are lumped together to illustrate that these two categories may be interchangeable depending on the goals of the stakeholders. A tourism-based watershed may intend to utilize natural resources much differently than an agrarian watershed. In any event, however, it is critical to understand all of the land uses and proportional distributions, as well as the distribution of natural resources in the watershed prior to examining potential stressors and ecological effects. The outcome of this exercise should be a spatial distribution of land use and critical natural resources within the watershed that should serve as the foundation for understanding the potential impact of present and future land use scenarios on ecosystems. This also facilitates the characterization of both stressors and ecological effects in a spatial manner within the watershed.

Stressor characteristics

Stressors are chemical, physical, or biological influences causing negative impact on populations and communities in the ecosystems at risk. Chemical stressors include a variety of inorganic and organic substances. Careful consideration must be given to characterizing secondary stressors that result from the interaction of primary stressors with the ecosystem. Examples of this include major breakdown products of organic contaminants, biotic transformations of inorganic contaminants (e.g. methylation), and the depletion of ozone caused by chlorofluorocarbons that could result in increased ultraviolet radiation exposure. Physical stressors include extremes of natural conditions and habitat alteration or destruction. Examples of these might be water temperature changes due to power generation, hydrologic alterations of the drainage network to prevent flooding or store water for municipalities, and siltation due to high suspended solids caused by heavy erosion during construction, mining, or forestry operations.

Biological stressors may often be overlooked but the impact may be devastating to a watershed. A good example is the gypsy moth (*Lymantria dispar* L. Lepidoptera: Lymantriidae) that has single-handedly destroyed both economies and natural resources within certain watersheds. Other introduced biological stresses included aquatic plants, rodents, fish, clams, fungi, and bacteria. In addition, competitive and predative relationships may change and ultimately redefine biological resources within the watershed.

It is important to note that a direct connection exists between land use and stressors; ultimately this connection must be extended to explain the relationship between land use and ecosystem impact. It is difficult, therefore, to isolate the stressors from the species response. The degree to which stressors influence the survivorship of species depends on the magnitude of the stress (intensity), the duration of the stress (how long the species is exposed relative to its own life history characteristics), the frequency (how often a stress of a particular intensity occurs), and the timing (when the stress occurs relative to critical life history stages of the species). This exposure scenario may be very complex, especially in watersheds where complex multiple stressor interactions occur as one moves down from the head waters through the hydrological network.

As in any risk assessment, it is important to consider both point and non-point source pollutants. Most watersheds will display a suite of point source discharges (identified through the permitting process) and non-point source pollutants (usually poorly characterized and loosely associated with various land uses). Both sources are important to stressor characterization. Ultimately understanding the inter-relationship between stressor sources may lay the framework for management scenarios including pollutant trading and land use exchanges.

The task of the risk assessor is to assemble and analyze a suite of previously compiled chemical, physical and biological data. Literature databases contain a variety of environmental toxicology and chemical fate data. These data facilitate the prediction, within certain constraints, of the interaction of contaminants with the physical and biological ecosystem components. This occurs during the analysis phase.

Another goal of the stressor characterization exercise is to begin to develop a spatial and temporal characterization of the stressor burden within the watershed. Climate and hydrology will play major roles as many stressors will be present during or just after storm events, and peak concentrations will move downstream through the watershed. In addition, duration of exposure will change according to storm location, size, and intensity as well as stream order (larger order streams usually have longer hydrographs while small order streams have short hydrographs with very high flow rates). The hydrology is very different in the headwaters versus the larger streams, rivers, lakes, reservoirs, wetlands, and estuaries that typify lower regions of watersheds. It is critical to characterize these processes as they influence stressor exposure scenarios and, in conjunction with land use, determine which portions of the watershed have the greatest stressor burden.

Ecological effects

All known ecological effects data concerning the stressors, watershed, or similar systems should be assembled. These effects data should include individual, population, and community responses. Sources for these data include historical databases, field observations (e.g. fish kills, changes in aquatic community structure, and endangered species distributions), laboratory toxicity tests (e.g. single species and microcosm bioassays), and chemical structure–activity relationships. It is important to note that the stakeholders may have anecdotal evidence of ecosystem stress that should be investigated. These observations may uncover problems not apparent to local regulatory agencies.

It is important to assemble natural history data (e.g. home range, feeding area, and migratory patterns) on the species of concern in the watershed. Much of this data can come from governmental agencies, environmental groups, and citizens' coalitions who have made natural resource inventories within the watershed. In addition, the open literature may contain information from other areas that is relevant to the watershed. These data, together with the spatial and temporal characterization of stressor exposures within the watershed can help predict potential ecological effects. Analysis of this information can help focus the assessment on specific stressors and on ecological components relevant to the watershed.

The utility of available ecological effects data for problem formulation may be limited by an inability to accurately extrapolate laboratory test data to the complex ecosystems that comprise a watershed. On the other hand, direct field observations of effects (e.g. fish kills and algal blooms) may be hard to interpret because of natural variation or the interaction of a suite of stressors. In many instances, the ecological effects portion of Watershed Assessment Plan Development will identify data gaps and help focus the ecological portion of the Analysis Phase of the assessment.

Endpoint selection

Once the stakeholders, risk assessor, and risk manager have established the goals of the assessment, relevant endpoints can be selected. These are dependent on the critical resources chosen for the watershed, stressors characterized, and ecological effects anticipated. An endpoint is a characteristic of an ecological component (e.g. increased mortality in fish) that may be affected by exposure to a stressor (Suter, 1993). Both assessment and measurement endpoints are used to determine risk in the watershed. Assessment endpoints generally refer to characteristics of populations and ecosystems defined over rather large scales (e.g. forest production over a large geographic area or fish populations in a reservoir). These endpoints are the closest reflection of the health of the critical resources to be protected and maintained in the watershed. Thus, it is critical that assessment endpoints have biological as well as societal value so that scientific information can be linked to the risk management process.

For an ecological risk assessment to produce sound, acceptable results, there are five criteria necessary for choosing endpoints (Suter, 1993): 1) societal relevance, 2) biological relevance, 3) unambiguous operational definition, 4) accessibility to prediction and measurement, and 5) susceptibility to the hazardous agent. In a watershed-based risk assessment it is critical that the assessment endpoints accurately meet the needs and concerns of the stakeholders as reflected in the identification of critical resources. Additionally assessment endpoints must be ecologically sound and susceptible to previously characterized stressors.

Obviously choice of assessment endpoints is neither trivial nor simple. These choices significantly influence the development of the conceptual model that is the product of the Watershed Plan Development Phase. Assessment endpoints must consider both the critical resources in the watershed and how effects will be defined. Some assessment endpoints are mandated legally or politically; however, it is important to determine what endpoints should be selected on technical grounds. Suter (1993) suggests performing one of the following formal analyses of the relationship of components of the action being assessed and components of the receiving environment:

1   Create a matrix of exposure alternatives (e.g. sediment loading from agricultural fields during rain events, nutrient loading from suburban lawns and golf courses, accidental industrial chemical spills, etc.) and environmental receptors (e.g. fish, algae, aquatic birds, etc.) that are potentially affected. Environmental receptors are then checked off and possibly scored for the intensity and duration of the exposure and relative sensitivity to the toxic material.

2   A receptor identification exercise is valuable to identify which organism will be most exposed to a chemical. This consists of two steps: 1) performing a rapid quantitative exposure assessment to determine what media are most contaminated (note: this may be from a fate model determination); and 2) determining what communities, trophic groups, populations, and life stages are most exposed to those media.

3   Indirect effects of stressors can be identified by developing models, including 'event trees' showing causal linkages between site contaminants and various environmental components.

4   Existing data can be reviewed to determine the sensitivity of species or processes to the contaminant or to similar contaminants. These may include data from toxicity testing or from biological monitoring of prior releases.

While assessment endpoints are explicit expressions of the actual environmental value that is to be predicted, measurement endpoints are measurable responses to a stressor that are related to the valued characteristics chosen as the assessment endpoints (Suter, 1993). Assessment endpoints are the ultimate focus in risk characterization and link measurement endpoints to the risk management process.

Each assessment endpoint requires a specific set of measurement endpoints. If an assessment endpoint can be directly measured then it is also a measurement endpoint. Usually, however, assessment endpoints and measurement endpoints are not the same. In many cases, measurement endpoints are necessary because the assessment endpoints are difficult to measure. Measurement endpoints must be quantitatively or qualitatively related to the assessment endpoint. Measurement endpoints are either single numbers (e.g. LC_{50}s) or, more preferably, multi-dimensional descriptive models (e.g. concentration-response functions) (Suter, 1993). Selection of measurement endpoints must be carefully thought out prior to undertaking an ecological risk assessment. The Framework document presents a useful synopsis of considerations in selecting measurement endpoints.

Conceptual model

Watersheds are complex assemblages of ecosystems interconnected by other ecosystems (hydrologic conduits). Each ecosystem may be impacted by adjacent land use or by any stressor exported from an upstream ecosystem. Results from the Identification of Critical Resources section can be used to draw a picture of the watershed. The major focus of the conceptual model is the development of working hypotheses regarding how stressors might affect ecological components of the natural environment. From a conceptual standpoint, the watershed can begin to appear as a series of compartments representing various land uses and aquatic ecosystems each connected by potential contaminant transport arrows (Figure 16.1). While this may appear oversimplified, it facilitates the formulation of hypotheses concerning spatial and temporal exposure scenarios and the impact of stressors on the ecosystems.

The hypothetical watershed represented in Figure 16.1 has several land uses contributing stressor burdens to different aquatic ecosystems. These different aquatic systems are hierarchically connected and contribute stressors through water flow. In addition, simple water flow can be considered a stressor. An important concept here is that the majority of stressors will enter the aquatic system through water flow: point source discharges, non-point source runoff during storm flow, ground water seepage during base flow, and simple downhill water flow from one aquatic ecosystem to another. This is not to discount direct chemical applications (purposefully or accidentally) as these may be major stressor burdens on the ecosystems into which they are discharged.

It is critical to understand hydrology within the watershed as this ultimately dictates exposure scenarios. Different land uses have different storm water hydro-graphs resulting in different exposure durations (Figure 16.2). In addition, runoff from different land uses may have different intensity resulting in different physical stresses (e.g. hydrology, erosion, and sedimentation) on the receiving stream ecosystem. Alterations of the hydrological conduit exacerbate or mediate some of these physical stresses (e.g. channelization increases the velocity of the water and impoundments decrease the water velocity). These alterations may significantly change exposure scenarios as well as critical ecological components potentially at risk. The incorporation of different land use hydrology into exposure scenarios

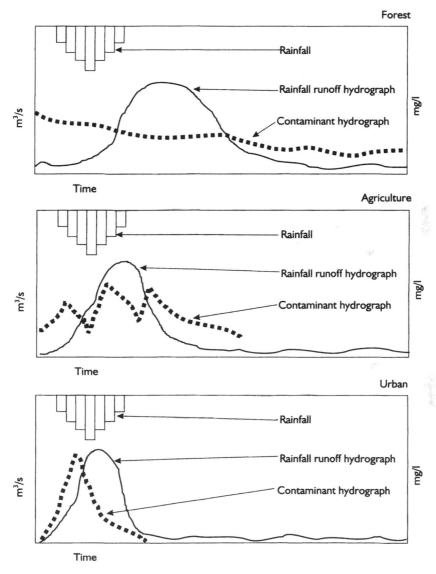

Figure 16.2 Relationship between rainfall, water runoff, and contaminant transport for different land uses

can be a significant aid to visualizing spatial and temporal relationships between stressors and ecological components within the watershed (Figure 16.3). In addition, visualizing these relationships in a diagram can help develop and prioritize hypotheses for the conceptual model.

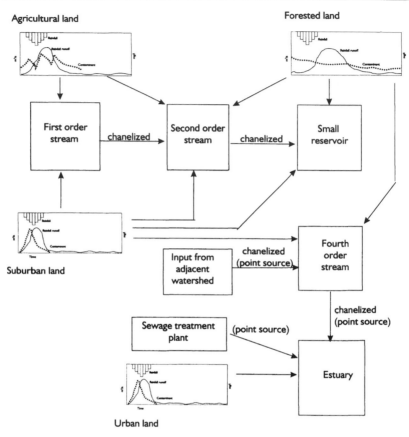

Figure 16.3 Incorporation of exposure scenarios into watershed model to facilitate hypothesis formulation and prioritization

Each hypothesis concerning an exposure scenario should be defined in terms of the stressor, the type of biological system and principal ecological components, how the stressor will contact or interact with the system, and the spatial and temporal scales (Table 16.6). Hypotheses are then prioritized for likelihood of contribution to risk. Hypotheses are selected for testing in the analysis phase. The number of hypotheses selected often depends on the financial constraints of the risk assessment. The approach for testing hypotheses that are carried forward into the analysis phase should be described in the conceptual model, including types of data and analytical tools needed; hypotheses not carried forward are saved for future consideration when uncertainty is addressed in risk characterization.

Before the analysis phases, a detailed work plan should be written that describes objectives, data requirements (including assessment and measurement endpoints), experimental design, procedures and methods, quality assurance objectives, and a time schedule to estimate duration and completion dates of various phases of the assessment. In formulating a work plan, it is critical to address how data gaps will

Table 16.6 Examples of aquatic resources in the ERA conceptual model

Aquatic system	Description of resources
First order stream	Little benthic community
	Low flow between storm events
Second order stream	Ripples, pools, diverse biota
	Established primary producers
	Benthic community
Small reservoir	Highly productive
	Fishing, boating
	Drinking water
	Mesotrophic
	Water fowl
Fourth order stream	Migratory fishes
	Rooted macrophyte beds along shore
	Diverse benthic community
Estuary	Migratory fishes
	Shellfish production
	Boating

be handled and to explicitly state data quality objectives (USEPA, 1992). The conceptual model describes the approach that will be used for the Analysis phase and the types of data and analytical tools that will be needed.

Analysis

The Analysis phase of watershed risk assessment consists of a quantitative determination of the probable fate and effects of multiple stressors in the watershed. The stressors and the assessment endpoints considered in the Analysis phase are those identified in the conceptual model during the planning phase. The analytical procedures must provide for both spatial and temporal dynamics of the stressors and their effects. Although the characterization of both exposure and effects are important in the risk characterization phase of a watershed risk assessment, in practice they are both parts of a single process.

The process of characterizing exposure and effects requires the application of computer simulation linked with GIS. Models of chemical transport, when linked with GIS, can show the spatial distribution of predicted concentrations within the watershed. These values provide the input to individual-based or aggregated population models to illustrate the predicted effects in both geographic space and time. The models must be capable of simulating multiple stressors (both chemical and non-chemical) and multiple endpoints simultaneously.

Watershed characterization

Watershed characterization includes the identification of ecosystems and other ecological components potentially at risk, hydrologic conduits, and land uses. Much of this is accomplished within the context of the conceptual model. In the analysis phase the interrelationships among the components of the watershed are re-examined as a part of hypothesis testing. As with any ecological risk assessment, interactions between stressors and watershed components must be considered. Microbial communities can modify chemical stressors through biotransformation. The impact of storm events, with their increased sediment and chemical loads, can be regulated by water management in reservoirs.

In a watershed risk assessment, the spatial and temporal distribution of ecological components relative to the comparable distribution of stressors is particularly important. In an area the size of a major watershed, it is likely that the juxtaposition of stressor and organism will change in complex ways. Animal species can move in and out of contaminated areas or move among areas of different levels of contamination. Land use changes will alter the distribution of plant species and wildlife habitats. In some cases, natural succession or silvicultural practices will change vegetative composition and structure over time.

As with the characterization of stressors, the characterization of the ecological components in a watershed will require the integration of simulation modeling and GIS. Models of plant and animal indicator species should start at the individual organism level (individual-based models) and then become aggregated if necessary (DeAngelis, 1994). These models should include a behavioral component to predict an organism's movements within the watershed. Radio telemetry studies can provide the data to develop and verify these models as well as monitoring exposure. Because of the number of species requiring monitoring and the extent of the area over which monitoring can occur, an automated telemetry system will be required.

Hydrologic modeling is a major component to watershed risk assessment. Because of the number of sub-basins in a watershed, the different orders of streams, and the overall length of the river channel, a hierarchical approach to hydrologic modeling is required. In this approach, individual sub-basins and stream reaches are modeled separately, with linkages established between them.

All modeling (hydrologic, individual, food-web, and ecosystem) should be integrated with GIS. Model parameters can be geographically referenced and maintained in a spatial database. Models linked to the database obtain the appropriate parameters for simulation of a given component. The GIS can display simulation results in real time or time-averaged results can be mapped for further analysis. Mapping of land use can utilize satellite imagery or digital aerial photography, depending upon the resolution required.

Characterization of exposure

Exposure requires the presence of a stressor and an ecological component in the same place at the same time. At the level of a watershed, exposure may occur over a wide geographic area and over a long period of time. To develop an accurate exposure profile, it is necessary to determine the spatial and temporal distribution of both the stressor and the ecological component. Exposure profiles are developed for each of the scenarios identified in the planning phase and considered with the associated effects in the risk characterization phase.

One aspect of stressor characterization is the determination of the spatial pattern of the stressor. This pattern often changes through time, so it is also necessary to follow stressor movement in both space and time. This approach will require the use of transport and fate simulation models that are linked to a GIS. For some chemical stressors and non-chemical stressors such as changes in land use, habitat alteration, or habitat fragmentation, spatial statistics can measure changes in a spatial pattern over time and a GIS can display the changes when these stressors are mapped. The scope of the stressor in the watershed will determine the scale of the sources of data used for mapping its distribution. For large areas, small scale data such as multispectral satellite imagery is a likely source. For smaller areas, the greater resolution provided by digital aerial photography may be required.

It is necessary to assess the effects of multiple stressors in watershed risk assessment. This requires considering the possibility of synergistic or antagonistic interactions among chemical stressors. There also could be interactions between chemical and non-chemical stressors. For example, clearing land could increase the runoff of both chemicals and suspended solids which may interact with existing chemicals in receiving waters.

The objective in watershed risk assessment is to characterize risk from exposure to a stressor for the duration of that exposure throughout the watershed. This requires continual monitoring and simulation of the stressor for the complete time of exposure.

Although the emphasis in watershed risk assessment is on population and eco-system effects, the proximal response is by individual organisms. Whether individual responses are aggregated or not, the spatial and temporal distributions of individuals must be combined with the spatial and temporal distributions of stressors to evaluate exposure. In the case of chemical stressors, exposure depends upon the chemical concentrations with which an organism comes into contact. An organismal response will depend upon uptake of that chemical, transport to a target organ, and both chemical concentration and residence time at that target organ. Exposure to non-chemical stressors such as habitat alteration results if an organism depends upon that habitat for any life support activities (e.g. nesting, feeding, resting, or hibernating) during its complete life cycle.

The analysis of the spatial and temporal distribution of the stressor must match that of the ecological component under assessment. Wide-ranging organisms will respond to habitat alterations over a large area. Less mobile organisms will respond to more local disturbances. Similarly individuals with a large home range will be exposed to a chemical stressor for only a short time unless the stressor overlaps most of the home range. Animals with small home ranges can receive large doses if their home range coincides with high concentrations of a chemical. The population effect of such an exposure will depend upon the spatial pattern of the chemical in the environment and the number of organisms exposed.

Although models are often used to describe patterns of exposure in ecological risk assessments, the watershed assessment process relies heavily on a modeling approach. Direct linkages need to be established between models of transport, fate, and uptake by individual organisms. Transport and fate models predict the spatial and temporal distribution of chemical stressors in the watershed. The output from these models provides input to individual-based models which predict the location of an organism in the watershed and its dose at that location.

Because of its complexity, a watershed risk assessment can be considered a continual process in which new scenarios concerning additional stressors can be assessed as additional data are added to the database. A permanent system needs to be established that links all appropriate models with a GIS. All stakeholders in the watershed should have access to the system through a computer network.

During the Watershed Plan Development phase, alternative scenarios are identified for evaluation and comparison for degree of risk. During exposure characterization and analysis, profiles of exposure can be developed for each scenario. Each ecological component identified in a scenario is evaluated for its exposure to a stressor (or combination of stressors). In practice, the development of an exposure profile can be a single step process or a multistep process depending upon the level of integration of transport and fate models with plant and animal uptake models. If both models are completely linked, an exposure (or a dose) profile can be obtained for an exposed individual by simulating the movement of that individual at the same time as simulating the spatial and temporal dynamics of the stressor. The exposure profile in this case represents the cumulative exposure over time and space. If the models must be exercised separately, the output from the transport models can be averaged over time to provide a geographic profile of the stressor. Output from the individual models also can be averaged – or measured home ranges can be used – to obtain an integrated exposure profile (Ott et al., 1986).

Uncertainty in the characterization of exposure is accounted for by measuring the variance in data collection and by incorporating the variance in stochastic models and Monte Carlo simulation. In a stochastic model, random variables representing state variables or model parameters (or both), will take on values according to some statistical distribution. In other words, there will be a probability associated with the value of the parameter or state variable. Monte Carlo simulation is a numerical technique of finding a solution to a stochastic model. For those random features of the model, values are chosen at random from a probability distribution. Repeated runs of the model then will result in different outcomes. Exposure profiles then can be presented as a probability function which gives the probability of the ecological component being exposed to a given level of the stressor for a given period of time.

Characterization of effects

Characterization of ecological effects involves the quantification of the functional relationship between the stressor(s) and the measurement and assessment endpoints identified in the Risk Planning phase. The relation needs to include both acute and chronic effects as well as possible interactions among stressors. Ecological response analysis involves the incorporation of the functional relations into the development of ecological effects models. Measurement and assessment endpoints are correlated and expressed as state variables in the models. Simulation with the ecological effects model provides a stressor–response profile for each of the risk planning scenarios. Because the effects models are linked to environmental transport models, the exposure and stressor–response profiles are developed in sequence rather than in parallel.

A watershed risk assessment requires an enormous amount of data. It is important, therefore, that the data so acquired are managed effectively. A database or set of relational databases must be created to store, manipulate, and update data used for all aspects of the risk assessment. These aspects include data used to characterize the watershed, data used to establish the relationship between the stressor and endpoint, and data used to develop model parameters. Because of the importance of spatial variability in a watershed risk assessment, the data that vary geographically should be geo-referenced and maintained as a GIS database.

The models used to predict ecological effects likely will be the same individual-based models used to predict exposure of organisms to the stressors. Parameters for the models can be derived from laboratory experiments, microcosm and mesocosm studies, and field research. Laboratory and microcosm experiments, because they are conducted under controlled conditions and therefore have relatively low random error, provide quality data for determining stressor–response relations and other model parameters. It is important, however, that the same range of possible environmental conditions and stressor levels be represented in the experiments.

Mesocosms can be used to increase the complexity, and therefore the realism, of experimental research. Mesocosm studies, however, typically produce data with more random error than laboratory experiments but less than field studies. They can be an important source of data, but perhaps the greatest function of mesocosms is to provide an intermediate validation step for effects models. Before too much time and expense is spent on a model, its predictive ability can be tested with mesocosm data.

Field studies are the only source of some data, although the data are likely to have relatively high random errors. Population and landscape parameters are available only from large field studies. For example, they are required to determine the relationship between habitat fragmentation and the probability of an animal's presence. Field studies also are the ultimate test of a model's predictive ability.

In model validation, the data set used to test the model must be independent from the one used to develop the model. A reasonable approach then would be to use laboratory experiments and microcosms to generate the data for model development, and data from mesocosm and field studies for model validation.

Because of the large number of stakeholders involved in a watershed risk assessment, the database must be accessible as part of the established computer network.

The computer system designed for a watershed risk assessment provides stakeholder access to all the data, including geographic data themes. The connection between the watershed database and stakeholder can work in both directions. Some stakeholders (e.g. state and federal agencies) have responsibility for managing data for their own purposes. These data can be imported into the watershed database, which then can be updated as new data are collected.

As in Exposure Analysis, the complexity of watershed risk assessment requires mathematical modeling and computer simulation for Ecological Response Analysis. Following data evaluation, the data are used to develop stressor–response relationships and estimate parameters for the effects models. Relationships also must be determined between measurement endpoints and assessment endpoints.

The same alternative scenarios used in Exposure Analysis are used to analyze the response of the ecological component to the given level of exposure. Each ecological component identified in a scenario is evaluated for its response to a stressor (or combination of stressors). As in Exposure Analysis, the characterization of ecological response can be a single step process or a multi-step process depending upon the level of integration of transport and fate models with plant or animal effects models. If both models are completely linked, an ecological response can be obtained for an individual by simulating the movements of that individual at the same time as simulating the spatial and temporal dynamics of the stressor. The simulation output in this case illustrates the response of the ecological component as a function of time. If the models must be exercised separately, the output from the transport models is stored in a file that can be used as input to the individual models.

Simulation output can show predicted values of the stressor levels and the response of measurement and assessment endpoints over time. This output has the advantage of indicating the time of peak exposure and effects as well as time to recovery. Because Monte Carlo simulations are used, confidence intervals can be produced for each state variable.

Stressor–response profile

The final step in the characterization in the analysis phase is the development of stressor–response profiles. Both stressor variables and the responses of ecological components to those stressors are included in simulation models as state variables. For each risk planning scenario, model output for these variables can generate the stressor–response profiles. Stressor levels and ecological responses can be plotted over time to evaluate timing of the response relative to the stressor. Alternatively response values can be plotted as a direct function of the stressor. Spatial output for stressor levels and effects are produced by running the models for sub-basins or other segments of the watershed. This output then can be mapped with the GIS.

Risk characterization

Risk characterization is the final phase of risk assessment. The deliverable of this phase is the foundation for the development of a watershed management plan. Hence, this phase must focus on the products of the analysis phase, stressor–response profile and exposure profile, and integrate them into a risk estimate. This is delivered to the risk manager along with a description of the major uncertainties and an interpretation of the ecological significance of anticipated effects.

As described in the Framework document (USEPA, 1992), there are at least three different approaches to integrating the stressor–response and exposure profiles. The first is a simple comparison of single effect and exposure values. This approach is based on the ratio of a single expected environmental concentration (EEC) to a single reference effect level of a contaminant. This ratio, known as the risk quotient, Q, is compared to the maximum acceptable risk quotient, otherwise known as the level of concern. This approach may be useful for a single point source and receiving stream within the watershed but is usually insufficient to describe the complex exposure scenarios and response distributions that occur throughout a watershed. In addition, for non-point source pollutants exposure is episodic followed by definite recovery periods where the contaminant(s) is either absent or below the threshold level. Hence, the results of laboratory bioassays may not accurately reflect the dynamics involved in the alternating stress and recovery periods. This approach is the least probabilistic of the three approaches described here and is heavily dependent on professional judgement.

The second approach is to compare the distributions of effects and exposure within the watershed. In this method, the results of the refined risk analysis are presented as a distribution of toxicity rather than a single point estimate. This approach uses all relevant single-species toxicity data and compares them to exposure distributions. Solomon et al. (1996) described this approach, summarized the assumptions, and provided examples. When both exposure (EEC) and effects (acute or chronic toxicity) are expressed in a probabilistic manner, decisions can be made to attempt to protect 95 percent of the species from all chronic effects. This also provides a good estimate of the maximum acceptable concentration and a target for mitigation measures.

An extension of this technique can be used to account for the episodic nature of non-point source contaminants, provided environmental monitoring data are sufficient to document the extent of exposure and recovery periods. Exposure distributions can be expressed not only in terms of concentration but also duration followed by recovery periods. This is particularly critical for contaminants at levels that induce sublethal effects that are reversible upon removal of the stressor burden.

The third approach for integrating the stressor–response and exposure distributions is through simulation modeling. A diversity of models is required to conduct a complete watershed risk assessment. These include the basic hydrologic model of the watershed and other transport models, models of individual organism responses to stressors, and individual-based and/or aggregated models to predict population effects. The models should be linked together and to a GIS to display simulation results.

As described above, models of exposure are linked to models of effects. This makes watershed risk assessment different from most ecological risk assessments in that a response can be related directly to the level of exposure. For a given exposure scenario, the response of the endpoints to that exposure can be predicted. Both exposure level of the stressor and the ecological response can be plotted over time to evaluate timing of the response relative to the stressor.

The simulation modeling approach in watershed risk assessment is different from site-specific risk assessment also in that the models are an integral part of a permanent watershed assessment plan. A given watershed is described by a set of models that is capable of representing the effects of multiple stressors on multiple endpoints. The models then can be used a decision-aiding tool under both current conditions and for assessing risk of possible future stressors in the watershed.

Uncertainty

Uncertainties are associated with the models used to conduct the assessment and the data used to develop and validate the models. In the Watershed Assessment Development Plan phase there is uncertainty associated with the conceptual model development. Poorly chosen endpoints or stressors as variables in the models will lead to an inadequate or inappropriate risk assessment. In the mathematical formulation of the models, errors or faulty assumptions about functional relationships in the model will yield inaccurate predictions of effects. Errors in parameter estimation also will lead to biased simulation results. The implementation of the model by programming the model for computer simulation is another possible source of uncertainty. Standard model validation procedures are available to guard against adverse impacts from these uncertainties (Balsi and Sargent, 1981; 1984).

Data are needed to suggest important variables to include in the models. Other data requirements include determining functional relationships among variables, estimating parameters, and comparing simulation results with observations from the real system for model validation. Where these data are lacking or where there are inaccuracies, the predicted effects of stressors on endpoints will be biased and lead to poor management decisions.

RESEARCH NEEDS

Watershed assessment plan development

1   Development of a model for risk assessment team formation and productivity.
2   Characterization of the relationship between land use and water quality.
3   Characterization of ecosystem response from chronic, episodic exposure scenarios with non-point source pollutants.

4   Development of a suite of biomonitoring tools for watersheds in the Gulf of Mexico.
5   Correlation between assessment and measurement endpoints.

Analysis

1   Characterization of the aquatic-terrestrial interface.
2   Development of automated telemetry systems for characterizing spatial and temporal movement of both terrestrial and aquatic organisms.
3   Linkage of chemical transport and fate models with GIS.
4   Linkage of individual effects models with GIS.
5   Utilization of mathematical models linked with GIS to generate and illustrate stressor–response profiles in watersheds.

Risk characterization

1   Characterization of the exposure scenarios for non-point source chemicals whose movement is closely associated with climatic events such as rainfall.
2   Development of better sampling procedures such as passive samplers for the detection of low levels of recalcitrant chemicals.
3   Characterization of the impact of channelization and other hydraulic modifications on the aquatic biota.
4   Characterization of how activities in one part of the watershed impact the other parts of the watershed.
5   Development of a better understanding of the effects of physical stressors on aquatic communities.
6   Development and refinement of biomarkers of stress for aquatic organisms.
7   Characterization of the assimilative capacity of aquatic communities. (When are aquatic ecosystems processing pollutants and when are they being stressed by pollutants?)
8   Characterization of the link between aquatic ecosystem stress and terrestrial ecosystem impacts.
9   Assessment and refinement of current approaches to biomonitoring at individual, population and community levels.
10   Further development of in-situ aquatic bioassays.

REFERENCES

Anderson, J.W. 1985. *Toxicity of Dispersed and Undispersed Prudhoe Crude Oil Fraction to Shrimp, Fish and their Larvae*. Washington, DC, USA: American Petroleum Institute.
Balsi, O. and Sargent, R.G. 1981. A methodology for cost-risk analysis in the statistical validation of simulation models. *Commun Assoc Comput Mach* (now *Commun ACM*). 24:190–7.

Balsi, O. and Sargent, R.G. 1984. Validation of simulation models via simultaneous confidence intervals. *Am J Math Manage Sci.* 4:375–406.

Brecken-Folse, J., Babikow, M.G. and Duke, T.W. 1993. Comparison of Gulf of Mexico drainage systems: input of toxic chemicals and potential for ecological effects. Draft Manuscript nr M1351. Gulf Breeze, FL, USA: USEPA.

Darnell, R.M. and Soniat, T.M. 1979. The estuary/continental shelf as an interactive system. In: Livingston, R.J. (ed.) *Ecological Processes in Coastal and Marine Systems.* New York, USA: Plenum.

Darnell, R.M., Defenbaugh, R.E. and Moore, D. 1983. Atlas of biological resources of the continental shelf, NW Gulf of Mexico. USDI/BLM Report nr 82-04. New Orleans, LA, USA: Bureau of Land Management (BLM), US Department of the Interior (USDI).

Darnell, R.M. and Kleypas, J.A. 1987. Eastern gulf shelf bio-atlas: a study of the distribution of demersal fishes and penaeid shrimp of soft bottoms of the continental shelf from the Mississippi River Delta to the Florida Keys. Outer Continental Shelf (OCS) Study/Minerals Management Service (MMS) 86-0041. New Orleans, LA, USA: USDI/MMS.

Darnell, R.M. 1988. Marine biology. In: Phillips, N.W. and James, B.M. (eds) *Offshore Texas and Louisiana Marine Ecosystems Data Synthesis,* Volume 2, *MMS Gulf of Mexico OCS Region.* New Orleans, LA, USA: USDI/MMS.

DeAngelis, D.L. 1994. What food web analysis can contribute to wildlife toxicology. In: Kendall, R.J. and Lacher, Jr T.E. (eds) *Wildlife Toxicology and Population Modeling: Integrated Studies of Agroecosystems.* Boca Raton, FL, USA: Lewis, CRC Press, pp. 365–82.

Durako, M.J. 1988. Turtle grass (*Thalassia testudinum* Banks ex König) – a seagrass Section III.9. In: Bajaj, Y.P.S. (ed.) *Biotechnology in Agriculture and Forestry 6 Crops II.* Berlin Heidelberg, Germany: Springer-Verlag, pp. 504–20.

Gosselink, J.G., Cordes, C.L. and Parsons, J.W. 1979. *An Ecological Characterization Study of the Chenier Plain Coastal Ecosystem of Louisiana and Texas,* Volume 1: Narrative report. FWB/OBS-78/9. Slidell, LA, USA: Office of Biological Services (OBS), US Fish and Wildlife Service (USFWS).

Iverson, R.L. and Bittaker, H.F. 1986. Seagrass distribution and abundance in eastern Gulf of Mexico coastal waters. *Estuarine Coastal and Shelf Science.* 22(5):577–602.

Kenworthy, W.J., Thayer, G.W. and Fonseca, M.S. 1988. The utilization of seagrass meadows by fishery organisms. In: Hook, D.D., McKee, Jr W.H., Smith, H.K., Gregory, J., Burrel, Jr V.G., DeVoe, M.R., Sojka, R.E., Gilbert, S., Banks, R., Stolzy, L.H., Brooks, C., Matthews. T.D. and Shear, T.H. (eds) *The Ecology and Management of Wetlands,* Volume 1, *Ecology of Wetlands.* Portland, OR, USA: Timber Press.

LaRoe, E.T., Farris, G.S., Puckett, C.E., Doran, P.D. and Mac, M.J. 1995. Our living resources: a report to the nation on the distribution, abundance, and health of U.S. plants, animals, and ecosystems. Washington, DC, USA: US Department of the Interior, National Biological Service (USDI/NBS).

Lewis, R.R., Durako, M.J., Moffler, M.D. and Phillips, R.C. 1985. Seagrass meadows of Tampa Bay: a review. In: Treat, S.F., Simon, J.L., Lewis, R.R. and Whitman, R.L. (eds) Proc, Tampa Bay Area Sci. Info Symp. Florida Sea Grant College Report 65. Minneapolis, MN, USA: Burgess Publishers, pp. 216–46.

Linton, T.L. 1988. Socioeconomics. In: Phillips, N.W. and James, B.M. (eds) *Offshore Texas and Louisiana Marine Ecosystems Data Synthesis,* Volumes 2 and 3. MMS Gulf of Mexico OCS Region. New Orleans, LA, USA: USDI/MMS.

Louisiana Department of Environmental Quality. 1988. State of Louisiana: Nonpoint source pollution assessment report (Draft of August 30, 1988). Baton Rouge, LA, USA: Louisiana Department of Environmental Quality.

Mager, A. and Ruebsamen, R. 1989. National Marine Fisheries Service habitat conservation efforts in the coastal southeastern United States. *Mar Fish Re.* 50(3):43–50.

Meade, N.F. and Leeworthy, V.R. 1986. Public expenditures on outdoor recreation in coastal areas of the USA. Washington, DC: US Department of Commerce, National Oceanographic and Atmospheric Agency (NOAA), National Ocean Service (NOS), Strategic Assessment Branch, Ocean Assessments Division.

National Marine Fisheries Service (NMFS). 1988a. Fisheries of the United States, 1987: Current fisheries statistics. Report nr 8700. Silver Spring, MD, USA: US Department of Commerce, NMFS.

NMFS. 1988b. Processed fishery products, annual summary, 1986: Current fisheries statistics. Report nr 8386. Silver Spring, MD, USA: US Department of Commerce, NMFS.

NMFS. 1989. Fisheries of the United States, 1988. National Marine Fisheries Service, Fisheries Statistics and Economics Division, Silver Spring, MD, USA: US Department of Commerce, NMFS.

NMFS. 2001. Fisheries of the United States, 2000. National Marine Fisheries Service, Fisheries Statistics and Economics Division, Silver Spring, MD, USA: US Department of Commerce, NMFS. http://www.st.nmfs.gov/.

NOAA. 1985. Gulf of Mexico coastal and ocean zones strategic assessment: Data atlases. Washington, DC, USA: US Department of Commerce, NOAA.

NOAA. 1988a. National estuarine inventory data atlas. Volume 4: Public recreation facilities in coastal areas. Washington, DC, USA: US Department of Commerce, NOAA.

NOAA. 1988b. National status and trends program for marine environmental quality. Progress report: A summary of selected data on chemical contaminants in sediments collected during 1984, 1985, 1986, and 1987. NOAA Technical Memo NOS OMA 44. Washington, DC, USA: US Department of Commerce, NOAA.

NOAA. 1991. The national shellfish register of classified estuarine waters. National Ocean Service. Washington, DC, USA: US Department of Commerce, NOAA/NOS.

Odum, W.E., McIvor, C.C. and Smith, T.J., III. 1982. *The Ecology of the Mangroves of South Florida: A Community Profile.* FWS/OBS-81/24. Washington, DC: USFWS, OBS.

Ott, W., Wallace, L., Mage, D., Akland, G., Lewis, R., Sauls, H., Rodes, C., Kleffman, D., Kuroda, D. and Morehouse, K. 1986. The Environmental Protection Agency's research program on total human exposure. *Environ Int.* 12:475–94.

Pait, A.S., Desouza, A.E. and Farrow, D.R.G. 1992. *Agricultural Pesticide Use in Coastal Areas: A National Summary.* Rockville, MD, USA: US Department of Commerce, NOAA.

Rabalais, N.N., Turner, R.E., Wiseman, W.J. Jr and Boesch, D.E. 1991. A brief summary of hypoxia on the northern Gulf of Mexico continental shelf: 1985–1988. In: Tyson, R.V. and Pearson, T.H. (eds) *Modern and Ancient Continental Shelf Anoxia.* Geological Society Special Publication nr 58. London, UK: The Royal Geological Society.

Roach, E.R. 1987. Wetland trends in coastal Alabama. In: Lowery, T.A. (ed.) Symposium on the natural resources of the Mobile Bay Estuary. Alabama Sea Grant Extension Service, Auburn University. MASGP-87-007. Auburn, AL, USA: Auburn University, pp. 92–101.

Schmied, R.L. and Burgess, E.D. 1987. Marine recreational fisheries in the southeastern United States: an overview. *Mar Fish Re.* 49(2):2–7.

Sports Fishing Institute. 1988. *Economic Impact of Sport Fishing in Louisiana, Mississippi, Alabama, Texas and Florida.* Washington, DC: Sports Fishing Institute.

Solomon, K.R., Baker, D.B., Richards, R.P., Dixon D.R., Klaine, S.J., LaPoint, T.W., Kendall, R.J., Weisskopf, C.P., Giddings, J.M., Giesy, J.P., Hall, L.W. and Williams, W.M. 1996. Ecological risk assessment of atrazine in North American surface waters. *Environ Toxicol Chem* 15(1):31–76.

Stevenson, J.C. 1988. Comparative ecology of submersed grass beds in freshwater, estuarine and marine environments. *Limnol Oceanogr.* 33:867–93.

Summers, J.K., Macauly, J.M., Heitmuller, P.T., Engle, V.D., Adams, A.M. and Brooks, G.T. 1993. Annual statistical summary: EMAP-Estuaries Louisianian province, 1991. USEPA, Environmental Research Laboratory, EPA 620/R-93/007. Gulf Breeze, FL, USA: USEPA.

Suter GW II. 1993. Ecological Risk Assessment. Chelsea, MI, USA: Lewis.

Texas Water Commission. 1986. *The State of Texas Water Quality Inventory*, 8th edn. LP86-07. Austin, TX, USA: Texas Water Commission.

Thayer, G.W., Kenworthy, W.J. and Fonseca, M.S. 1984. *The Ecology of Eelgrass Meadows of the Atlantic Coast: A Community Profile.* FWS/OBS-84-02. Washington, DC: USFWS, OBS.

Thornton, K.W., Saul, G.E. and Hyatt, D.E. 1994. *Environmental Monitoring and Assessment Program Assessment Framework.* EPA/620/R-94/016. Washington, DC: USEPA.

Turner, R.E. and Cahoon, D.R. 1987. *Causes of Wetland Loss in the Coastal Central Gulf of Mexico.* USDI, MMS, Gulf of Mexico OCS Region. OCS Study/MMS 87-0119. New Orleans, LA, USA: USDI, MMS.

USDI. 2000. Endangered and threatened wildlife and plants; 12-month finding for a petition to revise critical habitat for Alabama beach mouse, Perdido Key beach mouse, and Choctawhatchee beach mouse. *Federal Register.* 65(187):57800–2.

USEPA. 1990. Gulf facts. Gulf of Mexico Program, GMP-FS-001. John C Stennis Space Center, MS, USA: USEPA.

USEPA. 1992. Framework for ecological risk assessment. EPA/630/R-92/001. Washington, DC: USEPA.

USEPA. 1999. Ecological condition of estuaries in the Gulf of Mexico. EPA 620-R-98-004. Washington, DC, USA: USEPA.

USEPA. 2000. 1999 toxics release inventory (TRI) public data release report. EPA 260-R-01-001. Washington, DC, USA: USEPA, pp. 2-1–3-27.

Zieman, J.C. 1982. The Ecology of the Seagrasses of South Florida: A Community Profile. FWS/OBS-82-25. Washington, DC, USA: OBS, USFWS.

Chapter 17

Summary

Milton D. Taylor and Stephen J. Klaine

GOVERNMENTAL POLICIES

Existing environmental legislation, regulations, and standards concerning the coastal and marine environments must be enforced. However, there is a need in some countries to strengthen and broaden environmental protection either through creation of enforcement agencies where absent or by developing and expanding the legal framework for protecting coastal and marine environments. The legal framework for protecting the marine environment must be further developed on both the international level and within the legal context of individual countries. Comprehensive, enforceable laws dealing with pollution of surface and ground waters by pesticides and other pollutants are needed. Many of the less developed countries of the tropics have modeled their environmental institutions and laws after those of more industrialized and developed nations; however, those countries still in the process of establishing or revising their environmental protections should learn from the mistakes of others and seek to improve the basic model to fit local conditions, environmental, cultural, and political. As authorities responsible for regulatory control of pesticides become adequately conversant with control systems already in effect in other countries, they will be able to more effectively streamline the guidelines and standards currently in effect in various countries to ensure the future safe use of pesticides.

In those countries where pesticide regulatory or control systems have not yet been implemented, a request for registration of a pesticide product by a manufacturer should be accompanied by a comprehensive data set collected by the manufacturer's research group, which would include a wide range of toxicity data, persistence data, environmental fate and effects data (all of which should be collected under local conditions if economically feasible, or under suitable surrogate conditions, i.e. a similar tropical climate) and details of the nature and sensitivity of analytical techniques used to collect the data. Pesticides should also be re-registered periodically with a review of current data on the pesticide and a requirement that the registering company submit residue data collected under local conditions. Currently, in most countries once registration is given, a pesticide can be marketed forever until restricted or banned by a government agency because of an awakening recognition of its hazards. No pesticide product, or active ingre-dient, should be registered or reregistered without limits placed on its use. Pesticide products should be registered for a specific purpose on a particular crop with guidelines to describe the proper manner of application. Failure to do this can become a recipe for disaster in countries with poorly educated farmers if they use a product indiscriminately. Choosing a pesticide for a particular purpose is a highly skilled task and should not be performed by the uninformed. Thus, all stages of a pesticide's life cycle, e.g. import licenses, registration, storage, packing, labeling, transport, retailing, aerial spraying, and applicator training, must be addressed and regulated.

Many tropical countries need to develop programs to exercise control over residue levels present in food at the time it is offered for sale. Awareness among the general public about pesticide residues and their potential for contaminating the environment is also lacking. Because it is impossible to test all farm produce, a monitoring approach requires the establishment of regulations concerning maximum permissible residue limits that must not be exceeded in marketed food. Exceeding these limits must lead to legal action against the offending farmer or trader and destruction of the condemned produce. There is neither the political will nor sufficient money in most national budgets to police all local markets and imported products. Further, this approach would be meaningless unless the govern-ments concerned establish well-equipped laboratories of international caliber and reputation and staff them with teams of trustworthy analysts and inspectors. The capability of many tropical countries' governmental agencies to generate and analyze data on pesticide contamination in food, feedstuffs, and the environment is limited by a lack of sophisticated instrumentation, the requisite equipment main-tenance funding and staff, and adequate manpower with the expertise needed to run such nationwide programs. Funding for such programs, initiatives, and infra-structure is not available locally; if it is to be done, then the international community, through the United Nations, will have to step forward and provide both leadership and resources.

Environmental impact assessments of proposed large-scale projects must be given due consideration. Government funds for environmental impact studies of pesticides and other toxicants are limited, with many government's highest priority geared toward increasing production in the agriculture and fisheries sectors. Unfortunately, support from international sources is often limited or non-continuing in nature and, therefore, pesticide residue data are few and far between. Often, a lack of coordination and integration of government effort and the funding required for implementation is lacking, especially in the area of environmental management. Some countries, e.g. Vietnam and some parts of the Commonwealth Carribean, are in the initial stages of tourism development. For them, the challenge is to maintain the pristine condition of the ecosystems that attracted the development. They have a unique opportunity, in the planning, design, and construction phases, to develop chemical management and land use strategies to prevent or minimize ecosystem deterioration. Pristine coastal ecosystems can be quickly spoiled by untreated or inadequately treated sewage and wastewater discharges from the rapid expansion of tourism infrastructure, e.g. airports, hotels, marinas, and restaurants, encouraged by increased tourist interest in previously undiscovered areas. If legislation, policies, and regulations are already in place, regions can profit from becoming the 'new' and 'in' destination without harming the very asset that had attracted visitors initially.

The success of pest and pesticide management in tropical countries is directly related to the state of environmental awakening in the general public and a general appreciation for environmental problems developed among policy and decision makers in government and industry and the public. Many countries have established education and training programs to promote industry and public awareness of environmental problems and concerns and sound environmental practices. Citizen education and outreach to pesticide users may also reduce mismanagement and misuse of these chemicals. Other programs are designed to educate tropical countries' farmers and agricultural workers about IPM techniques and management practices that minimize excessive use of pesticides and chemical fertilizers. Similar programs should be adopted by other tropical countries with an effort to learn from others' successes and mistakes. The use of the farmer field school concept for teaching IPM to rice farmers in several of the countries has achieved notable success in most locations where it has been implemented. Its adoption for other crops could significantly reduce pollution from chemical pesticides and fertilizers.

In the tropics lindane, endosulfan, chlordane, and BHC are the remaining OCs in limited to widespread agricultural use. However, they are of primary concern with respect to the aquatic environment. Lindane and endosulfan are in the most widespread use and have proven highly toxic to aquatic life forms. While these compounds may be restricted or banned by more countries in the near future, the implementation of buffer zones in sensitive areas may help to minimize their entry into waterways. In areas where such an approach may not be practical, alternative pesticides with minimal toxic effects to aquatic organisms, while still maintaining field efficacy, should replace those currently being used. There seems to be a general consensus that the pesticide industry must energetically develop new pesticide classes and pesticide varieties with high performance, novel modes of action, low toxicity, and low residues to replace the older pesticide varieties that can cause serious pollution of agricultural ecosystems and leave high residue levels on farm produce. Concurrently, pesticide manufacturing countries must also pursue research, development, and production of biological pesticides, alternative pest control measures including biocontrol, natural predators and pathogens, pest resistant cultivars, and genetic engineering of crops and pursue subsidized use of biological and 'safe' chemical pesticides. This policy would force product structures to tend toward becoming more ecologically friendly.

Finally, it is essential that laboratories that produce data on pesticide interactions with environmental compartments and residues be required to have quality assurance (QA) and control (QC) procedures that meet the standard criteria of ISO-25. Good laboratory practices (GLP) and laboratory standard operating procedures (SOPs) are necessary for reliable and dependable analytical systems and include standardization of facilities for analysis. The reliability of data generated by these laboratories must be assured and internationally accepted.

RESEARCH NEEDS

More extensive studies are needed in many tropical countries on the extent and effect of pesticide pollution with particular emphasis on residues in marine, coastal, and estuarine environments. Although attempts have been made to determine residues in the marine environment and a large amount of data is available, a planned, systematic survey is lacking. Such a study would help in arriving at residue distribution among different compartments of marine ecosystems and elucidating the interactions between compartments. Sufficient data is still needed in many countries to properly develop and manage the marine environment. These studies would measure the level of contaminants, study contaminant accumulation that might lead to biological impacts, and record baseline data on the distribution of flora and fauna on beaches, in coastal lagoons and estuaries, and in other nearshore ecosystems. Additionally, documentation and evaluation of pesticide use in aquaculture and fishing activities is necessary to provide baseline information for managing this agricultural sector. Research is needed to determine the environmental consequences of chemical use in inland and coastal aquaculture and this research should examine pesticide impacts on non-target organisms, chemical fate and movement, effects, accumulation, degradation, and pest resistance development.

There is also a need for studies to assess the impact of inland drainage and land-based pollution sources on coastal lands and waters. Such studies would provide a scientific basis for legislative provisions to establish appropriate abatement and control measures. Additional studies are needed to characterize industrial effluents and to identify the most hazardous pollutants that might require implementation of immediate control measures. Also an estimate of the input rate of pollutants into estuaries and coastal zones from land-based sources, the distribution pathways of pollutants into estuarine and coastal waters, and long-term studies on the biological impact of pollution discharges into the coastal zone would be very useful.

Although much can be learned from studies conducted in temperate countries, there is clearly a need to conduct similar studies to elucidate the movement and fate, distribution, behavior, and bioavailability of pesticides in tropical ecosystems to assess the potential impacts of these chemicals in the tropics. Supervised field trials must be arranged to supplement a manufacturer's data and to ensure that local climatic and environmental factors are accounted for in registration deliberations. Safety in the use of pesticides is a dynamic challenge and locally generated data must cover the formulations in use, use patterns, and cropping systems from the tropics. Ecotoxicological aspects of pesticide use under a given ecological scenario are an essential requirement for safe use of pesticides. Knowledge about the environmental movement and fate, distribution, and bioavailability of pesticides and the development of pest resistance under tropical conditions is essential for understanding the consequences of pesticide use and misuse on tropical aquatic ecosystems, developing environmental impact statements and risk assessments, making prudent pest management decisions, and improving aquatic, estuarine, and coastal management policies. Also, special emphasis should be given to studies examining the potential impact on tropical aquatic ecosystems of repeated and continual low-level exposures to mixtures of pesticides.

Contemporary pesticides, e.g. ametryn, cadusafos, chlorothalonil, cypermethrin, propiconazole, quinclorac, and carbofuran among others, which have been found in aquatic ecosystems should be regarded as priority substances for future studies. A major need will be the development of sensitive methods to monitor pesticide effects on ecosystems. Criteria related to general water quality and specific criteria for tropical aquatic ecosystems must be developed. The concept of acceptable risk levels should be assessed relative to protection of valuable tropical aquatic ecosystems. It will be necessary to develop a 'tropical' definition of water quality, acceptable risk, and methods for environmental evaluation. Studies to develop pesticide reduction strategies especially for the more toxic pesticides are greatly needed.

The toxicity and effects of many of the currently used pesticides and their metabolites to aquatic organisms, especially invertebrates, need to be studied. A database developed from such a study would greatly assist efforts to conduct risk assessments of pesticides. A systematic study with well-defined short and long-range objectives would be of great value in evaluating and sustaining the health of the tropical marine environment.

While agricultural activity is the greatest contributor to pesticide pollution of the environment, there is a significant environmental contribution from the waste discharges of pesticide manufacturers. Characterization of these waste products is important for understanding their effects on ecosystems.

Considering the limitations of acute toxicity data, information on pesticide residue effects on ecosystems is essential to properly assess impacts. To achieve this, micro- and mesocosm studies may serve as a bridge between simple LC_{50} data and comprehensive ecosystem assessments. Ultimately field validation will be necessary to match predictions derived from laboratory, micro-, and mesocosm tests to observations of the responses of complex tropical ecosystems.

There is clearly an increasing need to develop and adopt IPM strategies for other tropical crops besides rice and the few major crops for which the information is available. This would require extensive research in various approaches to pest control in the specific tropical agroecosystems, including the introduction of multi-pest resistant cultivars, biological control methods, natural predators and pathogens, and effective training of farmers in the implementation of IPM strategies and techniques.

EDUCATION NEEDS AND INITIATIVES

Education and training programs, including workshops, seminars, pilot demon-stration parcels with farmers, extension training projects, and public service announcement campaigns, should be established to educate the general public about the environment in general, environmental problems, chemical and non-chemical pollutants and contaminants, and sound environmental practices. These nationwide campaigns are imperative and must include school children to provide continuity for the program. Furthermore, agricultural management practices that minimize excessive use of pesticides and chemical fertilizers should be developed and instituted throughout the tropics. There is a widespread need for developing and implementing training programs for pesticide handlers and applicators and their families. A major concern for the continued success and expansion of IPM in the tropics is the belief by farmers that pesticides are the only viable solution to their farming problems. For historical reasons, many tropical farmers' understanding of the scientific basis for the use of pesticides is incomplete and their concept of environmental protection is minimal. Because farmers directly use pesticides, it is very important to increase their knowledge of the reasons behind protecting the environment and minimizing pesticide use. Moreover, it is also essential to conduct technical and environmental awareness training for policy-makers at different political levels in addition to training the technicians and workers involved in pesticide production and application.

Application of pesticides in the field is the predominant cause for their conta-mination of the environment. Thus, an extensive program of public education to fuel public awareness on the proper uses of pesticides needs to be instituted where lacking and continued, improved, and reinforced where in place to minimize the indiscriminate and irresponsible use of pesticides. While countries readily accept the responsibility of promulgating relevant pesticide regulations, they must also assume the task of educating their people on safe pesticide use and establish an efficient means of supervising pesticide use to safeguard people's health. The side effects of pesticides caused by poor production and poor application techniques may include serious pollution and other environmental problems in addition to their toxic effects on wildlife and human beings. Continuous use of single pesticides leads to rapid resistance development in pests and, ultimately, to failure of the pesticide from pest resistance. Manufacturers and farmers seldom investigate the causes of such failures. Farmers often blindly increase the concentration of the pesticide or its frequency of use, further inducing resistance by pests and polluting the environment. A nationwide information dissemination and training program, especially for farmers and aquaculturists, on the development of resistance and the effects of pesticides in the environment can go a long way toward alleviating this problem.

Finally, the training of more environmental scientists, engineers, and managers, both locally and overseas, should be instituted throughout the tropics with adequate long-term funding and support from the international community.

MONITORING NEEDS

While the widespread use of pesticides continues, there is a need for extensive monitoring of their residues in the environment. Thus, where absent, effective monitoring programs must be established to monitor estuarine and marine pesticide levels, oil pollution, industrial and sewage pollution, and to encourage research on pesticide residues and their effects on aquatic organisms. Additionally, pollution monitoring of beaches and coastal waters should be instituted and bacteriological quality control of bathing waters should begin. Such monitoring programs must be supported by the necessary regulatory capacities, coupled with effective enforcement mechanisms to prevent contamination levels from exceeding locally established limits as stipulated by the appropriate legislatures.

Throughout these studies, the impact of pesticide residues contributed from agricultural activities could be discerned if river mouths of major rivers passing through the agricultural fields' drainage areas were monitored. As the capability of monitoring contaminants is strengthened, the environmental and social costs of inland and coastal aquaculture, rice fish culture, agriculture, and manufacturing can be assessed. Then, pesticides and other contaminants can be evaluated not only for localized effects but in the context of nationwide risk assessments. This will provide data for new legislative initiatives to protect the marine environment from unnecessary risks from pesticide use.

PROSPECTS

Tropical agriculture, like that in other countries, faces certain common problems. There is a continuous reduction in the amount of useful agricultural land due to growing cities, recreational areas, industrialization, and park preserves. Furthermore, every year valuable soils are lost to erosion from the deforestation that began more than 100 years ago but continues today. An increasing population, which demands higher quantities and better qualities of agroproducts, and a need for expanded agricultural exports to contribute to national economic development pose a tremendous challenge to tropical agriculture. It must produce more and better crops using less land and this necessarily implies an increase in productivity, achievable only through correct and timely application of science-based agricultural knowledge, including the prudent use of pesticides.

Despite worldwide efforts to find substitutes for agrochemicals, world food production will depend on the use of these chemicals for the foreseeable future. To relieve projected food shortages, Africa and Latin America are expected to increase their use of agrochemicals. Therefore, increased monitoring of pesticide residues will be necessary to preserve the environment and contribute to sustainable agriculture. This will require both an investment in scientific, regulatory, and enforcement infrastructure and the periodic upgrading of the knowledge and skill levels of environmental and agricultural scientists. More support for basic and applied research will be needed if enforcement of environmental regulations is to become easier so as to achieve the end result of better protection for the tropic's unique and diverse environment.

CONCLUSIONS AND RECOMMENDATIONS

Measurements of OCs in marine waters from the Arabian Sea showed low residue levels. Aldrin, HCHs, and DDTs were the most abundant and most commonly found pesticides in the Arabian Sea. Concentrations of DDT in the zooplankton showed a decreasing gradient from the near-shore to offshore. However, sediments from the Bay of Bengal contained an order of magnitude higher residues than sediments from the Arabian Sea. This was attributed to residues carried by the major rivers, which primarily flow east through heavily agricultural lands located there. Green mussels collected from the East Coast of the Indian Subcontinent had high levels of HCH while West Coast samples had high levels of DDT. This pattern is indicative of the different pesticide usage patterns for agriculture (HCH) and public health purposes (DDT). It appears that, in general, pesticide residues are low in the Indian Subcontinent marine environment, possibly due to the impact of semi-diurnal tides coupled with the influence of the biannual reversal of the direction of monsoon winds that ensures widespread dispersion of pollutants throughout the Arabian Sea, the Bay of Bengal, and the northern arms of the Indian Ocean.

Assessment of the fisheries sector in Asia has identified resource depletion, environmental damage, poverty among fisherfolk, low productivity, and limited utilization of offshore waters by commercial fishermen as major problems for the industry. Over-fishing and habitat degeneration has resulted in no substantial increase in fish capture in near-shore areas of some countries. However, many countries expect an increment of increased fish production to come from aquaculture, but aquaculture's use of chemicals may result in excessive environmental costs.

Monoculture-type agriculture for producing banana, coffee, sugarcane, rice, ornamentals, and fruits is one reason for the intense and predominant use of pesticides in the Caribbean, Central, and South America. This method of farming depends heavily on agrochemical use, which has many negative consequences. These include pest resistance development, soil deterioration, aquatic ecosystem degradation, the emergence and proliferation of secondary pests, adverse health effects on the general and agricultural labor populations, and various other environmental effects from the exposure of wildlife to pesticides residues.

1   In general, few large-scale effects have been documented given the ubiquitous OC pesticide residues in these tropical ecosystems. One caveat to this is that endocrine effects have not been investigated.

2 Toxicological investigations in these countries focus on the individual organism level and little attention has been paid to other organization levels including molecular, biochemical, population, community, and ecosystem levels.

3 Pesticide residues in sediments can be hydrologically connected to upstream land management practices.

4 International aid efforts must have education and outreach as integral components of their efforts.

5 IPM, involving a combination of chemical, biological, and cultural methods of pest control, has proven a realistic and viable means of decreasing the negative impacts of the excessive use of pesticides throughout the tropics.

6 Farmer and farm worker education is critical to the successful implementation of IPM programs.

7 Education should not be confined to the agricultural community but should embrace all sectors of society including school children and political leaders.

8 It is obvious that additional multi-nation monitoring programs are needed to document changes from the present pesticide residue levels. New chemistries and co-operation between regional nations

9 Continued influx of resources for pesticide monitoring must come from both internal and international government agencies.

10 The concept of sustainable resource management must be in the forefront of decision-making.

11 Better cause and effect relationships between land use and the deterioration of near coastal resources must be developed.

12 It is important to consider the environmental problems associated with pesticide use in association with the related agricultural, economic, political, and public health issues.

13 Best management strategies currently used in temperate climates must be successfully implemented and evaluated – with modification as needed – in tropical land use and development.

A FEW FINAL REMARKS

Both governmental and non-governmental institutions must work together with farmers, farmers' associations, and other players in the agricultural and food marketing and distribution sectors to conduct research and facilitate technology transfers with a goal of more rational and sustainable agricultural practices. The use of integrated pest management (IPM) programs and organic farming movements in a number of countries are excellent examples of positive movement toward sustainable agriculture.

Effective schemes for minimizing some of the risks associated with the use of pesticides already exist in countries such as the USA and the European Community. Agencies of the United Nations are extending cooperation, collaboration, and expert guidance to developing countries around the world in devising practical steps for the control of pesticides. They are assisting in maximizing pesticides' beneficial role while minimizing risks associated with undesirable levels of residues in abiotic compartments, biota, food chains, and foodstuffs and reducing untoward effects on non-target organisms in the environment.

Perceived personal benefits should not be the overriding factor in the decision to apply pesticides because every time they are used, a certain risk is involved especially if the usage is not judicious. People from many countries should begin to reorient their concept of the environment so that it is not limited to the house, the yard, the place of work, and the immediate community but focuses on the national and global scale. The increased productivity of countries' resources must proceed hand-in-hand with the conservation and preservation of those resources for future generations. The idea of sustainable development, natural capital, and the responsibility of the current generation to preserve its resources for the future should be ingrained in every citizen of the world.

Appendix

Listing of all pesticides mentioned, identification numbers and activities

Pesticide name(s)	Identification number	Activity
2,4-D	CAS Nr 94-75-7	phenoxyacetic herbicide; auxin
2,4-DB	CAS Nr 94-82-6	phenoxybutyric herbicide; auxin
2,4,5-T	CAS Nr 93-76-5	phenoxyacetic herbicide; auxin
2,4,5-TP, Silvex, fenoprop	CAS Nr 93-72-1	phenoxypropionic herbicide; auxin
Abate, temephos	CAS Nr 3383-86-8	phenyl thioOP insecticide
acephate	CAS Nr 30560-19-1	phosphoramidothioate OP insecticide
acetamiprid	CAS Nr 135410-20-7	pyridine insecticide
acifluorfen-sodium	CAS Nr 62476-59-9	nitrophenyl ether herbicide
Agrocide, hexachlorocyclohexane (HCH), benzene hexachloride (BHC)	CAS Nr 608-73-1	OC rodenticide and insecticide
Agroxone, MCPA	CAS Nr 94-74-6	phenoxyacetic herbicide
alachlor	CAS Nr 15972-60-8	chloroacetanilide herbicide
alar, daminozide	CAS Nr 1596-84-5	plant growth retardant
aldicarb	CAS Nr 116-06-3	oxime carbamate insecticide, acaricide, and nematicide
aldrin	CAS Nr 309-00-2	cyclodiene insecticide
Aliette, fosetyl, fosetyl-aluminium	CAS Nr 15845-66-6	OP fungicide
aluminum phosphide, gastoxin	CAS Nr 20859-73-8	fumigant insecticide
ametryn	CAS Nr 834-12-8	methylthiotriazine herbicide
amitraz	CAS Nr 33089-61-1	formamidine acaricide and insecticide
amobam	CAS Nr 3566-10-7	thiocarbamate fungicide
anilofos	CAS Nr 64249-01-0	OP herbicide
aramite	CAS Nr 140-57-8	chlorosulfite acaricide
arsenous oxide, arsenic trioxide	CAS Nr 1327-53-3	arsenic herbicide, rodenticide, and insecticide
atrazine	CAS Nr 1912-24-9	chlorotriazine herbicide
azadirachtin	CAS Nr 11141-17-6	unclassified insect growth regulator
azamethiphos	CAS Nr 35575-96-3	organothiophosphate (thioOP) insecticide
azinphos ethyl	CAS Nr 2642-71-9	thioOP acaricide; benzotriazine thioOP insecticide
Bacillus thuringiensis, Bt	Merek Index Nr 945	antifeedant insecticide
bellater	atrazine + cyanazine	chlorotriazine herbicides
benfuracarb	CAS Nr 82560-54-1	methylcarbamate insecticide

Pesticide name(s)	Identification number	Activity
benomyl	CAS Nr 17804-35-2	carbamate acaricide and nematicide; benzimidazolylcarbamate fungicide
bensulfuron	CAS Nr 99283-01-9	sulfonylurea herbicide
bentazone	CAS Nr 25057-89-0	unclassified herbicide
benthiocarb, thiobencarb(e)	CAS Nr 28249-77-6	thiocarbamate herbicide
BHC, Agrocide, HCH	CAS Nr 608-73-1	OC insecticide and rodenticide
blasticidin-S	CAS Nr 2079-00-7	antibiotic fungicide
borax, sodium tetraborate decahydrate, boric acid	CAS Nr 1303-96-4	insecticide
BPMC, fenobucarb	CAS Nr 3766-81-2	phenyl methylcarbamate insecticide
brodifacoum	CAS Nr 56073-10-0	coumarin rodenticide
bromacil	CAS Nr 314-40-9	uracil herbicide
bromadiolone	CAS Nr 28772-56-7	coumarin rodenticide
bromophos	CAS Nr 2104-96-3	thioOP acaricide; phenyl thioOP insecticide
bromopropylate	CAS Nr 18181-80-1	bridged biphenyl acaricide
buprofezin	CAS Nr 69327-76-0	chitin synthesis inhibitor insecticide
butachlor	CAS Nr 23184-66-9	chloroacetanilide herbicide
caacobre	not available	copper (II) oxide
cadusafos	CAS Nr 95465-99-9	thioOP insecticide and nematicide
calcium cyanide	CAS Nr 592-01-8	fumigant insecticide and rodenticide
callifan 50CE, endosulfan	CAS Nr 115-29-7	OC acaricide; cyclodiene insecticide
camphechlor, polychlorocamphene, toxaphene	CAS Nr 8001-35-2	OC acaricide and insecticide
captafol	CAS Nr 2425-06-1	dicarboximide fungicide
captan	CAS Nr 133-06-2	dicarboximide fungicide
carbaryl	CAS Nr 63-25-2	carbamate acaricide and insecticide; plant growth inhibitor
carbendazim	CAS Nr 10605-21-7	benzimidazolylcarbamate fungicide
carbetamide	CAS Nr 16118-49-3	carbanilate herbicide
carbofuran, furadan	CAS Nr 1563-66-2	carbamate acaricide and nematicide; benzofuranyl methylcarbamate insecticide
carbophenothion	CAS Nr 786-19-6	thioOP acaricide and phenyl thioOP insecticide
carbosulfan	CAS Nr 55285-14-8	benzofuranyl methylcarbamate insecticide; carbamate nematicide
carboxin	CAS Nr 5234-68-4	oxathiin fungicide
cartap	CAS Nr 15263-53-3	nereistoxin analogue insecticide
Champion, copper hydroxide, Kocide	CAS Nr 20427-59-2	copper fungicide
chinomethionat(e), oxythioquinox	CAS Nr 2439-01-2	quinoxaline acaricide and fungicide
chlorbenside	CAS Nr 103-17-3	bridged diphenyl acaricide
chlordane	CAS Nr 57-74-9	OC insecticide
chlordecone	CAS Nr 143-50-0	cyclodiene insecticide

Pesticide name(s)	Identification number	Activity
chlordimeform	CAS Nr 6164-98-3	formamidine acaricide; formamidine antifeedant insecticide
chlorfenson	CAS Nr 80-33-1	bridged diphenyl acaricide
chlorfenvinphos	CAS Nr 470-90-6	OP acaricide and insecticide
chlormequat	CAS Nr 703-89-6	plant growth regulator (retardant)
chlorobenzilate	CAS Nr 510-15-6	bridged diphenyl acaricide
chlorofos, trichlorfon	CAS Nr 52-68-6	phosphonate acaricide and insecticide
chloroneb	CAS Nr 2675-77-6	aromatic fungicide
chloropicrin	CAS Nr 76-06-2	unclassified fungicide, insecticide, and nematicide
chlorothalonil	CAS Nr 1897-45-6	aromatic fungicide
chlorpyrifos, diazinon	CAS Nr 2921-88-2	thioOP acaricide and nematicide; pyridine thioOP insecticide
chlorthal-dimethyl	CAS Nr 1861-32-1	benzoic acid herbicide
Chuchongjuzhu, pyrethrin	Merek Index Nr 7978	botanical insecticide
cianoga	not available	formicide
Compound 1080, sodium monofluor acetate, sodium fluoroacetate	CAS Nr 62-74-8	unclassified rodenticide
copper hydroxide, Champion, Kocide	CAS Nr 20427-59-2	copper fungicide
copper oxychloride, basic cupric chloride	CAS Nr 1332-40-7	bird repellant; copper fungicide
coumaphos	Merek Index Nr 2559	thioOP acaricide and heterocyclic thioOP insecticide
coumatetralyl	CAS Nr 5836-29-3	coumarin rodenticide
cufraneb	CAS Nr 11096-18-7	copper fungicide; dithiocarbamate fungicide
cyanamide	Merek Index Nr 2691	amidocyanogen
cyanazine	CAS Nr 21725-46-2	chlorotriazine herbicide
cyfluthrin	CAS Nr 68359-37-5	pyrethroid ester insecticide
cyhalofop butyl	CAS Nr 122008-85-9	aryloxyphenoxypropionic herbicide
cyhalothrin-lambda, Grenade, Karate	CAS Nr 91465-08-6	pyrethroid ester acaricide and insecticide
cyhexatin	CAS Nr 13121-70-5	organotin acaricide
cypermethrin	CAS Nr 52315-07-8	pyrethroid ester acaricide and insecticide
cyproconazole	CAS Nr 94361-06-5	conazole fungicide
dalapon	CAS Nr 75-99-0	halogenated aliphatic herbicide
daminozide, alar	CAS Nr 1596-84-5	plant growth retardant
dazomet	CAS Nr 533-74-4	unclassified herbicide, fungicide, and nematicide
DBCP (dibromochloropropane), Nemagon	CAS Nr 96-12-8	soil fumigant nematicide
DD	1,2-dichloropropane + 1,3-dichloropropene	fumigant insecticide; unclassified nematicide + soil fumigant nematicide
DDD	CAS Nr 72-54-8	OC insecticide
DDT	CAS Nr 50-29-3	OC acaricide and insecticide
declorane, declorano, mirex	CAS Nr 2385-85-5	cyclodiene insecticide
deltamethrin	CAS Nr 52918-63-5	pyrethroid ester insecticide
demeton	CAS Nr 8065-48-3	thioOP acaricide and insecticide
desmetryn	CAS Nr 1014-69-3	methylthiotriazine herbicide
di-allate	CAS Nr 2303-16-4	thiocarbamate herbicide

Pesticide name(s)	Identification number	Activity
diazinon	CAS Nr 333-41-5	thioOP acaricide; pyridine thioOP insecticide
diazoben, fenaminosulf	CAS Nr 140-56-7	bactericide; unclassified fungicide
difenzoquat	CAS Nr 49866-87-7	quaternary ammonium herbicide
diphacinone	CAS Nr 82-66-6	unclassified rodenticide
dicamba	CAS Nr 1918-00-9	benzoic acid herbicide; unclassified plant growth regulator
dichlorvos, DDVP	CAS Nr 62-73-7	OP acaricide and insecticide
dicofol	CAS Nr 115-32-2	bridged diphenyl acaricide
dicrotophos	CAS Nr 141-66-2	OP insecticide
dieldrin	CAS Nr 60-57-1	cyclodiene insecticide
diethylthiophosphate, DETP	CAS Nr 5871-17-0 (K$^+$ salt)	thioOP insecticide
diflubenzuron	CAS Nr 35367-38-5	chemosterilant; chitin synthesis inhibitor insecticide
Dimecron, phosphamidon	CAS Nr 13171-21-6	OP insecticide and nematicide
dimelon	not available	not available
dimethoate	CAS Nr 60-51-5	thioOP acaricide and nematicide; aliphatic amide thioOP insecticide
dinitrocresol, DNC, DNOC	CAS Nr 534-52-1	dinitrophenol herbicide; ovicidal insecticide, fungicide, and acaricide
dinocap	CAS Nr 39300-45-3	dinitrophenol acaricide and fungicide
dinoseb	CAS Nr 88-85-7	dinitrophenol herbicide
dioxathion	CAS Nr 78-34-2	thioOP acaricide; heterocyclic thioOP insecticide
diphacinone	CAS Nr 82-66-6	indandione rodenticide
diquat	CAS Nr 2764-72-9	quaternary ammonium herbicide
diuron	CAS Nr 330-54-1	phenylurea herbicide
DMAH, dimethylaluminumhydride	CAS Nr 865-37-2	formamidine insecticide
DSMA	CAS Nr 144-21-8	arsenical herbicide
EDB, ethylene dibromide	CAS Nr 106-93-4	fumigant insecticide
edifenphos, hinosan	CAS Nr 17109-49-8	OP fungicide
Ekalux, quinalphos	CAS Nr 13593-03-8	thioOP acaricide; quinoxaline thioOP insecticide
endosulfan, callifan 50CE	CAS Nr 115-29-7	OC acaricide; cyclodiene insecticide
endrin	CAS Nr 72-20-8	avicide; cyclodiene insecticide
EPTC, Eptam	CAS Nr 759-94-4	thiocarbamate herbicide
esfenvalerate	CAS Nr 56230-04-4	pyrethroid ester insecticide
ethachlor	not available	not available
ethephon	CAS Nr 16672-87-0	plant growth regulator; ethylene releasing defoliant
ethion	CAS Nr 563-12-2	thioOP acaricide; aliphatic thioOP insecticide
ethoprop, ethoprophos	CAS Nr 13194-48-4	aliphatic thioOP insecticide; thioOP nematicide
ethylene dibromide, EDB	CAS Nr 106-93-4	fumigant insecticide
ethylene oxide	CAS Nr 75-21-8	fumigant insecticide
etofenprox	CAS Nr 80844-07-1	pyrethroid ester insecticide
fenaminosulf, diazoben	CAS Nr 140-56-7	bactericide; unclassified fungicide
fenamiphos	CAS Nr 22224-92-6	phosphoramidate insecticide; OP nematicide
fenchlorphos	CAS Nr 299-84-3	phenyl thioOP insecticide

Pesticide name(s)	Identification number	Activity
fenbuconazole	CAS Nr 114369-43-6	conazole fungicide
fenbutatin oxide, hexakis	CAS Nr 13356-08-6	organotin acaricide
fenitrothion, sumithion	CAS Nr 122-14-5	phenyl thioOP insecticide
fenobucarb, BPMC	CAS Nr 3766-81-2	phenyl methylcarbamate insecticide
fenom C (Novartis)	profenophos + cypermethrin	phenyl thioOP + pyrethroid ester insecticide
fenoprop, 2,4,5-TP, Silvex	CAS Nr 93-72-1	phenoxypropionic herbicide; auxin plant growth regulator
fenoxaprop	CAS Nr 73519-55-8	aryloxyphenoxypropionic herbicide
fenpropathrin	CAS Nr 39515-41-8	pyrethroid ester acaricide and insecticide
fentin chloride	CAS Nr 639-58-7	organotin fungicide and molluscicide; antifeedant
fentin acetate	CAS Nr 900-95-8	organotin fungicide and molluscicide; antifeedant
fenthion	CAS Nr 55-38-9	avicide; phenyl thioOP insecticide
fenvalerate	CAS Nr 51630-58-1	pyrethroid ester acaricide and insecticide
ferbam	CAS Nr 14484-64-1	dithiocarbamate fungicide
flocoumafen	CAS Nr 90035-08-8	coumarin rodenticide
fluazifop-butyl	CAS Nr 69806-50-4	aryloxyphenoxypropionic herbicide
flumethrin	CAS Nr 69770-45-2	pyrethroid ester acaricide
fluometuron	CAS Nr 2164-17-2	phenylurea herbicide
fluoroacetamide, fussol	CAS Nr 640-19-7	unclassified rodenticide
fluroxypyr methyl heptyl ester	CAS Nr 69377-81-7	pyridyloxyacetic acid herbicide
flutolanil	CAS Nr 66332-96-5	benzanilide fungicide
folpet, phaltan	CAS Nr 133-07-3	dicarboximide fungicide
fonofos	CAS Nr 944-22-9	phenyl ethylphosphonothioate insecticide
formothion	CAS Nr 2540-82-1	thioOP acaricide; aliphatic amide thioOP insecticide
fosetyl, Aliette, fosetyl-aluminium (fosetyl-aluminium)	CAS Nr 15845-66-6 (CAS Nr 39148-24-8)	OP fungicide
foxim, phoxim	CAS Nr 14816-18-3	thioOP acaricide; oxime thioOP insecticide
gastoxin, aluminum phosphide	CAS Nr 20859-73-8	fumigant insecticide
Gesatop, simazine	CAS Nr 122-34-9	chlorotriazine herbicide
gibberellin	gibberellins	plant growth regulators
gliftor, DFP	1,3-difluoro-2-propanol	rodenticide
glufosinate, glufosinate-ammonium	CAS Nr 51276-47-2	OP herbicide
glyphosate	CAS Nr 1071-83-6	OP herbicide
gramaxone, paraquat	CAS Nr 4685-14-7	quaternary ammonium herbicide
Grofol 20-30-10	not available	foliar NPK + micronutrients
HCB, hexachlorobenzene	CAS Nr 118-74-1	aromatic fungicide
HCH, Agrocide, BHC	CAS Nr 608-73-1	OC rodenticide; OC insecticide
heptachlor	CAS Nr 76-44-8	cyclodiene insecticide
HETP, tetraethyl pyrophosphate (TEPP)	CAS Nr 107-49-3	OP acaricide and insecticide
hexaconazole ·	CAS Nr 79983-71-4	conazole fungicide
hinosan, edifenphos	CAS Nr 17109-49-8	OP fungicide
hostathion, triazophos	CAS Nr 24017-47-8	thioOP acaricide and nematicide; triazole thioOP insecticide
hydroprene	CAS Nr 41096-46-2	juvenile hormone analogue
imazalil	CAS Nr 35554-44-0	conazole fungicide

Pesticide name(s)	Identification number	Activity
imazapyr	CAS Nr 81334-34-1	imidazolinone herbicide
imidacloprid, marathon	CAS Nr 138261-41-3	pyridine insecticide
inacide (indomethacin)	CAS Nr 53-86-1	not available
iprobenfos, IBP, Kitazin-p	CAS Nr 26087-47-8	OP fungicide; synergist
iprodione	CAS Nr 36734-19-7	dicarboximide or imidazole fungicide
isazofos	CAS Nr 42509-80-8	triazole thioOP insecticide; thioOP nematicide
isodrin	CAS Nr 465-73-6	cyclodiene insecticide
isofenphos	CAS Nr 25311-71-1	phosphoramidothioate insecticide
isoprocarb, MIPC	CAS Nr 2631-40-5	phenyl methylcarbamate insecticide
isoprothiolane	CAS Nr 50512-35-1	unclassified fungicide and insecticide
isoproturon	CAS Nr 34123-59-6	phenylurea herbicide
Jiamijuzhi (methrothrin)	not available	pyrethroid pesticide
Jianganmycin, Jingan meisu	not available	biological fungicide
kasugamycin	CAS Nr 6980-18-3	bactericidal and fungicidal metabolite of *Streptomyces kasugaensis*; a microorganism protein biosynthesis inhibitor
Kocide, copper hydroxide, Champion	CAS Nr 20427-59-2	copper fungicide
Kuliansu (tooosederin)	not available	biological pesticide
Kusen (materine)	not available	biological pesticide
lead arsenate	CAS Nr 3687-31-8	inorganic insecticide
lead arsenite	CAS Nr 10031-13-7	inorganic insecticide
leptophos	CAS Nr 21609-90-5	phenyl phenylphosphonothioate insecticide
lindane, γ-HCH	CAS Nr 58-89-9	OC rodenticide, acaricide and insecticide
linuron	CAS Nr 330-55-2	phenylurea herbicide
macbal, XMC	CAS Nr 2655-14-3	phenyl methylcarbamate insecticide
MAFA	ferrous salt of methyl ammonium arsonic acid CH_5AsO_3	arsenical herbicide
magnesium phosphide (releases phosphine)	CAS Nr 12057-74-8 CAS Nr 7803-51-2	fumigant insecticide
malathion	CAS Nr 121-75-5	thioOP acaricide; aliphatic thioOP insecticide
mancozeb	CAS Nr 8018-01-7	dithiocarbamate fungicide
maneb	CAS Nr 12427-38-2	dithiocarbamate fungicide
marathon, imidacloprid	CAS Nr 138261-41-3	pyridine insecticide
MCPA, Agroxone	CAS Nr 94-74-6	phenoxyacetic herbicide
mebenil	CAS Nr 7055-03-0	benzanilide fungicide
menazon	CAS Nr 78-57-9	heterocyclic thioOP insecticide
metalaxyl	CAS Nr 57837-19-1	xylylalanine fungicide
metaldehyde	CAS Nr 108-62-3	molluscicide
methacrifos	CAS Nr 62610-77-9	thioOP acaricide; aliphatic thioOP insecticide
methamidophos	CAS Nr 10265-92-6	phosphoramidothioate acaricide andinsecticide
methidathion	CAS Nr 950-37-8	thiadiazole thioOP insecticide
methazole	CAS Nr 20354-26-1	unclassified herbicide
methomyl	CAS Nr 16752-77-5	oxime carbamate insecticide
methoxychlor	CAS Nr 72-43-5	OC insecticide

Pesticide name(s)	Identification number	Activity
methyl bromide	CAS Nr 78-57-9	unclassified fungicide and nematicide; halogenated aliphatic herbicide; fumigant insecticide
methyl parathion	CAS Nr 74-83-9	phenyl thioOP insecticide
methyl isothiocyanate	CAS Nr 556-61-6	unclassified fungicide, herbicide, and nematicide
metiram, Polyram	CAS Nr 9006-42-2	dithiocarbamate fungicide
metolachlor, primagram	CAS Nr 51218-45-2	chloroacetanilide herbicide
metolcarb, MTMC, tsumacide	CAS Nr 1129-41-5	carbamate acaricide; phenyl methylcarbamate insecticide
metsulfuron methyl	CAS Nr 74223-64-6	sulfonylurea herbicide
mirex, declorane, declorano	CAS Nr 2385-85-5	cyclodiene insecticide
molinate	CAS Nr 2212-67-1	thiocarbamate herbicide
monocrotophos	CAS Nr 6923-22-4	OP acaricide and insecticide
monorun	CAS Nr 150-65-5	phenylurea herbicide
MSMA, monosodium methanearsonate acid	CAS Nr 2163-80-6	arsenical herbicide
MTMC, metolcarb, tsumacide	CAS Nr 1129-41-5	carbamate acaricide; phenyl methylcarbamate insecticide
NAA, α-naphthaleneacetic acid	CAS Nr 86-87-3	auxin plant growth regulator
n-decanol, n-decyl alcohol	CAS Nr 13171-21-6	herbicide
Nemagon, DBCP	CAS Nr 96-12-8	soil fumigant nematicide
niclosamide	CAS Nr 50-65-7	molluscicide
nicotine, Yanjian	CAS Nr 54-11-5	botanical insecticide
nitrofen	CAS Nr 1836-75-5	nitrophenyl ether herbicide
Ofunack, pyridiphenthion	CAS Nr 119-12-0	heterocyclic thioOP insecticide
omethoate	CAS Nr 1113-02-6	thioOP acaricide; aliphatic amide thioOP insecticide
oxadiazon	CAS Nr 19666-30-9	unclassified herbicide
oxamyl	CAS Nr 23135-22-0	oxime carbamate acaricide, insecticide, and nematicide
oxydemeton methyl, demeton-methyl	CAS Nr 301-12-2	aliphatic thioOP insecticide
oxyfluorfen	CAS Nr 42874-03-3	nitrophenyl ether herbicide
oxythioquinox, chinomethionat(e)	CAS Nr 2439-01-2	quinoxaline acaricide and fungicide
paraquat, gramaxone	CAS Nr 4685-14-7	quaternary ammonium herbicide
parathion	CAS Nr 56-38-2	thioOP acaricide and phenyl thioOP insecticide
parathion-methyl	CAS Nr 298-00-0	thioOP acaricide and phenyl thioOP insecticide
pendimethalin	CAS Nr 40487-42-1	dinitroaniline herbicide
pentachlorophenol, PCP	CAS Nr 87-86-5	aromatic fungicide; unclassified herbicide; OC insecticide; molluscicide; plant defoliant
permethrin	CAS Nr 52645-53-1	pyrethroid ester acaricide and insecticide
phaltan, folpet	CAS Nr 133-07-3	dicarboximide fungicide
phenazine	CAS Nr 92-82-0	fungicide
phenothrin, Sumithrin	CAS Nr 26002-80-2	pyrethroid ester insecticide
phenoxyacetic acid	CAS Nr 122-59-8	fungicide
phenthoate	CAS Nr 2597-03-7	thioOP insecticide
phorate	CAS Nr 298-02-2	thioOP acaricide and nematicide; aliphatic thioOP insecticide
phosalone	CAS Nr 2310-17-0	thioOP acaricide; heterocyclic thioOP insecticide

Pesticide name(s)	Identification number	Activity
phosdrin (a.i. is mevinphos CAS Nr 26718-65-0)	CAS Nr 7786-34-7	OP acaricide and insecticide
phosmet	CAS Nr 732-11-6	thioOP acaricide; isoindole thioOP insecticide
phosphamidon, Dimecron	CAS Nr 13171-21-6	OP insecticide and nematicide
phoxim, foxim	CAS Nr 14816-18-3	thioOP acaricide; oxime thioOP insecticide
picloram	CAS Nr 1918-02-1	picolinic acid (pyridine) herbicide
piperophos	CAS Nr 24151-93-7	OP herbicide
pirimicarb	CAS Nr 23103-98-2	dimethylcarbamate insecticide
pirimiphos methyl	CAS Nr 29232-93-7	thioOP acaricide; pyrimidine thioOP insecticide
pirimiphos-ethyl	CAS Nr 23505-41-1	pyrimidine thioOP insecticide
polychlorocamphene, camphechlor, toxaphene	CAS Nr 8001-35-2	OC acaricide and insecticide
polythrion	not available	
pracol	not available	ampicillin trihydrateBantibacterial
primigram, metolachlor	CAS Nr 51218-45-2	chloroacetanilide herbicide
profenofos	CAS Nr 41198-08-7	phenyl thioOP insecticide
prometryn	CAS Nr 7287-19-6	methylthiotriazine herbicide
propachlor	CAS Nr 1918-16-7	chloroacetanilide herbicide
propanil	CAS Nr 709-98-8	anilide herbicide
propargite	CAS Nr 2312-35-8	sulfite ester acaricide
propiconazole	CAS Nr 60207-90-1	conazole fungicide
propineb	CAS Nr 9016-72-2	dithiocarbamate fungicide
propoxur	CAS Nr 114-26-1	carbamate acaricide; phenyl methylcarbamate insecticide
pyrethrin, Chuchongjuzhu	Merek Index Nr 7978	botanical insecticides
pyridiphenthion, Ofunack	CAS Nr 119-12-0	heterocyclic thioOP insecticide
pyroquilon	CAS Nr 57369-32-1	unclassified fungicide
quinalphos, Ekalux	CAS Nr 13593-03-8	thioOP acaricide; quinoxaline thioOP insecticide
quinclorac	CAS Nr 84087-01-4	quinolinecarboxylic acid herbicide
quintiofos	CAS Nr 1776-83-6	organothiophosphate acaricide
quizalofop-ethyl	CAS Nr 76578-12-6	aryloxyphenoxypropionic herbicide
rotenone, Yutenqin	CAS Nr 83-79-4	botanical insecticide
Samppi No. 3	not available	not available
schradan	CAS Nr 152-16-9	OP acaricide and insecticide
secto	not available	lindane + synergized pyrethroids (D-trans-allethrins)
sethoxydim	CAS Nr 74051-80-2	cyclohexene oxime herbicide
Shachonsuan	not available	OP
Silvex, fenoprop, 2,4,5-TP	CAS Nr 93-72-1	phenoxypropionic herbicide; auxin plant growth regulator
simazine, Gesatop	CAS Nr 122-34-9	chlorotriazine herbicide
sodium arsenite	CAS Nr 7784-46-5	arsenical herbicide and insecticide; unclassified rodenticide
sodium pentachlorophenoxide, sodium pentachlorophenate	CAS Nr 131-52-2	aromatic fungicide and molluscicide
sodium monofluor acetate, Compound 1080, sodium fluoroacetate	CAS Nr 62-74-8	unclassified rodenticide
sodium tetraborate decahydrate, borax, boric acid	CAS Nr 10043-35-3	insecticide
sodium chlorate	CAS Nr 7775-09-9	herbicide

Pesticide name(s)	Identification number	Activity
sodium trichloroacetate	CAS Nr 650-51-1	herbicide
strobane (terpene polychlorinates)	CAS Nr 8001-50-1	chloroterpene insecticide
sulfur, dusting	CAS Nr 7704-34-9	unclassified acaricide and fungicide
sulfotep	CAS Nr 3689-24-5	thioOP acaricide and insecticide
sulprofos	CAS Nr 35400-43-2	phenyl thioOP insecticide
sumicidin	CAS Nr 51630-58-1	pyrethroid insecticide
sumithion, fenitrothion	CAS Nr 122-14-5	phenyl thioOP insecticide
swep	CAS Nr 1918-18-9	carbanilate herbicide
tebuconazol	CAS Nr 80443-41-0	conazole fungicide
temephos, Abate	CAS Nr 3383-96-8	phenyl thioOP insecticide
TEPP (tetraethyl pyrophosphate), HETP	CAS Nr 107-49-3	OP acaricide and insecticide
terbufos	CAS Nr 13071-79-9	aliphatic thioOP insecticide; thioOP nematicide
terbumeton	CAS Nr 33693-04-8	methoxytriazine herbicide
terbuthylazine	CAS Nr 5915-41-3	chloro-triazine herbicide/algicide
terbutryn	CAS Nr 886500	methylthiotriazine herbicide
tetrachlorvinphos	CAS Nr 961-11-5	OP acaricide and insecticide
tetradifon	CAS Nr 116-29-0	bridged biphenyl acaricide
tetramethrin	CAS Nr 7696-12-0	pyrethroid ester insecticide
thallium salts	CAS Nr 7446-18-6	inorganic rodenticide
thiabendazole	CAS Nr 148-79-8	benzimidazole fungicide
thiobencarb(e), benthiocarb	CAS Nr 28249-77-6	thiocarbamate herbicide
thiodicarb	CAS Nr 59669-26-0	oxime carbamate insecticide
thiophanate methyl	CAS Nr 23564-05-8	carbamate fungicide
thiram	CAS Nr 127-26-8	dithiocarbamate fungicide
toxaphene, camphechlor, polychlorocamphene	CAS Nr 8001-35-2	OC acaricide and insecticide
tralomethrin	CAS Nr 66841-25-6	pyrethroid ester insecticide
triadimefon	CAS Nr 43121-43-3	conazole fungicide
triadimenol	CAS Nr 55219-65-3	conazole fungicide
triazophos	CAS Nr 24017-47-8	thioOP acaricide and nematicide; triazole thioOP insecticide
tribromophenol	CAS Nr 118-79-6	herbicide
tributyl tin naphthenate	CAS Nr 85409-17-2	organotin fungicide
trichlorfon, chlorofos	CAS Nr 52-68-6	phosphonate acaricide and insecticide
triclopyr	CAS Nr 55335-06-3	pyridine herbicide
tricyclazole	CAS Nr 41814-78-2	unclassified fungicide
tridemorph	CAS Nr 24602-86-6	morpholine fungicide
trifenmorph	CAS Nr 1420-06-0	molluscicide
trifluralin	CAS Nr 1582-09-8	dinitroaniline herbicide
triforine	CAS Nr 26644-46-2	unclassified fungicide
tsumacide, MTMC, metolcarb	CAS Nr 1129-41-5	carbamate acaricide; phenyl methylcarbamate insecticide
vamidothion	CAS Nr 002275-23-2	thioOP acaricide; aliphatic amide thioOP insecticide
warfarin	CAS Nr 81-81-2	coumarin rodenticide
XMC, macbal	CAS Nr 2655-14-3	phenyl methylcarbamate insecticide
Yanjian, nicotine	CAS Nr 54-11-5	botanical insecticide
Yinbieqin (diapropetryn)	not available	biological pesticide
Yutenqin, rotenone	CAS Nr 83-79-4	botanical insecticide
zinc phosphide	CAS Nr 1314-84-7	unclassified rodenticide
zineb	CAS Nr 12122-67-7	dithiocarbamate fungicide
ziram	CAS Nr 137-30-4	dithiocarbamate fungicide, bird and mammal repellent

Index

Printed and bound by CPI Group (UK) Ltd, Croydon, CR0 4YY

23/10/2024

01778238-0017